W9-CRL-628

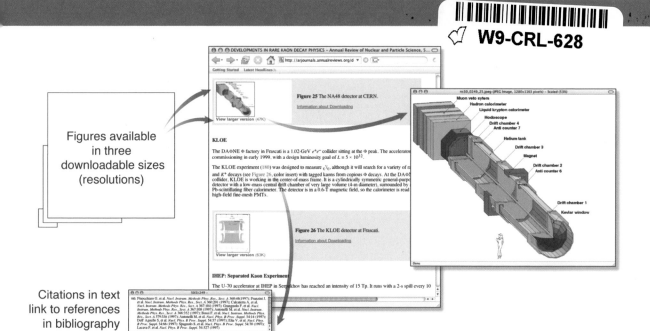

Figures available in three downloadable sizes (resolutions)

Citations in text link to references in bibliography

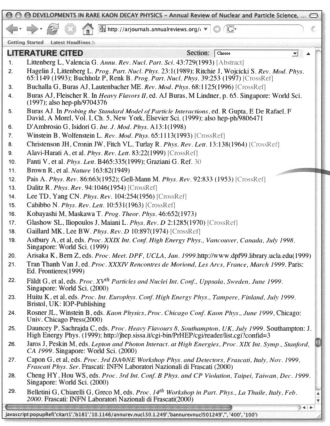

References in Annual Reviews article bibliography link out to sources of cited articles online

Annual Review of
Nuclear and Particle Science

Editorial Committee (2007)

Mirjam Cvetič, University of Pennsylvania
Wick Haxton, University of Washington
Barry R. Holstein, University of Massachusetts
Abolhassan Jawahery, University of Maryland
Boris Kayser, Fermi National Accelerator Laboratory
Georg G. Raffelt, Max-Planck-Institute für Physik
Paul L. Tipton, Yale University
Stanley G. Wojcicki, Stanford University

Responsible for the Organization of Volume 57
(Editorial Committee, 2005)

Barry R. Holstein
Abolhassan Jawahery
Boris Kayser
Steven Ritz
Hamish Robertson
Matthew Strassler
Paul L. Tipton
William Zajc
Mirjam Cvetič (Guest)

Production Editor: Joyce Tang
Managing Editor: Veronica Dakota Padilla
Bibliographic Quality Control: Mary A. Glass
Electronic Content Coordinator: Suzanne K. Moses
Illustration Editor: Douglas Beckner

Annual Review of
Nuclear and Particle Science

Volume 57, 2007

Boris Kayser, *Editor*
Fermi National Accelerator Laboratory

Barry R. Holstein, *Associate Editor*
University of Massachusetts

Abolhassan Jawahery, *Associate Editor*
University of Maryland

www.annualreviews.org • science@annualreviews.org • 650-493-4400

Annual Reviews
4139 El Camino Way • P.O. Box 10139 • Palo Alto, California 94303-0139

QC
770
A5
v.57
phys

Annual Reviews
Palo Alto, California, USA

COPYRIGHT © 2007 BY ANNUAL REVIEWS, PALO ALTO, CALIFORNIA, USA. ALL RIGHTS RESERVED. The appearance of the code at the bottom of the first page of an article in this serial indicates the copyright owner's consent that copies of the article may be made for personal or internal use, or for the personal or internal use of specific clients. This consent is given on the condition that the copier pay the stated per-copy fee of $20.00 per article through the Copyright Clearance Center, Inc. (222 Rosewood Drive, Danvers, MA 01923) for copying beyond that permitted by Section 107 or 108 of the U.S. Copyright Law. The per-copy fee of $20.00 per article also applies to the copying, under the stated conditions, of articles published in any *Annual Review* serial before January 1, 1978. Individual readers, and nonprofit libraries acting for them, are permitted to make a single copy of an article without charge for use in research or teaching. This consent does not extend to other kinds of copying, such as copying for general distribution, for advertising or promotional purposes, for creating new collective works, or for resale. For such uses, written permission is required. Write to Permissions Dept., Annual Reviews, 4139 El Camino Way, P.O. Box 10139, Palo Alto, CA 94303-0139 USA.

International Standard Serial Number: 0163-8998
International Standard Book Number: 978-0-8243-1557-3
Library of Congress Catalog Card Number: 53-995

All Annual Reviews and publication titles are registered trademarks of Annual Reviews.

⊗ The paper used in this publication meets the minimum requirements of American National Standards for Information Sciences—Permanence of Paper for Printed Library Materials, ANSI Z39.48-1992.

Annual Reviews and the Editors of its publications assume no responsibility for the statements expressed by the contributors to this *Annual Review*.

TYPESET BY APTARA, INC.
PRINTED AND BOUND BY FRIESENS CORPORATION, ALTONA, MANITOBA, CANADA

Annual Review of
Nuclear and
Particle Science

Volume 57, 2007

Contents

The Indiana Cooler: A Retrospective
    Hans-Otto Meyer 1

Chiral Perturbation Theory
    Véronique Bernard and Ulf-G. Meißner 33

What Do Electromagnetic Plasmas Tell Us about the Quark-Gluon Plasma?
    Stanisław Mrówczyński and Markus H. Thoma 61

Viscosity, Black Holes, and Quantum Field Theory
    Dam T. Son and Andrei O. Starinets 95

Physics of String Flux Compactifications
    Frederik Denef, Michael R. Douglas, and Shamit Kachru 119

Systematic Errors
    Joel Heinrich and Louis Lyons 145

Two-Photon Physics in Hadronic Processes
    Carl E. Carlson and Marc Vanderhaeghen 171

Glauber Modeling in High-Energy Nuclear Collisions
    Michael L. Miller, Klaus Reygers, Stephen J. Sanders, and Peter Steinberg 205

The Cosmic Microwave Background for Pedestrians: A Review for
    Particle and Nuclear Physicists
    Dorothea Samtleben, Suzanne Staggs, and Bruce Winstein 245

Cosmic-Ray Propagation and Interactions in the Galaxy
    Andrew W. Strong, Igor V. Moskalenko, and Vladimir S. Ptuskin 285

An Introduction to Effective Field Theory
    C.P. Burgess 329

Recent Developments in the Fabrication and Operation of
    Germanium Detectors
    Kai Vetter 363

Searching for New Physics in $b \rightarrow s$ Hadronic Penguin Decays
    Luca Silvestrini 405

Quantum Communication
Judy Jackson and Neil Calder 441

Primordial Nucleosynthesis in the Precision Cosmology Era
Gary Steigman 463

Indexes

Cumulative Index of Contributing Authors, Volumes 48–57 493

Cumulative Index of Chapter Titles, Volumes 48–57 496

Errata

An online log of corrections to *Annual Review of Nuclear and Particle Science* articles may be found at http://nucl.annualreviews.org/errata.shtml

Related Articles

From the *Annual Review of Astronomy and Astrophysics*, Volume 45 (2007)

An Accidental Career
Geoffrey Burbidge

Theory of Star Formation
Christopher F. McKee and Eve C. Ostriker

Relativistic X-Ray Lines from the Inner Accretion Disks Around Black Holes
J.M. Miller

From the *Annual Review of Materials Research*, Volume 37 (2007)

Single-Molecule Micromanipulation Techniques
K.C. Neuman, T. Lionnet, and J.-F. Allemand

The Study of Nanovolumes of Amorphous Materials Using Electron Scattering
David J.H. Cockayne

Electron Holography: Applications to Materials Questions
Hannes Lichte, Petr Formanek, Andreas Lenk, Martin Linck, Christopher Matzeck, Michael Lehmann, and Paul Simon

Atom Probe Tomography of Electronic Materials
Thomas F. Kelly, David J. Larson, Keith Thompson, Roger L. Alvis, Joseph H. Bunton, Jesse D. Olson, and Brian P. Gorman

Electron Holography: Phase Imaging with Nanometer Resolution
Martha R. McCartney and David J. Smith

From the *Annual Review of Physical Chemistry*, Volume 58 (2007)

Molecular Motors: A Theorist's Perspective
Anatoly B. Kolomeisky and Michael E. Fisher

Theory of Structural Glasses and Supercooled Liquids
Vassiliy Lubchenko and Peter G. Wolynes

Femtosecond Stimulated Raman Spectroscopy
Philipp Kukura, David W. McCamant, and Richard A. Mathies

Density-Functional Theory for Complex Fluids
Jianzhong Wu and Zhidong Li

Annual Reviews is a nonprofit scientific publisher established to promote the advancement of the sciences. Beginning in 1932 with the *Annual Review of Biochemistry*, the Company has pursued as its principal function the publication of high-quality, reasonably priced *Annual Review* volumes. The volumes are organized by Editors and Editorial Committees who invite qualified authors to contribute critical articles reviewing significant developments within each major discipline. The Editor-in-Chief invites those interested in serving as future Editorial Committee members to communicate directly with him. Annual Reviews is administered by a Board of Directors, whose members serve without compensation.

2007 Board of Directors, Annual Reviews

Richard N. Zare, *Chairman of Annual Reviews, Marguerite Blake Wilbur Professor of Chemistry, Stanford University*

John I. Brauman, *J.G. Jackson–C.J. Wood Professor of Chemistry, Stanford University*

Peter F. Carpenter, *Founder, Mission and Values Institute, Atherton, California*

Karen S. Cook, *Chair of Department of Sociology and Ray Lyman Wilbur Professor of Sociology, Stanford University*

Sandra M. Faber, *Professor of Astronomy and Astronomer at Lick Observatory, University of California at Santa Cruz*

Susan T. Fiske, *Professor of Psychology, Princeton University*

Eugene Garfield, *Publisher, The Scientist*

Samuel Gubins, *President and Editor-in-Chief, Annual Reviews*

Steven E. Hyman, *Provost, Harvard University*

Joshua Lederberg, *University Professor, The Rockefeller University*

Sharon R. Long, *Professor of Biological Sciences, Stanford University*

J. Boyce Nute, *Palo Alto, California*

Michael E. Peskin, *Professor of Theoretical Physics, Stanford Linear Accelerator Center*

Harriet A. Zuckerman, *Vice President, The Andrew W. Mellon Foundation*

Management of Annual Reviews

Samuel Gubins, President and Editor-in-Chief
Richard L. Burke, Director for Production
Paul J. Calvi Jr., Director of Information Technology
Steven J. Castro, Chief Financial Officer and Director of Marketing & Sales
Jeanne M. Kunz, Human Resources Manager and Secretary to the Board

Annual Reviews of

Analytical Chemistry
Anthropology
Astronomy and Astrophysics
Biochemistry
Biomedical Engineering
Biophysics and Biomolecular Structure
Cell and Developmental Biology
Clinical Psychology
Earth and Planetary Sciences
Ecology, Evolution, and Systematics
Economics
Entomology
Environment and Resources
Fluid Mechanics
Genetics
Genomics and Human Genetics
Immunology
Law and Social Science
Marine Science
Materials Research
Medicine
Microbiology
Neuroscience
Nuclear and Particle Science
Nutrition
Pathology: Mechanisms of Disease
Pharmacology and Toxicology
Physical Chemistry
Physiology
Phytopathology
Plant Biology
Political Science
Psychology
Public Health
Sociology

SPECIAL PUBLICATIONS
Excitement and Fascination of Science, Vols. 1, 2, 3, and 4

The Indiana Cooler: A Retrospective

Hans-Otto Meyer

Department of Physics, Indiana University, Bloomington, Indiana;
email: meyer1@indiana.edu

Annu. Rev. Nucl. Part. Sci. 2007. 57:1–31

First published online as a Review in Advance on March 5, 2007

The *Annual Review of Nuclear and Particle Science* is online at http://nucl.annualreviews.org

This article's doi:
10.1146/annurev.nucl.57.090506.123102

Copyright © 2007 by Annual Reviews.
All rights reserved

0163-8998/07/1123-0001$20.00

Key Words

storage ring, internal target, electron cooling, polarized beams, polarized targets

Abstract

From 1983 to 2002, the Indiana Cooler was constructed and operated at the Indiana University Cyclotron Facility. During that period, a relatively small group of people built an accelerator complex, explored the new technology of electron cooling, and demonstrated its usefulness in nuclear and particle physics. This review recounts the history of the project, describes the facility, and summarizes the scientific results in atomic, nuclear, and particle physics, and in the physics of beams.

Contents

1. INTRODUCTION 2
2. THE FACILITY 3
    2.1. Construction History 3
    2.2. Facility Description 4
    2.3. The Cooler as a User Facility 8
3. ACCELERATOR PHYSICS 8
    3.1. Unpolarized Beam Studies 9
    3.2. Polarized Beam Studies 11
    3.3. Beam-Target Interaction 14
4. TECHNOLOGY 14
    4.1. Internal Targets 14
    4.2. Detectors and Experimental Techniques 17
5. NUCLEAR AND ATOMIC PHYSICS 18
    5.1. Pion Production 18
    5.2. Nucleon Scattering 21
    5.3. Three-Nucleon System 23
    5.4. Atomic Physics 25
6. AUTHOR'S COMMENTS 26

1. INTRODUCTION

The Indiana Cooler was built in response to the emerging new technology of electron cooling. Electron cooling (Section 3.1.3) was first demonstrated in Novosibirsk in 1974. Five years later, a proposal for the Low-Energy Antiproton Ring at CERN, featuring a cooled beam in a storage ring, promised unprecedented experimental possibilities. By that time, the scientists at the Indiana University Cyclotron Facility (IUCF) began to consider the application of cooling methods to intermediate-energy nuclear physics. Finally, when electron cooling of 200-MeV protons was demonstrated at Fermilab, the scientists at the IUCF decided that they were on safe ground, submitting a proposal entitled "The IUCF Cooler-Tripler: Proposal for an Advanced Light-Ion Physics Facility" to the National Science Foundation in December 1980. The plan included a new cyclotron, the Tripler, which was dropped in a second version of the proposal (and later realized in Osaka, Japan). Start-up funds ($1 million) were received in 1982, and funding for the Cooler construction ($6.5 million) began in 1983. The state of Indiana paid for the necessary building addition ($2.7 million).

The Cooler was leading the way, but was joined quickly by similar projects at a number of laboratories in Europe and Japan, which also began planning and building storage rings that employed either electron or stochastic cooling (1). This review describes the Cooler facility, recounts its history, and summarizes the scientific accomplishments that encompassed the fields of the physics of beams and of atomic, nuclear, and particle physics. For a review of the role of storage rings in nuclear physics, see References 2–4.[1]

2. THE FACILITY

This section recounts briefly the construction of the Cooler from 1982 to 1988 and describes in some detail the accelerator complex and its components. Because the Cooler was a national-user facility, we also include information about the composition of the user community and its interaction with the facility.

2.1. Construction History

2.1.1. Construction. In 1982, a Cooler working party consisting of 14 engineers, technicians, and physicists was formed and began holding weekly meetings. In 1983, the construction of a new building was initiated, and the first funds for major equipment purchases became available. By the end of the next year, most major components to complete the ring were procured and started to appear in the new building. The years 1985 to 1987 were marked by an intense assembly and installation effort. There were 28 full-time employees working on the project. Lamination stamping and stacking, coil winding, and fabrication and mapping of the magnetic elements were in full swing. The high-voltage platform for the electron cooler was erected and its 300-kV supply was tested. Concrete support blocks were cast, magnets were mounted and surveyed, and the beam pipe was installed and sections of it put under vacuum. By the end of July 1987, the first beam was injected and the beam path commissioned, section by section, until the orbit was completed. The functioning of the storage ring was demonstrated when a beam lifetime of more than 1 s was observed.

2.1.2. Commissioning. Early in 1988, the ring was operated as a synchrotron accelerator for the first time, and events from a nuclear reaction were observed by the ce01 detector (Section 4.2.1) in the G region. Finally, on April 16, 1988, electron cooling of a stored 45-MeV proton beam was achieved. The next day, the first interaction of a cooled beam with an internal hydrogen target was observed, and an equilibrium between cooling and target heating was established as expected, opening the way to experiments with internal targets. Shortly thereafter, deuterons and polarized protons were stored and cooled. The construction phase ended in May 1988. A detailed account of the progress of the Cooler construction is provided in Reference 5 and in a series of quarterly reports produced to inform the funding agency of the progress made.

This was the beginning of an exciting period, marked by many firsts and rapid progress in demonstrating the capabilities of the new technology. On June 2, 1988,

---

[1]The references cited throughout this review amount to a (it is hoped) complete list of all Cooler-related journal articles. Contributions to conference proceedings are included if they are the sole record on a given subject. For reasons of space, the reference list was limited to the scientific output of the Cooler. For links to supporting work, the reader is referred to the citations in these Cooler-related publications.

a formal dedication of the new facility took place. This event coincided with a celebration of 50 years of nuclear physics research at Indiana University.

2.1.3. Later Additions.

For the first few years, beam for the Cooler was supplied by the existing cyclotron. In 1994, the source terminal of the cyclotron was equipped with a new polarized ion source (6) to improve the intensity of polarized beam.

To overcome the limitations inherent in the cyclotron as an injector, a new accelerator complex, dedicated to Cooler injection, was constructed between 1994 and 1997. It consisted of a new ion source, a pre-accelerator, and a small accumulator synchrotron (Section 2.2.2). Funding ($3.5 million) was provided jointly by Indiana University and the National Science Foundation; beam delivery to the Cooler started in September 2000. With this addition, the Cooler facility reached the peak of its performance. The first Cooler experiment had been conducted with less than 25 µA of stripping-injected unpolarized protons on a simple H_2 gas jet target. By the time operations ended, 1.5 mA of polarized proton or deuteron beam could be stored, and polarized proton and deuteron targets were available (7).

In 1999, the National Science Foundation announced that it would no longer support the operation of the Cooler. An exit strategy for a final research program was implemented, and the last beam was orbiting in the ring at the end of July 2002.

2.2. Facility Description

2.2.1. Cooler ring.

At the time of decommissioning, the Cooler facility presented itself as shown in **Figure 1**. The lattice of the ring had the shape of a hexagon. Breaking with the conventional wisdom of ring design, which calls for perfect symmetry, compromises were made to enhance beam-parameter flexibility and to accommodate

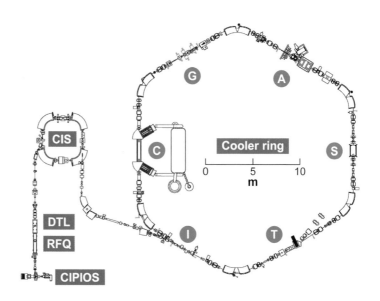

Figure 1

Layout of the Cooler facility as it presented itself in 2001. Shown are the Cooler Injector Polarized Ion Source (CIPIOS), the radio-frequency quadrupole accelerator (RFQ), the drift-tube linac (DTL), the Cooler Injector Synchrotron (CIS), and the storage ring itself.

the requirements imposed by nuclear physics experiments. The six straight sections were labeled C, I, T, S, A, and G for cooling, injection, time-of-flight, spectroscopy, adjustment, and general purpose, respectively.

The main bends at the corners of the hexagon (four of 60° and two of 54°) were accomplished by pairs of dipole magnets with a maximum field of 1.51 T. Additional horizontal bending magnets were located in the straight sections C, I, and T, as discussed below. The beam position was controlled by 2 horizontal and 16 vertical steerers, together with extra windings in the corner dipoles and the additional bending magnets. The magnetic elements needed for focusing and chromatic control (36 quadrupoles and 12 sextupoles) were grouped near the corners to provide more space for experimental equipment (6.1 m in S, I, and G; 5.1 m in A and T; and 7.1 m in C). The optics design called for tight waists with dispersion (i.e., the orbit depends locally on beam momentum) in S, I, and G, whereas in the other straight sections the beam was not dispersed.

Beam from the injector entered the Cooler at the center of the dedicated I region between two 3° bending magnets. Initially, the ring was filled by stripping H_2^+, D_2^+, or $^3He^+$ ions on a 20 µg cm^{-2} carbon foil with one unsupported edge. For beams for which no strippable ions were available, such as polarized beams, injection was achieved by a pair of full-aperture kickers, followed by rf stacking. For both injection modes, cooling was employed to increase the stored proton current (8). For polarized beams, spin-precession solenoids in the beam line between the injector and the Cooler were used to match the spin alignment of the incoming beam to the stable spin direction (Section 3.2.1) at the injection point.

The C region contained a 2.8-m-long solenoid with a maximum field of 15 T, which was needed to prevent the spread of the cooling electron beam due to its own space charge. Adjacent to the solenoid, two toroidal field coils served to inflect and extract the electron beam. The electron beam had a diameter of 2.5 cm, overlapping with the stored beam. Its largest energy was 275 keV, and currents of up to 2 A could be provided. After its extraction, the electron beam was decelerated and efficiently collected. To compensate for the steering of the stored beam by the toroids, the C region contained four strong steerer pairs. To offset spin precession in the cooling solenoid, two compensation solenoids with opposite fields were placed before the first toroid and after the second one, respectively.

The straight sections T, S, A, and G were all used for experiments. The T region contained a 6° bending magnet with a large gap, supplied by the University of Pittsburgh. This magnet served to separate low-rigidity reaction products from the beam for a number of experiments, and made possible the observation of neutrons at 0°. **Figure 2** shows a panoramic view of the S region, and part of the A and T regions.

The maximum rigidity of the stored beam was 3.6 T m (corresponding to a proton energy of 490 MeV). An rf cavity in the A region (obtained from DESY, Hamburg, and built originally for the Princeton-Penn accelerator) was used to bunch and accelerate the beam. A second rf cavity was added to widen the choice of the beam time structure. During acceleration, the rigidity was ramped at a rate of up to 1.0 T m s^{-1}. To suppress eddy currents, all magnetic elements were laminated. Electron cooling was employed

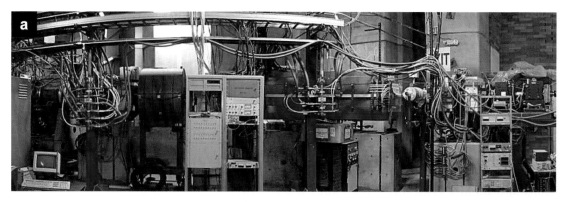

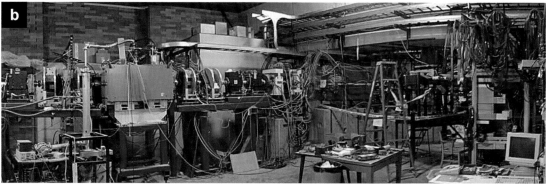

Figure 2

Panoramic view of approximately one-third of the Cooler circumference at the time of decommissioning. (*a*) The pd elastic scattering experiment in the A region is visible on the left. The orange corner dipole pair at the end of the S region is visible to the right of the center of the picture. The view continues in panel *b*. (*b*) The blue box houses the spin-precession solenoid of the Siberian snake. Another dipole pair is also visible, and the equipment that was used to measure the dd$\rightarrow\alpha\pi^0$ cross section in the T region can be seen on the right. Photo courtesy of R.E. Pollock.

before and after the acceleration. The rf cavities were also used to decelerate the beam. The ramping protocol differed between acceleration and deceleration because of magnet hysteresis. The circumference of the ring was 86.77 m (9). The transverse design acceptance of the lattice was 25π mm mrad, and the momentum acceptance was $\Delta p/p = \pm 0.2\%$. The betatron tunes in the horizontal and vertical directions were $Q_x = 3.86$ and $Q_y = 4.86$, respectively.

The beam pipe was constructed of welded stainless steel, prebaked at 900°C, with a few nonconducting sections made from ceramic. After installation and pump down, the beam pipe was baked at approximately 150°C. In the corners of the ring, the vacuum was maintained at 10^{-10} torr. The stringent vacuum constraints were often relaxed near tight waists in the interest of more flexibility in the use of internal targets.

2.2.2. New injector and ion source.
After the completion of a new injector (10) in 2000, the beam was produced by the newly constructed Cooler Injector Polarized Ion Source (CIPIOS) (11). This negative ion source featured a pulsed hydrogen or deuterium gas jet emerging from a cooled nozzle. The appropriate magnetic substates were selected by a combination of permanent sextupole magnets and radio-frequency transitions, as necessary to produce polarized protons or vector- or tensor-polarized deuterons. The atomic beam was ionized by a plasma (H^-, D^-) charge-exchange ionizer. The source potential was -25 kV, and the extracted peak current was up to 12 mA with an emittance of 1.6π mm mrad.

For proton operation, the beam was first accelerated to 3 MeV by a radio-frequency quadrupole and then to 7 MeV by a commercial drift-tube linac. The beam was then accumulated in a small Cooler Injector Synchrotron (CIS) by stripping on a 4.5 µg cm^{-2} carbon foil. After acceleration to 203 MeV, the entire CIS content (approximately 5×10^9 protons) was transferred and kick injected into the Cooler ring as a single bunch. This process was repeated several times with a repetition rate of 0.8 Hz. For deuteron operation, the vane structure of the radio-frequency quadrupole was replaced with a different assembly, a process that took approximately five days (mainly to recover the vacuum). In this mode, the radio-frequency quadrupole accelerated deuterons to 4 MeV, while the drift-tube linac was inactive. The deuteron energy was ramped to 90 MeV in the CIS before transfer to the Cooler (12).

The beam usage for different tasks of the Cooler over its lifetime is illustrated in **Figure 3**. The category labeled overhead includes regular maintenance and the setups for new runs. The category labeled research is detailed further in the following section.

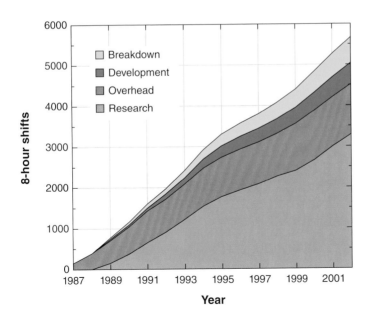

Figure 3

Accumulated beam production during the lifetime of the Indiana Cooler in units of eight-hour shifts. The total beam usage is subdivided into equipment breakdown, development of new capabilities, overhead for maintenance and changeover, and research time during which the experimenters had control of the beam.

2.3. The Cooler as a User Facility

Planning for the first Cooler experiments started in 1984 with weekly meetings among the Cooler experimenters. The group quickly realized that a measurement with an internal target is quite different from the traditional situation, where the beam passes through the target only once, because in a ring the presence of the internal target affects the beam properties and background generated after the target may show up in front of it.

While the Cooler was under construction, the experimenters were preparing for the challenge ahead. A Monte Carlo simulation of how an internal target affects the beam properties was carried out (13), and various methods to prepare sufficiently thin internal targets were explored (Section 3.4). In addition, a workshop on "Nuclear Physics with Stored Cooled Beams" was organized. It took place at McCormick's State Park and was attended by 114 participants from 10 countries (14).

In the fall of 1984, the nineteenth regular biannual IUCF Program Advisory Committee for the first time considered two Cooler proposals. On the basis of its recommendation, procurement for a major detector system in the G region was initiated. The components of this detector were tested concurrently with Cooler commissioning activities. The detector system was operational by the time that cooled, and accelerated beams became available in the fall of 1989. Outside users from the Universities of Michigan, Pittsburgh, and Wisconsin and from Northwestern University contributed strongly to the initial experimental effort, which consisted of most of the first 10 proposed experiments.

The Cooler research effort soon became notable for its worldwide visibility. One example of this is the fact that in the first five years after commissioning, Cooler scientists received 55 invitations for talks at international conferences. The IUCF was also the host for the Fourth International Conference on Nuclear Physics at Storage Rings (STORI99) (15).

Several Cooler proposals appeared on the agenda of each of the thirty Program Advisory Committee sessions that followed the nineteenth Program Advisory Committee, resulting in a total of 75 new proposals and 9 updates of active experiments. The authors on these proposals were from 116 different institutions in 16 countries, and included 54 graduate students. Approximately half of the 38 spokespersons were from the IUCF. Most Cooler experiments represented relatively modest projects that could be completed within 40 to 60 eight-hour shifts of beam time. However, some more involved research efforts consumed up to 300 shifts, while 11 experiments did not go beyond the proposal stage. Approximately two-thirds of the 3200 shifts of beam time spent on research was devoted to nuclear and particle physics. The remainder was split between accelerator physics, atomic physics, and tests of new techniques (**Figure 4**).

3. ACCELERATOR PHYSICS

The exploratory nature of the Cooler project made the accelerator structure accessible to changes or additions. It thus provided an ideal laboratory for investigating accelerator physics topics related to storage rings.

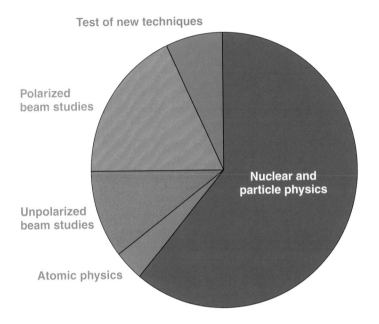

Figure 4

Fractional expenditure of research beam time for different fields of study.

3.1. Unpolarized Beam Studies

3.1.1. Phase space. The coordinates of the four-dimensional transverse phase space include the position and angle (or transverse momentum) of the stored particles relative to the equilibrium orbit, in both the vertical and horizontal directions. To first order, the particles execute betatron oscillations about the equilibrium point (owing to the focusing elements in the ring). The number of such oscillations per turn is termed the betatron tune. The occupied phase volume is known as the emittance.

The coordinates of the two-dimensional longitudinal phase space include the phase and momentum relative to an orbiting fixed point. When an rf cavity is present, the beam is bunched and the fixed point is taken to be the center of the bunch phase space distribution. In the absence of drag forces (from an internal target or residual gas), an equilibrium particle at the fixed point encounters the cavity when the field is zero. All other particles execute synchrotron oscillations about that point. The number of oscillations per turn is termed the synchrotron tune.

3.1.2. Nonlinear beam dynamics. In rings with long storage times, the small, nonlinear part of the forces acting on the stored particles may play a significant role. For instance, during the design of the Superconducting Super Collider in the years after 1990, it became urgent to verify the nonlinear beam motion in accelerators and to model the long-term beam stability in the presence of nonlinear forces.

A beam bunch of small emittance marks the path of a single particle, and a turn-by-turn measurement of the bunch position at two locations a quarter of a betatron oscillation apart then yields the transverse phase space coordinates. The small-emittance beam of the Cooler thus offered the opportunity to investigate nonlinear dynamics

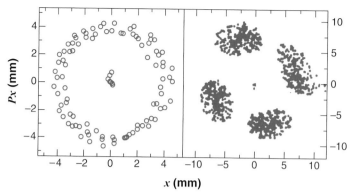

Figure 5

Measured Poincaré map at a fourth-order betatron resonance (*right panel*). The left panel shows the same, slightly off the resonance. The axes correspond to the horizontal phase space coordinates. Reprinted with permission from Lee SY, et al. *Phys. Rev. Lett.* 67:3768 (1991). Copyright 1991 by the American Physical Society (16).

experimentally. From a study of a Poincaré map (a plot of phase space coordinates for subsequent turns, as in **Figure 5**), the Hamiltonian was derived that described the complex particle motion at third- and fourth-order resonances (16, 17), linear coupling resonances (18), and nonlinear sextupolar coupling resonances (19). Knowledge of the Hamiltonian allows one to devise measures to compensate for the ring nonlinearity. The same technique also found use in measuring the betatron tune (20).

To manipulate the beam tune, a ferrite window-frame magnet was developed that could function as a dipole or as a quadrupole (21). Tune modulation was used to lock particles onto resonance islands to establish chaotic dynamics (22), and by employing the magnet as a quadrupole for rapid tune jumps, the beam dynamics near overlapping resonances was explored (21).

The longitudinal phase space coordinates for individual turns can be deduced from a measurement of the time of passage of a beam bunch at a given point and its position in a dispersed location. This has been used to construct longitudinal Poincaré maps of a beam driven by rf modulation (23–25), and to study the coupling between synchrotron and betatron motion and its effect on beam emittance (26–28). The latter is important because the synchrotron motion in large rings is of a low enough frequency that it can be excited by ground movements. Demonstrations of the use of nonlinear beam dynamics in manipulating beam distributions included bunch compression (29), bunch phase space dilution experiments (30, 31), and investigations of parametric resonances in quasi-isochronous systems (32).

3.1.3. Electron cooling. The use of a ring for beam storage gained practical interest with the invention of cooling, that is, when it became possible to control the phase space occupied by the beam particles. There are essentially two methods to achieve this. In electron cooling, the stored beam is immersed in a comoving electron beam for part of the ring circumference. The random motion of the beam particles in the electron rest frame is then transferred to the cold electrons. The beam energy is limited by the energy range for which one can generate an intense electron beam. In the other method, stochastic cooling, the departure of the stored particles from some reference point in phase space is sensed at one location of the ring, and a corresponding corrective kick is applied at an appropriate second location. Because the kick affects all particles, the correction has to be decreased as the number of particles increases, which results in a weaker cooling force. This method, therefore, works best for low-intensity beams over a wide range of energies, and has mainly been applied in antiproton accumulation. However, electron cooling is largely independent of the number of stored particles and was thus the method of choice for the Cooler.

In a ring with an rf cavity, the velocity of an equilibrium particle may differ from the electron velocity. When this difference is large enough, the fixed point abruptly changes into a limit cycle and the beam distribution in longitudinal phase space becomes ring like. Mathematically, this corresponds to a Hopf bifurcation. This phenomenon was observed and investigated in the Cooler (33–35), and it was demonstrated that one could make use of it to measure the temperature of the electron beam.

Electron cooling at the IUCF Cooler provided beams with an emittance of less than 1π mm mrad and a momentum spread $\Delta p/p$ of less than a few times 10^{-4}. Electron cooling also proved crucial in injecting beam into the ring (8) because it made it possible to overcome the limitation of a constant phase space distribution, required by Liouville's theorem.

3.1.4. Intense beams. With beam cooling working well, it was possible for the Cooler beam intensity to reach the space charge limit (36). This limit is of general importance in accelerator physics, and the Cooler provided a welcome opportunity to systematically study the dynamics of space-charge-dominated beams (37–39). Together with the earlier results from nonlinear beam dynamics studies, this activity contributed to a better fundamental understanding of beams near the space charge limit.

Intense beams may also be vulnerable to collective instabilities caused by the impedance in the accelerator. A common remedy is a double rf system that increases the tune spread of the beam and enhances Landau damping. A series of experiments was carried out to understand the dynamics of double rf systems (40–42).

Alternatively, the collective beam instability can be alleviated by modulating the rf amplitude at twice the synchrotron frequency. A detailed study of such a dynamical system showed the bifurcation of the nonlinear resonance and the suppression of noise in the beam spectrum, and established this technique as a method for providing active damping of collective beam instabilities (43, 44).

3.2. Polarized Beam Studies

3.2.1. Depolarizing resonances. The behavior of particles with spin in a storage ring is determined by the interaction of their magnetic moment with the magnetic fields along the orbit. In a synchrotron, vertical fields guide the particles around a closed orbit. With only these fields present, the vertical component of a spin vector is stable, whereas any in-plane component precesses. The number of precessions per orbit is termed the spin tune v_s.

Sometimes magnetic field components in the plane of the ring are present that do not average to zero over one orbit. Such fields inflict a kick on the spin vector away from its stable direction, once per revolution. When the spin tune is an integer, these periodic kicks add up coherently and the beam can be depolarized. This condition is known as an imperfection resonance.

Particles in a ring carry out betatron oscillations (Section 3.1.1) owing to the focusing quadrupole fields. The horizontal fields encountered by a vertically oscillating particle also provide precession kicks. An intrinsic resonance occurs if the spin tune is an integer multiple of the vertical betatron tune.

Other depolarizing resonances arise from coupling between the vertical and horizontal betatron tunes (45), or between the transverse and the synchrotron motion of the stored particles. It is also possible to induce a depolarizing resonance on purpose by generating precession kicks at the proper frequency, using an rf solenoid or a dipole.

Depolarizing resonances can make it difficult to accelerate a stored beam to high energy. Tune-jumping techniques can be used to solve this problem when only a small number of resonances are encountered, but this method is no longer feasible with many partly overlapping resonances. To overcome this difficulty, a local spin rotator has been proposed (e.g., a longitudinal solenoid field) that precesses the spin by 180° (known as full snake) or by less than 180° (partial snake). With a full snake, the spin tune is exactly a half integer, independent of beam energy. The experimental proof that a full snake indeed removes intrinsic, as well as imperfection, resonances (**Figure 6**) was obtained soon after the Cooler was commissioned (46, 47), attracting worldwide attention.

Partial snakes, which have been studied extensively (48–51), also remove imperfection resonances, but in the case of intrinsic resonances, merely shift the beam energy at which these occur. These displaced resonances are then termed snake resonances (52). Overlapping resonances (53), which are obviously important in high-energy machines, were studied in the Cooler by introducing an induced resonance either on top of the imperfection resonance at 108.4 MeV (54) or in the form of synchrotron sidebands that occur with imperfection or induced resonances (55–57).

The polarization lifetime in the vicinity of a depolarizing resonance was investigated for an intrinsic (58) and for an induced resonance (59). The result showed that the polarization lifetime is usually much longer than the beam lifetime, unless one is quite close to a resonance.

3.2.2. Polarized beam properties. When the stored-beam parameters cross the condition for a depolarizing resonance at the appropriate rate, the spins are reversed in a process called adiabatic fast passage. The ability to flip the spin of the stored beam is clearly important to nuclear experiments, and thus extensive tests of this technique were carried out (60–66). In these studies, an induced resonance was moved by ramping its driving rf frequency. A spin flip efficiency of better than 99% was achieved, and several nuclear physics experiments made use of the method.

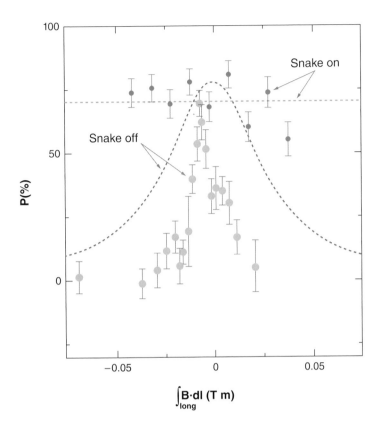

Figure 6

Proton polarization as a function of a longitudinal residual field integral. This imperfection field was varied by means of the cooling compensating solenoids. Without snake, the $G\gamma = 2$ imperfection resonance destroys the polarization unless the residual field is exactly compensated. With a full snake (*small red circles*), the depolarizing resonance has vanished. Reprinted with permission from Krisch AD, et al. *Phys. Rev. Lett.* 63:1137 (1989). Copyright 1989 by the American Physical Society (46).

In a ring with only vertical fields, the spin tune is given by $\nu_s = G\gamma$, where G is the anomalous moment of the particles and γ is the relativistic Lorentz factor. The location of the imperfection resonance then depends only on the beam energy and can be used for energy calibration. This is not quite true in a ring that contains a series of vertical bends. Even if the net bending angle is zero, there is still net spin precession because the respective spin rotations do not commute. The horizontal fields of the toroids in the C region of the Cooler and of the related correction dipoles represent just such an arrangement. The existence of this type III snake has been proposed (67) and experimentally verified (68, 69) at the Cooler.

The stable spin direction is determined by the magnetic fields present along the orbit and does not depend on the polarization direction prior to injection. Snakes not only change the spin tune but also affect the stable spin direction. For example, at a point opposite from a full snake, the stable spin direction is along the beam. Several experiments in the A region required longitudinal beam polarization, which was produced by a superconducting solenoid in the T region and a solenoid field in the C region. The latter was obtained by simply reversing the current in the solenoids that normally are used to compensate the electron confinement field, resulting in a field integral of twice that of the confinement solenoid.

Spin-1 particles such as deuterons are more complicated because the description of their polarization requires a vector as well as a tensor, which arises from an imbalance of the $m = \pm 1$ substates, relative to the $m = 0$ substate. When operating the ring with a stored deuteron beam of mixed vector and tensor polarization close enough to an induced resonance (so that the polarization lifetime was short enough to be measured), the experiment showed that the vector polarization lifetime was twice that of tensor polarization (70). In addition, spin flipping of a vector beam was demonstrated, and it was shown that, in principle, tensor polarization could be manipulated as well (66).

Some researchers have suggested that a stored beam may be polarized by spatially separating stored particles of opposite spin by the Stern-Gerlach effect. There has never been any evidence of this, and an exploratory measurement with the Cooler (71) did not lead to more activity.

3.3. Beam-Target Interaction

Electron cooling makes the use of internal targets possible. As a prerequisite to experiments with internal targets, it was necessary to understand the interplay between such targets and the properties of the stored beam (lifetime, emittance, energy spread).

At low energy, multiple scattering in an internal target causes beam emittance growth, or heating, an effect counteracted by cooling. Beam loss occurs mostly by Rutherford scattering by an angle large enough for the particle to leave the ring acceptance. At higher energies, beam loss is dominated by nuclear scattering. For protons on a H_2 target, the transition between the two regimes takes place at approximately 1 GeV, that is, well above the Cooler energy range. The lifetime of a stored cooled beam over a range of Cooler energies was systematically studied in the presence of targets of varying thickness and species, and models were developed to explain the results (72).

The target thickness is limited by the cooling power available to compensate for multiple scattering. For electron-cooled protons, this limit is approximately $10^{16}/[Z(Z+1)]$ atoms per cm^2, where Z is the atomic number of the target material. Beam loss is caused by large-angle Rutherford scattering, and thus the beam lifetime scales roughly as $(T/Z)^2$, where T is the beam kinetic energy. The lifetime also depends on the machine acceptance. A short beam lifetime means that more unproductive time is spent refilling the ring. The target thickness and the timing of the experimental cycle, which consists of ring filling, beam manipulation, data taking, filling, and so on, can be chosen such that the average luminosity is optimized (72). Researchers concluded that the average luminosity is roughly proportional to $1/Z^2$, indicating that heavier targets come with a penalty of lower luminosity.

4. TECHNOLOGY

4.1. Internal Targets

The limit on the thickness of internal targets rules out self-supporting target foils or windows of gas cells. Various target schemes that avoided this restriction were developed in preparation for Cooler experiments, including a pure electron beam target that was used as a diagnostic tool (73, 74). In the following, the development of targets for nuclear physics experiments will be recapitulated. Typical internal targets are windowless. Thus, the beam encounters only the nuclei of interest, and undesired background is minimized. The targets are thin (Section 3.3), and low-energy outgoing reaction products are detectable. The small target thickness is offset by a large stored current, and the resulting luminosity is comparable to that of a conventional one-pass experiment.

4.1.1. Gas targets. The ideal thin internal target is a localized volume occupied by a rarefied gas. For Cooler experiments, this was achieved by a gas jet emerging from a cooled glass nozzle, positioned a few millimeters from the stored beam (75). To maintain the ring vacuum, differential pumping was added on each side of the target region. This jet target, which served many experiments, was operated with H_2, D_2, HD, N_2, argon gas, and water vapor (to produce an oxygen target). A target thickness of up to 10^{16} atoms cm^{-2} was achieved, 85%–90% of which was in the jet itself, with the remainder distributed along 16 cm inside the first pumping stage (76).

4.1.2. Solid targets. Even though target foils are ruled out, solid targets are still acceptable if they intercept only a small fraction of the circulating beam, so that a given beam particle misses the target most of the time while being cooled all of the time. Early experiments with the Cooler demonstrated that, averaged over many beam revolutions, the effect on beam lifetime and energy spread is the same as that of a uniform target of equivalent thickness. Despite the effort to develop solid targets, there turned out to be little demand for them, mainly because of the severe luminosity penalty incurred by heavier targets.

A solid internal target may take the form of a stream of micron-sized particles. Such a target is versatile because a wide range of materials is commercially available in particulate form. In the Cooler dust target, a laminar flow of a carrier gas in a capillary was seeded with the particles. This produced a narrow beam of particles exiting the capillary, where the carrier gas was removed by differential pumping. The dust beam then crossed the stored beam (77). A test with graphite particles and target thicknesses from 3×10^{13} to 2×10^{15} atoms cm^{-2} proved that such a target is feasible (78).

Thin fibers represent another choice of a nonuniform solid target. Extremely thin carbon fibers were produced by evaporating carbon through a wire grid onto a glass substrate and floating off the resulting microribbons onto a water surface, from where they could be picked up and mounted on a frame. The thinnest usable ribbons produced in this manner were 7 μg cm^{-2} thick, 12 μm wide, and 5 cm long (79). The time-averaged target thickness could be decreased further by periodically sweeping the fiber across the beam. The interaction of fiber targets with stored beams was investigated in detail (80). This study included the effect of fiber heating on the interaction energy spread, and of fiber charging (by secondary electron emission) on the stability of the beam. Thin fibers like these have found use in Coulomb-nuclear interference polarimeters in high-energy rings.

Skimmer targets consist of a thick slab of target material that intercepts just the fringe of the stored beam. Because the intercepted beam is lost, such a target acts as a form of slow extraction rather than a true internal target. The position of the skimmer edge has been adjusted remotely to keep the interaction rate constant. In most polarized beam studies (Section 3.2) a graphite skimmer was used to measure proton polarization.

4.1.3. Polarized targets. The thickness of polarized gas targets is limited by the production rate of polarized atoms. Despite impressive advances made over 40 years of development, polarized gas targets are usually too thin for conventional experiments. However, with the intense beam accumulated in a storage ring, the use of such targets became feasible (81). Polarized internal targets have unprecedented properties: They are pure, not susceptible to radiation damage, and offer rapid reversal of the sign of the polarization and free choice of its direction.

All polarized targets in the Cooler relied on a storage cell to enhance the target thickness. A storage cell is a narrow open-ended tube through which the stored-beam passes. Polarized atoms arrive at the center of the cell via a capillary or a feed tube. The cell confines the atoms to the vicinity of the beam while they drift toward the open ends of the tube. To minimize depolarization when the atoms collide with the cell wall, the cell must consist of (or be coated with) a suitable material (82, 83). Effects that influence the target polarization include the formation of molecules (Section 5.4.2), and spin-exchange collisions (Section 5.4.3). When choosing the diameter of the storage cell, one has to deal with a trade-off between target thickness and ring acceptance (which affects the beam lifetime). Experimental requirements such as minimization of background from cell walls and detectability of recoil nuclei must also be taken into account (84). Two typical storage cell targets are shown in **Figure 7**.

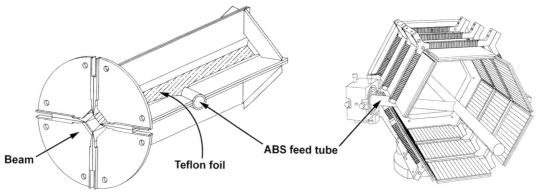

Beam   **Teflon foil**   **ABS feed tube**

Figure 7

Two typical storage cells. (*Left*) The walls of this cell consist entirely of very thin Teflon foils (83), facilitating the observation of low-energy recoils. (*Right*) This arrangement offered full azimuthal coverage for recoil particles. The storage cell in the center is made from 25-μm-thick aluminum; it is surrounded by 18 microstrip detectors. A polarized atomic deuteron beam from the atomic beam source (ABS) entered through the feed tubes indicated in the figure.

The orientation of the spin alignment axis is determined by a magnetic field present at the cell location. Usually, this holding field merely has to overcome the ambient Earth field, and is generated with coils outside the target chamber. For a typical cell of 1 cm diameter and 25 cm length, as used in conjunction with the atomic beam target, the target thickness was between 10^{13} and 10^{14} atoms cm^{-2} (85, 86).

The first polarized target installed in the Cooler used ^{3}He atoms obtained by creating metastable atoms by optically pumping the $^3S_1 - {}^3P_0$ transition, and then transferring the nuclear polarization by metastability-exchange collisions to the ground state. A flow into the storage cell of 10^{17} atoms s^{-1} with a polarization of $P = 0.4$ was reported (87). The proof of principle for spin-correlation experiments in a ring was the measurement of p^{3}He scattering at 45 MeV and 197 MeV (88).

The second source of polarized atoms that was operated in the Cooler was an atomic beam source (ABS). In this source, polarized H and D atoms were produced by dissociating molecules, forming a beam, and then selecting a single hyperfine state (for H) or the appropriate combination of hyperfine states (for D). The selection was achieved by a combination of inhomogeneous magnetic fields and rf-induced transitions between magnetic substates. Typically, the ABS generated a beam of 1 cm diameter with a fluence of approximately 3×10^{16} polarized H atoms s^{-1} in a pure spin state (89). The proton polarization was approximately $P = 0.75$. This target was used by 13 experiments. The capability to produce polarized deuterons was added in 2002 (90).

An alternative method for producing polarized hydrogen by spin exchange with optically pumped potassium was also tested in the Cooler (91). The fluence from such a laser-driven source (approximately 10^{18} s^{-1}) is larger than that from an ABS, but the nuclear polarization is low (approximately $P = 0.15$), and there is a contamination of K atoms of a few percent.

A comprehensive review of the development of polarized gas targets for the Cooler in the context of similar work at other facilities can be found in Reference 92.

4.2. Detectors and Experimental Techniques

4.2.1. Forward detectors.
In the many Cooler experiments dedicated to meson production near threshold (Section 5.1.1), it was sufficient to cover a relatively small forward cone with a detector with moderate energy and angle resolution. These requirements governed the design of the ce01 detector (93), which was in operation throughout the life of the Cooler and served numerous experiments. It consisted of two scintillator arrays, able to stop 200-MeV protons, and four wire chamber planes. The wire chambers were of a new design and featured a hole in the center (for the stored beam to pass through) bounded by a narrow ring that was supported solely by the wires and thin foils (94). The ce01 detector was capable of measuring the momenta of multiple charged particles, making it possible to reconstruct the kinematics of each recorded event. The cylindrical symmetry of the detector was essential in disentangling polarization observables.

4.2.2. Recoil detection.
Because of the low beam halo of a stored beam, solid-state detectors could be placed just a few centimeters from the beam, traversing a

thin, windowless target, to detect low-energy recoil particles (95). When operated in coincidence with a forward detector, position-sensitive recoil detectors of this type were used to localize the interaction point and to define the events of interest. Recoil detection was also used in the production of a tagged neutron beam (96, 97) (Section 5.2.2), and an array of microstrip detectors with associated electronics (98) was developed for a measurement of pion production from ^{12}C via the detection of the recoil nuclei (99).

4.2.3. Polarimetry. All polarization measurements in the Cooler were related to the analyzing power of proton-proton (pp) elastic scattering at a fixed energy (183.1 MeV) and angle (8.64°_{lab}). This calibration point ($A_y = 0.2122 \pm 0.0017$) was established by a careful measurement, relative to an absolute standard (100). This calibration could be exported to other energies by making use of energy ramping. To this aim, first the polarization of a beam stored at the calibration energy was established. The stored beam was then accelerated to the new energy, and the pp analyzing power was measured, becoming the new standard. The beam was then decelerated to the injection energy to verify that no beam depolarization was incurred while changing the beam energy (101). Thus, when used with a hydrogen target, the ce01 detector (Section 4.2.1) provided an absolute proton polarimeter for the energy range from 200 MeV to 500 MeV. For most of the polarized beam studies (Section 3.2), where absolute measurements were not essential, a carbon skimmer target (Section 4.1.2) was employed because it was more efficient and easier to use than a gas target.

The polarization of stored deuterons (12) was deduced from proton-deuteron (pd) elastic scattering, using analyzing powers from the literature. To measure the output polarization from the CIS (Section 2.2.2), a novel polarimeter had to be developed that could cope with a luminosity duty factor of 10^{-8} (102).

5. NUCLEAR AND ATOMIC PHYSICS

The intended purpose of the Cooler was to explore how a stored cooled beam interacting with an internal target might benefit nuclear physics experiments. It turned out that some of the attempted experiments were spectacularly successful, leading to data of unprecedented precision that could not have been measured in any other way.

5.1. Pion Production

5.1.1. Heavy-meson exchange. The first nuclear physics measurement with the newly commissioned Cooler was devoted to the pp→ppπ^0 total cross section near threshold. Originally viewed as a warm-up experiment, it demonstrated dramatically how new technology often leads to surprising results and new physics.

Pion production benefits from the internal target environment in a number of ways. A pure and windowless target is crucial because it avoids heavier nuclei for which the pion-production threshold is much lower. Furthermore, with a thin target, the momenta of the emerging baryons can be measured, the mass of the produced particle can be reconstructed, and thus the events of interest can be cleanly selected.

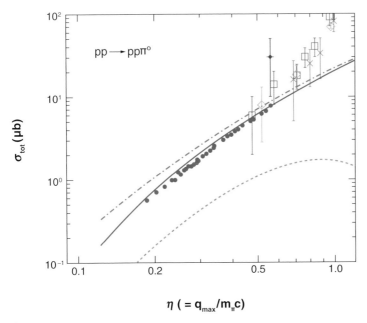

Figure 8

Total cross section for pp→ppπ^0 versus the maximum pion momentum in units of m_π. The solid blue dots are the Cooler data (9); all other symbols represent the data available prior to the Cooler experiment. A calculation with heavy-meson exchange is shown as a solid blue line. Also shown is the same calculation without the Coulomb interaction (*dash-dotted green line*) and without heavy-meson exchange (*dotted orange line*) (105). Reprinted with permission from Horowitz CJ, Meyer HO, Griegel DK. *Phys. Rev. C* 49:1337 (1994). Copyright 1994 by the American Physical Society.

Despite the thin target, the luminosity is reasonable because intense orbiting beams can be obtained by accumulation. Emittance and energy spread of a cooled beam are small, and the absolute beam energy can be determined to a few hundred keV from measuring the orbit frequency. This is important because, near threshold, the cross section varies rapidly with bombarding energy.

The pp→ppπ^0 total cross section data from the first Cooler experiment (103) are shown in **Figure 8**. The energy dependence of the cross section is well explained by phase space, the Coulomb repulsion between the final-state nucleons, and the final-state nucleon-nucleon (NN) interaction. [In a remeasurement in small energy steps (9), no evidence of the opening of the competing channels pp→dπ$^+$ and pp→pnπ$^+$ was found.] Surprisingly however, the magnitude of the cross section turned out to be five times larger than expected. At that time, the theoretical treatment included the impulse term (where the pion is emitted by one of the nucleons), with negligible contribution from rescattering (where the pion is emitted by one of the nucleons and then rescattered by the other). These findings were presented at a Workshop on Particle Production near Threshold that was hosted by the IUCF (104). It took a couple years for the theorists to realize that a process as fundamental

as this may not be understood. As is illustrated in **Figure 8**, the main missing ingredient turned out to be the exchange of σ and ω mesons (in a Z-graph with an
intermediate negative-energy state), but rescattering, driven by the off-shell isoscalar
πN amplitude, may also be more important than originally thought. The ensuing
theoretical activity included the work of local theorists (105), and is ongoing today.

The impact of the first pion-production experiment was ample justification for
subsequent measurements of the $pp \rightarrow pn\pi^+$ (106) and $pp \rightarrow d\pi^+$ (107) cross section
and analyzing-power angular distributions, and of the total $pd \rightarrow pd\pi^0$ cross section
(108, 109). Near threshold, the angular momenta in the final state are either 0 or
1. Together with the Pauli principle, and the conservation of angular momentum,
parity, and isospin, this greatly limits the number of partial waves that can contribute.
With polarized proton beam and target, spin-dependent total cross sections and spin-
correlation coefficients could be measured, and it became possible to isolate individual
partial waves. This was demonstrated for $pp \rightarrow pp\pi^0$ (110, 111), for $pp \rightarrow pn\pi^+$ (112,
113), and for $pp \rightarrow d\pi^+$ (114). The culmination of this program was a complete map
of all $pp \rightarrow pp\pi^0$ observables that can be measured with both initial-state protons
polarized, as a function of two angles needed to describe the three-body final state
(115). This data set is in principle sufficient to determine all contributing partial-wave
amplitudes, providing a stringent test of models as they are being refined. This measurement also showed that the longitudinal analyzing $pp \rightarrow pp\pi^0$ (with three particles
in the final state) can be quite large (116), even though this observable vanishes by
parity conservation in reactions with a two-body final state.

5.1.2. Chiral symmetry.
Double-pion production, $pd \rightarrow {}^3He\pi^+\pi^-$, has a threshold
just within the energy range of the Cooler. It was studied with the expectation that
some of the pion pairs in the final state would form a bound state (pionium). Measuring
the branching ratio for pionium decay into two photons or two pions, respectively,
would then determine the $\pi^+\pi^-$ scattering length, which in turn is related to the
breaking of chiral symmetry. The experiment was carried out just 1 MeV above
threshold, where the recoil ${}^3He$ nuclei as well as pions from unbound pairs fall into a
narrow cone in the forward direction and can be detected with near 100% efficiency.
A 67-pb cross section for the production of free pions was observed, but unfortunately
no evidence for pionium was found (117).

5.1.3. Charge symmetry.
The goal of the very last Cooler experiment was to mea-
sure the total cross section for the reaction $dd \rightarrow \alpha\pi^0$. This reaction is forbidden,
unless charge symmetry is broken. The measurement was carried out at two energies
close to threshold using a deuterium gas jet as a target. **Figure 9** shows missing-mass
spectra with a distinct peak from the $dd \rightarrow \alpha\pi^0$ reaction. At the lower energy, the
measured cross section was 12.7 ± 2.2 pb (118). This result has the potential to
constrain the contributions to charge symmetry breaking by the up-down quark mass
difference and by meson mixing mechanisms.

Previous attempts to measure this cross section had failed, and it is remarkable
that this experiment was successful in establishing a nonzero value while facing a hard

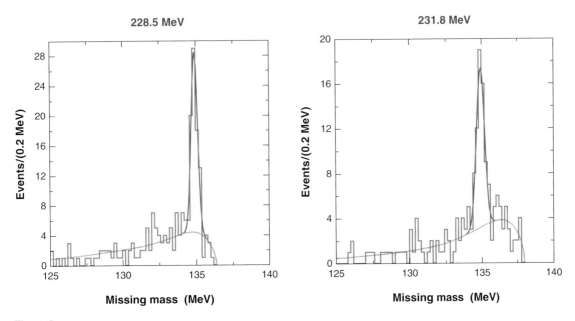

228.5 MeV

231.8 MeV

Figure 9

Missing-mass spectra from the reaction $dd \rightarrow \alpha + X$ at two bombarding energies. The peak corresponds to the charge-symmetry-violating $dd \rightarrow \alpha \pi^0$ reaction (118). The background is due to the reaction $dd \rightarrow \alpha \gamma \gamma$. Reprinted with permission from Stephenson EJ, et al. *Phys. Rev. Lett.* 91:142302 (2003). Copyright 2003 by the American Physical Society.

deadline owing to the looming permanent shutdown. Thus, the last Cooler experiment, like the first one, provided another surprise, attracting worldwide attention.

5.2. Nucleon Scattering

5.2.1. Proton-proton partial waves. Much of our understanding of the NN force has been gained from the study of free NN scattering. The observables are parameterized in terms of empirical partial-wave phase shifts, which serve as the basis for tests of NN interaction models. Below 1 GeV, the main features of the pp scattering phase shifts are known quite well, and it is clear that new data can have an impact only if they have a large weight (small experimental uncertainties) and if they comprise observables that are not yet well represented in the database. These conditions were met by a new generation of analyzing-power and spin-correlation measurements in pp scattering that became possible with the Cooler.

Scattering experiments in a storage ring use a pure target, yield background-free data, and offer the possibility to measure cleanly at small angles, where the nuclear and the Coulomb amplitudes interfere. This was demonstrated by a measurement of the pp analyzing power from 5° to 20° at 183 MeV with a hydrogen jet target (119). The data proved sensitive to the small higher-order terms of the Coulomb potential (120). The start-up experiment with the polarized internal ABS target in the A region

(Section 4.1.3) was a measurement at 198-MeV bombarding energy of three of the four possible spin-correlation coefficients (A_{xx}, A_{yy}, A_{xz}) from 4° to 17° (lab angle) (121). After some detector improvements that allowed the measurement to cover nearly the full angle range, this study was extended to a number of energies between 200 MeV and 450 MeV (122, 123), where previous measurements of spin-correlation data were quite sparse. The data were collected in less than one week of running time. Using a snake (Section 3.2) to make the polarization at the A-region target longitudinal, the fourth correlation coefficient, A_{zz}, was also measured at 198 MeV (124). The achieved statistical uncertainty was approximately ± 0.01, while systematic errors were less than that, and the error of the absolute overall normalization was 2.4%. These uncertainties were considerably smaller than some of the differences between different phase shift analyses, and thus the new data improved our knowledge of the free pp interaction, in particular the higher pp partial waves and the tensor splitting parameters.

At a given polar angle, 16 yields were measured (four azimuths, with positive or negative polarization for both, the beam, and the target). A new method was developed to reduce these yields to observables, making use of a mathematical method known as diagonal scaling (125). Diagonal scaling represents a generalization of the cross-ratio method used traditionally in analyzing-power measurements.

5.2.2. Pion-nucleon coupling constant. The ease with which low-energy reaction products could be detected was utilized to produce a secondary, tagged neutron beam by observing the two protons from the pd→ppn breakup reaction. An internal deuterium jet target was used with position-sensitive solid-state detectors in close proximity of the target, to provide precise reconstruction of the neutron four momentum for each event. This secondary neutron beam was used for a careful, absolute measurement of the neutron-proton (np) elastic scattering cross section near 180° (126, 127). From an extrapolation of the large-angle np cross section to the pion pole, the πNN coupling constant could be deduced. The result of the Cooler measurement was consistent with the Nijmegen 1993 partial-wave analysis, and disagreed with a previous np scattering experiment from CELSIUS in Uppsala.

5.2.3. Off-shell interaction. An important aspect of the force between nucleons in nuclei is its off-shell component. Models of the NN force predict different off-shell behavior. Since the early days of nuclear physics, it has been hoped that the bremsstrahlung process pp→ppγ would provide empirical constraints of the off-shell NN interaction. In the 1980s, a number of new calculations became available, creating a demand for more data. The Cooler inadvertently made a contribution to this effort when it was realized that missing-mass spectra, obtained during pp→ppπ^0 measurements, also featured a peak at zero mass, corresponding to pp→ppγ. With a 310-MeV cooled proton beam on a hydrogen jet target, some 70,000 ppγ events were collected (128). Because both outgoing protons were inside a 20° cone, the collected data were far off the kinematics for elastic scattering. Comparison with theory, however, was hampered by the difficulty of providing calculations that reflected the actual coverage of the three-body final-state phase space.

5.3. Three-Nucleon System

5.3.1. Spin-structure function of ^{3}He.

The ^{3}He nucleus is of interest because it is a nuclear system with properties that can be calculated and precisely compared with data. It is also thought that polarized ^{3}He can be used as an effective polarized neutron target in nuclear and particle physics experiments. Quasi-elastic knockout of the constituent nucleons of polarized ^{3}He by polarized protons offers a direct method to constrain the spin dependence of the single-particle wave functions of ^{3}He. Such a measurement was carried out with 197-MeV protons and an optically pumped polarized ^{3}He target (87). This was the first Cooler experiment that made use of a polarized target. The measured spin-correlation coefficients for ^{3}He(p,2p) and ^{3}He(p,np) are described in Reference 129, while the results in terms of spin-dependent momentum distributions are discussed in Reference 130. The experiment revealed that with increasing nucleon momentum, the polarization of the neutron in polarized ^{3}He decreases, and that the two protons have nonzero net polarization (**Figure 10**). These findings are in agreement with the theoretical expectation for a nucleon momentum of up to approximately 300 MeV/c.

5.3.2. Three-nucleon force.

In March 2000, the polarized ABS in the A region was upgraded to supply a vector- or tensor-polarized deuterium target, and in early 2001, the CIPIOS (Section 2.2.2) started to deliver a polarized deuteron beam to the Cooler. Having the choice of either polarized protons or polarized deuterons for either the beam or the target provided a unique opportunity to measure an almost complete set of polarization observables in the three-nucleon system. Subsequently, angular distributions for the proton analyzing power, four target analyzing powers, five vector correlation coefficients, and seven tensor correlation coefficients in pd elastic scattering were measured at 135- and 200-MeV proton bombarding energy (131).

The data were compared with Faddeev calculations that predict three-nucleon observables on the basis of a given NN potential. It is usually assumed that these calculations accurately describe how nature would behave without a three-nucleon force (3NF). Thus, discrepancies with the data are attributed to the effect of a 3NF. However, including various versions of a 3NF in the calculation did not significantly improve the overall agreement with the Cooler data. This leads to the conclusion that either the theoretically constructed 3N potentials are not realistic, or that the difference between the data and the 2N calculation is not really dominated by 3NF effects. In either case, there is no empirical evidence for a 3NF. Thus, the long-standing quest to understand the 3NF is still open, and new theoretical input is required.

A measurement of the breakup reaction dp→ppn, carried out with a polarized 270-MeV deuteron beam on a polarized proton target, focused on the axial polarization observables A_z (longitudinal analyzing power), $C_{y,x}$–$C_{x,y}$ (vector correlation coefficient), and $C_{zz,z}$ (tensor correlation coefficient) (132). In reactions with a two-body final state, these observables are zero by parity conservation, but in the three-body breakup, they may be sizeable. A theoretical argument suggested that axial observables might be particularly sensitive to 3NF effects. To compare the data to theory, a novel method was developed to carry out Faddeev calculations in a way that

Figure 10

Experimentally determined polarization of the neutron (*upper panel*) and a proton (*lower panel*) in ^{3}He versus the momentum of the nucleon. The curves indicate the limits of a PWIA model. Reprinted with permission from Milner RG, et al. *Phys. Lett. B* 379:67 (1996). Copyright 1996 by Elsevier (130).

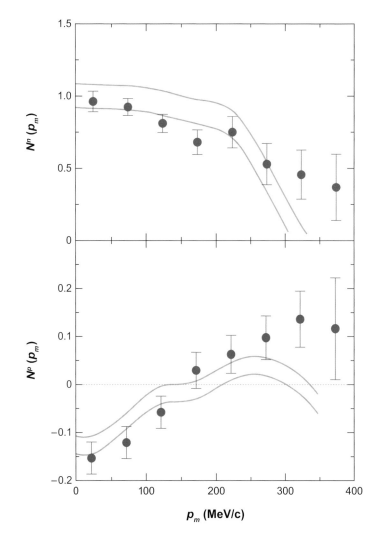

accurately reflects the phase space covered by the experiment (133). As with elastic scattering, including a 3NF in the calculation did not systematically improve the overall agreement with the data (**Figure 11**).

In the past, evidence for a 3NF has often been claimed on the basis of a limited data set for which including a 3NF in the calculation happened to improve the agreement with the data. An example is the measurement of the proton and deuteron analyzing powers and the correlation coefficient $C_{y,y}$ at 197 MeV for center-of-mass angles from 70° to 120° (134). The experiment was carried out with a laser-driven polarized proton target (Section 4.1.3) in the G region.

In connection with a pion-production experiment, the small-angle differential cross section of pd elastic scattering was measured with a 200-MeV proton beam (135). An internal target of HD gas was used, making it possible to relate the cross

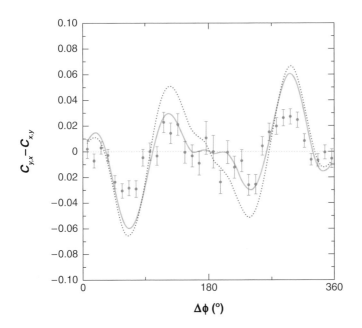

Figure 11

The vector correlation coefficient $C_{y,x}-C_{x,y}$ in pd breakup versus the coplanarity angle $\Delta\varphi$ of the two protons in the final state, measured with a vector polarized beam and target. For $\Delta\varphi = 0$ or π, or for a two-body final state, this observable is zero by parity conservation. The dotted and solid curves represent a Faddeev calculation with and without a three-nucleon force. For details, see Reference 132. Reprinted with permission from Meyer HO, et al. *Phys. Rev. Lett.* 93:112502 (1994). Copyright 1994 by the American Physical Society.

section normalization to the well-known pp cross section. The experiment revealed that Faddeev calculations underestimate the forward cross section by up to 20%, independent of the force model or whether a 3NF is included.

5.4. Atomic Physics

5.4.1. Di-electronic recombination.
When a positive, but not fully stripped, ion captures an electron, the binding energy may be used to excite an already present atomic electron. Calculations of this di-electronic recombination (DR) depend strongly on how the electron-electron interaction is treated. At the time of Cooler commissioning, new models were being developed, and it was realized that the new technology made an electron target (the cooling beam) readily available and could yield new data to test these models (136). Because the electron excitation energy is discrete, the DR process has resonant character, with the corresponding peaks occurring at relative electron-to-ion bombarding energies of up to 50 eV. In the Cooler measurements, a beam of non-fully stripped ions orbited the storage ring once and then was stopped in a Faraday cup (the vacuum was not good enough to actually store such a beam). Recombination took place with the cooling electrons in the C region and led to neutral atoms that did not round the next corner and could be easily detected. By controlling the cooling beam energy, the electron energy in the rest frame of the atoms could be ramped over the region of interest, resulting in a measurement of DR in $^3\mathrm{He}^+$ (137, 138). Similar measurements of DR were carried out with $^6\mathrm{Li}^+$ ions (139, 140).

5.4.2. Nuclear polarization of molecules.
A fraction of the polarized atoms in a storage cell recombine to form H_2 molecules, and one would like to know to what

extent the protons in these molecules are still polarized. Recombination takes place when a polarized atom colliding with the cell wall meets an unpolarized atom that has been adsorbed on the wall. The nuclear polarization of the resulting molecule is thus 1/2. The internal molecular field B_c of these molecules precesses about the external field B. At each subsequent wall collision, the polarization decreases by B_c/B^2. The predicted nuclear polarization of the molecules then depends on the external field and the number of wall collisions. A dedicated measurement of the polarization of H_2 molecules was carried out with a longitudinally polarized proton beam on a longitudinally polarized target (141). The target polarization was measured by observing the pp spin-correlation coefficient A_{zz}. The formation of molecules was enhanced by admitting the target atoms into a recombination cell with copper walls. The external field dependence of the remaining nuclear polarization of H_2 was in agreement with the theoretical expectation.

5.4.3. Spin exchange in polarized deuterium. When two atoms with antiparallel electron spins collide, both spins flip with a high probability. In a gas where mixed hyperfine states are present, such as in tensor-polarized deuterium targets, this effect modifies the populations of the substates such that they approach the spin temperature equilibrium. The rate at which this happens depends on the collision rate and thus on the density. This effect was demonstrated with a deuterium target with mixed vector and tensor polarization. Elastic scattering of unpolarized 135-MeV protons was used to measure both polarizations. The observed tensor depolarization was measured as a function of the gas density in the target cell, and was consistent with theoretical expectations (142).

6. AUTHOR'S COMMENTS

The Indiana Cooler represents a unique technological and intellectual achievement by a comparatively small group of dedicated people, working together with the desire to reach their common goal. I well remember the joy and pride of those involved when a cooled beam passed through an internal target for the first time. Studying how the new tool may be used for nuclear science was like opening the door to an unknown land, and tackling the technical challenges that presented themselves was a fascinating task for the scientists involved. During the following decade, the modest-sized Cooler project contributed a number of landmark experiments to nuclear and accelerator science, and became recognized as a world-class research facility. It was certainly a privilege for me and my Cooler colleagues to be a part of this effort.

SUMMARY POINTS

1. The Indiana Cooler was constructed and operated between 1983 and 2002 at the IUCF. The facility consisted of an ion source, an injector, and a storage ring. Its declared goal was to explore the new technology of electron cooling and to demonstrate its usefulness in nuclear and particle physics.

2. Electron cooling is a method to contract the phase volume of a stored beam. It makes the use of internal targets possible and provides a beam of small emittance and momentum spread for beam dynamics studies.

3. Manipulation of beam properties becomes possible in a storage ring. Experiments made use of the possibilities to accelerate or decelerate the beam, to select its time structure, and to choose the polarization direction of the stored beam and flip its sign.

4. The Cooler research program is an example of how a new technology, applied to an existing field of research, can lead to significant advances.

5. Nuclear physics research with the Cooler benefited from the unique experimental environment offered by internal targets in a storage ring. This was demonstrated, for instance, by the study of reactions near threshold. The large, precise data sets of spin-correlation coefficients in pp and pd scattering and reactions involving polarized proton and deuteron beams and targets could not have been measured anywhere else in the world.

6. The Cooler provided a unique laboratory to explore stored-beam dynamics, including nonlinear behavior and beam instabilities. The understanding gained, as well as the related technical developments, has influenced the design of new accelerators. Examples include the management of depolarizing resonances encountered when accelerating a stored beam to high energy and the development of methods to counter intensity-limiting phenomena.

LITERATURE CITED

1. Pollock RE. *Nucl. Instrum. Methods A* 287:313 (1990)
2. Pollock RE. *Comments Nucl. Part. Phys.* 12:73 (1983)
3. Pollock RE. *Annu. Rev. Nucl. Part. Sci.* 41:357 (1991)
4. Meyer HO, et al. *Annu. Rev. Nucl. Part. Sci.* 47:235 (1997)
5. Pollock RE. *Proc. IEEE Part. Accel. Conf.*, Chicago, 1989, p. 17, **http://accelconf.web.cern.ch/AccelConf/p89/PDF/PAC1989_0017.PDF** (1989)
6. Wedekind M, et al. *Proc. IEEE Part. Accel. Conf.*, Washington, D.C., 1993, p. 3184, **http://accelconf.web.cern.ch/AccelConf/p93/PDF/PAC1993_3184.PDF** (1993)
7. Friesel DL, Derenchuk VL, Sloan T, Stephenson EJ. *Proc. Eur. Part. Accel. Conf.*, Vienna, 2000, p. 539, **http://accelconf.web.cern.ch/AccelConf/e00/PAPERS/MOP5B05.pdf** (2000)
8. Friesel DL, Pollock RE, Ellison T, Jones WP. *Nucl. Instrum. Methods B* 10/11:864 (1985)
9. Meyer HO, et al. *Nucl. Phys. A* 539:633 (1992)
10. Friesel DL, East G, Sloan T. *Nucl. Instrum. Methods A* 441:134 (2000)

11. Derenchuk VP, Belov AS. *Proc. IEEE Part. Accel. Conf.*, Chicago, 2001, p. 2093, **http://accelconf.web.cern.ch/AccelConf/p01/PAPERS/WPAH008.PDF** (2001)

12. Stephenson EJ, et al. *Proc. 9th Int. Workshop Polariz. Sources Targets*, eds. V Derenchuk, B von Przewoski, p. 304. Hackensack, NJ: World Sci. (2002)

13. Meyer HO. *Nucl. Instrum. Methods B* 10/11:342 (1985)

14. Schwandt P, Meyer HO, eds. *Proc. Workshop Nucl. Phys. Stored Cooled Beams, McCormick's Creek, Indiana*, Vol. 128. New York: AIP. 349 pp. (1985)

15. Meyer HO, Schwandt P, eds. *Proc. 4th Int. Conf. Nucl. Phys. Storage Rings (STORI99), Bloomington, Indiana*, Vol. 512. New York: AIP. 436 pp. (2000)

16. Lee SY, et al. *Phys. Rev. Lett.* 67:3768 (1991)

17. Caussyn DD, et al. *Phys. Rev. A* 46:7942 (1992)

18. Liu YJ, et al. *Phys. Rev. E* 49:2347 (1994)

19. Ellison M, et al. *Phys. Rev. E* 50:4051 (1994)

20. Hamilton BJ, et al. *Nucl. Instrum. Methods A* 342:314 (1994)

21. Budnick J, et al. *Nucl. Instrum. Methods A* 368:572 (1996)

22. Wang Y, et al. *Phys. Rev. E* 49:5697 (1994)

23. Ellison M, et al. *Phys. Rev. Lett.* 70:591 (1993)

24. Huang H, et al. *Phys. Rev. E* 48:4678 (1993)

25. Chu CM, et al. *Nucl. Instrum. Methods A* 381:215 (1996)

26. Syphers M, et al. *Phys. Rev. Lett.* 71:719 (1993)

27. Wang Y, et al. *Phys. Rev. E* 49:1610 (1994)

28. Lee SY. *Phys. Rev. E* 49:5706 (1994)

29. Fung KM, et al. *Phys. Rev. Spec. Top. Accel. Beams* 3:100101 (2000)

30. Jeon D, et al. *Phys. Rev. Lett.* 80:2314 (1998)

31. Chu CM, et al. *Phys. Rev. E* 60:6051 (1999)

32. Jeon D, et al. *Phys. Rev. E* 54:4192 (1996)

33. Caussyn DD, et al. *Phys. Rev. Lett.* 73:2696 (1994)

34. Caussyn DD, et al. *Phys. Rev. E* 51:4947 (1995)

35. Lee SY, et al. *Phys. Rev. E* 53:1287 (1996)

36. Ellison TJP, et al. *Phys. Rev. Lett.* 70:790 (1993)

37. Riabko A, et al. *Phys. Rev. E* 51:3529 (1995)

38. Lee SY, Riabko A. *Phys. Rev. E* 51:1609 (1995)

39. Lee SY, Okamoto H. *Phys. Rev. Lett.* 80:5133 (1998)

40. Liu YJ, et al. *Part. Accel.* 49:221 (1995)

41. Lee SY, et al. *Phys. Rev. E* 49:5717 (1994)

42. Liu YJ, et al. *Phys. Rev. E* 50:R3349 (1994)

43. Li D, et al. *Phys. Rev. E* 48:R1638 (1993)

44. Li D, et al. *Nucl. Instrum. Methods A* 364:205 (1995)

45. Ohmori C, et al. *Phys. Rev. Lett.* 75:1931 (1995)

46. Krisch AD, et al. *Phys. Rev. Lett.* 63:1137 (1989)

47. Goodwin JE, et al. *Phys. Rev. Lett.* 64:2779 (1990)

48. Anferov VA, et al. *Phys. Rev. A* 46:R7383 (1992)

49. Blinov BB, et al. *Phys. Rev. Lett.* 73:1621 (1994)

50. Phelps RA, et al. *Phys. Rev. Lett.* 72:1479 (1994)

51. Alexeeva LA, et al. *Phys. Rev. Lett.* 76:2714 (1996)
52. Phelps RA, et al. *Phys. Rev. Lett.* 78:2772 (1997)
53. Lee SY. *Phys. Rev. E* 47:3631 (1993)
54. Baiod R, et al. *Phys. Rev. Lett.* 70:2557 (1993)
55. Lee SY, Berglund M. *Phys. Rev. E* 54:806 (1996)
56. Chu CM, et al. *Phys. Rev. E* 58:4973 (1998)
57. Blinov BB, et al. *Phys. Rev. Spec. Top. Accel. Beams* 2:064001 (1999)
58. Meyer HO, et al. *Phys. Rev. E* 56:3578 (1997)
59. von Przewoski B, et al. *Rev. Sci. Instrum.* 69:3146 (1998)
60. Caussyn DD, et al. *Phys. Rev. Lett.* 73:2857 (1994)
61. Crandell DA, et al. *Phys. Rev. Lett.* 77:1763 (1996)
62. von Przewoski B, et al. *Rev. Sci. Instrum.* 67:165 (1996)
63. Blinov BB, et al. *Phys. Rev. Lett.* 81:2906 (1998)
64. Derbenev YS, et al. *Phys. Rev. Spec. Top. Accel. Beams* 3:094001 (2000)
65. Blinov BB, et al. *Phys. Rev. Lett.* 88:014801 (2002)
66. Morozov VS, et al. *Phys. Rev. Lett.* 91:214801 (2003)
67. Pollock RE. *Nucl. Instrum. Methods A* 300:210 (1991)
68. Minty MG, et al. *Phys. Rev. D* 44:R1361 (1991)
69. Minty MG, et al. *Part. Accel.* 41:71 (1993)
70. von Przewoski B, et al. *Phys. Rev. E* 68:046501 (2003)
71. Akchurin N, et al. *Phys. Rev. Lett.* 69:1753 (1992)
72. Pollock RE, et al. *Nucl. Instrum. Methods A* 330:380 (1993)
73. Pollock RE, Klasen M, Lash J, Sloan T. *Nucl. Instrum. Methods A* 330:27 (1993)
74. Stoller D, Dueck T, Muterspaugh M, Pollock RE. *Nucl. Instrum. Methods A* 423:489 (1999)
75. Doskow JE, Sperisen F. *Nucl. Instrum. Methods A* 362:20 (1995)
76. Sperisen F, et al. *Nucl. Instrum. Methods A* 274:604 (1989)
77. Meyer HO, et al. *Nucl. Instrum. Methods A* 295:53 (1990)
78. Berdoz A, et al. *Nucl. Instrum. Methods B* 40/41:455 (1989)
79. Lozowski WR, Hudson JD. *Nucl. Instrum. Methods A* 303:34 (1991)
80. von Przewoski B, et al. *Nucl. Instrum. Methods A* 328:435 (1993)
81. Haeberli W. *Proc. Workshop Nucl. Phys. Stored Cooled Beams, McCormick's Creek, Indiana*, eds. P Schwandt, HO Meyer, 128:251. New York: AIP (1985)
82. Price JS, Haeberli W. *Nucl. Instrum. Methods A* 349:321 (1994)
83. Lozowski WR, Hudson JD, Dezarn WA. *Nucl. Instrum. Methods A* 362:189 (1995)
84. Ross MA, et al. *Nucl. Instrum. Methods A* 326:424 (1993)
85. Ross MA, et al. *Nucl. Instrum. Methods A* 344:307 (1994)
86. Dezarn WA, et al. *Nucl. Instrum. Methods A* 362:36 (1995)
87. Bloch C, et al. *Nucl. Instrum. Methods A* 354:437 (1995)
88. Lee K, et al. *Phys. Rev. Lett.* 70:738 (1993)
89. Wise T, Roberts AD, Haeberli W. *Nucl. Instrum. Methods A* 336:410 (1993)
90. von Przewoski B, et al. *Proc. 9th Int. Workshop Polariz. Sources Targets, Nashville, Indiana*, eds. V Derenchuk, B von Przewoski, p. 57. Hackensack, NJ: World Sci. (2002)

91. Poelker M, et al. *Nucl. Instrum. Methods A* 3264:58 (1995)
92. Steffens E, Haeberli W. *Rep. Prog. Phys.* 66:1887 (2003)
93. Rinckel T, et al. *Nucl. Instrum. Methods A* 439:117 (2000)
94. Solberg K, et al. *Nucl. Instrum. Methods A* 281:283 (1989)
95. Pitts WK, et al. *Nucl. Instrum. Methods A* 302:382 (1991)
96. Peterson T, et al. *IEEE Trans. Nucl. Sci.* 47:768 (2000)
97. Peterson T, et al. *Nucl. Instrum. Methods A* 527:432 (2004)
98. Yu Z, et al. *Nucl. Instrum. Methods A* 351:460 (1994)
99. Jacobsen ER, et al. *Nucl. Instrum. Methods A* 371:439 (1996)
100. von Przewoski B, et al. *Phys. Rev. C* 44:44 (1991)
101. Pollock RE, et al. *Phys. Rev. E* 55:7606 (1997)
102. Stephenson EJ, et al. *Proc. 9th Int. Workshop Polariz. Sources Targets, Nashville, Indiana*, eds. V Derenchuk, B von Przewoski, p. 294. Hackensack, NJ: World Sci. (2002)
103. Meyer HO, et al. *Phys. Rev. Lett.* 65:2846 (1990)
104. Nann H, Stephenson E, eds. *Proc. Workshop Part. Prod. Near Thresh., Nashville, Indiana*, Vol. 221. New York: AIP. 471 pp. (1991)
105. Horowitz CJ, Meyer HO, Griegel DK. *Phys. Rev. C* 49:1337 (1994)
106. Daehnick WW, et al. *Phys. Rev. Lett.* 74:2913 (1995)
107. Heimberg P, et al. *Phys. Rev. Lett.* 77:1012 (1996)
108. Rohdjess H, et al. *Phys. Rev. Lett.* 70:2864 (1993)
109. Meyer HO, Niskanen JA. *Phys. Rev. C* 47:2474 (1993)
110. Meyer HO, et al. *Phys. Rev. Lett.* 81:3096 (1998)
111. Meyer HO, et al. *Phys. Rev. Lett.* 83:5439 (1999)
112. Saha SK, et al. *Phys. Lett. B* 461:175 (1999)
113. Daehnick WW, et al. *Phys. Rev. C* 65:024003 (2002)
114. von Przewoski B, et al. *Phys. Rev. C* 61:064604 (2000)
115. Meyer HO, et al. *Phys. Rev. C* 63:064002 (2001)
116. Meyer HO, et al. *Phys. Lett. B* 480:7 (2000)
117. Betker AC, et al. *Phys. Rev. Lett.* 77:3510 (1996)
118. Stephenson EJ, et al. *Phys. Rev. Lett.* 91:142302 (2003)
119. Pitts WK, et al. *Phys. Rev. C* 45:R1 (1992)
120. Pitts WK. *Phys. Rev. C* 45:455 (1992)
121. Haeberli W, et al. *Phys. Rev. C* 55:597 (1997)
122. Rathmann F, et al. *Phys. Rev. C* 58:658 (1998)
123. von Przewoski B, et al. *Phys. Rev. C* 58:1897 (1998)
124. Lorentz B, et al. *Phys. Rev. C* 61:054002 (2000)
125. Meyer HO. *Phys. Rev. C* 56:2074 (1997)
126. Sarsour M, et al. *Phys. Rev. Lett.* 94:082303 (2005)
127. Sarsour M, et al. *Phys. Rev. C* 74:044003 (2006)
128. von Przewoski B, et al. *Phys. Rev. C* 45:2001 (1992)
129. Miller MA, et al. *Phys. Rev. Lett.* 74:502 (1995)
130. Milner RG, et al. *Phys. Lett. B* 379:67 (1996)
131. von Przewoski B, et al. *Phys. Rev. C* 74:064003 (2006)
132. Meyer HO, et al. *Phys. Rev. Lett.* 93:112502 (1994)

133. Kuros-Kolnierczuk J, et al. *Few Body Syst.* 34:259 (2004)
134. Cadman RV, et al. *Phys. Rev. Lett.* 86:967 (2001)
135. Rohdjess H, et al. *Phys. Rev. C* 57:2111 (1998)
136. Tanis JA, et al. *Nucl. Instrum. Methods B* 43:290 (1989)
137. Tanis JA, et al. *Nucl. Instrum. Methods B* 56/57:337 (1991)
138. Haar RR, et al. *Phys. Rev. A* 47:R3472 (1993)
139. Zavodszky PA, et al. *Nucl. Instrum. Methods B* 99:35 (1995)
140. Zavodszky PA, et al. *Phys. Rev. A* 58:2001 (1998)
141. Wise T, et al. *Phys. Rev. Lett.* 87:042701 (2001)
142. von Przewoski B, et al. *Phys. Rev. A* 68:042705 (2003)

Chiral Perturbation Theory

Véronique Bernard[1] and Ulf-G. Meißner[2]

[1]Laboratoire de Physique Théorique, Université Louis Pasteur, F-67084 Strasbourg, France; email: bernard@lpt6.u-strasbg.fr

[2]Helmholtz Institut für Strahlen und Kernphysik, Universität Bonn, D-53115 Bonn, Germany; and Forschungszentrum Jülich, Institut für Kernphysik, D-52425 Jülich, Germany; email: meissner@itkp.uni-bonn.de

Annu. Rev. Nucl. Part. Sci. 2007. 57:33–60

First published online as a Review in Advance on March 6, 2007

The *Annual Review of Nuclear and Particle Science* is online at http://nucl.annualreviews.org

This article's doi: 10.1146/annurev.nucl.56.080805.140449

Copyright © 2007 by Annual Reviews. All rights reserved

0163-8998/07/1123-0033$20.00

Key Words

effective field theory, quantum chromodynamics

Abstract

This review gives a brief introduction to chiral perturbation theory in its various settings. We discuss some applications of recent interest, including chiral extrapolations for lattice gauge theory.

Contents

1. INTRODUCTION AND DISCLAIMER 34
2. QCD SYMMETRIES AND THEIR REALIZATION 35
3. EFFECTIVE CHIRAL LAGRANGIAN AND POWER
   COUNTING 39
4. CHIRAL LOOPS AND LOW-ENERGY CONSTANTS 41
5. A SPECIFIC ONE-LOOP CALCULATION 42
6. EXPLORING ANALYTICITY AND UNITARITY 44
7. APPLICATIONS 48
   7.1. Goldstone Boson Masses 48
   7.2. Goldstone Boson Scattering 49
   7.3. Neutral Pion Photoproduction 51
   7.4. Connection to LQCD 53

1. INTRODUCTION AND DISCLAIMER

Quantum chromodynamics (QCD) is the theory of the strong interactions. Although asymptotic freedom allows for a perturbative analysis at large energies, the low-energy domain is characterized by the appearance of hadrons that contain (and hide) the fundamental QCD degrees of freedom—the quarks and gluons. This property of QCD is termed confinement; it is beyond the reach of perturbation theory, calling for a nonperturbative treatment. Furthermore, QCD also exhibits spontaneous, explicit, and anomalous symmetry breaking—and the consequences of these broken symmetries can be analyzed in terms of an appropriately formulated effective field theory (EFT). This EFT is chiral perturbation theory (CHPT), and in the following we review its salient properties, together with some phenomenological applications and its connection to the lattice formulation of QCD. Lattice QCD (LQCD) promises exact solutions, utilizing a formulation on a discretized space-time and solving the pertinent path integral with the help of large computers. Most lattice calculations are, however, done at unphysically large quark masses, and CHPT offers a model-independent scheme to perform the necessary chiral extrapolations.

We end this introduction with a disclaimer: This is not an all-purpose review, but rather one that stresses some fundamentals and selected applications. In what follows, we supply a sufficient amount of references for the reader to immerse himself or herself more deeply into the subject.

In Section 2, we discuss the symmetries of QCD and their realization underlying the EFT. The corresponding effective Lagrangian and the pertinent power counting are given in Section 3, while Section 4 contains some remarks on chiral loops and the meaning of the low-energy coupling constants of the EFT. As a specific example, the scalar form factor of the pion is analyzed to one-loop accuracy in Section 5. The role of unitarity and analyticity is discussed in Section 6. In particular, we discuss the dispersive representation of the scalar pion form factor and show how dispersion relations and CHPT can be combined to give accurate predictions of low-energy observables. A few selected applications that represent the state of the art in CHPT and a discussion of the interplay between LQCD and CHPT are given in Section 7.

2. QCD SYMMETRIES AND THEIR REALIZATION

First, we must discuss chiral symmetry in the context of QCD. Chromodynamics is a nonabelian $SU(3)_{color}$ gauge theory with N_f flavors of quarks, three of them light (u, d, s), and the other three heavy (c, b, t). Here, light and heavy refer to a typical hadronic scale of approximately 1 GeV. In what follows, we consider light quarks only (the heavy quarks are considered as decoupled). The QCD Lagrangian reads

$$\mathcal{L}_{QCD} = -\frac{1}{2g^2} \text{Tr} \left(G_{\mu\nu} G^{\mu\nu} \right) + \bar{q} i \gamma^\mu D_\mu q - \bar{q} \mathcal{M} q = \mathcal{L}_{QCD}^0 - \bar{q} \mathcal{M} q, \qquad 1.$$

where we have absorbed the gauge coupling in the definition of the gluon field and where color indices are suppressed. The three-component vector q collects the quark fields: $q^T(x) = (u(s), d(x), s(x))$. As far as the strong interactions are concerned, the different quarks u, d, and s have identical properties, except for their masses. The quark masses are free parameters in QCD; the theory can be formulated for any value of the quark masses. In fact, light-quark QCD can be well approximated by a fictitious world of massless quarks, denoted $\mathcal{L}_{QCD}^0$ in Equation 1. Remarkably, this theory contains no adjustable parameter; the gauge coupling g merely sets the scale for the renormalization-group-invariant scale Λ_{QCD}. Furthermore, in the massless world, left- and right-handed quarks are completely decoupled. These are defined via

$$q_L = P_L q, \quad q_R = P_R q, \qquad 2.$$

in terms of the projection operators

$$P_L = \frac{1}{2}(1 + \gamma_5), \quad P_R = \frac{1}{2}(1 - \gamma_5),$$

$$P_L^2 = P_L, \quad P_R^2 = P_R, \quad P_L + P_R = \mathbf{1}, \quad P_L \cdot P_R = 0. \qquad 3.$$

The Lagrangian of massless QCD is invariant under separate unitary global transformations of the left- and right-handed quark fields, the so-called chiral rotations,

$$q_I \to V_I q_I, \quad V_I \in U(3), \quad I = L, R, \qquad 4.$$

leading to $3^2 = 9$ conserved left- and 9 conserved right-handed currents by virtue of Noether's theorem. These can be expressed in terms of vector ($V \sim L + R$) and axial-vector ($A \sim L - R$) currents:

$$V_0^\mu = \bar{q} \gamma^\mu q, \quad V_a^\mu = \bar{q} \gamma^\mu \frac{\lambda_a}{2} q,$$

$$A_0^\mu = \bar{q} \gamma^\mu \gamma_5 q, \quad A_a^\mu = \bar{q} \gamma^\mu \gamma_5 \frac{\lambda_a}{2} q. \qquad 5.$$

Here, $a = 1, \ldots 8$, and the λ_a are Gell-Mann's $SU(3)$ flavor matrices. As discussed below, the singlet axial current is anomalous, and thus not conserved. The actual symmetry group of massless QCD is generated by the charges of the conserved currents: $G_0 = SU(3)_R \times SU(3)_L \times U(1)_V$. The $U(1)_V$ subgroup of G_0 generates conserved baryon number because the isosinglet vector current counts the number of quarks minus antiquarks in a hadron. The remaining group, $SU(3)_R \times SU(3)_L$, is often referred to as chiral $SU(3)$. Note that one also considers the light u and d quarks only (with the strange quark mass fixed at its physical value). In that case, one speaks of chiral $SU(2)$ and must replace the generators in Equation 5 by the Pauli matrices. QCD is also invariant under the discrete symmetries of parity (P), charge conjugation (C) and time reversal (T). Although interesting in itself, we do not consider strong CP violation and the related θ-term in what follows (see Reference 1).

The chiral symmetry is a symmetry of the Lagrangian of QCD, but not of the ground state or the particle spectrum. To describe the strong interactions in nature, it is crucial that chiral symmetry is spontaneously broken. This can be seen most easily from the fact that hadrons do not appear in parity doublets. If chiral symmetry were exact, from any hadron one could generate by virtue of an axial transformation another state of exactly the same quantum numbers except of opposite parity. The spontaneous symmetry breaking leads to the formation of a quark condensate in the vacuum $\sim \langle 0|\bar{q}q|0 \rangle = \langle 0|\bar{q}_L q_R + \bar{q}_R q_L|0 \rangle$, thus connecting the left- with the right-handed quarks. In the absence of quark masses, this expectation value is flavor independent: $\langle 0|\bar{u}u|0 \rangle = \langle 0|\bar{d}d|0 \rangle = \langle 0|\bar{q}q|0 \rangle$. More precisely, the vacuum is invariant only under the subgroup of vector rotations times the baryon number current: $H_0 = SU(3)_V \times U(1)_V$. This is the commonly accepted picture that is supported by general arguments (2) as well as lattice simulations of QCD (for a recent study, see Reference 3 and references therein). In fact, the vacuum expectation value of the quark condensate is only one of the many possible order parameters characterizing the spontaneous symmetry violation. All operators that share the invariance properties of the vacuum (Lorentz invariance, parity, invariance under $SU(3)_V$ transformations) qualify as order parameters. The quark condensate nevertheless enjoys a special role. It is related to the density of small eigenvalues of the QCD Dirac operator (see Reference 4 and more recent discussions in References 5 and 6),

$$\lim_{\mathcal{M} \to 0} \langle 0|\bar{q}q|0 \rangle = -\pi \rho(0). \qquad 6.$$

For free fields, $\rho(\lambda) \sim \lambda^3$ near $\lambda = 0$. Only if the eigenvalues accumulate near zero does one obtain a nonvanishing condensate. This scenario is indeed supported by lattice simulations and many model studies involving topological objects such as instantons or monopoles.

Before discussing the implications of spontaneous symmetry breaking for QCD, we briefly remind the reader of Goldstone's theorem (7, 8). To every generator of a spontaneously broken symmetry corresponds a massless excitation of the vacuum. This can be understood in a nut-shell (ignoring subtleties such as the normalization of states and alike, the argument also goes through in a more rigorous formulation): Assume that $\mathcal{H}$ is some Hamiltonian invariant under some charges Q^i, i.e., $[\mathcal{H}, Q^i] = 0$ with $i = 1, \ldots, n$. Assume further that m of these charges ($m \leq n$) do not annihilate the vacuum, i.e., $Q^j|0 \rangle \neq 0$ for $j = 1, \ldots, m$. Define a single-particle state via $|\psi \rangle = Q^j|0 \rangle$. This is an energy eigenstate with eigenvalue zero, as $H|\psi \rangle = HQ^j|0 \rangle =$

$Q^j H|0\rangle = 0$. Thus, $|\psi\rangle$ is a single-particle state with $E = \vec{p} = 0$, i.e., a massless excitation of the vacuum. These states are the Goldstone bosons, collectively denoted as pions $\pi(x)$ in what follows. Through the corresponding symmetry current, the Goldstone bosons couple directly to the vacuum:

$$\langle 0| J^0(0)|\pi\rangle \neq 0. \qquad \qquad 7.$$

In fact, the nonvanishing of this matrix element is a necessary and sufficient condition for spontaneous symmetry breaking. In QCD, we have eight (three) Goldstone bosons for $SU(3)$ $(SU(2))$ with spin-zero and negative parity. The latter property is a consequence of these Goldstone bosons being generated by applying the axial charges on the vacuum. The dimensionful scale associated with the matrix element (Equation 7) is the pion decay constant (in the chiral limit)

$$\langle 0 \left| A^a_\mu(0) \right| \pi^b(p)\rangle = i\delta^{ab} F p_\mu, \qquad \qquad 8.$$

which is a fundamental mass scale of low-energy QCD. In the world of massless quarks, the value of F differs from the physical value by terms proportional to the quark masses, to be introduced later, $F_\pi = F[1 + \mathcal{O}(\mathcal{M})]$. The physical value of F_π is 92.4 MeV, determined from pion decay, $\pi \to \upsilon\mu$. For a discussion of F_π in the context of the Standard Model and beyond, see Reference 9.

Of course, in QCD the quark masses are not exactly zero. The quark mass term leads to the so-called explicit chiral symmetry breaking. Consequently, the vector and axial-vector currents are no longer conserved (with the exception of the baryon number current):

$$\partial_\mu V^\mu_a = \frac{1}{2} i\bar{q}[\mathcal{M}, \lambda_a]q, \quad \partial_\mu A^\mu_a = \frac{1}{2} i\bar{q}\{\mathcal{M}, \lambda_a\}\gamma_5 q. \qquad \qquad 9.$$

However, the consequences of the spontaneous symmetry violation can still be analyzed systematically because the quark masses are small. QCD possesses what is known as approximate chiral symmetry. In that case, the mass spectrum of the unperturbed Hamiltonian and the one including the quark masses cannot be significantly different. Stated differently, the effects of the explicit symmetry breaking can be analyzed in perturbation theory. This perturbation generates the remarkable mass gap of the theory; the pions (and, to a lesser extent, the kaons and the eta) are much lighter than all other hadrons. To be more specific, consider chiral $SU(2)$. The second formula of Equation 9 is nothing but a Ward identity that relates the axial current $A^\mu = \bar{d}\gamma^\mu\gamma_5 u$ with the pseudoscalar density $P = \bar{d}i\gamma_5 u$,

$$\partial_\mu A^\mu = (m_u + m_d)P. \qquad \qquad 10.$$

Taking on-shell pion matrix elements of this Ward identity, one arrives at

$$M^2_\pi = (m_u + m_d)\frac{G_\pi}{F_\pi}, \qquad \qquad 11.$$

where the coupling G_π is given by $\langle 0| P(0)|\pi(p)\rangle = G_\pi$. This equation leads to some intriguing consequences: In the chiral limit, the pion mass is exactly zero, in accordance with Goldstone's theorem. More precisely, the ratio G_π/F_π is a constant in the chiral limit and the pion mass grows as $\sqrt{(m_u + m_d)}$ as the quark masses are turned on. A more detailed discussion of the Goldstone boson masses and their relation to the quark masses is given in Section 7.1.

There is even further symmetry related to the quark mass term. Hadrons appear in isospin multiplets, characterized by very tiny splittings of the order of a few MeV. These are generated by the small quark mass difference $m_u - m_d$ (small with respect to the typical hadronic mass scale of a few hundred MeV), and also by electromagnetic effects of the same size (with the notable exception of the charged to neutral pion mass difference that is almost entirely of electromagnetic origin). This can be made more precise. For $m_u = m_d$, QCD is invariant under $SU(2)$ isospin transformations:

$$q \rightarrow q' = Uq, \quad q = \begin{pmatrix} u \\ d \end{pmatrix}, \quad U = \begin{pmatrix} a^* & b^* \\ -b & a \end{pmatrix}, \quad |a|^2 + |b|^2 = 1. \qquad 12.$$

In this limit, up and down quarks cannot be disentangled as far as the strong interactions are concerned. Rewriting of the QCD quark mass term allows one to make the strong isospin violation explicit:

$$\mathcal{H}_{QCD}^{SB} = m_u \bar{u}u + m_d \bar{d}d = \frac{1}{2}(m_u + m_d)(\bar{u}u + \bar{d}d) + \frac{1}{2}(m_u - m_d)(\bar{u}u - \bar{d}d), \qquad 13.$$

where the first (second) term is an isoscalar (isovector). Extending these considerations to $SU(3)$, one arrives at the eightfold way of Gell-Mann & Ne'eman (10), which played a decisive role in our understanding of the quark structure of the hadrons. The $SU(3)$ flavor symmetry is also an approximate one, but the breaking is much stronger than is the case for isospin. From this, one can directly infer that the quark mass difference $m_s - m_d$ must be much bigger than $m_d - m_u$. Again, this will be made more precise in Section 7.1.

There is one further source of symmetry breaking, which is best understood in terms of the path-integral representation of QCD. The effective action contains an integral over the quark fields that can be expressed in terms of the so-called fermion determinant. Invariance of the action under chiral transformations requires not only the action to be left invariant, but also the fermion measure (11). Symbolically,

$$\int [d\bar{q}][dq] \ldots \rightarrow |\mathcal{J}| \int [d\bar{q}'][dq'] \ldots. \qquad 14.$$

If the Jacobian is not equal to one, $|\mathcal{J}| \neq 1$, one encounters an anomaly. Of course, such a statement has to be made more precise because the path integral requires regularization and renormalization. Still, it captures the essence of the chiral anomalies of QCD. One can show in general that certain 3-, 4-, and 5-point functions with an odd number of external axial-vector sources are anomalous. Particular examples are the famous triangle anomalies of Adler, Bell, and Jackiw and the divergence of the singlet axial current,

$$\partial_\mu(\bar{q}\gamma^\mu \gamma_5 q) = 2iq m\gamma_5 q + \frac{N_f}{8\pi} G^a_{\mu\mu} \tilde{G}^{\mu\mu,a}, \qquad 15.$$

which is related to the generation of the η' mass. There are many interesting aspects of anomalies in the context of QCD and CHPT. Space does not allow us to discuss these, so we refer the reader to Reference 12.

We end this section by providing a list of reviews on the foundations and applications of CHPT. See References 13–22; a recent status report is Reference 23. The state-of-the-art two-loop calculations are reviewed in Reference 24.

3. EFFECTIVE CHIRAL LAGRANGIAN AND POWER COUNTING

The appropriate workhorse to analyze the consequences of spontaneous, explicit, and anomalous symmetry breaking in QCD is the chiral effective Lagrangian (25). The relevant degrees of freedom are the Goldstone bosons coupled to external fields. Two remarks are in order: (*a*) Extensions of this scheme to include, for example, matter fields are discussed briefly below, and (*b*) one can work equally well with the generating functional [see, e.g., the classical papers (26, 27)]. In the chiral limit, we are dealing with a theory without mass gap. Consequently, S-matrix elements and transition currents are dominated by pion-exchange contributions. The QCD Lagrangian can thus be mapped onto an effective Lagrangian,

$$\mathcal{L}_{QCD}[\bar{q}, q, G] \rightarrow \mathcal{L}_{eff}[U, \partial_\mu U, \ldots, \mathcal{M}], \qquad 16.$$

where $U(x)$ is a matrix-valued $SU(3)$ field that collects the Goldstone bosons. Furthermore, $\partial_\mu U$ reminds us that all interactions are of derivative nature owing to Goldstone's theorem, and $\mathcal{M}$ keeps track of the explicit symmetry breaking owing to the finite quark masses. Leutwyler (28) and Weinberg (29) have given a formal proof of this equivalence based on the analysis of the chiral Ward identities, but space does not allow us to discuss these beautiful papers in more detail here.

The effective Lagrangian leads to a well-defined quantum field theory in which gauge and chiral symmetries as well as the chiral anomaly are manifest. The best strategy to construct the most general $\mathcal{L}_{eff}$ consistent with the QCD symmetries is to consider QCD in the presence of locally chiral-invariant external fields,

$$\begin{aligned}
\mathcal{L}_{QCD} &= \mathcal{L}_{QCD}^0 - \bar{q}\gamma_\mu(v^\mu + \gamma_5 a^\mu)q + \bar{q}(s - i\gamma_5 p)q, \\
&= \mathcal{L}_{QCD}^0 - \bar{q}_L \gamma_\mu P_L l^\mu q_L - \bar{q}_R \gamma_\mu P_R r^\mu q_R + \cdots \qquad 17.
\end{aligned}$$

in terms of scalar $s(x)$, pseudoscalar $p(x)$, vector $v^\mu(x)$, and axial-vector $a^\mu(x)$ sources. Explicit symmetry breaking is included in the scalar source, $s(x) = \mathcal{M} + \cdots = \mathrm{diag}(m_u \cdot m_d, m_s) + \cdots$. Electroweak interactions are easily incorporated via

$$r_\mu = e Q \mathcal{A}_\mu, \quad l_\mu = e Q \mathcal{A}_\mu + \frac{e}{\sqrt{2}\sin^2\theta_W}(W_\mu^+ T_+ + \mathrm{h.c.})$$

$$Q = \mathrm{diag}\left(\frac{2}{3}, -\frac{1}{3}, -\frac{1}{3}\right), \quad T_+ = \begin{pmatrix} 0 & V_{ud} & V_{us} \\ 0 & 0 & 0 \\ 0 & 0 & 0 \end{pmatrix}, \qquad 18.$$

where $\mathcal{A}_\mu$ is the photon field, W_μ is the charged massive vector boson field, θ_W is the weak mixing angle, and V_{ud}, V_{us} are the pertinent elements of the CKM matrix.

The most important ingredient to making the EFT a useful tool is the power counting. In CHPT, we have a dual expansion in small external momenta and small quark masses (with a fixed ratio to have a well-defined chiral limit). The corresponding small parameter is denoted by q, where small refers to the typical hadronic scale of approximately 1 GeV. First, one assigns a chiral dimension to all building blocks of $\mathcal{L}_{\text{eff}}$: $U(x) = \mathcal{O}(1)$, $\partial_\mu U(x)$, $l_\mu(x)$, $r_\mu(x) = \mathcal{O}(q)$ and $s(x)$, $p(x) = \mathcal{O}(q^2)$. The last assignment is a consequence of Equation 11. The Goldstone boson masses are non-analytic in the quark masses. (For an alternative power counting, see Reference 30.) The lowest-order effective Lagrangian then takes the form

$$\mathcal{L}^{(2)} = \frac{F^2}{4} \langle D_\mu U D^\mu U^\dagger + \chi U^\dagger + \chi^\dagger U \rangle, \qquad 19.$$

where the brackets denote the trace in flavor space, $D_\mu U = \partial_\mu U + i l_\mu U - i U r_\mu$ is the chiral covariant derivative, and $\chi = 2B(s + ip)$ parameterizes the explicit chiral symmetry breaking. The Lagrangian (Equation 19) is consistent with the strictures from Goldstone's theorem. To this order, the theory is completely specified by two parameters: the pion decay constant in the chiral limit F (see Equation 8), and B, which measures the strength of the quark condensate in the chiral limit, $B = |\langle 0|\bar{q}q|0\rangle|/F^2$. At next-to-leading-order $\mathcal{O}(q^4)$, the effective Lagrangian contains 10 (7) local operators for $SU(3)$ ($SU(2)$). These are accompanied by coupling constants not determined by chiral symmetry, the so-called low-energy constants (LECs). For the explicit form of $\mathcal{L}^{(4)}$, see References 26 and 27. However, at this order there are further contributions. Interactions generate loops; for example, closing two external lines in a tree-level pion-pion scattering graph leads to the one-loop pion tadpole (pion mass shift), and the chiral anomaly is formally of order q^4. At two-loop order, one has further contributions from tree graphs with dimension-six insertions, from one-loop graphs with exactly one insertion from $\mathcal{L}^{(4)}$, and from two-loop graphs with insertions from $\mathcal{L}^{(2)}$. The complete structure of the effective Lagrangian at two-loop order is given in Reference 31 (where one can also find references to earlier work on that topic). All this is captured in Weinberg's (25) power-counting formula, which orders the various contributions to any S-matrix element for pion interactions according to the chiral dimension D (the inclusion of external fields is straightforward),

$$D = 2 + \sum_d N_d(d - 2) + 2L, \qquad 20.$$

with N_d the number of vertices with dimension d (derivatives and/or pion mass insertions) and L the number of pion loops. Chiral symmetry gives a lower bound for D, $D \geq 2$. These are exactly the tree graphs with lowest-order $d = 2$ vertices and $L = 0$ [giving the soft-pion (current algebra) predictions of the 1960s].

To address issues such as isospin violation or the extraction of quark mass ratios, one must include virtual photons and leptons in the EFT. Space forbids us from discussing the many interesting aspects of these extensions. The interested reader should consult some of the classics (see References 32–37).

Matter fields such as, for example, nucleons can also be included in CHPT. In that case, special care has to be taken of the new (hard) mass scale introduced by the matter field (such as the nucleon mass in the chiral limit). This can be treated in various ways for baryons (heavy fermion approach, infrared regularization, extended on-mass-shell scheme, and so on). A detailed review on this topic can be found in

Reference 17, and more recent updates can be found in References 38–40. Virtual photons in baryon CHPT are addressed in References 41 and 42.

4. CHIRAL LOOPS AND LOW-ENERGY CONSTANTS

Beyond tree level, any observable calculated in CHPT receives contributions from tree and loop graphs. The loops not only generate the imaginary parts, but are also in most cases divergent, requiring regularization and renormalization. In CHPT, one usually chooses a mass-independent regularization scheme to avoid power divergences. (There are, however, instances in which other regulators are more appropriate or physically intuitive; for a beautiful discussion of this and related issues, see References 43 and 44.) The method of choice in CHPT is dimensional regularization, which introduces the scale λ. Varying this scale has no influence on any observable O (renormalization scale invariance),

$$\frac{d}{d\lambda} O(\lambda) = 0, \qquad \qquad 21.$$

but this also means that it makes little sense to assign a physical meaning to the separate contributions from the contact terms and the loops. Physics, however, dictates the range of scales appropriate for the process under consideration. Describing the pion vector radius (at one loop) by chiral loops alone would necessitate a scale of approximately 1/2 TeV (as Leutwyler stressed long ago). In this case, the coupling of the ρ meson generates the strength of the corresponding one-loop counter term that gives most of the pion radius (more on this below).

The most intriguing aspects of chiral loops are the chiral logarithms (chiral logs). In the chiral limit, the pion cloud becomes long ranged and there is no more Yukawa factor $\sim \exp(-M_\pi r)$ to cut it off. This generates terms such as $\log M_\pi^2$, $1/M_\pi$, ..., that is, contributions that are nonanalytic in the quark masses. Such statements can be applied to all hadrons surrounded by a cloud of pions, which by virtue of their small masses can move very far away from the object that generates them. Stated differently, in QCD the approach to the chiral limit is nonanalytic in the quark masses and the low-energy structure of QCD can therefore not be analyzed in terms of a simple Taylor expansion. (An early paper that deals with the subtleties of approaching the chiral limit in QCD is Reference 45 and literature quoted therein.) The exchange of the massless Goldstone bosons generates poles and cuts starting at zero-momentum transfer, such that the Taylor series expansion in powers of the momenta fails. This is a general phenomenon of theories that contain massless particles. The Coulomb scattering amplitude due to photon exchange is proportional to e^2/t, with $t = (p' - p)^2$ the momentum transfer squared between the two charged particles.

As stated above, most loops are divergent. In dimensional regularization, all one-loop divergences are simple poles in $1/(d - 4)$, where d is the number of space-time dimensions (for renormalization at two loops, see Reference 46). Consequently, these divergences can be absorbed in the pertinent LECs:

$$L_i \to L_i^{\text{ren}} + \beta_i L(\lambda), \quad L(\lambda) = \frac{\lambda^{d-4}}{16\pi^2} \left(\frac{1}{d-4} - \frac{1}{2}(\ln(4\pi) + \Gamma'(1) + 1) \right), \qquad 22.$$

where β_i is the corresponding β function and where the renormalized and finite L_i^{ren} must be determined by a fit to data (or calculated eventually using LQCD). Having determined the values of the LECs from experiment, one is faced with the issue of trying to understand these numbers. Not surprisingly, the higher-mass states of QCD leave their imprint in the LECs. Consider again the ρ-meson contribution to the vector radius of the pion. Expanding the ρ propagator in powers of t/M_ρ^2—its first term is a contact term of dimension four, with the corresponding finite LEC L_9 given by $L_9 = F_\pi^2/2M_\rho^2 \simeq 7.2 \times 10^{-3}$—close to the empirical value, $L_9 = 6.9 \times 10^{-3}$ at $\lambda = M_\rho$. This so-called resonance saturation (pioneered in References 47–49) holds generally for most LECs at one loop and is used frequently in two-loop calculations to estimate the $\mathcal{O}(p^6)$ LECs (for a recent study on this issue, see Reference 50). More precisely, there are two types of LECs, the so-called dynamical LECs and the symmetry breakers. The contributions proportional to the LECs of the first type are nonvanishing in the chiral limit and can be determined from phenomenology. The symmetry breakers, however, are much more difficult to pin down from data and are also difficult to model. Here, recent progress in LQCD promises a determination of the contribution from these LECs at unphysical values of the quark masses. Much progress has been made in the field of resonance saturation in the past few years. For a state-of-the-art calculation, see Reference 51 and the many references therein. For extensions of the idea of resonance saturation of the LECs in the pion-nucleon and two-nucleon sectors, see References 52 and 53.

Let us end this section with a short remark on the pion cloud of the nucleon, a topic that has gained some prominence in recent years; the literature abounds with incorrect statements. Consider as an example the isovector Dirac radius of the proton (54). At third order in the chiral expansion, it takes the form

$$\langle r^2 \rangle_1^V = \left(0.61 - (0.47 \text{ GeV}^{-2})\tilde{d}(\lambda) + 0.47 \log \frac{\lambda}{1 \text{ GeV}} \right) \text{fm}^2, \qquad 23.$$

where $\tilde{d}(\lambda)$ is a dimension-three pion-nucleon LEC that parameterizes the nucleon core contribution. Comparing Equation 23 with the empirical value for the Dirac radius, $\langle r^2 \rangle_1^V = (0.585 \text{ fm})^2$, one finds that even the sign of the core contribution $\sim \tilde{d}(\lambda)$ is not fixed if λ is varied within the sensible range from 600 MeV to 1 GeV. Only the sum of the core and the cloud contribution constitute a meaningful quantity that should be discussed.

5. A SPECIFIC ONE-LOOP CALCULATION

Let us now consider a specific example of a one-loop calculation based on Feynman diagrams, the scalar form factor of the pion [the same calculation using the generating functional can be found in the classical paper (26)]. We consider this simple three-point function because it allows us to make our arguments with the least amount of algebra. In chiral $SU(2)$, the coupling of the pion to a scalar-isoscalar source defines the scalar form factor $F_S(t)$,

$$\delta^{ik} F_S(t) = \langle \pi^i(p') | \bar{q}q | \pi^k(p) \rangle, \quad t = (p' - p)^2, \qquad 24.$$

where i, k are isospin indices and t the invariant four-momentum transfer squared. Because there are no scalar-isoscalar sources, this form factor can only be inferred indirectly, making use, for example, of dispersive techniques (see Section 6 or Reference 55).

The important role of the scalar form factor stems from the observation that its value at $t = 0$ is proportional to the expectation value of the quark mass term in the QCD Hamiltonian,

$$\frac{\partial M_\pi^2}{\partial \hat{m}} = \langle \pi | \bar{q} q | \pi \rangle. \tag{25.}$$

To one-loop accuracy, one finds (the pertinent tree and one-loop graphs are shown in **Figure 1**)

$$f(t) = 1 + b(t) + \mathcal{O}(p^4),$$
$$b(t) = b_0 + b_1 t + \frac{1}{2 F_\pi^2} \left(2t - M_\pi^2 \right) \bar{J}(t). \tag{26.}$$

Here, $f(t) = F_S(t)/2B$ is the normalized scalar form factor,

$$\bar{J}(t) = \frac{1}{16\pi^2} \left(\sigma \ln \frac{\sigma - 1}{\sigma + 1} + 2 \right), \quad \sigma = \sqrt{1 - \frac{4M_\pi^2}{t}} \quad (t < 0) \tag{27.}$$

is the fundamental meson loop integral (the so-called fundamental bubble), and b_0 and b_1 are polynomials in the pion mass (modulo logs) that depend on the one-loop renormalized LECs ℓ_3 and ℓ_4,

$$b_0 = \frac{M_\pi^2}{16\pi^2 F_\pi^2} \left(\ln \frac{M_\pi^2}{\mu^2} + 64\pi^2 \ell_3(\lambda) + \frac{1}{2} \right),$$
$$b_1 = \frac{1}{16\pi^2 F_\pi^2} \left(-\ln \frac{M_\pi^2}{\mu^2} + 16\pi^2 \ell_4(\lambda) - 1 \right). \tag{28.}$$

These LECs are universal and relate various Green functions. For the case at hand, ℓ_4 can be obtained from the ratio F_K/F_π [extending the theory to $SU(3)$ and then matching the corresponding LEC to ℓ_4 by integrating out the kaons and the eta], and ℓ_3 from the expansion of M_π in powers of the quark masses. The chiral logarithms in b_0 and b_1 are generated by some of the loop graphs depicted in **Figure 1**. Note that

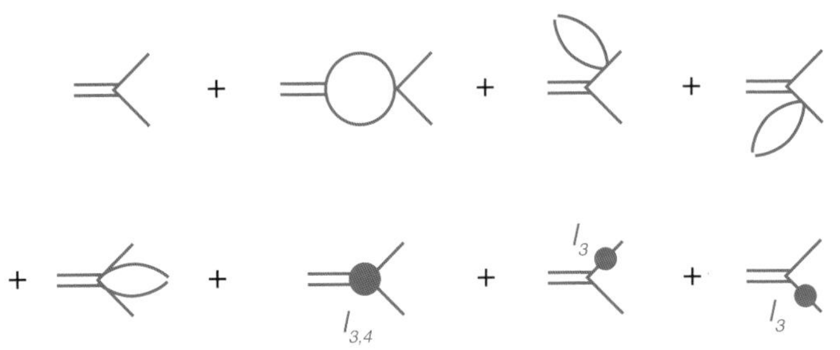

Figure 1

Graphs contributing to the pion scalar form factor at one loop. The double line denotes the scalar-isoscalar source; solid lines are pions. The filled circle depicts an insertion from the next-to-leading-order effective Lagrangian.

the scalar form factor is finite in the chiral limit. It is instructive to study its expansion at low-momentum transfer,

$$f(t) = 1 + \frac{1}{6}\langle r_S^2 \rangle t + \mathcal{O}(t^2),\qquad\qquad 29.$$

which defines the scalar radius r_S. Its low-energy representation can be read off from Equations 26 and 28:

$$\langle r_S^2 \rangle = 6b_1 - \frac{1}{192\pi^2 F_\pi^2}.\qquad\qquad 30.$$

Remarkably, the scalar radius is considerably larger than the corresponding vector radius [that can be extracted from $e^+e^- \to \pi^+\pi^-$ data (see Reference 56)]:

$$\langle r_S^2 \rangle = (0.61 \pm 0.02)\ \mathrm{fm}^2 \gg \langle r_V^2 \rangle = (0.452 \pm 0.013)\ \mathrm{fm}^2,\qquad\qquad 31.$$

with the values for the scalar radius taken from Reference 55. This difference is understood; it is generated by the strong pion-pion interaction in the isospin-zero S wave. This can also be seen from the fact that the coefficient of the chiral logarithm contained in $\langle r_S^2 \rangle$ is six times larger than the time-honored corresponding coefficient in $\langle r_V^2 \rangle$ (57).

6. EXPLORING ANALYTICITY AND UNITARITY

In CHPT, imaginary parts of vertex functions and scattering amplitudes are generated by loop diagrams so that unitarity is obeyed perturbatively. The nontrivial unitarity effects generated by the loops are generated by the propagation of on-shell intermediate states in the loop diagrams. Causality implies certain properties of the analytic structure of amplitudes that allow one to relate real and imaginary parts in the form of dispersion relations. Consider, for example, the dispersion relation for the normalized scalar form factor,

$$f(s) = \frac{1}{\pi}\int_0^\infty ds'\,\frac{\mathrm{Im}\,f(s')}{s - s' - i\varepsilon}.\qquad\qquad 32.$$

Knowledge of $\mathrm{Im}\,f(s)$ for all s thus allows one to reconstruct $f(s)$ in the low-energy region (and above). It is evident that subtractions of the dispersion relation can soften the dependence of this integral on large s. The contents of the chiral loops and the content of the dispersive integral must therefore be related. Even more, possible subtraction constants in the dispersive representation must be mapped onto combinations of LECs because they represent the most general polynomial contribution consistent with the underlying symmetries. It was therefore argued early (see References 58–60) that dispersion relations may be used to extend the range of applicability of CHPT. However, one has to make the relation between the chiral and the dispersive representations more precise. To appreciate the content of the dispersive compared with the chiral representation, consider again the normalized scalar form factor of the pion, $f(t)$. It is given in terms of an analytic function in the complex t-plane, cut along the real axis for $t \le 4M_\pi^2$ (**Figure 2**):

$$\frac{1}{2i}[f(t+i\varepsilon) - f(t-i\varepsilon)] = |f(t)|\sin\delta_0^0(t).\qquad\qquad 33.$$

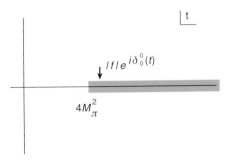

Figure 2

Analytic properties of the scalar form factor.

making use of Watson's theorem, $f(t) = |f(t)| \exp(i\delta_0^0(t))$, with δ_0^0 the elastic $\pi\pi$ isospin-zero S-wave scattering phase shift. This holds to very good approximation up to the $\bar{K}K$ threshold. At order p^2, the scalar form factor has a discontinuity given by

$$\delta_0^0(t) = \frac{1}{32\pi^2 F_\pi^2} \sqrt{1 - \frac{4M_\pi^2}{t}} \left(2t - M_\pi^2\right). \qquad 34.$$

We now want to construct a function that has exactly this discontinuity,

$$k(t) = \frac{1}{2F_\pi^2} \left(2t - M_\pi^2\right) \bar{J}(t), \qquad 35.$$

so that

$$f(t) = a + bt + \frac{1}{2F_\pi^2} \left(2t - M_\pi^2\right) \bar{J}(t), \qquad 36.$$

with a and b unknown coefficients. However, we are able to draw some conclusions about the pion mass dependence of these coefficients. Because the form factor is normalized to one at $t = 0$, we know that

$$a = 1 + a_1 M_\pi^2 \quad \text{(modulo logs)}. \qquad 37.$$

As the pion mass tends to zero, we can expand the function $\bar{J}(t)$,

$$\bar{J}(t) \xrightarrow{M_\pi \to 0} \frac{1}{16\pi^2} \ln M_\pi^2 + \cdots, \qquad 38.$$

where the ellipsis denotes terms not relevant for the following. To have a finite scalar form factor in the chiral limit, we must require that

$$b = -\frac{1}{16\pi^2} \ln M_\pi^2. \qquad 39.$$

Combining Equations 36, 37, and 39, we have obtained a representation that is algebraically equivalent to the one-loop representation given in Equation 26, without having calculated a single loop diagram. This is truly a pleasure. The procedure we have performed to relate the LECs with the subtraction constants is referred to as matching. Matching can be done at various orders. At tree level, the form factor is real, and one thus only needs to ensure that the normalization is correct (see Equation 37). Matching at the one-loop level allows one to reconstruct the chiral representation at that order, with the subtraction constants taking over the role of certain LECs.

It is instructive to take a closer look at this dispersive representation of $f(s)$ in comparison with the chiral one. First, note that the two results have the same algebraic structure. The dispersive representation, however, contains less information. The subtraction constants a and b take the role of the LECs ℓ_3 and ℓ_4 in Equation 28, but such subtraction constants are process dependent and cannot be related to other Green functions. Nevertheless, the chiral log in b lets one understand the enhancement of the corresponding LEC. These arguments can easily be carried out to higher orders. The dispersive representation of the scalar form factor based on the one-loop CHPT amplitude is worked out in Reference 61, and the full two-loop result was later given in Reference 62 (for more recent work on the scalar form factors of the pion and the kaon, see References 63–67). To summarize this little exercise, as long as one is interested in the algebraic structure of a given observable (matrix element), the dispersive method is fine and also easy to apply. However, to relate Green functions to other quantities or to analyze the complete pion mass dependence of observables or LECs, a one- (or higher) loop calculation in CHPT is mandatory.

Historically, the first use of analyticity and unitarity dates back long before the advent of CHPT, namely, to Lehmann's calculation of massless pion-pion scattering to fourth order in the pion momenta (68). Because his arguments are so elegant, it is worth repeating them here. For massless pions, the leading-order $\pi\pi$ scattering amplitude is given in terms of a single invariant function, $A^{(2)}(s, t, u) = s/F^2$, where F is the pion decay constant in the chiral limit and the Mandelstam variables obey $s + t + u = 0$. Furthermore, $A(s, t, u)$ must be symmetric in t, u (Bose symmetry). Thus, at fourth order one has only two independent combinations, $s^2 = (t + u)^2$ and tu. This fixes the polynomial parts at fourth order. However, at this order the scattering amplitude is no longer real. If one employs elastic unitarity of the scattering amplitude, $\mathrm{Im}\, T = |T|^2$, it follows that the imaginary part of the invariant function $A^{(4)}$ takes the form

$$\mathrm{Im}\, A^{(4)} = \frac{1}{16\pi F^4} \left\{ \frac{1}{2} s^2 \theta(s) + \frac{1}{6} t(t - u)\theta(t) + \frac{1}{6} u(u - t)\theta(u) \right\}, \qquad 40.$$

where the three terms in the brackets are due to the s-, t-, and u-channel cuts, respectively. Next, one makes use of analyticity to construct a function that produces this imaginary part. Such a function is

$$A^{(4)} = \frac{1}{16\pi^2 F^4} \left\{ -\frac{1}{2} s^2 \log \frac{-s}{s_0} - \frac{1}{6} t(t - u) \log \frac{-t}{t_0} - \frac{1}{6} u(u - t) \log \frac{-u}{t_0} \right\}, \qquad 41.$$

where s_0 and t_0 are constants, and in the last term t_0 appears because of Bose symmetry. This result is quite remarkable; at fourth order the $\pi\pi$ scattering amplitude depends solely on two constants that cannot be determined from analyticity. If we now introduce two scale-dependent parameters $G_1(\mu^2)$ and $G_2(\mu^2)$, we obtain for the amplitude at fourth order

$$A(s, t, u) = A^{(2)}(s, t, u) + A^{(4)}(s, t, u) + \mathcal{O}(s^3, \ldots)$$

$$A^{(4)}(s, t, u) = G_1(\mu^2)s^2 + G_2(\mu^2)tu$$

$$+ \frac{1}{16\pi^2 F^4}\left\{ -\frac{1}{2}s^2 \log\frac{-s}{\mu^2} - \frac{1}{6}t(t-u)\log\frac{-t}{\mu^2} - \frac{1}{6}u(u-t)\log\frac{-u}{\mu^2}\right\},$$

$$G_1(\mu) = \frac{1}{32\pi^2 F^4}\left(\log\frac{s_0}{\mu^2} + \frac{1}{3}\log\frac{t_0}{\mu^2}\right),$$

$$G_2(\mu^2) = -\frac{1}{24\pi^2 F^4}\left(\log\frac{t_0}{\mu^2}\right).$$

42.

This dispersive representation can be matched to the one-loop CHPT representation for massless pions, which contains the two LECs ℓ_1 and ℓ_2—completely analogous to the case of the scalar form factor (see Equation 28).

In the meantime, this program has been carried much further by combining the Roy equations (69)—special dispersion relations utilizing the high degree of crossing symmetry of the elastic pion scattering amplitude—with the two-loop chiral representation for the $\pi\pi$ scattering amplitude. This has led to the remarkable prediction of Reference 70 for the S-wave scattering lengths,

$$a_0^0 = 0.220 \pm 0.05, \quad a_0^2 = -0.0444 \pm 0.0010.$$

43.

Such a precision is rarely achieved in low-energy QCD. Space forbids us from describing these wonderful calculations in detail. The interested reader should consult the original literature (see References 71–73 for the two-loop representation of the $\pi\pi$ amplitude, Reference 74 for a review on Roy equation studies of $\pi\pi$ scattering, and References 75 and 76 for other calculations of this type; see also Reference 77). The comparison of the chiral prediction with experimental determinations of the scattering lengths is discussed in Section 7.2.

Another interesting consequence of unitarity are the so-called (threshold) cusps that appear in scattering processes when a new channel opens. More precisely, owing to kinematical reasons, such cusps are visible only in S waves. One example is the already discussed scalar form factor of the pion, which exhibits a cusp at the two-pion threshold (see figure 2 in Reference 61). A similar effect in the vector form factor is washed out by the kinematical P-wave prefactors. Another cusp effect that was long considered an academic curiosity appears in $\pi^0\pi^0 \to \pi^0\pi^0$ scattering owing to the opening of the $\pi^+\pi^-$ threshold just 9.2 MeV about the $\pi^0\pi^0$ threshold, as first found in Reference 34 (and shown graphically in Reference 78). It was realized years later that the same cusp effect appears in the decay $K^+ \to \pi^+\pi^0\pi^0$ and allows one to accurately extract the scattering-length combination $a_0^0 - a_0^2$ (79). Further theoretical work on this issue can be found in References 80, 80a, and 81. In the last reference, a field-theoretical method is developed to analyze this effect consistently. For the present state of experimental determinations of $a_0^0 - a_0^2$ from kaon decays, see Section 7.2. Another manifestation of this phenomenon appears in neutral pion production off the nucleon, as discussed in Section 7.3.

7. APPLICATIONS

7.1. Goldstone Boson Masses

One of the most interesting applications of the CHPT machinery is the extraction of the light-quark mass ratios, for example, from the chiral expansion of the Goldstone boson masses. Only with additional input, say from QCD sum rules or LQCD, can one extract values of the quark masses in a given scheme at a given scale. In CHPT the quark masses always appear together with the LEC B. In 1989 Donoghue reviewed the basic ideas of how to apply CHPT in the analysis of the quark mass ratios and the state of the art of such extractions in this journal (82). Since then, theoretical activity has focused on Leutwyler's ellipse that relates the quark mass ratios m_s/m_d and m_u/m_d via

$$\frac{1}{Q^2}\left(\frac{m_s}{m_d}\right)^2 + \left(\frac{m_u}{m_d}\right)^2 = 1, \qquad\qquad 44.$$

modulo corrections of $\mathcal{O}(m_d^2/m_s^2)$, where $Q^2 = (m_s^2 - \hat{m}^2)/(m_d^2 - m_u^2)$ and $\hat{m} = (m_u + m_d)/2$. The numerical value of $Q \simeq 23$ is difficult to pin down accurately because it depends sensitively on the next-to-leading-order electromagnetic corrections to the Goldstone boson masses (the corrections to Dashen's theorem) (see, for example, the discussion in Reference 83 and references therein). Furthermore, in the EFT, a redefinition of the quark condensate and certain LECs allows one to move freely on the ellipse—the famous Kaplan-Manohar ambiguity (84). Historically, the ellipse was first defined in the seminal work of Gasser & Leutwyler (27), and it was first drawn by Kaplan & Manohar (84). It can also be shown that this ambiguity persists at two-loop level. Consequently, some additional information such as, for example, quark mass ratios from the baryon mass splittings is needed to pin down the physical values of the quark mass ratios. **Figure 3** shows the state of the art of such determinations. All determinations agree on one result: The up-quark mass is nonzero ($m_u = 0$ would trivially solve the strong CP problem). For orientation, we give here the results from Reference 83:

$$\frac{m_u}{m_d} = 0.553 \pm 0.043, \quad \frac{m_s}{m_d} = 18.9 \pm 0.8, \quad \frac{m_s}{\hat{m}} = 24.4 \pm 1.5. \qquad 45.$$

It is remarkable that the up- and down-quark masses are so different. Naively, one would thus expect very sizeable strong isospin violation. However, this difference is effectively masked because it is so small compared to any hadronic scale, may it be Λ_{QCD}, m_ρ, or Λ_χ. The role of $\bar{s}s$ fluctuations on the ratio $m_s/\hat{m}$ is discussed in Reference 88.

Another interesting recent result concerns the chiral expansion of the pion mass, which to leading order is given by the Gell-Mann–Oakes–Renner relation (89)

$$M_\pi^2 = B\,(m_u + m_d) + \mathcal{O}\left(m_{u,d}^2\right), \qquad\qquad 46.$$

and the corrections quadratic in the quark masses are parameterized by one LEC (known as $\bar{\ell}_3$) that can also be obtained from data on $K_{\ell 4}$ decays (see the following section). Reference 90 showed that $|\bar{\ell}_3| \leq 16$, which implies that the Gell-Mann–Oakes–Renner relation represents a good approximation—more than 94% of the

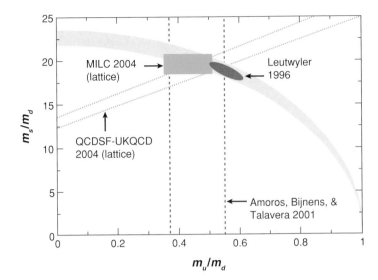

Figure 3

Light-quark mass ratios. The light blue shaded band is the first quadrant of Leutwyler's ellipse for $Q = 22.7 \pm 0.8$. Depicted are results from Leutwyler (83) and from the Lund group (85), as well as from two recent lattice calculations (86, 87). Figure courtesy of Jürg Gasser.

pion mass originates from the first term in its quark mass expansion. This also shows that the quark condensate is indeed the leading-order parameter.

7.2. Goldstone Boson Scattering

The purest reaction to test the chiral dynamics of QCD is elastic pion-pion scattering at threshold. Because the pion three momentum vanishes at threshold, the dual expansion of CHPT is given by one single small parameter, $M_\pi^2/(4\pi F_\pi)^2 \simeq 0.014$. The chiral expansion for the S-wave scattering lengths takes the form

$$a_0^0 = \frac{7 M_\pi^2}{32\pi F_\pi^2} \left[1 + \Delta_4 + \Delta_6\right] + \mathcal{O}\left(M_\pi^8\right),$$

$$a_0^2 = -\frac{M_\pi^2}{16\pi F_\pi^2} \left[1 + \tilde{\Delta}_4 + \tilde{\Delta}_6\right] + \mathcal{O}\left(M_\pi^8\right), \qquad 47.$$

where Δ_4 and Δ_6 collect the one- and two-loop corrections, first given in References 91 and 72, respectively. The numerical evaluation of these corrections gives as central values (using $F_\pi = 92.4$ MeV)

$$a_0^0 = 0.159[1+0.26+0.10] = 0.216, \quad a_0^2 = 0.0454[1-0.02+0.00] = 0.0445. \quad 48.$$

The corrections in the isospin-zero channel are remarkably large. This effect is understood completely in terms of the very strong final-state interactions that effectively generate a very broad pole at $\sqrt{s} \simeq 440$ MeV, far off the real axis. Reference 92 provides a layman's discussion of this effect. In contrast, for isospin two, the corrections have the expected small size. As remarked above, the most precise determination of these fundamental quantities of QCD comes from a combination of CHPT with dispersion relations (see Equation 43). **Figure 4** collects the presently available theoretical and experimental information on the S-wave scattering lengths. The universal

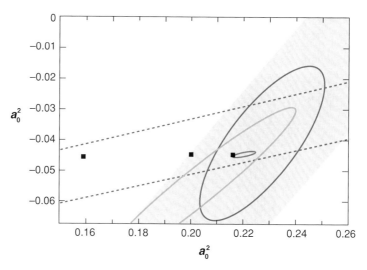

Figure 4

S-wave $\pi\pi$ scattering lengths: theory versus experiment. The black squares give the central value of the tree (97), one-loop (91), and two-loop (72) calculations, respectively. The small red ellipse is the result from Reference 70. The dotted lines denote the universal curve (74). The gray area represents the results obtained from the DIRAC experiment (pionium lifetime) (95), the orange ellipse is the $K_{\ell 4}$ result from E865 (94), and the purple ellipse is the result obtained by the NA48/2 Collaboration from analyzing the cusp in $K \to 3\pi$ (96). Figure courtesy of Heiri Leutwyler.

curve was already established from dispersion theoretical studies of $\pi\pi$ scattering in the 1960s. All solutions that give phase shifts compatible with the ρ meson and Regge behavior at high energies lead to a narrow band for a_0^0 and a_0^2 (93). Shown is the updated universal curve from Reference 74. The experimental information stems from the analysis of $K_{\ell 4}$ decays (94), the measurement of the lifetime of pionium (an electromagnetic bound state of a $\pi^+\pi^-$ pair) (95), and the analysis of the aforementioned cusp in $K \to 3\pi$ decays (96). Note that the preliminary analysis of K_{e4} data from NA48/2 seems to be in conflict with these values.

The next simple and pure Goldstone boson process including also strange quarks is elastic pion-kaon scattering in the threshold region. Here, the situation is less satisfactory. First, the expansion parameter is now $M_K^2/(4\pi F_\pi)^2 \simeq 0.18$, which is sizably larger than in the $SU(2)$ case. The one- and two-loop corrections have been worked out in References 98 and 99, and 100, respectively. Furthermore, the Roy-Steiner equations for pion-kaon scattering with constraints from CHPT have been developed and analyzed in References 101 and 102. Most of the low-energy data are in poor agreement with the solutions of the Roy-Steiner equations. Nevertheless, for the S-wave scattering lengths in the basis of physical isospin 1/2 and 3/2, the chiral expansion seems to converge reasonably well. A typical two-loop result from Reference 100 is (note, however, that the paper also contains other fits with very different values)

$$a_0^{1/2} = 0.220[0.224 \pm 0.022], \quad 10a_0^{3/2} = -0.47[-0.448 \pm 0.077], \qquad 49.$$

where the numbers in the brackets are the dispersive results of Reference 102.

However, there persists a real puzzle. One can formulate an $SU(2)$ low-energy theorem (LET) for the isovector scattering length $a_0^- = (a_0^{1/2} - a_0^{3/2})/3$ (103),

$$a_0^- = \frac{M_\pi^2}{8\pi F_\pi^2(1 + M_\pi/M_K)}\left(1 + \mathcal{O}\left(M_\pi^2\right)\right), \qquad 50.$$

which is not affected by kaon loop effects at next-to-leading order. [Note that there is a small caveat in this statement because the pertinent LECs have not been properly adapted from $SU(3)$ to $SU(2)$.] Because the final-state interactions in $K\pi$ scattering are weaker than in $\pi\pi$, one expects smaller corrections to a_0^- than to a_0^0. This is also borne out by the one-loop calculation; the correction is approximately 12% (103, 104). However, matching the $SU(3)$ two-loop representation of Reference 100 to $SU(2)$, it was found that the subleading corrections to the LET (Equation 50) are of the same size as the leading ones (105). This may be related to the fact that a poor convergence was also found for some of the subthreshold parameters that parameterize the πK scattering amplitude inside the Mandelstam triangle (see Reference 100). More work is needed to clarify the situation.

7.3. Neutral Pion Photoproduction

The chiral structure of QCD can also be analyzed in the presence of matter fields, in particular for nucleons. Neutral pion photoproduction off nucleons in the threshold region, $\gamma N \to \pi^0 N$ ($N = p, n$), exhibits one of the most intriguing realizations of pion loop effects. In the threshold region, this process can be parameterized in terms of one complex S-wave and three complex P-wave multipoles, known as E_{0+} and $P_{1,2,3}$, respectively (for precise definitions, see Reference 106 and references therein). To disentangle these multipoles, one has to measure differential cross sections and one polarization observable, such as the photon asymmetry in $\gamma p \to \pi^0 p$. Historically, an LET for the electric dipole amplitude was derived in the heyday of current algebra under certain analyticity (smoothness) assumptions (107, 108). Measurements at Mainz, Saclay, and Saskatoon seemed in conflict with this LET. However, Reference 109 showed that the presence of the pions in loop graphs (more precisely in the so-called triangle diagram) generates infrared singularities in the Taylor coefficients of the invariant amplitude for E_{0+}, invalidating the smoothness assumption made in References 107 and 108. The correct form of the LET thus reads

$$\begin{Bmatrix} E_{0+}^{\pi^0 p} \\ E_{0+}^{\pi^0 n} \end{Bmatrix} = -\frac{e g_{\pi N}}{8\pi m}\left[\mu F_1 + \mu^2 F_2 + \mathcal{O}(\mu^3)\right],$$

$$F_1 = \begin{pmatrix} 1 \\ 0 \end{pmatrix}, \quad F_2 = -\frac{1}{2}\begin{pmatrix} 3 + \kappa^p \\ -\kappa^n \end{pmatrix} - \Delta\begin{pmatrix} 1 \\ 1 \end{pmatrix}, \quad \Delta = \frac{m^2}{16 F_\pi^2} \simeq 6.4, \qquad 51.$$

in terms of the small parameter $\mu = M_\pi/m \simeq 1/7$, and m is the nucleon mass. Furthermore, $g_{\pi N} \simeq 13$ is the strong pion-nucleon coupling constant and $\kappa_{p,n}$ the

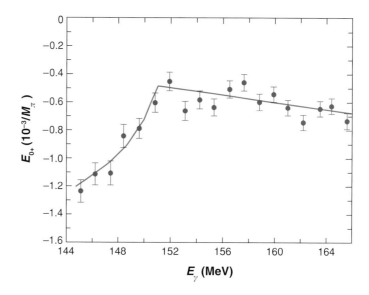

Figure 5

The real part of the electric dipole amplitude E_{0+} in the threshold region. The red line is the one-loop CHPT prediction (106, 112), and the blue data points are the most recent measurements from MAMI (113).

anomalous magnetic moment of the proton and neutron, respectively. The contribution $\sim\Delta$ is the novel pion loop effect first found in Reference 109. In fact, the terms $\sim\mu^3$ have also been worked out. Besides loop contributions, one has two polynomial terms that in the threshold region can be combined in one structure accompanied by one LEC.

Consider now, for definiteness, neutral pion production off the proton. Fixing this LEC at the opening of the π^+n threshold just ≈ 7 MeV above the $\pi^0 p$ threshold, one obtains an excellent description of the threshold data for Re E_{0+}, as shown in **Figure 5**. Also clearly visible is the cusp at the π^+n threshold—its strength is directly proportional to the isovector (charge exchange) scattering length. Thus, a more precise measurement of this cusp would give additional experimental information on zero-energy pion-nucleon scattering (see also Reference 110 for further discussion). Also, CHPT predicts the counterintuitive result that at threshold $|E_{0+}^{\pi^0 n}| > |E_{0+}^{\pi^0 p}|$. This prediction is vindicated by the measurement of coherent neutral pion production off the deuteron and by utilizing CHPT for few-nucleon systems (see Reference 111). Note, however, that the chiral expansion for the electric dipole amplitude is not converging very well. This is another manifestation of strong final-state interactions, this time in the πN system. Quite unexpectedly, in Reference 106, novel LETs for the P-wave multipoles $P_{1,2}$ were found (including terms of order μ), and the corrections of $\mathcal{O}(\mu^2)$ were analyzed in Reference 112. These LETs are in good agreement with the data from Mainz (which performed so far the only polarization measurement that allows disentangling P_1 from P_2) (see Reference 112 or 113 for detailed discussions). The comparison of the CHPT predictions with the most recent data from Mainz reads

$$E_{0+} = -1.19[-1.23 \pm 0.08 \pm 0.03],$$
$$\bar{P}_1 = 9.67[9.46 \pm 0.05 \pm 0.28], \qquad\qquad 52.$$
$$\bar{P}_2 = -9.6[-9.5 \pm 0.09 \pm 0.28],$$

in the conventional units of $10^{-3}/M_{\pi^+}$ and $10^{-3}/M_{\pi^+}^2$, respectively. Note also that the third P-wave multipole P_3 is given in terms of one third-order LEC. P_3 is completely dominated by the excitation of the Δ resonance. In fact, free fits to the data and estimating the strength of the LEC from resonance saturation give almost identical results. Recently, the relation between dispersion relations and the chiral representation for neutral pion photoproduction has been analyzed in the context of the so-called Fubini-Furlan-Rosetti sum rule (see References 114–116). In particular, relations between the LECs and the subtraction constants of the dispersion relations were derived and used to pin down some of the LECs with an unprecedented accuracy (see the discussion in Section 6). For similar studies combining dispersion relations and CHPT for elastic pion-nucleon scattering, see References 117 and 118. In particular, the Roy-type equations for πN scattering are given in Reference 118—they just remain to be solved.

7.4. Connection to LQCD

QCD matrix elements can be calculated from first principles in the framework of LQCD. Space-time is approximated by a box of size $V = L_s \times L_s \times L_s \times L_t$, where L_s (L_t) refers to the length in the spatial (temporal) directions. Furthermore, a UV cutoff is given by the inverse lattice spacing a. At present, typical lattice sizes and spacings are $L_s \simeq L_t \simeq 2 \dots 3$ fm and $a \simeq 0.07 \dots 0.15$ fm, respectively. In addition, it is very difficult to perform numerical simulations with light fermions, so the state-of-the-art calculations using various (improved) actions and modern algorithms have barely reached pion masses of approximately 250 MeV. In addition, exact chiral symmetry can be implemented only in the computer-time-intensive domain wall (119) or overlap (120) formulations. Most results obtained so far are based on simulations using considerably heavier pions (and no exact chiral symmetry). To connect lattice results to the real world of continuum QCD, one has to perform the continuum limit ($a \to 0$), the thermodynamic limit ($V \to \infty$), and chiral extrapolations from the unphysical to the physical quark masses. All this can be performed in the framework of suitably tailored effective field theories, which are variations of continuum CHPT discussed so far, such as staggered CHPT, Wilson CHPT, and so on, via the intermediate step of the Symanzik effective action (121, 122). Space does not allow us to review all these interesting developments; we refer the reader to the recent comprehensive review by Sharpe (123) (for an early review on the CHPT treatment of finite volume effects, see Reference 14).

Obviously, the lattice practitioners need CHPT. For truly ab initio calculations, the quark mass dependence of the lattice results must be analyzed in terms of a model-independent approach, such as CHPT, in a regime of quark masses where it is applicable. Using resummation schemes or models to try to extend the range of applicability of CHPT inevitably induces an uncontrolled systematic uncertainty that should be avoided. It is indeed astonishing how often one finds in the literature statements of precise determinations of hadron properties based on extrapolation functions that are not rooted in QCD or are at best models with a questionable relation to QCD. Below we address the issue of to what extent CHPT representations and LQCD results already overlap.

CHPT provides unambiguous chiral extrapolation functions, parameterized in terms of the pertinent LECs. Ideally, one would like to perform global fits to a large variety of observables because these are interrelated through the appearance of certain LECs. At present, however, this is not yet possible, and for a variety of applications it is mandatory to include phenomenological input for some of the LECs; an example is given below. Note that frequently in the literature, one-loop extrapolation functions are used for pion masses well outside the regime of their applicability. As a matter of fact, most pion and kaon Green functions have been worked out at two-loop accuracy in the continuum (for a review, see Reference 24), and considerable progress has been reported for many of these quantities for partially quenched QCD, in which one allows for different values of the valence and sea quarks (see References 124–127). Clearly, with increasing quark masses, the CHPT representations become increasingly inaccurate, and this is in fact a strength of the EFT in that it provides a measure of the theoretical uncertainty.

Let us illustrate these issues for the pion decay constant F_π. Its special role for spontaneous symmetry breaking was already explained in Section 2. Its analytic form at two-loop order in the continuum is (62, 128)

$$
\begin{aligned}
F_\pi &= F\left[1 + X\tilde{\Delta}^{(4)} + X^2\tilde{\Delta}^{(6)}\right] \\
&= F\left\{1 + X[\tilde{L} + \tilde{\ell}_4] + X^2\left[-\frac{3}{4}\tilde{L}^2 + \tilde{L}\left(-\frac{7}{6}\tilde{\ell}_1 - \frac{4}{3}\tilde{\ell}_2 + \tilde{\ell}_4 - \frac{29}{12}\right)\right.\right. \\
&\quad \left.\left. + \frac{1}{2}\tilde{\ell}_3\tilde{\ell}_4 + \frac{1}{12}\tilde{\ell}_1 - \frac{1}{3}\tilde{\ell}_2 - \frac{13}{192} + \tilde{r}(\mu)\right]\right\}, \\
X &= \frac{M_\pi^2}{16\pi^2 F^2}, \quad \tilde{L} = \log\frac{\mu^2}{M_\pi^2}, \quad \tilde{\ell}_i = \log\frac{\Lambda_i^2}{\mu^2},
\end{aligned}
$$
53.

where $\tilde{r}(\mu)$ is a combination of dimension-six LECs. Note that, in contrast to common lore, the number of new LECs does not explode when going to higher orders when one considers specific observables. With $\Lambda_1 = 0.12^{+0.04}_{-0.03}$ GeV, $\Lambda_2 = 1.20^{+0.06}_{-0.06}$ GeV, $\Lambda_3 = 0.59^{+1.40}_{-0.41}$ GeV, $\Lambda_4 = 1.25^{+0.04}_{-0.03}$ GeV, and $\tilde{r}(\mu) = 0 \pm 3$ from resonance saturation, varying the scale of dimensional regularization μ between 500 MeV and 1 GeV and using $F_\pi = 92.4 \pm 0.3$ MeV, one obtains the (yellow) band between the solid lines in **Figure 6**. For comparison, the dashed lines correspond to the one-loop result. Note that for pion masses below 300 MeV, the one- and two-loop representations are essentially equivalent, but for higher pion masses it is important to include the two-loop corrections for a realistic assessment of the theoretical uncertainty. This is consistent with expectations based on naive dimensional analysis; the expansion parameter is $X = 0.01, 0.07, 0.18$ for $M_\pi = 139.57, 300, 500$ MeV, respectively. For orientation, we also show the recent lattice results from Reference 129. In that paper, the one-loop representation for M_π and F_π is used and the LEC $\tilde{\ell}_4$ is extracted with an accuracy of a few percent. This appears optimistic if one accounts for the two-loop corrections at these pion masses, as shown in **Figure 6**. Other collaborations such as QCDSF, ETM, and CERN-Roma have also obtained results for F_π at low pion

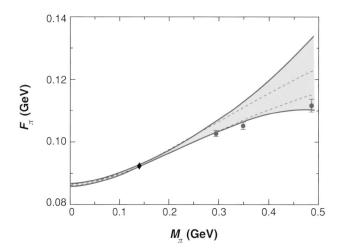

Figure 6

Pion (quark) mass dependence of the pion decay constant. Lattice results from Reference 129 (*purple circles*) in comparison with the two-loop CHPT result from Reference 128 (*yellow area*). The one-loop band obtained by varying $\tilde{\ell}_4$ is given by the dashed red lines. The black diamond is the physical point.

masses. However, these data had not been fully analyzed by the time this review was written (see Reference 130 for some of these preliminary results).

Matters are somewhat different in the baryon (nucleon) sector for two reasons. Although a multitude of ground-state (and some excited state) properties have been simulated, most CHPT calculations have been performed to one-loop accuracy. Furthermore, in these extrapolation functions, one has even and odd powers of the expansion parameter(s) and correspondingly more LECs. For these reasons, chiral extrapolations can be performed only over a smaller range of pion masses with a tolerable uncertainty as compared with the meson sector. Space does not allow for an overall review here; thus, we refer the reader to Reference 40. We briefly discuss the quark mass dependence of the nucleon axial-vector coupling g_A here, as a variety of lattice data exists for pion masses ranging from 340 MeV to 1 GeV. Also, the two-loop representation of g_A was recently published (131). This allows one to discuss some subtleties that can arise in the chiral expansion of certain observables. The pion mass expansion of g_A takes the form

$$g_A = g_0 \left\{ \underbrace{1}_{\text{tree}} + \underbrace{\Delta^{(2)} + \Delta^{(3)}}_{\text{1-loop}} + \underbrace{\Delta^{(4)} + \Delta^{(5)}}_{\text{2-loop}} \right\} + \mathcal{O}\left(M_\pi^6\right), \qquad 54.$$

where g_0 is the chiral limit value of g_A and $\Delta^{(n)}$ collects the corrections proportional to M_π^n. At one-loop order, the chiral representation of g_A contains g_0, one combination of dimension-two LECs ($c_2 + c_3$), and one dimension-three LEC (d_{16}). The latter two can be determined from the analysis of $\pi N \rightarrow \pi N$ and $\pi N \rightarrow \pi \pi N$, respectively. This is one case where it is mandatory to include such phenomenological information to analyze the chiral expansion of a given observable. The one-loop representation is dominated by the M_π^3 term as the pion mass increases. One is therefore unable to connect to the lattice results, which show a very weak dependence of g_A on the pion mass. At two loops, one has further operators accompanied by certain combinations of LECs. The dominant contributions to these stem from $1/m$ corrections to the

Figure 7

Pion mass dependence of
the axial-vector coupling.
Lattice results from
LHPC/MILC (132)
(*orange triangles*) and
QCDSF (Göckeler M,
private communication)
(*inverted purple triangles*) in
comparison with the
two-loop CHPT result
from Reference 131
(*yellow area*). The black
circle is the physical point.

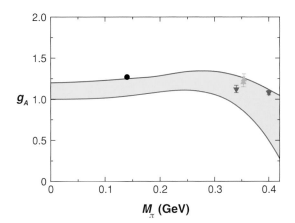

lower-order LECs. The remaining pieces can be estimated assuming naturalness only. Demanding further that the trend of the lattice data is followed, one arrives at the (yellow) band shown in **Figure 7**. The width of the band is a consequence of this condition. There are strong cancellations between the various contributions. Also, one sees that the range of applicability of the chiral extrapolation function and the lattice results barely overlap. To arrive at precision results for g_A, pion masses of less than 300 MeV are mandatory.

In view of this, the claim in Reference 132 that g_A has been calculated and determined precisely in the chiral regime appears overly optimistic. More data at lower pion masses are called for to substantiate such claims. Also note that these strong cancellations between various orders were first found in an EFT approach including the $\Delta(1232)$ as an active degree of freedom and counting the nucleon-delta mass splitting $m_\Delta - m_N \simeq 3 F_\pi$ as an additional small parameter at leading one-loop order (see Reference 133 and the recent update, Reference 134). The claims in these papers that one can truthfully represent the chiral expansion of g_A in leading one-loop order for pion masses up to 700 MeV once the delta is included are unfounded. The resummation of some subset of corrections induces an uncontrolled uncertainty because other higher-order effects are ignored. A beautiful example for this is provided by the cancellations of delta contributions and higher-order πN loop effects in the nucleon's magnetic polarizabilities (135). Note also that finite volume corrections for g_A, which are essential for connecting the lattice results with the real world, are discussed in References 136–138.

Finally, we note that CHPT practitioners also need the lattice. In particular, for nonleptonic, weak meson interactions and in the baryon sector, the amount of precise phenomenological information is limited. Therefore, it appears impossible to pin down all LECs, in particular the symmetry breakers (see Section 4). Since the pioneering work in Reference 139 in the mid-1990s, the ALPHA Collaboration has reported considerable progress in the lattice determination of the next-to-leading-order LECs in the meson sector (140). Further work on the leading-order LECs for $\Delta S = 1$ transitions are reported in Reference 141, and a method to determine the pion-nucleon LEC c_3 was suggested in Reference 142. In our opinion, more effort should be invested in the determination of LECs from lattice studies, and this would further tighten the interplay between the CHPT and the lattice communities. This is important because it will take a long time before complicated processes or reactions with many external probes will be amenable to lattice simulations.

ACKNOWLEDGMENTS

We are very grateful to Jürg Gasser and Heiri Leutwyler for many useful comments and communications. We have also profited from discussions with Hans Bijnens, Gilberto Colangelo, John Donoghue, Gerhard Ecker, Bastian Kubis, Akaki Rusetsky, and Jan Stern. Partial financial support under the European Union Integrated Infrastructure Initiative Hadron Physics Project (contract number RII3-CT-2004-506078), by the Deutsche Forschungsgemeinschaft (TR 16, "Subnuclear Structure of Matter"), and by the Bundesministerium für Bildung und Forschung (research grant number 06BN411) is gratefully acknowledged. This work was supported in part by the European Union contract number MRTN-CT-2006-035482, "FLAVIAnet."

LITERATURE CITED

1. Peccei RD, Quinn HR. *Phys. Rev. Lett.* 38:1440 (1970)
2. Vafa C, Witten E. *Nucl. Phys. B* 234:173 (1984)
3. McNeile C. *Phys. Lett. B* 619:124 (2005)
4. Banks T, Casher A. *Nucl. Phys. B* 169:103 (1980)
5. Leutwyler H, Smilga A. *Phys. Rev. D* 46:5607 (1992)
6. Stern J. hep-ph/9801282 (1998)
7. Goldstone J. *Nuovo Cim.* 19:154 (1961)
8. Goldstone J, Salam A, Weinberg S. *Phys. Rev.* 127:965 (1962)
9. Stern J. hep-ph/0611127 (2006)
10. Gell-Mann M, Ne'eman Y. *The Eightfold Way*. New York: W.A. Benjamin (1964)
11. Fujikawa K. *Phys. Rev. D* 29:285 (1984)
12. Bijnens J. *Int. J. Mod. Phys. A* 8:3045 (1993)
13. Donoghue JF. Presented at Int. Sch. Low Energy Antiprotons, Erice, Italy, Jan. 25–31, 1990. CERN-TH-5667–90 (1990)
14. Meißner U-G. *Rep. Prog. Phys.* 56:903 (1993)
15. Leutwyler H. hep-ph/9406283 (1994)
16. Ecker G. *Prog. Part. Nucl. Phys.* 35:1 (1995)
17. Bernard V, Kaiser N, Meißner U-G. *Int. J. Mod. Phys. E* 4:193 (1995)
18. Pich A. *Rep. Prog. Phys.* 58:563 (1995)
19. Gasser J. *Nucl. Phys. Proc. Suppl.* 86:257 (2000)
20. Leutwyler H. hep-ph/0008124 (2000)
21. Scherer S. *Adv. Nucl. Phys.* 27:277 (2003)
22. Gasser J. *Lect. Notes Phys.* 629:1 (2004)
23. Ecker G. *Am. Inst. Phys. Conf. Proc.* 806:1 (2006)
24. Bijnens J. hep-ph/0604043 (2006)

25. Weinberg S. *Physica A* 96:327 (1979)
26. Gasser J, Leutwyler H. *Ann. Phys.* 158:142 (1984)
27. Gasser J, Leutwyler H. *Nucl. Phys. B* 250:465 (1985)
28. Leutwyler H. *Ann. Phys.* 235:165 (1994)
29. Weinberg S. hep-ph/9412326 (1994)
30. Fuchs NH, Sazdjian H, Stern J. *Phys. Lett. B* 269:183 (1991)
31. Bijnens J, Colangelo G, Ecker G. *JHEP* 9902:020 (1999)
32. Gasser J, Leutwyler H. *Phys. Rep.* 87:77 (1982)
33. Urech R. *Nucl. Phys. B* 433:234 (1995)
34. Meißner U-G, Müller G, Steininger S. *Phys. Lett. B* 406:154 (1997)
35. Knecht M, Neufeld H, Rupertsberger H, Talavera P. *Eur. Phys. J. C* 12:469 (2000)
36. Schweizer J. *JHEP* 0302:007 (2003)
37. Gasser J, Rusetsky A, Scimemi I. *Eur. Phys. J. C* 32:97 (2003)
38. Meißner U-G. *At the Frontier of Particle Physics*, Vol. 1, ed. M Shifman, p. 417. Singapore: World Sci. (2001)
39. Scherer S. *Int. J. Mod. Phys. A* 21:881 (2006)
40. Bernard V. Chiral perturbation theory and baryon properties. *Prog. Nucl. Part. Phys.* In press (2007)
41. Müller G, Meißner UG. *Nucl. Phys. B* 556:265 (1999)
42. Gasser J, et al. *Eur. Phys. J. C* 26:13 (2002)
43. Georgi H. *Annu. Rev. Nucl. Part. Sci.* 43:209 (1993)
44. Espriu D, Matias J. *Nucl. Phys. B* 418:494 (1994)
45. Gasser J, Zepeda A. *Nucl. Phys. B* 174:445 (1980)
46. Bijnens J, Colangelo G, Ecker G. *Ann. Phys.* 280:100 (2000)
47. Ecker G, Gasser J, Pich A, de Rafael E. *Nucl. Phys. B* 321:311 (1989)
48. Ecker G, et al. *Phys. Lett. B* 223:425 (1989)
49. Donoghue JF, Ramirez C, Valencia G. *Phys. Rev. D* 39:1947 (1989)
50. Kampf K, Moussallam B. *Eur. Phys. J. C* 47:723 (2006)
51. Cirigliano V, et al. *Nucl. Phys. B* 753:139 (2006)
52. Bernard V, Kaiser N, Meißner U-G. *Nucl. Phys. A* 615:483 (1997)
53. Epelbaum E, Meißner UG, Glöckle W, Elster C. *Phys. Rev. C* 65:044001 (2002)
54. Bernard V, Hemmert TR, Meißner UG. *Nucl. Phys. A* 732:149 (2004)
55. Donoghue JF, Gasser J, Leutwyler H. *Nucl. Phys. B* 343:341 (1990)
56. Bijnens J, Talavera P. *JHEP* 0203:046 (2002)
57. Beg MAB, Zepeda A. *Phys. Rev. D* 6:2912 (1972)
58. Truong TN. CERN-TH-4748/87 (1987)
59. Donoghue JF, Holstein BR. *Phys. Rev. D* 48:137 (1993)
60. Donoghue JF. hep-ph/9506205 (1995)
61. Gasser J, Meißner U-G. *Nucl. Phys. B* 357:90 (1991)
62. Bijnens J, Colangelo G, Talavera P. *JHEP* 9805:014 (1998)
63. Moussallam B. *Eur. Phys. J. C* 14:111 (2000)
64. Meißner U-G, Oller JA. *Nucl. Phys. A* 679:671 (2001)
65. Frink M, Kubis B, Meißner U-G. *Eur. Phys. J. C* 25:259 (2002)
66. Ananthanarayan B, et al. *Phys. Lett. B* 602:218 (2004)

67. Lähde TA, Meißner U-G. *Phys. Rev. D* 74:034021 (2006)
68. Lehmann H. *Phys. Lett. B* 41:529 (1972)
69. Roy SM. *Phys. Lett. B* 36:353 (1971)
70. Colangelo G, Gasser J, Leutwyler H. *Phys. Lett. B* 488:261 (2000)
71. Knecht M, Moussallam B, Stern J, Fuchs NH. *Nucl. Phys. B* 457:513 (1995)
72. Bijnens J, et al. *Phys. Lett. B* 374:210 (1996)
73. Bijnens J, et al. *Nucl. Phys. B* 508:263 (1997)
74. Ananthanarayan B, Colangelo G, Gasser J, Leutwyler H. *Phys. Rep.* 353:207 (2001)
75. Descotes-Genon S, Fuchs NH, Girlanda L, Stern J. *Eur. Phys. J. C* 24:469 (2002)
76. Kaminski R, Pelaez JR, Yndurain FJ. *Phys. Rev. D* 74:014001 (2006)
77. Pelaez JR, Yndurain FJ. *Phys. Rev. D* 68:074005 (2003)
78. Meißner U-G. *Nucl. Phys. A* 629:72c (1998)
79. Cabibbo N. *Phys. Rev. Lett.* 93:121801 (2004)
80. Cabibbo N, Isidori G. *JHEP* 0503:021 (2005)
80a Gamiz E, Prades J, Scimemi I. hep-ph/0602023 (2006)
81. Colangelo G, Gasser J, Kubis B, Rusetsky A. *Phys. Lett. B* 638:187 (2006)
82. Donoghue JF. *Annu. Rev. Nucl. Part. Sci.* 39:1 (1989)
83. Leutwyler H. *Phys. Lett. B* 378:313 (1996)
84. Kaplan DB, Manohar AV. *Phys. Rev. Lett.* 56:2004 (1986)
85. Amoros G, Bijnens J, Talavera P. *Nucl. Phys. B* 602:87 (2001)
86. Aubin C, et al. (MILC Collab.) *Phys. Rev. D* 70:114501 (2004)
87. Göckeler M, et al. (QCDSF Collab.) *Phys. Lett. B* 639:307 (2006)
88. Descotes-Genon S, Fuchs NH, Girlanda L, Stern J. *Eur. Phys. J. C* 34:201 (2004)
89. Gell-Mann M, Oakes RJ, Renner B. *Phys. Rev.* 175:2195 (1968)
90. Colangelo G, Gasser J, Leutwyler H. *Phys. Rev. Lett.* 86:5008 (2001)
91. Gasser J, Leutwyler H. *Phys. Lett. B* 125:325 (1983)
92. Meißner U-G. *Comments Nucl. Part. Phys.* 20:119 (1991)
93. Morgan D, Shaw G. *Nucl. Phys. B* 10:261 (1969)
94. Pislak S, et al. *Phys. Rev. D* 67:072004 (2003)
95. Adeva B, et al. (DIRAC Collab.) *Phys. Lett. B* 619:50 (2005)
96. Batley JR, et al. (NA48/2 Collab.) *Phys. Lett. B* 633:173 (2006)
97. Weinberg S. *Phys. Rev. Lett.* 17:616 (1966)
98. Bernard V, Kaiser N, Meißner U-G. *Phys. Rev. D* 43:2757 (1991)
99. Bernard V, Kaiser N, Meißner U-G. *Nucl. Phys. B* 357:129 (1991)
100. Bijnens J, Dhonte P, Talavera P. *JHEP* 0405:036 (2004)
101. Ananthanarayan B, Buettiker P, Moussallam B. *Eur. Phys. J. C* 22:133 (2001)
102. Buettiker P, Descotes-Genon S, Moussallam B. *Eur. Phys. J. C* 33:409 (2004)
103. Roessl A. *Nucl. Phys. B* 555:507 (1999)
104. Kubis B, Meißner U-G. *Phys. Lett. B* 529:69 (2002)
105. Schweizer J. *Phys. Lett. B* 625:217 (2005)
106. Bernard V, Kaiser N, Meißner U-G. *Z. Phys. C* 70:483 (1996)
107. De Baenst P. *Nucl. Phys. B* 24:633 (1970)
108. Vainshtein AI, Zakharov VI. *Nucl. Phys. B* 36:589 (1972)
109. Bernard V, Kaiser N, Gasser J, Meißner U-G. *Phys. Lett. B* 268:291 (1991)

110. Bernstein AM, et al. *Phys. Rev. C* 55:1509 (1997)

111. Beane SR, et al. *Nucl. Phys. A* 618:381 (1997)

112. Bernard V, Kaiser N, Meißner U-G. *Eur. Phys. J. A* 11:209 (2001)

113. Schmidt A, et al. *Phys. Rev. Lett.* 87:232501 (2001)

114. Pasquini B, Drechsel D, Tiator L. *Eur. Phys. J. A* 23:279 (2005)

115. Bernard V, Kubis B, Meißner U-G. *Eur. Phys. J. A* 25:419 (2005)

116. Pasquini B, Drechsel D, Tiator L. *Eur. Phys. J. A* 27:231 (2006)

117. Büttiker P, Meißner U-G. *Nucl. Phys. A* 668:97 (2000)

118. Becher T, Leutwyler H. *JHEP* 0106:017 (2001)

119. Kaplan DB. *Phys. Lett. B* 288:342 (1992)

120. Narayanan R, Neuberger H. *Nucl. Phys. B* 443:305 (1995)

121. Symanzik K. *Nucl. Phys. B* 226:187 (1983)

122. Symanzik K. *Nucl. Phys. B* 226:205 (1983)

123. Sharpe SR. hep-lat/0607016 (2006)

124. Bijnens J, Danielsson N, Lähde TA. *Phys. Rev. D* 70:111503 (2004)

125. Bijnens J, Lähde TA. *Phys. Rev. D* 71:094502 (2005)

126. Bijnens J, Lähde TA. *Phys. Rev. D* 72:074502 (2005)

127. Bijnens J, Danielsson N, Lähde TA. *Phys. Rev. D* 73:074509 (2006)

128. Colangelo G, Dürr S. *Eur. Phys. J. C* 33:543 (2004)

129. Beane SR, Bedaque PF, Orginos K, Savage MJ. (NPLQCD Collab.) *Phys. Rev. D* 73:054503 (2006)

130. Meißner U-G, Schierholz G. hep-lat/0611072 (2006)

131. Bernard V, Meißner U-G. *Phys. Lett. B* 639:278 (2006)

132. Edwards RG, et al. (LHPC Collab.) *Phys. Rev. Lett.* 96:052001 (2006)

133. Hemmert TR, Procura M, Weise W. *Phys. Rev. D* 68:075009 (2003)

134. Procura M, Musch BU, Hemmert TR, Weise W. hep-lat/0610105 (2006)

135. Bernard V, Kaiser N, Schmidt A, Meißner U-G. *Phys. Lett. B* 319:269 (1993)

136. Beane SR, Savage MJ. *Phys. Rev. D* 70:074029 (2004)

137. Colangelo G, Fuhrer A, Haefeli C. *Nucl. Phys. Proc. Suppl.* 153:41 (2006)

138. Khan AA, et al. hep-lat/0603028 (2006)

139. Myint S, Rebbi C. *Nucl. Phys. B* 421:241 (1994)

140. Heitger J, Sommer R, Wittig H. (ALPHA Collab.) *Nucl. Phys. B* 588:377 (2000)

141. Giusti L, et al. *JHEP* 0404:013 (2004)

142. Bedaque PF, Grießhammer HW, Rupak G. *Phys. Rev. D* 71:054015 (2005)

What Do Electromagnetic Plasmas Tell Us about the Quark-Gluon Plasma?

Stanisław Mrówczyński[1] and Markus H. Thoma[2]

[1] Institute of Physics, Świętokrzyska Academy, PL-25-406 Kielce, Poland; and Sołtan Institute for Nuclear Studies, PL-00-681 Warsaw, Poland; email: mrow@fuw.edu.pl

[2] Max-Planck-Institute for Extraterrestrial Physics, D-85741 Garching, Germany; email: mthoma@mpe.mpg.de

Annu. Rev. Nucl. Part. Sci. 2007. 57:61–94

First published online as a Review in Advance on April 4, 2007

The *Annual Review of Nuclear and Particle Science* is online at http://nucl.annualreviews.org

This article's doi:
10.1146/annurev.nucl.57.090506.123124

Copyright © 2007 by Annual Reviews.
All rights reserved

0163-8998/07/1123-0061$20.00

Key Words

relativistic heavy-ion collisions

Abstract

Because the quark-gluon plasma (QGP) reveals some obvious similarities to the well-known electromagnetic plasma (EMP), an accumulated knowledge on EMPs can be used in QGP studies. After discussing similarities and differences of the two systems, we present the theoretical tools used to describe the plasmas. The tools include kinetic theory, hydrodynamic approach, and diagrammatic perturbative methods. We consider collective phenomena in the plasma, with a particular emphasis on instabilities that crucially influence the temporal evolution of the system. Finally, properties of strongly coupled plasma are discussed.

Contents

1. INTRODUCTION 62
2. THEORETICAL TOOLS 64
    2.1. Transport Theory 64
    2.2. Hydrodynamic Approach 68
    2.3. Diagrammatic Methods 73
3. COLLECTIVE PHENOMENA 74
    3.1. Screening 75
    3.2. Collective Modes 76
    3.3. Instabilities 79
    3.4. Energy Loss 84
4. STRONGLY COUPLED PLASMAS 86

1. INTRODUCTION

Plasma—the ionized gas of electrons and ions—has been actively studied since its discovery in a discharge tube at the end of nineteenth century. The term plasma was introduced by Irving Langmuir in 1929. Prospects to get a practically unlimited source of energy due to nuclear fusion reactions in a hot ionized gas of hydrogen isotopes have stimulated a large-scale program to study plasmas in terrestrial experiments for more than half a century. Plasmas are also actively studied by astrophysicists, as it appears to be the most common phase of matter. Approximately 99% of the entire visible Universe is in the plasma phase. Not only are stars formed of ionized gas, but the interstellar and intergalactic mediums are also plasmas, although very sparse ones. Principles of plasma physics can be found in, for example, well-known textbooks (1, 2).

The quark-gluon plasma (QGP) is the system of quarks and gluons that are not confined in the hadron's interiors but can move freely in a whole volume occupied by the system. A broad presentation of the whole field of QGP physics is contained in three volumes of review articles (3–5); the lectures (6) can serve as an elementary introduction. Active studies of the QGP started in the mid-1980s when relativistic heavy-ion collisions offered an opportunity to create a drop of the QGP in a laboratory. The experimental programs at CERN and BNL provided evidence of the QGP production at the early stage of nucleus-nucleus collisions, when the system is extremely hot and dense, but properties of the QGP remain enigmatic. So, one can ask, what do electromagnetic plasmas (EMPs) tells us about the QGP?

The QGP reveals some obvious similarities to the well-known EMP, as quantum chromodynamics (QCD) describing the interactions of the quarks and gluons resembles quantum electrodynamics (QED), which governs interactions of charged objects. Thus, some lessons from EMPs should be useful in the exploration of the QGP. The aim of this review is to discuss what QGP physicists can actually learn from their EMP colleagues, and how the huge accumulated knowledge on EMPs can be used in QGP studies. However, we must be aware not only of similarities but also of important differences between EMPs and the QGP. Some differences are of rather trivial origin, but some are deeply rooted in the dynamical foundations of the two systems.

Let us enumerate these trivial dissimilarities. The QGP is usually relativistic or even ultrarelativistic, whereas the EMP is mostly nonrelativistic in laboratory experiments. The differences between the nonrelativistic and relativistic plasmas go far beyond the kinematics of motion of plasma particles. For example, let us consider the plasma's composition. In the nonrelativistic system, there are particles but no antiparticles, and the particle's number is conserved. In the relativistic system, we have both particles and antiparticles (as electrons and positrons in EMPs), and the lepton number—not the particle's number—is conserved. Particle number density is not a proper way to characterize the system. For this reason, QGP physicists use the baryon and strangeness densities.

Another trivial but very important distinctive feature of the EMP is the huge mass difference between electrons and ions, which is responsible for a specific dynamic role of heavy ions. The ions are usually treated as a passive background, which merely compensates the charge of electrons, but electro-ion collisions drive the system toward equilibrium and maintain the equilibrium. However, the energy transfer between electrons and ions is very inefficient, and their mutual equilibration is very slow. Therefore, we have electron and ion fluids of different temperatures for a relatively long time. There is nothing similar in the QGP. There are heavy quarks—charm, bottom, and top—that are, however, much less populated than the light quarks and gluons, and their lifetime is short. Therefore, the heavy quarks hardly influence the QGP dynamics.

The EMP, which is the closest analog of the QGP, is the relativistic system of electrons, positrons, and photons. Such a plasma is actually studied in the context of some astrophysical applications, for example, supernovae explosions. The differences between the QGP and the EMP are then of dynamical origin: The first one is governed by QCD and the second one by QED. The latter theory is Abelian, whereas the former one is non-Abelian with a prominent role for gluons that carry color charges and, thus, not only mediate the interaction among colored quarks and antiquarks but interact among themselves. Gluons, in contrast to photons, also contribute to the density of color charges and to the color current.

The most important common feature of the EMP and the QGP is the collective character of the dynamics. The range of electrostatic interaction is, in spite of the screening, usually much larger than the interparticle spacing. There are many particles in the Debye sphere—the sphere of the radius equal to the effective interaction range—and motion of these particles is highly correlated. There is a similar situation in the deconfined perturbative phase of QCD (7). The Debye mass is of order gT, where g is the QCD constant and T is the temperature. Because the particle density in QGP is of order T^3, the number of partons in the Debye sphere, which is roughly $1/g^3$, is large in the weakly coupled ($1/g \gg 1$) QGP.

In various laboratory experiments, the EMP is embedded in an external electromagnetic field. For example, the magnetic field is used to trap the plasma, and there are numerous fascinating phenomena occurring in such a situation. In the case of QGPs produced in relativistic heavy-ion collisions, it is hard to imagine any external chromodynamic field applied to the plasma. Therefore, we consider here only the systems in which fields are generated self-consistently in the plasma.

Our review is organized as follows: Theoretical tools, which are used to describe the plasmas, are presented in Section 2. The tools include the kinetic theory, the hydrodynamic approach, and diagrammatic methods of field theory. In Section 3 we discuss collective phenomena that are the most characteristic feature of plasmas. After explaining the phenomenon of screening, quasi-particle modes in the equilibrium and nonequilibrium plasma are presented. We pay much attention to instabilities that crucially influence plasma dynamics. The problem of a particle's energy loss in a plasma is also discussed. Section 4 is devoted to the strongly coupled plasma, which reveals particularly interesting properties.

Throughout the review we use the natural units, with $c = \hbar = k_B = 1$ and the metric $(1, -1, -1, -1)$. However, this gets a bit complicated. Plasma physicists usually use the Gauss (CGS) units, where the fine structure constant equals $\alpha = e^2 \approx 1/137$, and the electromagnetic counterpart of the units usually applied in QCD is the so-called Heaviside-Lorentz system, where the 4π factor does not show up in the Maxwell equations but $\alpha = e^2/4\pi$. We stick to the traditionally used units in the two fields of physics, and thus the factor of 4π must be additionally taken into account when comparing EMP and QGP formulas.

2. THEORETICAL TOOLS

2.1. Transport Theory

Transport theory provides a natural framework for studying equilibrium and nonequilibrium plasmas. The central object of the theory is the distribution function, which describes a time-dependent distribution of particles in a phase-space spanned by the particle's momenta and positions. The distribution function of each plasma component evolves owing to the interparticle collisions and the interaction with an external and/or self-consistently generated mean field. The two dynamical effects give rise to the collision and mean-field terms of a transport equation satisfied by the distribution function.

2.1.1. Electromagnetic plasma. A formulation of the kinetic theory of relativistic plasma can be found in Reference 8. The distribution function is denoted as $f_n(\mathbf{p}, x)$, with the index n labeling plasma components: electrons, positrons, ions. Spin is usually treated as an internal degree of freedom. The function depends on the four-position $x \equiv (t, \mathbf{x})$ and the three-momentum $\mathbf{p}$. The four-momentum p obeys the mass-shell constraint $p^2 = m^2$, where m is the particle mass. Then $p \equiv (E_p, \mathbf{p})$, with $E_p \equiv \sqrt{m^2 + \mathbf{p}^2}$.

The distribution function satisfies the transport equation

$$\left(p_\mu \partial^\mu + q_n p^\mu F_{\mu\nu} \partial_p^\nu \right) f_n(\mathbf{p}, x) = C[f_n], \qquad 1.$$

where $C[f_n]$ denotes the collision term, q_n is the charge of the plasma species n, and $F^{\mu\nu}$ is the electromagnetic strength tensor that either represents an external field applied to the system and/or is generated self-consistently by the four currents present in the plasma:

$$\partial^\mu F_{\mu\nu} = 4\pi j_\nu,$$

where

$$j^\mu(x) = \sum_n q_n \int \frac{d^3 p}{(2\pi)^3} \frac{p^\mu}{E_p} f_n(\mathbf{p}, x). \qquad 2.$$

The transport equation can be easily solved in the linear-response approximation when the collision term is neglected. The equation is linearized around the stationary and homogeneous state described by the distribution $\bar{f}_n(\mathbf{p})$. The state is also assumed to be neutral, and there are no currents. The distribution function is then decomposed as

$$f_n(\mathbf{p}, x) = \bar{f}_n(\mathbf{p}) + \delta f_n(\mathbf{p}, x),$$

where $\bar{f}_n(\mathbf{p}) \gg \delta f_n(\mathbf{p}, x)$.

The transport equation is linearized in δf_n, and $F^{\mu\nu}$ can be exactly solved after the Fourier transformation, which is defined as

$$f(k) = \int d^4 x e^{ikx} f(x), \quad f(x) = \int \frac{d^4 k}{(2\pi)^4} e^{-ikx} f(k). \qquad 3.$$

Then, one finds $\delta f_n(\mathbf{p}, k)$, which is the Fourier transform of $\delta f_n(\mathbf{p}, x)$, and the induced current, which can be written as

$$\delta j^\mu(k) = -\Pi^{\mu\nu}(k) A_\nu(k), \qquad 4.$$

with the polarization tensor equal to

$$\Pi^{\mu\nu}(k) = 4\pi \sum_n q_n^2 \int \frac{d^3 p}{(2\pi)^3} \bar{f}_n(\mathbf{p}) \frac{(p \cdot k)^2 g^{\mu\nu} + k^2 p^\mu p^\nu - (p \cdot k)(k^\mu p^\nu + k^\nu p^\mu)}{(p \cdot k)^2}. \qquad 5.$$

The tensor is symmetric $[\Pi^{\mu\nu}(k) = \Pi^{\nu\mu}(k)]$ and transverse $[k_\mu \Pi^{\mu\nu}(k) = 0]$, which guarantees that the current given by Equation 4 is gauge independent.

For isotropic plasmas, the polarization tensor has only two independent components, which are usually chosen as

$$\Pi_L(k) = \Pi_{00}(k),$$

$$\Pi_T(k) = \frac{1}{2} \left(\delta_{ij} - \frac{k_i k_j}{\mathbf{k}^2} \right) \Pi_{ij}(k), \qquad 6.$$

where the indices $i, j = 1, 2, 3$ label three-vector and tensor components. In the case of an ultrarelativistic $(T \gg m)$ electron-positron equilibrium plasma, the momentum integral in Equation 5 can be performed analytically in the high-temperature limit

$(T \gg \omega, |\mathbf{k}|)$, and the result already derived by Silin (9) in 1960 reads

$$\Pi_L(k) = -3m_\gamma^2 \left[1 - \frac{\omega}{2|\mathbf{k}|} \ln \frac{\omega + |\mathbf{k}|}{\omega - |\mathbf{k}|}\right],$$

$$\Pi_T(k) = \frac{3}{2} m_\gamma^2 \frac{\omega^2}{\mathbf{k}^2} \left[1 - \left(1 - \frac{\mathbf{k}^2}{\omega^2}\right) \frac{\omega}{2|\mathbf{k}|} \ln \frac{\omega + |\mathbf{k}|}{\omega - |\mathbf{k}|}\right], \qquad\qquad 7.$$

where $k \equiv (\omega, \mathbf{k})$ and $m_\gamma \equiv e\,T/3$ denotes the thermal photon mass generated by the interaction of the photons with the electrons and positrons.

The above polarization tensor was found in the collisionless limit of the transport equation. The effect of collisions can be easily taken into account if the so-called Bhatnagar-Gross-Krook (BGK) collision term is used in the transport equation (10). The result for an ultrarelativistic equilibrium plasma is given in Reference 11.

2.1.2. Quark-gluon plasma.

The transport theory of the QGP (12, 13) appears to be much more complicated than its electromagnetic counterpart. The distribution function of quarks $Q(\mathbf{p}, x)$ is a hermitian $N_c \times N_c$ matrix in color space [for an SU(N_c) color gauge group]. The distribution function is gauge dependent, and it transforms under a local gauge transformation $U(x)$ as

$$Q(\mathbf{p}, x) \rightarrow U(x) Q(\mathbf{p}, x) U^\dagger(x). \qquad\qquad 8.$$

Here and in most cases below, the color indices are suppressed. The distribution function of antiquarks, which we denote by $\tilde{Q}(\mathbf{p}, x)$, is also a hermitian $N_c \times N_c$ matrix, and it transforms according to Equation 8. The distribution function of gluons is a hermitian $(N_c^2 - 1) \times (N_c^2 - 1)$ matrix, which transforms as

$$G(\mathbf{p}, x) \rightarrow \mathcal{U}(x) G(\mathbf{p}, x) \mathcal{U}^\dagger(x), \qquad\qquad 9.$$

where

$$\mathcal{U}_{ab}(x) = 2\mathrm{Tr}[\tau^a U(x) \tau^b U^\dagger(x)],$$

with $\tau^a, a = 1, \ldots, N_c^2 - 1$ being the SU(N_c) group generators in the fundamental representation with $\mathrm{Tr}(\tau^a \tau^b) = \frac{1}{2}\delta^{ab}$.

The color current is expressed in the fundamental representation as

$$j^\mu(x) = -\frac{g}{2} \int \frac{d^3p}{(2\pi)^3} p^\mu \left[Q(\mathbf{p}, x) - \tilde{Q}(\mathbf{p}, x)\right]$$

$$- \frac{1}{N_c} \mathrm{Tr}\left[Q(\mathbf{p}, x) - \tilde{Q}(\mathbf{p}, x)\right] + 2\tau^a \mathrm{Tr}\left[T^a G(\mathbf{p}, x)\right], \qquad 10.$$

where g is the QCD coupling constant. A sum over helicities, two per particle, and over quark flavors N_f is understood in Equation 10, even though it is not explicitly written down. The SU(N_c) generators in the adjoint representation are expressed through the structure constants $T_{bc}^a = -i f_{abc}$, and are normalized as $\mathrm{Tr}[T^a T^b] = N_c \delta^{ab}$. The current can be decomposed as $j^\mu(x) = j_a^\mu(x)\tau^a$, with $j_a^\mu(x) = 2\mathrm{Tr}(\tau_a j^\mu(x))$. The distribution functions, which are proportional to the unit matrix in color space, are gauge independent, and they provide the color current (Equation 10) that vanishes identically.

Gauge-invariant quantities are given by the traces of the distribution functions. Thus, the baryon current and the energy-momentum tensor read

$$b^\mu(x) = \frac{1}{3} \int \frac{d^3 p}{(2\pi)^3} \, p^\mu \text{Tr} \left[Q(\mathbf{p}, x) - \tilde{Q}(\mathbf{p}, x) \right],$$

$$T^{\mu\nu}(x) = \int \frac{d^3 p}{(2\pi)^3} \, p^\mu p^\nu \text{Tr} \left[Q(\mathbf{p}, x) + \tilde{Q}(\mathbf{p}, x) + G(\mathbf{p}, x) \right],$$

where we use the same symbol $\text{Tr}[\cdots]$ for the trace both in the fundamental and adjoint representations.

The distribution functions of quarks, antiquarks, and gluons satisfy the transport equations

$$p^\mu D_\mu Q(\mathbf{p}, x) + \frac{g}{2} p^\mu \left\{ F_{\mu\nu}(x), \partial_p^\nu Q(\mathbf{p}, x) \right\} = C[Q, \tilde{Q}, G], \qquad 11.$$

$$p^\mu D_\mu \tilde{Q}(\mathbf{p}, x) - \frac{g}{2} p^\mu \left\{ F_{\mu\nu}(x), \partial_p^\nu \tilde{Q}(\mathbf{p}, x) \right\} = \tilde{C}[Q, \tilde{Q}, G], \qquad 12.$$

$$p^\mu D_\mu G(\mathbf{p}, x) + \frac{g}{2} p^\mu \left\{ \mathcal{F}_{\mu\nu}(x), \partial_p^\nu G(\mathbf{p}, x) \right\} = C_g[Q, \tilde{Q}, G], \qquad 13.$$

where $\{\ldots\ldots\}$ denotes the anticommutator and ∂_p^ν the four-momentum derivative.[1] The covariant derivatives D_μ and $\mathcal{D}_\mu$ act as

$$D_\mu = \partial_\mu - ig[A_\mu(x), \ldots], \quad \mathcal{D}_\mu = \partial_\mu - ig[\mathcal{A}_\mu(x), \ldots],$$

with A_μ and $\mathcal{A}_\mu$ being four-potentials in the fundamental and adjoint representations, respectively:

$$A^\mu(x) = A_a^\mu(x)\tau^a, \quad \mathcal{A}^\mu(x) = T^a A_a^\mu(x).$$

The strength tensor in the fundamental representation is $F_{\mu\nu} = \partial_\mu A_\nu - \partial_\nu A_\mu - ig[A_\mu, A_\nu]$, whereas $\mathcal{F}_{\mu\nu}$ denotes the field strength tensor in the adjoint representation. C, $\tilde{C}$ and C_g represent the collision terms.

The transport equations are supplemented by the Yang-Mills equation describing generation of the gauge field:

$$D_\mu F^{\mu\nu}(x) = j^\nu(x), \qquad 14.$$

where the color current is given by Equation 10. As in the case of the electromagnetic plasma, the transport equations, which are linearized around a stationary, homogeneous, and colorless state, can be solved. Because of the color neutrality assumption, the analysis is rather similar to that of the Abelian plasma, and it ends up with the polarization tensor that is proportional to the unit matrix in the color space and has the form of Equation 5.

---

[1] As the distribution functions do not depend on p_0, the derivative over p_0 is identically zero.

As in the case with EMPs, the collisions can be easily taken into account using the approximate BGK collision terms (14, 15). Within a more realistic approach, color charges are treated in a similar way as spin degrees of freedom, and one uses the so-called Waldmann-Snider collision terms (16, 17), which are usually applied to study spin transport.

2.2. Hydrodynamic Approach

Within the hydrodynamic approach, the plasma is treated as a liquid and described in terms of macroscopic variables that obey the equations of motion resulting from the conservation laws. The fluid equations are applied to a large variety of plasma phenomena, but, depending on the timescale of interest, the actual physical content of the equations is rather different.

Real hydrodynamics deals with systems in local equilibrium, and thus it is applicable only at sufficiently long timescales. The continuity and the Euler or Navier-Stokes equations are supplemented by the equation of state to form a complete set of equations. The equations can be derived from kinetic theory, using the distribution function of local equilibrium, which by definition maximizes the entropy density and thus cancels the collision terms of the transport equations.

In the electron-ion plasma, there are several timescales of equilibration. The electron component of the plasma reaches the equilibrium in the shortest time. Then ions are equilibrated, but for a relatively long time the electron and ion temperatures remain different from each other, as the energy transfer between electrons to ions is rather inefficient. This happens owing to the huge mass difference between electrons and ions.

When the electrons have reached local equilibrium with their own temperature and hydrodynamic velocity, the collision terms of the kinetic equations representing electron-electron collisions vanish, while the collision terms due to electron-ion collisions can be neglected because they influence the electron distribution function only at a sufficiently long timescale. Then, one obtains hydrodynamic equations of an electron fluid. When the ion component is also equilibrated, we have two fluids with different temperatures and hydrodynamic velocities. At the timescales when the fluid equations are applicable, the plasma can be treated as locally neutral. Charge fluctuations are obviously possible, but they disappear rapidly because the electric field generated by the local charges induces the currents, which in turn neutralize the charges. Because the plasma is nearly an ideal conductor, the process of plasma neutralization is very fast. Owing to the charge neutrality of the plasma, the electric field is not present in the fluid equations and we end up with magnetohydrodynamics, where the pressure gradients and magnetic field drive the plasma dynamics.

As explained above, the regime of magnetohydrodynamics appears because there is a heavy positive component (ions) and a light negative component (electrons) of the plasma. There is no QCD analog of magnetohydrodynamics, as every quark or gluon can carry opposite color charges. Therefore, when local equilibrium is reached, various color components of the plasma have the same temperatures and hydrodynamic velocities (17). Because the quark-gluon system becomes color neutral even before the local equilibration is reached (14, 16), we deal with hydrodynamics of a neutral fluid where the chromodynamic fields are absent. Such a relativistic hydrodynamics of colorless QGP has been actively studied over the past two decades (18, 19).

The hydrodynamic equations, which actually express macroscopic conservation laws, hold not only for systems in local equilibrium but for systems out of equilibrium as well. The equations can then be applied at timescales significantly shorter than that of local equilibrium. At such a short timescale, the collision terms of the transport equations can be neglected entirely. However, extra assumptions are then needed to close the set of equations, as the (equilibrium) equation of state cannot be used. Plasma physicists developed several methods to close the set of equations, and thus fluid equations are used to study bulk features of short timescale phenomena in the plasmas. To get more detailed information, kinetic theory is needed. Because the fluid equations are noticeably simpler than the kinetic ones, the hydrodynamic approach is used frequently in numerical simulations of plasma evolution, studies of nonlinear dynamics, and so on.

Below, we derive the fluid equations for the EMP and the QGP from the respective kinetic theory. Because the fluid approach under consideration is supposed to hold at sufficiently short timescales, we use the collisionless transport equations.

2.2.1. Electromagnetic plasma. We assume here that there are several streams in the relativistic plasma system and that the distribution functions of each plasma component (electrons, positrons, ions) belonging to each stream satisfy the collisionless transport equation. The equations are coupled only through the electromagnetic mean field, which is generated by the current coming from all streams. The field in turn interacts with every stream.

Integrating the collisionless transport equation (Equation 1) over momentum, one finds the continuity equation

$$\partial_\mu n_\alpha^\mu = 0, \tag{15.}$$

where the four-flow is

$$n_\alpha^\mu(x) \equiv \int \frac{d^3 p}{(2\pi)^3} p^\mu f_\alpha(\mathbf{p}, x). \tag{16.}$$

The index α simultaneously labels the streams and plasma components.

Multiplying the transport equation (Equation 1) by the four-momentum p and integrating over momentum, we get

$$\partial_\mu T_\alpha^{\mu\nu} + q_\alpha n_\alpha^\mu F_\mu^\nu = 0, \tag{17.}$$

where the energy-momentum tensor is

$$T_\alpha^{\mu\nu}(x) \equiv \int \frac{d^3 p}{(2\pi)^3} p^\mu p^\nu f_\alpha(\mathbf{p}, x). \tag{18.}$$

The structure of n_α^μ and $T_\alpha^{\mu\nu}$ is assumed to be that of an ideal fluid in local thermodynamic equilibrium. Thus, one has

$$n_\alpha^\mu(x) = n_\alpha(x) u_\alpha^\mu(x), \tag{19.}$$

$$T_\alpha^{\mu\nu}(x) = [\varepsilon_\alpha(x) + p_\alpha(x)] u^\mu(x) u^\nu(x) - p_\alpha(x) g^{\mu\nu}. \tag{20.}$$

To obtain the relativistic version of the Euler equation, Equation 17 needs to be manipulated following Reference 20. Substituting the energy-momentum tensor of the form of Equation 20 into Equation 17 and projecting the result on the direction of u_α^μ, one finds

$$u_{\alpha\nu}\partial_\mu T_\alpha^{\mu\nu} = u_\alpha^\mu \partial_\mu \varepsilon_\alpha + (\varepsilon_\alpha + p_\alpha)\partial_\mu u_\alpha^\mu = 0. \qquad 21.$$

Computing $\partial_\mu T_\alpha^{\mu\nu} - u_\alpha^\nu u_{\alpha\rho}\partial_\mu T_\alpha^{\mu\rho}$, one gets the Lorentz covariant form of the Euler equation

$$M_\alpha^\nu \equiv (\varepsilon_\alpha + p_\alpha)u_{\alpha\mu}\partial^\mu u_\alpha^\nu + \left(u_\alpha^\mu u_\alpha^\nu \partial_\mu - \partial^\nu\right) p - q_\alpha n_\alpha u_{\alpha\mu} F^{\mu\nu} = 0. \qquad 22.$$

In a more familiar form, the equation is given by $\mathbf{M}_\alpha - \mathbf{v}_\alpha M_\alpha^0 = 0$. Namely,

$$(\varepsilon_\alpha + p_\alpha)\gamma_\alpha^2 \left(\frac{\partial}{\partial t} + \mathbf{v}_\alpha \cdot \nabla\right)\mathbf{v}_\alpha + \left(\nabla + \mathbf{v}_\alpha \frac{\partial}{\partial t}\right)p_\alpha - q_\alpha n_\alpha \gamma_\alpha \left[\mathbf{E} - \mathbf{v}_\alpha(\mathbf{v}_\alpha \cdot \mathbf{E}) + \mathbf{v}_\alpha \times \mathbf{B}\right] = 0, \qquad 23.$$

where the four-velocity u_α^μ was expressed as $u_\alpha^\mu = (\gamma_\alpha, \gamma_\alpha \mathbf{v}_\alpha)$, with $\gamma_\alpha \equiv (1 - \mathbf{v}_\alpha^2)^{-1/2}$.

In the nonrelativistic limit (which is easily obtained when the velocity of light c is restored in the equation), Equation 23 gets the well-known form

$$\left(\frac{\partial}{\partial t} + \mathbf{v}_\alpha \cdot \nabla\right)\mathbf{v}_\alpha + \frac{1}{m_\alpha n_\alpha}\nabla p_\alpha - \frac{q_\alpha}{m_\alpha}\left(\mathbf{E} + \mathbf{v}_\alpha \times \mathbf{B}\right) = 0. \qquad 24.$$

The fluid Equations 15 and 17 with n_α^μ and $T_\alpha^{\mu\nu}$ given by Equations 19 and 20, respectively, do not constitute a closed set of equations. There are five equations and six unknown functions: n_α, p_α, ε_α, and three components of u_α^μ (because of the constraint $u_\alpha^\mu u_{\mu\alpha} = 1$, one component of u_α^μ can be eliminated). There are several methods to close the set of equations. In particular, assuming that the system's dynamics is dominated by the mean-field interaction, one can neglect the pressure gradients. One can also add an equation that relates p_α to ε_α. The relation is usually known as the equation of state, but one should be aware that the plasma system is not in equilibrium, and in general the thermodynamic relations do not hold.

In the ultrarelativistic limit when the characteristic particle's energy (the temperature of the equilibrium system) is much larger than the particle's mass, and thus $p^2 \cong 0$, the energy-momentum tensor is traceless ($T_{\mu\alpha}^\mu = 0$), as follows from Equation 18 for $p^2 = 0$. Then, Equation 20 combined with the constraint $u_\alpha^\mu(x)u_{\alpha\mu}(x) = 1$ provides the desired relation

$$\varepsilon_\alpha(x) = 3 p_\alpha(x), \qquad 25.$$

which coincides with the equation of state of an ideal gas of massless particles.

Because the distribution functions of every plasma component belonging to every stream are assumed to obey the collisionless transport equation, we have a separated set of fluid equations for every plasma component of every stream. The equations are coupled only through the electromagnetic mean field. More precisely, the electrons, positrons, and ions of every stream contribute to the current generating the field, which in turn interacts with the streams.

The fluid equations can be solved in the linear-response approximation. The equations are linearized around the stationary and homogeneous state described by $\bar{n}_\alpha$ and

$\bar{u}_\alpha^\mu$. This state is neutral and there are no currents, i.e.,

$$\sum_\alpha \bar{n}_\alpha \bar{u}_\alpha^\mu = 0. \qquad 26.$$

The charge density is decomposed as

$$n_\alpha(x) = \bar{n}_\alpha + \delta n_\alpha(x), \qquad 27.$$

where $\bar{n}_\alpha \gg \delta n_\alpha$. The fully analogous decomposition of the hydrodynamic velocity u_α^μ, pressure p_α, and energy density ε_α is also adopted.

The set of the continuity and Euler equations linearized in δn_α, δu_α^μ, δp_α, $\delta \varepsilon_\alpha$, and $F^{\mu\nu}$ can be exactly solved after they are Fourier transformed. Thus, one finds $\delta n_\alpha(k)$ and $\delta u_\alpha^\mu(k)$ when the set of fluid equations is closed by neglecting the pressure gradients. If the equation of state is used, one also finds $\delta \varepsilon_\alpha(k)$.

Keeping in mind that the induced current equals

$$\delta j^\mu = \sum_\alpha \left(q_\alpha \bar{n}_\alpha \delta u_\alpha^\mu + q_\alpha \delta n_\alpha \bar{u}_\alpha^\mu \right),$$

one finds from Equation 4

$$\Pi^{\mu\nu}(k) = \sum_\alpha \frac{4\pi q_\alpha^2 \bar{n}_\alpha^2}{\bar{\varepsilon}_\alpha + \bar{p}_\alpha} \frac{1}{(\bar{u}_\alpha \cdot k)^2} \Bigg[k^2 \bar{u}_\alpha^\mu \bar{u}_\alpha^\nu + (\bar{u}_\alpha \cdot k)^2 g^{\mu\nu} - (\bar{u}_\alpha \cdot k)\left(k^\mu \bar{u}_\alpha^\nu + k^\nu \bar{u}_\alpha^\mu \right)$$

$$+ \frac{(\bar{u}_\alpha \cdot k)k^2 \left(k^\mu \bar{u}_\alpha^\nu + k^\nu \bar{u}_\alpha^\mu \right) - (\bar{u}_\alpha \cdot k)^2 k^\mu k^\nu - k^4 \bar{u}_\alpha^\mu \bar{u}_\alpha^\nu}{k^2 + 2(\bar{u}_\alpha \cdot k)^2} \Bigg]. \qquad 28.$$

The first term gives the polarization tensor when the pressure gradients are neglected, and the second term gives the effect of the pressure gradients due to the equation of state given by Equation 25. The first term is symmetric $[\Pi^{\mu\nu}(k) = \Pi^{\nu\mu}(k)]$ and transverse $[k_\mu \Pi^{\mu\nu}(k) = 0]$. The second term is symmetric and transverse as well. Thus, the whole polarization tensor (Equation 28) is symmetric and transverse. The first term of Equation 28 can be obtained from the kinetic theory result (Equation 5) with the distribution function $\bar{f}_n(\mathbf{p})$ proportional to $\delta^{(3)}(\mathbf{p} - (\bar{\varepsilon}_\alpha + \bar{p}_\alpha)\mathbf{u}_\alpha/\bar{n}_\alpha)$. Thus, the first term neglects the thermal motion of plasma particles, whereas the second term takes this effect into account.

2.2.2. Quark-gluon plasma. The fluid approach presented here follows the formulation given in Reference 21. As in the EMP case, we assume that there are several streams in the plasma system and that the distribution functions of quarks, antiquarks, and gluons of each stream satisfy the collisionless transport equation. The streams are labeled with the index α.

Further analysis is limited to quarks, but inclusion of antiquarks and gluons is straightforward. The distribution function of quarks belonging to the stream α is denoted as $Q_\alpha(\mathbf{p}, x)$. Integrating over momentum the collisionless transport (Equation 11) satisfied by Q_α, one finds the covariant continuity equation

$$D_\mu n_\alpha^\mu = 0, \qquad 29.$$

where n_α^μ is an $N_c \times N_c$ matrix defined as

$$n_\alpha^\mu(x) \equiv \int \frac{d^3p}{(2\pi)^3} p^\mu Q_\alpha(\mathbf{p}, x). \qquad 30.$$

The four-flow n_α^μ transforms under gauge transformations as the quark distribution function, that is, according to Equation 8.

Multiplying the transport equation (Equation 11) by the four-momentum and integrating the product over momentum, we obtain

$$D_\mu T_\alpha^{\mu\nu} - \frac{g}{2} \left\{ F_\mu^\nu, n_\alpha^\mu \right\} = 0, \qquad 31.$$

where the energy-momentum tensor is

$$T_\alpha^{\mu\nu}(x) \equiv \int \frac{d^3p}{(2\pi)^3} p^\mu p^\nu Q_\alpha(\mathbf{p}, x). \qquad 32.$$

We assume further that the structure of n_α^μ and $T_\alpha^{\mu\nu}$ is

$$n_\alpha^\mu(x) = n_\alpha(x) u_\alpha^\mu(x), \qquad 33.$$

$$T_\alpha^{\mu\nu}(x) = \frac{1}{2} \left(\varepsilon_\alpha(x) + p_\alpha(x) \right) \left\{ u_\alpha^\mu(x), u_\alpha^\nu(x) \right\} - p_\alpha(x) g^{\mu\nu}, \qquad 34.$$

where the hydrodynamic velocity u_α^μ is, as n_α, ε_α, and p_α, an $N_c \times N_c$ matrix. The anticommutator of u_α^μ and u_α^ν is present in Equation 34 to guarantee the symmetry of $T_\alpha^{\mu\nu}$ with respect to $\mu \leftrightarrow \nu$, which is evident in Equation 32.

In the case of an Abelian plasma, the relativistic version of the Euler equation is obtained from Equation 31 by removing from it the component parallel to u_α^μ. An analogous procedure is not possible for the non-Abelian plasma because in general the matrices n_α, u_α^μ, and u_α^ν do not commute with each other. Thus, one has to work directly with Equations 29 and 31 with n_α^μ and $T_\alpha^{\mu\nu}$ defined by Equations 33 and 34, respectively. The equations have to be supplemented by the Yang-Mills equation (Equation 14) with the color current of the form

$$j^\mu(x) = -\frac{g}{2} \sum_\alpha \left(n_\alpha u_\alpha^\mu - \frac{1}{N_c} \text{Tr} \left[n_\alpha u_\alpha^\mu \right] \right), \qquad 35.$$

where only the quark contribution is taken into account.

The fluid Equations 29 and 31, as their electromagnetic counterpart, do not form a closed set of equations, but can be closed analogously. The only difference is that the equation of state (Equation 25) relates the matrix value functions ε_α and p_α to each other.

As in the case of the electromagnetic plasma, the fluid Equations 29 and 31, which are linearized around a stationary, homogeneous, and colorless state described by $\bar{n}_\alpha$, $\bar{\varepsilon}_\alpha$, $\bar{p}_\alpha$, and $\bar{u}_\alpha^\mu$, can be solved (21). Because of the color-neutrality assumption—$\bar{n}_\alpha$, $\bar{\varepsilon}_\alpha$, $\bar{p}_\alpha$, and $\bar{u}_\alpha^\mu$ are all proportional to the unit matrix in the color space—the analysis is rather similar to that of the Abelian plasma, and one ends up with the polarization tensor from Equation 28, which is proportional to the unit matrix in the color space.

2.3. Diagrammatic Methods

Various characteristics of the weakly coupled plasma can be calculated using the perturbative expansion, that is, diagrammatic methods of field theory. It requires a generalization of the Feynman rules applicable to processes, which occur in vacuum, to many-body plasma systems. When the plasma is in thermodynamic equilibrium, one can either follow the so-called imaginary-time formalism (see, e.g., References 22 and 23) or the real-time (Schwinger-Keldysh) formalism (24, 25). The latter can also be extended to nonequilibrium situations (26, 27).

The perturbative expansion expressed in terms of Feynman diagrams allows a systematic computation of various quantities. However, to obtain a gauge-invariant finite result, one often has to re-sum a class of diagrams, as required by the hard loop approach (28–30) (the real-time formulation is discussed in Reference 31). The approach, which was first developed for equilibrium systems (28–30) (for a review, see Reference 32) and then extended to the nonequilibrium case (33–35), distinguishes soft from hard momenta. In the case of ultrarelativistic QED plasmas in equilibrium, the soft momenta are of order eT, whereas the hard momenta are of order T, with T being the plasma temperature. One obviously assumes that $1/e \gg 1$. The hard loop approach deals with soft collective excitations generated by hard plasma particles that dominate the distribution functions.

As an example, we consider the polarization tensor given by Equation 5, which was obtained within the kinetic theory in Section 2.1. We restrict ourselves to ultrarelativistic QED plasmas. In the lowest order of the perturbative expansion, the polarization tensor or photon self-energy is given by the diagram shown in **Figure 1**. The tensor can be decomposed into vacuum and medium contributions. The first one requires a usual renormalization because of a UV divergence, whereas the medium part appears to be UV finite. One reproduces Equation 5 by applying to the diagrammatic result the hard loop approximation, which requires that the energy and momentum $(\omega, \mathbf{k})$ of the external photon line are much smaller than the momentum $(\mathbf{p})$ of the electron loop. Then, it appears that the vacuum part can be neglected, as it is much smaller than the medium part. In the case of an ultrarelativistic equilibrium EMP, Equation 7 was derived diagrammatically in References 36 and 37. In the QGP, the lowest-order polarization tensor (gluon self-energy) includes one-loop diagrams with internal gluon and ghost lines. The final result for the gluon-polarization tensor in the high-temperature approximation essentially coincides with the QED expression. The color degrees of freedom enter through the trivial color factor δ_{ab}. In the case of equilibrium QGP, one additionally replaces in Equation 7 the thermal photon

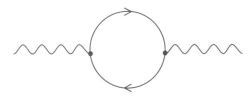

Figure 1

The lowest-order contribution to the QED polarization tensor.

mass by a thermal gluon mass given by

$$m_g^2 = \frac{g^2 T^2}{3}\left(1 + \frac{N_f}{6}\right), \qquad\qquad 36.$$

where N_f indicates the number of light-quark flavors.

The hard loop approach can be formulated nicely in terms of an effective action. Such an action for an equilibrium system was derived diagrammatically in Reference 29 and in the explicitly gauge-invariant form in Reference 30. The equilibrium hard loop action was also found within the semiclassical kinetic theory (38, 39). The action was generalized (33, 35) for nonequilibrium systems, which are, on average, locally color neutral, stationary, and homogeneous.

The starting point was the effective action, which describes an interaction of classical fields with currents induced by these fields in the plasma. The Lagrangian density is quadratic in the gluon and quark fields, and it equals

$$\mathcal{L}_2(x) = -\int d^4 y \left[\frac{1}{2} A^a_\mu(x)\Pi^{\mu\nu}_{ab}(x-y)A^b_\nu(y) + \bar\Psi(x)\Sigma(x-y)\Psi(y)\right], \qquad 37.$$

where $\Pi^{\mu\nu}_{ab}$ and Σ are the gluon-polarization tensor and the quark self-energy, respectively, while A^μ and Ψ denote the gluon and quark fields. Following Braaten & Pisarski (30), the Lagrangian from Equation 37 was modified to comply with the requirement of gauge invariance. The final result, which is nonlocal but manifestly gauge invariant, is

$$\mathcal{L}_{\mathrm{HL}}(x) = \frac{g^2}{2}\int \frac{d^3 p}{(2\pi)^3}\left[f(\mathbf{p})F^a_{\mu\nu}(x)\left(\frac{p^\nu p^\rho}{(p\cdot D)^2}\right)_{ab} F^{b\mu}_\rho(x)\right.$$
$$\left. + i\,\frac{N_c^2 - 1}{4N_c}\tilde f(\mathbf{p})\bar\Psi(x)\frac{p\cdot\gamma}{p\cdot D}\Psi(x)\right], \qquad 38.$$

where $F^{\mu\nu}_a$ is the strength tensor and D denotes the covariant derivative; $f(\mathbf{p})$ and $\tilde f(\mathbf{p})$ are the effective parton distribution functions defined as $f(\mathbf{p}) \equiv n(\mathbf{p}) + \bar n(\mathbf{p}) + 2N_c n_g(\mathbf{p})$ and $\tilde f(\mathbf{p}) \equiv n(\mathbf{p}) + \bar n(\mathbf{p}) + 2n_g(\mathbf{p})$, respectively; $n(\mathbf{p})$, $\bar n(\mathbf{p})$, and $n_g(\mathbf{p})$ are the distribution functions of quarks, antiquarks, and gluons, respectively, of a single-color component in a homogeneous and stationary plasma, which is locally and globally colorless; the spin and flavor are treated as parton internal degrees of freedom. The quarks and gluons are assumed to be massless. The effective action given by Equation 38 generates n-point functions, which obey the Ward-Takahashi identities. Equation 38 holds under the assumption that the field amplitude is much smaller than T/g, where T denotes the characteristic momentum of (hard) partons.

3. COLLECTIVE PHENOMENA

The most characteristic feature of the electromagnetic and QCD plasmas, which results from a long-range interaction governing both systems, is a collective behavior that leads to specific plasma phenomena such as screening, plasma oscillations, instabilities, and so on.

Because the electromagnetic and chromodynamic polarization tensors, which are obtained in the linear-response analysis, are essentially the same, the collective effects in EMPs and QGPs are very similar in the linear-response regime. As our discussion is limited to this regime, mostly the EMP is considered in this section.

3.1. Screening

We start with screening of electric charges in the plasma. To discuss the effect, let us consider an electric field generated by a point-like charge q moving with velocity $\mathbf{v}$ in the plasma. The problem is studied in numerous plasma handbooks, for example, in Reference 2. The induction vector obeys the Maxwell equation

$$\nabla \cdot \mathbf{D}(x) = 4\pi q \, \delta^{(3)}(\mathbf{r} - \mathbf{v}t) \,,$$

with $x \equiv (t, \mathbf{r})$. After the Fourier transformation, which is defined by Equation 3, the induction vector reads

$$i\mathbf{k} \cdot \mathbf{D}(k) = 8\pi^2 q \, \delta(\omega - \mathbf{k} \cdot \mathbf{v}), \qquad\qquad 39.$$

where $k \equiv (\omega, \mathbf{k})$. The induction vector $\mathbf{D}(k)$ is related to the electric field $\mathbf{E}(k)$ through the dielectric tensor $\varepsilon^{ij}(k)$ as

$$D^i(k) = \varepsilon^{ij}(k) E^j(k). \qquad\qquad 40.$$

We note that the dielectric tensor $\varepsilon^{ij}(k)$, which carries information on the electromagnetic properties of a medium, can be expressed through the polarization tensor as

$$\varepsilon^{ij}(k) = \delta^{ij} + \frac{1}{\omega^2} \Pi^{ij}(k). \qquad\qquad 41.$$

In an isotropic plasma, there are only two independent components of the dielectric tensor ε_T and ε_L, which are related to ε^{ij} as

$$\varepsilon^{ij}(k) = \varepsilon_T(k)(\delta^{ij} - k^i k^j /\mathbf{k}^2) + \varepsilon_L(k) k^i k^j /\mathbf{k}^2. \qquad\qquad 42.$$

Using Equations 40 and 42, and expressing the electric field $\mathbf{E}$ through the scalar ϕ and vector $\mathbf{A}$ potentials $[\mathbf{E}(k) = -i\mathbf{k}\phi(k) + i\omega\mathbf{A}(k)]$ in the Coulomb gauge $[\mathbf{k} \cdot \mathbf{A}(k) = 0]$, one finds the electric potential in a medium (the wake potential):

$$\phi(x) = 4\pi q \int \frac{d^3 k}{(2\pi)^3} \frac{e^{i\mathbf{k} \cdot (\mathbf{r} - \mathbf{v}t)}}{\varepsilon_L(\omega = \mathbf{v} \cdot \mathbf{k}, \mathbf{k})\mathbf{k}^2}. \qquad\qquad 43.$$

Let us first consider the simplest case of the potential generated by a static ($\mathbf{v} = 0$) charge. Using Equations 6 and 7, $\varepsilon_L(0, \mathbf{k})$ of an ultrarelativistic electron-positron plasma is found as

$$\varepsilon_L(0, \mathbf{k}) = 1 + \frac{m_D^2}{\mathbf{k}^2} \,, \qquad\qquad 44.$$

where m_D is the so-called Debye mass given by $m_D^2 = e^2 T^2 /3 = 3m_\gamma^2 = \Pi_L(0, \mathbf{k})$. Then, Equation 43 gives the well-known screened potential

$$\phi(\mathbf{r}) = \frac{q}{r} e^{-m_D r}, \qquad\qquad 45.$$

with $r \equiv |\mathbf{r}|$. Thus, the inverse Debye mass has the interpretation of the screening length of the potential. Because the average interparticle spacing in the ultrarelativistic plasma is of order T^{-1}, the number of particles in the Debye sphere (the sphere of the radius m_D^{-1}) is of order e^{-3}, which is, as already mentioned in the Introduction, much larger than unity in the weakly coupled plasma ($1/e^2 \gg 1$). This explains the collective behavior of the plasma, since motion of particles from the Debye sphere is highly correlated.

For $\mathbf{v} \neq 0$, the potential given by Equation 43 has a rich structure. It has been discussed in the context of QGPs in References 40–42, showing that it can exhibit attractive contributions even between like-sign charges in certain directions (40). For a supersonic particle, the potential can reveal a Mach cone structure associated with Cerenkov radiation when electromagnetic properties of the plasma are appropriately modeled (41, 42).

3.2. Collective Modes

Let us consider a plasma in a homogenous, stationary state with no local charges and no currents. As a fluctuation or perturbation of this state, there appear local charges or currents generating electric and magnetic fields, which in turn interact with charged plasma particles. Then, the plasma reveals a collective motion, which classically is termed plasma oscillations. Quantum mechanically we deal with quasi-particle collective excitations of the plasma.

The collective modes are solutions of the dispersion equation obtained from the equation of motion of the Fourier-transformed electromagnetic potential $A^\mu(k)$, which is

$$[k^2 g^{\mu\nu} - k^\mu k^\nu - \Pi^{\mu\nu}(k)]A_\nu(k) = 0, \qquad 46.$$

where the polarization tensor $\Pi^{\mu\nu}$ contains all dynamical information about the system. The general dispersion equation is then

$$\det[k^2 g^{\mu\nu} - k^\mu k^\nu - \Pi^{\mu\nu}(k)] = 0. \qquad 47.$$

Owing to the transversality of $\Pi^{\mu\nu}(k)$, not all components of $\Pi^{\mu\nu}(k)$ are independent of each other. Consequently, the dispersion equation (Equation 47), which involves a determinant of a 4×4 matrix, can be simplified to the determinant of a 3×3 matrix. For this purpose, one usually introduces the dielectric tensor $\varepsilon^{ij}(k)$, which is related to the polarization tensor by Equation 41. Then, the dispersion equation gets the form

$$\det[\mathbf{k}^2 \delta^{ij} - k^i k^j - \omega^2 \varepsilon^{ij}(k)] = 0. \qquad 48.$$

The relationship between Equation 47 and Equation 48 is most easily seen in the Coulomb gauge when $\phi = 0$ and $\mathbf{k} \cdot \mathbf{A}(k) = 0$. Then, $\mathbf{E} = i\omega\mathbf{A}$ and Equation 46 is immediately transformed into an equation of motion for $\mathbf{E}(k)$, which further provides the dispersion equation (Equation 48).

As expressed by Equation 42, there are only two independent components of the dielectric tensor [$\varepsilon_T(k)$ and $\varepsilon_L(k)$] in an isotropic plasma. Then, the dispersion

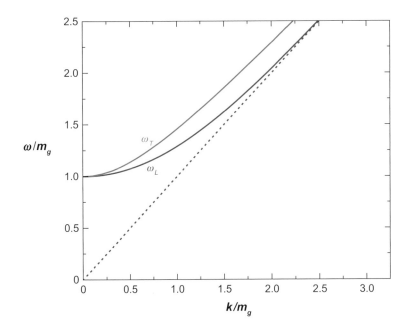

Figure 2

Dispersion relation of longitudinal and transverse plasma waves.

equation (Equation 48) splits into two equations:

$$\varepsilon_T(k) = \mathbf{k}^2/\omega^2, \quad \varepsilon_L(k) = 0. \qquad 49.$$

Solutions of the dispersion equations $\omega(\mathbf{k})$, with a generally complex frequency ω, represent plasma modes, which classically are, as already mentioned, the waves of electric and/or magnetic fields in the plasma, whereas quantum mechanically the modes are quasi-particle excitations of the plasma system. If the imaginary part of the mode's frequency $\Im\omega$ is negative, the mode is damped. Its amplitude decays exponentially in time as $e^{\Im\omega t}$. When $\Im\omega = 0$, we have a stable mode with a constant amplitude. Finally, if $\Im\omega > 0$, the mode's amplitude grows exponentially in time; there is an instability.

When the electric field of a mode is parallel to its wave vector $\mathbf{k}$, the mode is termed longitudinal. A mode is termed transverse when the electric field is transverse to the wave vector. The Maxwell equations show that the longitudinal modes, also known as electric, are associated with electric-charge oscillations. The transverse modes, also known as magnetic, are associated with electric-current oscillations.

The collective (boson) modes in the equilibrium ultrarelativistic plasma are shown in **Figure 2**. There are longitudinal modes, also termed plasmons, and transverse modes. Both start at zero momentum at the plasma frequency, which is identical to the thermal photon (or gluon) mass, $\omega_p = m_\gamma$. The dispersion relations lie above the light cone ($\omega > |\mathbf{k}|$), showing that the plasma waves are undamped (no Landau damping) in the high-temperature limit. As explained in Section 3.3.1, the Landau damping, which arises formally from the imaginary part of the polarization tensor given by Equation 7 at $\omega^2 < \mathbf{k}^2$, occurs when the energy of the wave is transferred to plasma particles moving with velocity equal to the phase velocity ($\omega/|\mathbf{k}|$) of the wave.

If the phase velocity is larger than the speed of light, such a transfer is obviously not possible.

3.2.1. Two-stream system.

As an example of the rich spectrum of collective modes, we consider the two-stream system within the hydrodynamic approach when the effect of pressure gradients is neglected. Details of the analysis can be found in Reference 21. The dielectric tensor provided by the polarization tensor from Equation 28 is

$$\varepsilon^{ij}(\omega, \mathbf{k}) = \left(1 - \frac{\omega_p^2}{\omega^2}\right)\delta^{ij} - \frac{4\pi}{\omega^2}\sum_\alpha \frac{q_\alpha^2 \bar{n}_\alpha^2}{\bar{\varepsilon}_\alpha + \bar{p}_\alpha}\left[\frac{\bar{v}_\alpha^i k^j + \bar{v}_\alpha^j k^i}{\omega - \mathbf{k}\cdot\bar{\mathbf{v}}_\alpha} - \frac{(\omega^2 - \mathbf{k}^2)\bar{v}_\alpha^i \bar{v}_\alpha^j}{(\omega - \mathbf{k}\cdot\bar{\mathbf{v}}_\alpha)^2}\right], \quad 50.$$

where $\bar{\mathbf{v}}_\alpha$ is the hydrodynamic velocity related to the hydrodynamic four-velocity and $\bar{u}_\alpha^\mu$. ω_p is the plasma frequency given as

$$\omega_p^2 \equiv 4\pi\sum_\alpha \frac{q_\alpha^2 \bar{n}_\alpha^2}{\bar{\varepsilon}_\alpha + \bar{p}_\alpha}. \quad 51.$$

The index α, which labels the streams and plasma components, has four values, $\alpha = L-, L+, R-, R+$. The first character labels the stream (R for right and L for left), while the second one labels the plasma component ($+$ for positive and $-$ for negative charges). For simplicity we assume here that the streams are neutral and identical to each other and their velocities, which are chosen along the z-axis, and are opposite to each other. Then,

$$\bar{n} \equiv \bar{n}_{L-} = \bar{n}_{L+} = \bar{n}_{R-} = \bar{n}_{R+}, \quad \bar{\varepsilon} \equiv \bar{\varepsilon}_{L-} = \bar{\varepsilon}_{L+} = \bar{\varepsilon}_{R-} = \bar{\varepsilon}_{R+},$$

$$\bar{p} \equiv \bar{p}_{L-} = \bar{p}_{L+} = \bar{p}_{R-} = \bar{p}_{R+}, \quad \bar{v} \equiv \bar{v}_{L-} = \bar{v}_{L+} = -\bar{v}_{R-} = -\bar{v}_{R+},$$

$$e = q_{L-} = -q_{L+} = q_{R-} = -q_{R+}, \quad 52.$$

and the plasma frequency is $\omega_p^2 = 16\pi e^2 \bar{n}^2/(\bar{\varepsilon} + \bar{p})$.

The wave vector is first chosen to be parallel to the x-axis, $\mathbf{k} = (k, 0, 0)$. Owing to Equation 52, the off-diagonal elements of the matrix in Equation 48 vanish and the dispersion equation with the dielectric tensor given by Equation 50 is

$$\left(\omega^2 - \omega_p^2\right)\left(\omega^2 - \omega_p^2 - k^2\right)\left(\omega^2 - \omega_p^2 - k^2 - \lambda^2\frac{k^2 - \omega^2}{\omega^2}\right) = 0, \quad 53.$$

where $\lambda^2 \equiv \omega_p^2 \bar{v}^2$. As solutions of the equation, one finds a stable longitudinal mode with $\omega^2 = \omega_p^2$ and a stable transverse mode with $\omega^2 = \omega_p^2 + k^2$. There are also transverse modes with

$$\omega_\pm^2 = \frac{1}{2}\left[\omega_p^2 - \lambda^2 + k^2 \pm \sqrt{\left(\omega_p^2 - \lambda^2 + k^2\right)^2 + 4\lambda^2 k^2}\right]. \quad 54.$$

As seen, $\omega_+^2 > 0$, but $\omega_-^2 < 0$. Thus, the mode ω_+ is stable and there are two modes with pure imaginary frequency corresponding to $\omega_-^2 < 0$. The first mode is overdamped, whereas the second one is the well-known unstable Weibel mode (43), leading to the filamentation instability. A physical mechanism of the instability is explained in Section 3.3.2.

The wave vector, as well as the stream velocities, is now chosen along the z-axis, i.e., $\mathbf{k} = (0, 0, k)$. Then, the matrix in Equation 48 is diagonal. With the dielectric tensor given by Equation 50, the dispersion equation reads

$$\left(\omega^2 - \omega_p^2 - k^2\right)^2 \left\{\omega^2 - \omega_p^2 - \omega_p^2 \left[\frac{k\bar{v}}{\omega - k\bar{v}} + \frac{(k^2 - \omega^2)\bar{v}^2}{2(\omega - k\bar{v})^2} - \frac{k\bar{v}}{\omega + k\bar{v}} + \frac{(k^2 - \omega^2)\bar{v}^2}{2(\omega + k\bar{v})^2}\right]\right\} = 0.$$
55.

There are two transverse stable modes with $\omega^2 = \omega_p^2 + k^2$. The longitudinal modes are solutions of the above equation, which can be rewritten as

$$1 - \omega_0^2 \left[\frac{1}{(\omega - k\bar{v})^2} + \frac{1}{(\omega + k\bar{v})^2}\right] = 0,$$
56.

where $\omega_0^2 \equiv \omega_p^2 / 2\bar{\gamma}^2$ with $\bar{\gamma} = (1 - \bar{v}^2)^{-1/2}$. With the dimensionless quantities $x \equiv \omega/\omega_0$ and $y \equiv k\bar{v}/\omega_0$, Equation 56 is

$$(x^2 - y^2)^2 - 2x^2 - 2y^2 = 0,$$
57.

and is solved by

$$x_{\pm}^2 = y^2 + 1 \pm \sqrt{4y^2 + 1}.$$
58.

As seen, x_+^2 is always positive and thus gives two real (stable) modes, x_-^2 is negative for $0 < y < \sqrt{2}$, and so there are two pure imaginary modes. The unstable one corresponds to the two-stream electrostatic instability. A physical mechanism of the instability is explained in Section 3.3.1.

3.3. Instabilities

The presence of unstable modes in a plasma system crucially influences its dynamics. Huge difficulties encountered by the half-century program to build a thermonuclear reactor are related to various instabilities experienced by a plasma, which make the system's behavior very turbulent, hard to predict, and hard to control.

There exists a large variety of instabilities; the history of plasma physics is said to be a history of discoveries of new instabilities. Plasma instabilities can be divided into two general groups: (*a*) hydrodynamic instabilities, caused by coordinate space inhomogeneities, and (*b*) kinetic instabilities due to the nonequilibrium momentum distribution of plasma particles.

The hydrodynamic instabilities are usually associated with phenomena occurring at the plasma boundaries. In the case of the QGP, this is the domain of highly nonperturbative QCD, where the non-Abelian nature of the theory is of crucial importance. Then, the behavior of the QGP is presumably very different from that of the EMP, and thus we will not speculate about possible analogies.

The kinetic instabilities are simply the collective modes with positive $\Im\omega$, introduced in Section 3.2 and found in Section 3.2.1 in the specific case of the two-stream system. Thus, we have longitudinal (electric) and transverse (magnetic) instabilities. In the nonrelativistic plasma, the electric instabilities are usually much more important than the magnetic ones, as the magnetic effects are suppressed by the factor v^2/c^2, where v is the particle's velocity. In the relativistic plasma, both types of instabilities are of similar strength. As we discuss below, the electric instabilities occur when the momentum distribution of plasma particles has more than one maximum, as in the two-stream system. A sufficient condition for the magnetic instabilities appears to be an anisotropy of the momentum distribution.

3.3.1. Mechanism of electric instability.

Let us consider a plane wave of the electric field, with the wave vector along the z-axis. For a charged particle, which moves with a velocity $v = p_z/E_p$ equal to the phase velocity of the wave $v_\phi = \omega/k$, the electric field does not oscillate, but is constant. The particle is then either accelerated or decelerated depending on the field's phase. For an electron with $v = v_\phi$, chances of being accelerated and decelerated are equal, as the time intervals spent by the particle in the acceleration zone and in the deceleration zone are the same.

Let us now consider electrons with velocities somewhat smaller than the phase velocity of the wave. Such particles spend more time in the acceleration zone than in the deceleration zone, and the net result is that the particles with $v < v_\phi$ are accelerated. Consequently, energy is transferred from the electric field to the particles. The particles with $v > v_\phi$ spend more time in the deceleration zone than in the acceleration zone, and thus they are effectively decelerated. Energy is transferred from the particles to the field. If the momentum distribution is such that there are more electrons in a system with $v < v_\phi$ than with $v > v_\phi$, the wave loses energy that is gained by the particles, as shown in the left graph of **Figure 3**. This is the mechanism of the famous collisionless Landau damping of the plasma oscillations. If there are more particles with $v > v_\phi$ than with $v < v_\phi$, the particles lose energy that is gained by the wave, as in the right graph of **Figure 3**. Consequently, the wave amplitude grows. This is the mechanism of electric instability. As explained above, it requires the existence of the momentum interval where $f_n(\mathbf{p})$ grows with $\mathbf{p}$. Such an interval appears when the momentum distribution has more than one maximum. This happens in the two-stream system discussed in Section 3.2.1 or in the system of a plasma and a beam, shown in **Figure 4**.

Figure 3

Mechanism of energy transfer between particles and fields.

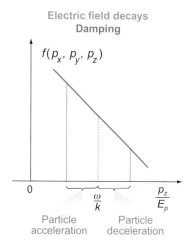

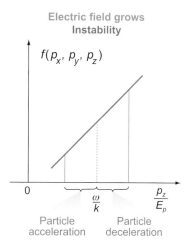

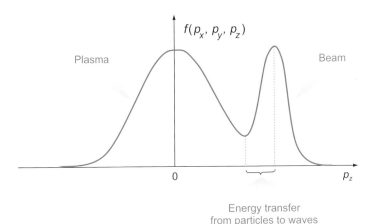

Figure 4

Momentum distribution of
the plasma-beam system.

3.3.2. Mechanism of magnetic instability. Because the magnetic instabilities appear to be relevant for QGPs produced in relativistic heavy-ion collisions (see below), we discuss them in more detail. Let us first explain following Reference 44 how the unstable transverse modes are initiated. For this purpose we consider a plasma system that is homogeneous, but where the momentum distribution of particles is not of equilibrium form—it is not anisotropic. The system is on average locally neutral $[\langle j^\mu(x)\rangle = 0]$ but current fluctuations are possible, and thus in general the correlator $\langle j^\mu(x_1) j^\nu(x_2)\rangle$ is nonzero. Because the plasma is assumed to be weakly coupled, the correlator can be estimated neglecting the interaction entirely. Then, when the effects of quantum statistics are also neglected, the correlator is

$$M^{\mu\nu}(t,\mathbf{x}) \overset{\text{def}}{=} \langle j^\mu(t_1,\mathbf{x}_1) j^\nu(t_2,\mathbf{x}_2)\rangle = \sum_n q_n^2 \int \frac{d^3p}{(2\pi)^3} \frac{p^\mu p^\nu}{E_p^2} f_n(\mathbf{p})\delta^{(3)}(\mathbf{x}-\mathbf{v}t), \qquad 59.$$

where $\mathbf{v} \equiv \mathbf{p}/E_p$ and $(t,\mathbf{x}) \equiv (t_2 - t_1, \mathbf{x}_2 - \mathbf{x}_1)$. Owing to the average space-time homogeneity, the correlator given by Equation 59 depends only on the difference $(t_2 - t_1, \mathbf{x}_2 - \mathbf{x}_1)$. The space-time points $(t_1, \mathbf{x}_1)$ and $(t_2, \mathbf{x}_2)$ are correlated in the system of noninteracting particles if a particle travels from $(t_1, \mathbf{x}_1)$ to $(t_2, \mathbf{x}_2)$. For this reason, the delta function $\delta^{(3)}(\mathbf{x} - \mathbf{v}t)$ is present in Equation 59. The sum and momentum integral represent summation over all particles in the system. The fluctuation spectrum is found as the Fourier transform of Equation 59; that is,

$$M^{\mu\nu}(\omega,\mathbf{k}) = \sum_n q_n^2 \int \frac{d^3p}{(2\pi)^3} \frac{p^\mu p^\nu}{E_p^2} f_n(\mathbf{p})2\pi\,\delta(\omega - \mathbf{k}\mathbf{v}). \qquad 60.$$

To further study the fluctuation spectrum, the particle's momentum distribution must be specified. We present here only a qualitative discussion of Equations 59 and 60, assuming that the momentum distribution is strongly elongated in one direction, which is chosen to be along the z-axis. Then, the correlator M^{zz} is larger than M^{xx} or M^{yy}. It is also clear that M^{zz} is at its largest when the wave vector $\mathbf{k}$ is along the direction of the momentum deficit. In such a case the delta function $\delta(\omega - \mathbf{k}\mathbf{v})$ does not much constrain the integral in Equation 60. Because the momentum distribution is elongated in the z-direction, the current fluctuations are at their largest when the wave vector $\mathbf{k}$ is in the x-y plane. Thus, we conclude that some fluctuations in the anisotropic system are large, much larger than in the isotropic one. An anisotropic system has a natural tendency to split into the current filaments parallel to the direction of the momentum surplus. These currents are seeds of the transverse unstable mode known as the filamentation, or Weibel, instability (43), which was found in the two-stream system discussed in Section 3.2.1.

Let us now explain in terms of elementary physics why the fluctuating currents, which flow in the direction of the momentum surplus, can grow in time. The form of the fluctuating current is chosen to be

$$\mathbf{j}(x) = j\,\hat{\mathbf{e}}_z\,\cos(k_x x), \qquad\qquad 61.$$

where $\hat{\mathbf{e}}_z$ is a unit vector in the z-direction. As seen in Equation 61, there are current filaments of the thickness $\pi/|k_x|$, with the current flowing in opposite directions in the neighboring filaments. The magnetic field generated by the current from Equation 61 is given as

$$\mathbf{B}(x) = 4\pi\,\frac{j}{k_x}\,\hat{\mathbf{e}}_y\,\sin(k_x x),$$

and the Lorentz force acting on the particles, which fly along the z-direction, is

$$\mathbf{F}(x) = q\mathbf{v} \times \mathbf{B}(x) = -4\pi q\,v_z\frac{j}{k_x}\hat{\mathbf{e}}_x \sin(k_x x),$$

where q is the particle's electric charge. One observes (see **Figure 5**) that the force distributes the particles in such a way that those that contribute positively to the current in a given filament are focused in the filament center, whereas those that contribute negatively are moved to the neighboring one. Thus, the initial current is growing and the magnetic field generated by this current is growing as well. The instability is driven by the energy transferred from the particles to fields. More specifically, the kinetic energy related to the motion along the direction of the momentum surplus is used to generate the magnetic field.

3.3.3. Role of instabilities. As mentioned above, there exists a large variety of plasma instabilities that strongly influence numerous plasma characteristics. Not much is known of the hydrodynamic instabilities of the QGP, and if they exist, they belong to the highly nonperturbative sector of QCD, which is still poorly understood. As explained in Section 3.3.1, the electric instabilities occur in a two-stream system, or more generally, in systems with a momentum distribution having more than one maximum. Although such a distribution is common in EMPs, it is rather irrelevant for QGPs produced in relativistic heavy-ion collisions, where the global as well as local momentum distributions are expected to monotonously decrease in every direction from the maximum. The electric instabilities are absent in such a system, but a magnetic unstable mode, which has been discussed in Section 3.3.2, is possible. The filamentaion instability was first argued to be relevant for the QGP produced in relativistic heavy-ion collisions in References 44–46. A characteristic time of

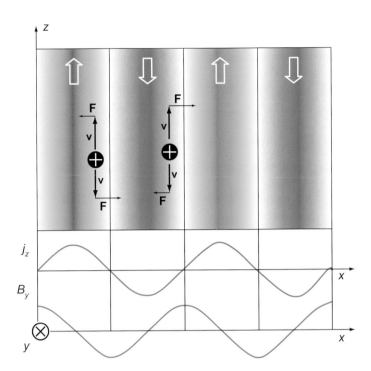

Figure 5

The mechanism of filamentation instability. See text for a description.

instability growth was estimated (45, 46) to be shorter or at least comparable to other timescales of the parton-system evolution. The mechanism of instability growth was also clarified (44). The early arguments were substantiated in the forthcoming analytic calculations (47–49) and numerical simulations (50–54).

A main consequence of instabilities is a fast equilibration of the weakly coupled plasma. The problem is of particular interest because the experimental data on heavy-ion collisions, where QGP production is expected, suggest that an equilibration time of the parton system is below 1 fm/c (55). A whole scenario of instabilities-driven equilibration is reviewed in Reference 56. Here we mention only the main points, starting with an observation that collisions of charged particles are not very efficient in redistributing particle momenta, as the Rutherford cross section is strongly peaked at a small momentum transfer. One needs either many frequent but soft collisions or a few rare but hard collisions to substantially change a particle's momentum. As a result, the inverse time of collisional equilibration of the QGP is of the order $g^4\ln(1/g)T$ (16), where T is the characteristic momentum of quarks or gluons. It appears that the momentum distribution approaches isotropy owing to instabilities within the inverse time of order gT (57). If $1/g \gg 1$, the collisional equilibration is obviously much slower. As discussed in Section 4, the situation changes in strongly coupled plasmas.

When the instabilities grow, the system becomes more and more isotropic (46, 57) because the Lorentz force changes the particle's momenta, and the growing fields carry an extra momentum. To explain the mechanism, let us assume that initially there is a momentum surplus in the z-direction. The fluctuating current flows in the

z-direction, with the wave vector pointing in the x-direction. Because the magnetic field has a y-component, the Lorentz force—which acts on partons flying along the z-axis—pushes the partons in the x-direction where there is a momentum deficit. The numerical simulation (52) shows that growth of the instabilities is indeed accompanied by the system's fast isotropization.

The system isotropizes not only owing to the effect of the Lorentz force, but also owing to the momentum carried by the growing field. When the magnetic and electric fields are oriented along the y- and z-axes, respectively, the Poynting vector points in the x-direction along the wave vector. Thus, the momentum carried by the fields is oriented in the direction of the momentum deficit of particles.

Although the scenario of instabilities-driven equilibration looks very promising, the problem of thermalization of QGPs produced in heavy-ion collisions is far from being settled. Schenke et al. (58) showed that interparton collisions, which have been modeled using the BGK collision term (10), reduce the growth of instabilities and thus slow down the process of equilibration. The equilibration is also slowed down owing to expansion of the QGP into vacuum (59, 60), which is a characteristic feature of QGPs produced in relativistic heavy-ion collisions. Finally, the late stage of instability development, when non-Abelian effects are crucially important, appears to be very complex (61, 62), and it is far from being understood.

As already mentioned, instabilities influence various plasma characteristics. In particular, turbulent magnetic fields generated in the systems, which are unstable with respect to transverse modes, are responsible for a reduction of plasma viscosity (63). Then, an anomalously small viscosity, which is usually associated with strongly coupled systems, can occur in weakly coupled plasmas as well. Recently, it has been argued (64, 65) that the mechanism of viscosity reduction is operative in the unstable QGP.

3.4. Energy Loss

A charged particle that moves across the plasma changes its energy owing to several processes (2). When the particle's energy (E) is comparable to the plasma temperature (T), the particle can gain energy owing to interactions with field fluctuations. (In context of the QGP, the problem was studied in Reference 66.) A fast particle with $E \gg T$ loses energy, and dominant contributions come from collisions with other plasma particles and from radiation. In the following we discuss the energy loss of a fast particle, as the problem is closely related to jet quenching, which was suggested long ago as a signature for QGP formation in relativistic heavy-ion collisions (67, 68).

Let us start with the collisional energy loss. The particle's collisions are split into two classes: hard, with high-momentum transfer, corresponding to the collisions with plasma particles, and soft, with low-momentum transfer dominated by the interactions with plasma collective modes. The momentum is called soft when it is of order of the Debye mass, m_D, or smaller, and it is hard when it is larger than m_D.

The soft contribution to the energy loss, which can be treated classically, is often known as plasma polarization. It leads to the energy loss per unit time given by the formula

$$\left(\frac{dE}{dt}\right)_{\text{soft}} = \int d^3x\, \mathbf{j}(x)\mathbf{E}(x), \qquad\qquad 62.$$

where $\mathbf{E}$ is the electric field induced in the plasma by the particle's current $\mathbf{j}$, which is of the form $\mathbf{j}(x) = q\mathbf{v}\delta^{(3)}(\mathbf{x} - \mathbf{v}t)$. The field can be calculated by means of the Maxwell equations. After eliminating the magnetic field, one finds the equation

$$\left[\varepsilon^{ij}(k) - \frac{\mathbf{k}^2}{\omega^2}\left(\delta^{ij} - \frac{k^i k^j}{\mathbf{k}^2}\right)\right] E_j(k) = \frac{4\pi}{i\omega} j^i(k).$$

Because we consider equilibrium plasma, which is isotropic, one introduces the longitudinal (ε_L) and transverse (ε_L) components of ε^{ij}. Then, Equation 62 can be manipulated to

$$\left(\frac{dE}{dx}\right)_{\text{soft}} = -\frac{4\pi i e^2}{v}\int \frac{d^3k}{(2\pi)^3}\left\{\frac{\omega}{\mathbf{k}^2 \varepsilon_L(k)} + \frac{\mathbf{v}^2 - \omega^2/\mathbf{k}^2}{\omega[\varepsilon_T(k) - \mathbf{k}^2/\omega^2]}\right\}, \qquad 63.$$

which gives the energy loss per unit length. This formula describes the effect of medium polarization. However, three comments are in order here:

1. Equation 63 includes the charge self-interaction signaled by the UV divergence of the integral from Equation 63. The self-interaction is removed by subtracting from Equation 63 the vacuum expression with $\varepsilon_L = \varepsilon_T = 1$.
2. Poles of the function under the integral from Equation 63 correspond to the plasma collective modes as given by the dispersion relation from Equation 49. Therefore, the explicit expressions of ε_L and ε_L are not actually needed to compute the integral in Equation 63. The knowledge of the spectrum of quasi-particles appears to be sufficient.
3. Equation 63 is derived in the classical approximation, which breaks down for a sufficiently large $\mathbf{k}$. Therefore, an upper cutoff is needed. The interaction with $\mathbf{k}$ above the cutoff, which, as already mentioned, is of order of the Debye mass, should be treated as hard collisions with plasma particles.

The energy loss per unit length due to hard collisions is

$$\left(\frac{dE}{dx}\right)_{\text{hard}} = \sum_i \int \frac{d^3k}{(2\pi)^3} n_i(k)\,[\text{flux factor}] \int d\Omega \frac{d\sigma^i}{d\Omega} v, \qquad 64.$$

where the sum runs over particle species distributed according to $n_i(k)$, $\nu \equiv E - E'$ is the energy transfer, and $d\sigma^i/d\Omega$ is the respective differential cross section. Combining Equations 63 and 64, one finds the complete collisional energy loss. The calculations of the energy loss of a fast parton in the QGP along the lines presented above were performed in References 69 and 70. Systematic calculations of the collisional energy loss using the hard thermal loop resummation technique were given in References 71 and 72, with a result that is infrared finite, gauge invariant, and complete to leading order. Recently, the calculations of the collisional energy loss have been extended to anisotropic QGPs (73).

It was realized that a sizeable contribution to the quark's energy loss comes from radiative processes (74). The problem, however, appeared to be very complex because the quark's successive interactions in the plasma cannot be treated as independent from each other, and there is a destructive interference of radiated gluons known as the Landau-Pomeranchuk-Migdal effect (75). There are numerous papers devoted to the radiative energy loss, and the whole problem is reviewed in Reference 76. A general conclusion of these studies is that the energy lost by a fast light quark depends quadraticaly (not linearly) on the path traversed in the QGP, since the radiative energy loss dominates over the collisional loss. Recent experiments at RHIC show (77, 78), however, that heavy quarks, whose radiative energy loss is significantly suppressed, are strongly decelerated in the QGP medium. It may suggest that the collisional energy loss should actually be enhanced as predicted theoretically in Reference 79.

4. STRONGLY COUPLED PLASMAS

Our discussion of the collective phenomena presented in Section 3 was limited to weakly interacting plasmas, with the coupling constant much smaller than unity. However, the QGP produced in ultrarelativistic heavy-ion collisions is presumably strongly coupled quark-gluon plasma (sQGP), as the temperature is never much larger than Λ_{QCD}, and the regime of asymptotic freedom is not reached. The QGP is certainly a strongly interacting system close to the confinement phase transition. There are indeed hints in the extensive experimental material collected at RHIC (80–83) that the matter produced at the early stage of nucleus-nucleus collisions is in the form of sQGP for an interval of time of a few femtometers per speed of light. In particular, the characteristics of elliptic flow and particle spectra, which are well described by ideal hydrodynamics, seem to indicate a fast thermalization and small viscosity of the plasma. Both features are naturally explained assuming a strong coupling of the plasma (55, 84–87).

Although a fast thermalization (56) as well as a small viscosity (64, 65) can also be explained by instabilities, the idea of sQGP must be examined. However, the theoretical tools presented in Section 2 implicitly or explicitly assume a small coupling constant, and they are of limited applicability. A powerful approach, which can be used to study sQGP is the lattice formulation of QCD (for a review, see Reference 88). However, lattice QCD calculations, which are mostly numerical, encounter serious problems in incorporating quark degrees of freedom. It is also very difficult to analyze time-dependent plasma characteristics.

Strongly coupled conformal field theories such as supersymmetric QCD can be studied by means of the so-called AdS/CFT duality (89). Although some very interesting results on the conformal QGP were obtained in this way (see, e.g., Reference 90 and references therein), the relevance of these results for the QGP governed by QCD—not supersymmetric QCD—is unclear. Thus, the question arises: What can we learn about sQGP from strongly coupled EMPs?

We first note that most EMPs in nature and technological applications are weakly coupled. That is, the interaction energy between the plasma particles is much smaller than their thermal (kinetic) energy. This is because strongly coupled plasmas require a high particle density and/or low temperature, at which usually strong recombination occurs and the plasma state vanishes. Exceptions are the ion component in white dwarfs, metallic hydrogen, and other states of dense warm matter in the interior of giant planets; short-living dense plasmas produced by intense laser or heavy-ion beams or in explosive shock tubes; dusty (or complex) plasmas; and two-dimensional electron systems in liquid helium (91–93). Therefore, it is a real challenge to study strongly coupled EMPs both theoretically and experimentally.

In nonrelativistic EMP, the interaction energy is given by the (screened) Coulomb potential. The Coulomb coupling parameter defined by

$$\Gamma = \frac{q^2}{a\,T} \qquad\qquad 65.$$

distinguishes between weakly coupled, $\Gamma \ll 1$, and strongly coupled, $\Gamma \gtrsim 1$, plasmas. Here, q is the particle charge, a the interparticle distance, and T the kinetic temperature of the plasma component (electrons, ions, charged dust grains) under consideration. In the case of a degenerate plasma, for example, the electron component in a white dwarf, the kinetic energy T is replaced by the Fermi energy. Owing to the strong interaction, the plasma can behave either as a gas or a liquid, or even a solid (crystalline) system.

The case of a one-component plasma (OCP) with a pure Coulomb interaction (a single species of charged particles in a uniform, neutralizing background) has been studied in great detail as a reference model for strongly coupled plasmas using simple models as well as numerical simulations (91). For $\Gamma > 172$, the plasma was shown to form regular structures (Coulomb crystallization) (94). Below this critical value, the OCP is in the supercritical state. For values of Γ larger than $\sim$50, it behaves like an ordinary liquid, whereas for small values below unity, it behaves like a gas. Only if Γ is large enough does the usual liquid behavior (Arrhenius' law for the viscosity, Stokes-Einstein relation between self-diffusion and shear viscosity, etc.) appear owing to caging of the particles (a single particle is trapped for some period of time in the cage formed by its nearest neighbors). For values of Γ smaller than approximately 50, caging is not sufficiently strong and the system shows complicated, not yet understood transport properties. However, the short-range ordering typical for liquids shows up already for $\Gamma > 3$ (95). A gas-liquid transition requiring a long-range attraction and a short-range repulsion, for example, Lennard-Jones potential, does not exist in the OCP with particles of like-sign charges.

In realistic systems with a screened Coulomb interaction (Yukawa potential), the phase diagram can be shown in the Γ-κ plane, where $\kappa = a/\lambda_D$ is the distance parameter, with λ_D being the Debye screening length. Numerical simulations based on molecular dynamics lead to the phase diagram shown in **Figure 6** (96).

The first quantity of interest of sQGP is the coupling parameter. In analogy to nonrelativistic EMP, it is defined as (97)

$$\Gamma = \frac{2Cg^2}{4\pi a\,T} = 1.5 - 5, \qquad\qquad 66.$$

where C is the Casimir invariant ($C = 4/3$ for quarks and $C = 3$ for gluons), $a \simeq$ 0.5 fm is the interparticle distance, and $T \simeq 200$ MeV is the QGP temperature corresponding to a strong-coupling constant $g \simeq 2$. The factor 2 in the numerator comes from taking into account the magnetic interaction in addition to the static electric (Coulomb) interaction, which are of the same magnitude in ultrarelativistic

Figure 6

Phase diagram of a
strongly coupled Yukawa
system. Figure reprinted
with permission from
Reference 96. Copyright
(1997) by the American
Physical Society.

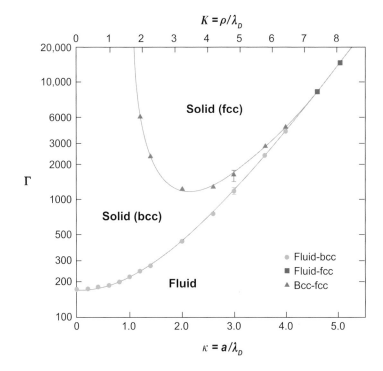

plasmas. The factor 4π in the denominator comes from using the Heaviside-Lorentz system in QCD, as discussed in the Introduction. The distance parameter κ of the QGP under the above conditions is rather small, typically between 1 and 3 (98).

Note that we have assumed here a classical interaction potential corresponding to one-gluon exchange. However, an effective potential that takes into account higher-order and nonperturbative effects may be much larger. This may be related to the fact that experimental data suggest a cross-section enhancement for the parton interaction by more than an order of magnitude (see below). Hence, the effective coupling parameter may be up to an order of magnitude larger than Equation 66.

As discussed above, comparison to the OCP model as well as experimental data suggests that QGPs close to the confinement phase transition could be in a liquid phase. The question arises whether there is a phase transition from a liquid to a gaseous QGP, as sketched in **Figure 7**. For such a transition, a Lennard-Jones-type interaction between the partons is required. However, the parton interaction in perturbative QCD is either purely repulsive or attractive in the various interaction channels, for example, in the quark-antiquark or diquark channels. Owing to nonlinear effects caused by the strong coupling, however, attractive interactions can arise even in the case of like-sign charges (see, e.g., Reference 99), leading to Lennard-Jones-type potentials. Hence, a gas-liquid transition in the QGP with a critical point, proposed in Reference 87, cannot be excluded and deserves further investigation.

An important quantity, which is very useful in theoretical and experimental studies of strongly coupled systems on the microscopic level, in particular in fluid physics

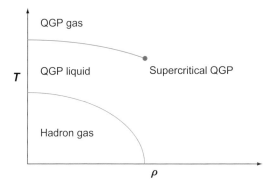

Figure 7

Sketch of a phase diagram of strongly interacting matter with a possible gas-liquid transition in the QGP phase. T denotes the temperature, while ρ is the baryon density.

(100), is correlation function. In particular, the pair-correlation function and the static-structure function provide valuable information on the equation of state of the system (100). The extension of the approach to the QGP has been proposed in Reference 98.

The static density-density autocorrelation function is defined for a classical system as (91, 100)

$$G(\mathbf{r}) = \frac{1}{N} \int d^3 r' \langle \rho(\mathbf{r} + \mathbf{r}', t) \rho(\mathbf{r}', t) \rangle,$$

where N is the total number of particles and

$$\rho(\mathbf{r},\ t) = \sum_{i=1}^{N} \delta^{(3)}(\mathbf{r} - \mathbf{r}_i(t))$$

is the local density of point particles, with $\mathbf{r}_i(t)$ denoting the position of i-th particle at time t. The density-density autocorrelation function is related to the pair-correlation function, which is defined as

$$g(\mathbf{r}) = \frac{1}{N} \left\langle \sum_{i,j,i \neq j}^{N} \delta^3(\mathbf{r} + \mathbf{r}_i - \mathbf{r}_j) \right\rangle,$$

by the relation $G(\mathbf{r}) = g(\mathbf{r}) + \delta^{(3)}(\mathbf{r})$. The static-structure function, defined by

$$S(\mathbf{p}) = \frac{1}{N} \langle \rho(\mathbf{p}) \rho(-\mathbf{p}) \rangle,$$

with the Fourier-transformed particle density

$$\rho(\mathbf{p}) = \int d^3 r \rho(\mathbf{r}) e^{-i\mathbf{p}\cdot\mathbf{r}},$$

is the Fourier transform of the density-density autocorrelation function

$$S(\mathbf{p}) = \int d^3 r e^{-i\mathbf{p}\cdot\mathbf{r}} G(\mathbf{r}).$$

The static-structure function $S(\mathbf{p})$ is constant for $\mathbf{p} \neq 0$ for uncorrelated particles (2). The typical behavior of the function in an interacting gas and in a liquid is sketched in **Figure 8**. The oscillatory behavior is caused by short-range correlations,

Figure 8

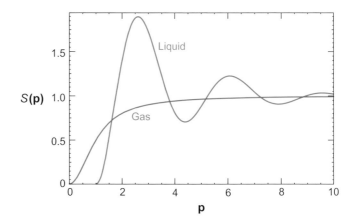

Sketch of the
static-structure functions
versus momentum in the
gas and liquid phases in
arbitrary units.

corresponding to a short-range ordering typical for liquids. In the case of an OCP, the oscillations appear for $\Gamma > 3$, indicating a liquid behavior (albeit with nonstandard transport properties) of the supercritical phase already for rather low values of Γ (95).

The static-quark structure function can be related to the longitudinal gluon-polarization tensor containing only the quark loop (**Figure 1**) via

$$S(\mathbf{p}) = -\frac{12}{\pi g^2 n} \int_0^\infty d\omega \Im \Pi_L(\omega, \mathbf{p}) \coth\frac{\omega}{2T},$$

where $n = N/V = \langle \rho(\mathbf{r}) \rangle$ is the average particle density in a homogeneous system. As a reference for the strong-coupling regime, the static-quark structure function has been calculated in the weak-coupling limit by resumming the polarization tensor in the high-temperature limit (Equation 7), leading to (98)

$$S(\mathbf{p}) = \frac{2N_f T^3}{n} \frac{\mathbf{p}^2}{\mathbf{p}^2 + m_D^2}, \qquad\qquad 67.$$

where $m_D^2 = N_f g^2 T^2/6$ is the quark contribution to the Debye screening mass. The static-structure function given by Equation 67 starts at zero for $|\mathbf{p}| = 0$ and saturates at the uncorrelated structure function $S(\mathbf{p}) = 2N_f T^3/n$ for large $|\mathbf{p}|$. Such a structure function corresponds to an interacting Yukawa system in the gas phase (see **Figure 8**). Indeed, the pair-correlation function, following from the Fourier transform of $S(\mathbf{p}) - 1$, is

$$g(r) = -\frac{N_f T^3}{2\pi n} \frac{m_D^2}{r} e^{-m_D r},$$

and it reproduces the Yukawa potential.

To compute the structure function in strongly coupled plasmas, molecular dynamics is used (91). Although QGP is not a classical system, as the thermal de Broglie wave length is of the same order as the interparticle distance, molecular dynamics may be useful as a first estimate (87). Using molecular dynamics for a classical sQGP (101–103), the expected behavior described above has been qualitatively verified (101). In strongly coupled dense matter, where quantum effects are important, quantum molecular dynamics based on a combination of classical molecular dynamics and density functional theory has been applied successfully (104). A generalization to the relativistic QGP has not been attempted so far. As an ultimate choice, lattice QCD could be used to calculate the structure or pair-correlation functions. This would provide a test for the state of sQGP as well as for the importance of quantum effects by comparing lattice results to classical molecular simulations.

As a last application, we consider the influence of strong coupling on the cross sections entering transport coefficients (shear viscosity), stopping power, and other dynamical quantities of the plasma. Besides higher-order and nonperturbative quantum effects, there is already a cross-section enhancement on the classical level. The reason is that the Coulomb radius, defined as $r_c = q^2/E$ for particle energy E, is of order of the Debye screening length—or r_c is even larger than λ_D in a strongly coupled plasma. Hence, the standard Coulomb scattering formula must be modified because the interaction with particles outside of the Debye sphere contributes significantly. Consequently, the inverse screening length cannot be used as an infrared cutoff. This modification leads, for example, to the experimentally observed enhancement of the so-called ion drag force in complex plasmas, which is caused by the ion-dust interaction (105).

In the QGP at $T \simeq 200$ MeV, the ratio r_c/λ_D equals 1–5. It may enhance a parton cross section by a factor of 2–9 (97) compared with perturbative results. An enhanced cross section reduces the mean free path λ and consequently reduces the viscosity η as $\eta \sim \lambda$. An enhancement of the elastic parton cross section by more than an order of magnitude compared with perturbative results also explains the elliptic flow and particle spectra observed at RHIC (106). An infrared cutoff smaller than the Debye mass provides a natural explanation for this enhancement. Also note that if the cross section is enhanced, the collisional energy loss grows. However, the radiative energy loss is expected to be suppressed in the sQGP by the Landau-Pomeranchuk-Migdal effect (75).

Finally, we mention two examples of strongly coupled systems that have not been considered in QGP physics but that may be of relevance. Strongly coupled plasmas such as two-dimensional Yukawa liquids (107) and dusty plasmas are non-Newtonian fluids. That is, the shear viscosity depends on the shear rate (flow velocity), as is observed in ketchup (shear thinning). The second example concerns nanofluidics. The expanding fireball in ultrarelativistic heavy-ion collisions has a transverse dimension of approximately 20 interparticle distances (approximately 10 fm). Fluids consisting of such a low number of layers exhibit properties different from large fluid systems. For example, the shear flow does not show a continuous velocity gradient but jumps owing to the adhesive forces between two layers. Such a behavior has been observed, for example, in complex plasmas.

LITERATURE CITED

1. Krall NA, Trivelpiece AW. *Principles of Plasma Physics*. New York: McGraw-Hill (1973)
2. Ichimaru S. *Basic Principles of Plasma Physics*. Reading: Benjamin (1973)

3. Hwa RC, ed. *Quark-Gluon Plasma*. Singapore: World Sci. (1990)

4. Hwa RC, ed. *Quark-Gluon Plasma 2*. Singapore: World Sci. (1995)

5. Hwa RC, Wang XN, eds. *Quark-Gluon Plasma 3*. Singapore: World Sci. (2004)

6. Mrówczyński St. *Acta Phys. Pol. B* 29:3711 (1998)

7. Blaizot JP, Iancu E. *Phys. Rep.* 359:355 (2002)

8. deGroot SR, Van Leeuwen WA, van Weert ChG. *Relativistic Kinetic Theory*. Amsterdam: North-Holland (1980)

9. Silin VP. *Sov. Phys. JETP* 11:1136 (1960)

10. Alexandrov AF, Bogdankovich LS, Rukhadze AA. *Principles of Plasma Electrodynamics*. Berlin: Springer (1984)

11. Carrington ME, Fugleberg T, Pickering D, Thoma MH. *Can. J. Phys.* 82:671 (2004)

12. Elze HT, Heinz UW. *Phys. Rep.* 183:81 (1989)

13. Mrówczyński St. *Phys. Rev. D* 39:1940 (1989)

14. Manuel C, Mrówczyński St. *Phys. Rev. D* 70:094019 (2004)

15. Schenke B, Strickland M, Greiner C, Thoma MH. *Phys. Rev. D* 73:125004 (2006)

16. Arnold P, Son DT, Yaffe LG. *Phys. Rev. D* 59:105020 (1999)

17. Manuel C, Mrówczyński St. *Phys. Rev. D* 68:094010 (2003)

18. Kolb PF, Heinz UW. In *Quark Gluon Plasma 3*, ed. RC Hwa, XN Wang, p. 634. Singapore: World Sci. (2004)

19. Huovinen P, Ruuskanen PV. *Annu. Rev. Nucl. Part. Sci.* 56:163 (2006)

20. Landau LD, Lifshitz EM. *Fluid Mechanics*. Oxford: Pergamon (1963)

21. Manuel C, Mrówczyński St. *Phys. Rev. D* 74:105003 (2006)

22. Kapusta JI. *Finite-Temperature Field Theory*. Cambridge: Cambridge Univ. Press (1989)

23. Le Bellac M. *Thermal Field Theory*. Cambridge: Cambridge Univ. Press (1996)

24. Schwinger JS. *J. Math. Phys.* 2:407 (1961)

25. Keldysh LV. *Zh. Eksp. Teor. Fiz.* 47:1515 (1964) [*Sov. Phys. JETP* 20:1018 (1965)]

26. Kadanoff LP, Baym G. *Quantum Statistical Mechanics*. New York: Benjamin (1962)

27. Bezzerides B, Dubois DF. *Ann. Phys.* 70:10 (1972)

28. Braaten E, Pisarski RD. *Nucl. Phys. B* 337:569 (1990)

29. Taylor JC, Wong SMH. *Nucl. Phys. B* 346:115 (1990)

30. Braaten E, Pisarski RD. *Phys. Rev. D* 45:1827 (1992)

31. Carrington ME, Hou D, Thoma MH. *Eur. Phys. J. C* 7:347 (1999)

32. Thoma MH. In *Quark-Gluon Plasma 2*, ed. RC Hwa, p. 51. Singapore: World Sci. (1995)

33. Pisarski RD. hep-ph/9710370 (1997)

34. Mrówczyński St, Thoma MH. *Phys. Rev. D* 62:036011 (2000)

35. Mrówczyński St, Rebhan A, Strickland M. *Phys. Rev. D* 70:025004 (2004)

36. Klimov VV. *Sov. Phys. JETP* 55:199 (1982)

37. Weldon HA. *Phys. Rev. D* 26:1394 (1982)

38. Blaizot JP, Iancu E. *Nucl. Phys. B* 417:608 (1994)

39. Kelly PF, Liu Q, Lucchesi C, Manuel C. *Phys. Rev. D* 50:4209 (1994)

40. Mustafa MG, Thoma MH, Chakraborty P. *Phys. Rev. C* 71:017901 (2005)
41. Ruppert J, Müller B. *Phys. Lett. B* 618:123 (2005)
42. Chakraborty P, Mustafa MG, Thoma MH. *Phys. Rev. D* 74:094002 (2006)
43. Weibel ES. *Phys. Rev. Lett.* 2:83 (1959)
44. Mrówczyński St. *Phys. Lett. B* 393:26 (1997)
45. Mrówczyński St. *Phys. Lett. B* 314:118 (1993)
46. Mrówczyński St. *Phys. Rev. C* 49:2191 (1994)
47. Randrup J, Mrówczyński St. *Phys. Rev. C* 68:034909 (2003)
48. Romatschke P, Strickland M. *Phys. Rev. D* 68:036004 (2003)
49. Arnold P, Lenaghan J, Moore GD. *JHEP* 0308:002 (2003)
50. Rebhan A, Romatschke P, Strickland M. *Phys. Rev. Lett.* 94:102303 (2005)
51. Arnold P, Lenaghan J. *Phys. Rev. D* 70:114007 (2004)
52. Dumitru A, Nara Y. *Phys. Lett. B* 621:89 (2005)
53. Arnold P, Moore GD, Yaffe LG. *Phys. Rev. D* 72:054003 (2005)
54. Rebhan A, Romatschke P, Strickland M. *JHEP* 0509:041 (2005)
55. Heinz U. *AIP Conf. Proc.* 739:163 (2005)
56. Mrówczyński St. *Acta Phys. Pol. B* 37:427 (2006)
57. Arnold P, Lenaghan J, Moore GD, Yaffe LG. *Phys. Rev. Lett.* 94:072302 (2005)
58. Schenke B, Strickland M, Greiner C, Thoma MH. *Phys. Rev. D* 73:125004 (2006)
59. Romatschke P, Venugopalan R. *Phys. Rev. Lett.* 96:062302 (2006)
60. Romatschke P, Rebhan A. *Phys. Rev. Lett.* 97:252301 (2006)
61. Arnold P, Moore GD. *Phys. Rev. D* 73:025013 (2006)
62. Dumitru A, Nara Y, Strickland M. hep-ph/0604149 (2006)
63. Abe T, Niu K. *J. Phys. Soc. Jpn.* 49:717 (1980)
64. Asakawa M, Bass SA, Müller B. *Phys. Rev. Lett.* 96:252301 (2006)
65. Asakawa M, Bass SA, Müller B. hep-ph/0608270 (2006)
66. Chakraborty P, Mustafa MG, Thoma MH. hep-ph/0611355 (2006)
67. Bjorken JD. *Fermilab preprint* 82/59-THY (1982)
68. Gyulassy M, Plümer M. *Phys. Lett. B* 243:432 (1990)
69. Thoma MH, Gyulassy M. *Nucl. Phys. B* 351:491 (1991)
70. Mrówczyński St. *Phys. Lett B* 269:383 (1991)
71. Braaten E, Thoma MH. *Phys. Rev. D* 44:R2625 (1991)
72. Braaten E, Thoma MH. *Phys. Rev. D* 44:1298 (1991)
73. Romatschke P, Strickland M. *Phys. Rev. D* 71:125008 (2005)
74. Gyulassy M, Wang XN. *Nucl. Phys. B* 420:583 (1994)
75. Landau LD, Pomeranchuk IY. *Dokl. Akad. Nauk. SSR* 92:535 (1953); Landau LD, Pomeranchuk IY. *Dokl. Akad. Nauk. SSR* 92:735 (1953); Migdal AB. *Phys. Rev.* 103:1811 (1956)
76. Baier R, Schiff D, Zakharov BG. *Annu. Rev. Nucl. Part. Sci.* 50:37 (2000)
77. Adler SS, et al. (PHENIX Collab.) *Phys. Rev. Lett.* 96:032301 (2006)
78. Abelev BI, et al. (STAR Collab.) nucl-ex/0607012 (2006)
79. Mustafa MG, Thoma MH. *Acta Phys. Hung. A* 22:93 (2005)
80. Arsene I, et al. (BRAHMS Collab.) *Nucl. Phys. A* 757:1 (2005)
81. Back BB, et al. (PHOBOS Collab.) *Nucl. Phys. A* 757:28 (2005)

82. Adams J, et al. (STAR Collab.) *Nucl. Phys. A* 757:102 (2005)
83. Adcox K, et al. (PHENIX Collab.) *Nucl. Phys. A* 757:184 (2005)
84. Gyulassy M, McLerran L. *Nucl. Phys. A* 750:30 (2005)
85. Shuryak EV. *Nucl. Phys. A* 774:387 (2006)
86. Cassing W, Peshier A. *Phys. Rev. Lett.* 94:172301 (2005)
87. Thoma MH. *Nucl. Phys. A* 774:307 (2006)
88. Karsch F. *Lect. Notes Phys.* 583:209 (2002)
89. Maldacena JM. *Adv. Theor. Math. Phys.* 2:231 (1998) [*Int. J. Theor. Phys.* 38:1113 (1999)]
90. Kovtun P, Son DT, Starinets AO. *Phys. Rev. Lett.* 94:111601 (2005); Janik RA, Peschanski R. *Phys. Rev. D* 74:046007 (2006)
91. Ichimaru S. *Rev. Mod. Phys.* 54:1017 (1982)
92. Fortov VE, et al. *Phys. Rep.* 421:1 (2005)
93. Tahir NA, et al. *J. Phys. A* 39:4755 (2006); Hoffmann DHH, et al. *J. Phys. IV* 133:49 (2006)
94. Slattery WL, Doolen GD, Witt HE. *Phys. Rev. A* 21:2087 (1980)
95. Daligault J. *Phys. Rev. Lett.* 96:065003 (2006)
96. Hamaguchi S, Farouki RT, Dubin DHE. *Phys. Rev. E* 56:4671 (1997)
97. Thoma MH. *J. Phys. G* 31:L7 (2005). Erratum. *J. Phys. G* 31:539 (2005)
98. Thoma MH. *Phys. Rev. D* 72:094030 (2005)
99. Tsytovich V. *Contrib. Plasma Phys.* 45:533 (2005)
100. Hansen JP, McDonald IR. *Theory of Simple Liquids*. London: Acad. Press (1986)
101. Gelman BA, Shuryak EV, Zahed I. *Phys. Rev. C* 74:044908 (2006)
102. Gelman BA, Shuryak EV, Zahed I. *Phys. Rev. C* 74:044909 (2006)
103. Hartmann P, Donko Z, Levai P, Kalman GJ. *Nucl. Phys. A* 774:881 (2006)
104. Car R, Parrinello M. *Phys. Rev. Lett.* 55:2471 (1985)
105. Yaroshenko V, et al. *Phys. Plasmas* 12:093503 (2005)
106. Molnar D, Gyulassy M. *Nucl. Phys. A* 697:495 (2002)
107. Donko Z, Goree J, Hartmann P, Kutasi K. *Phys. Rev. Lett.* 96:145003 (2006)

Viscosity, Black Holes, and Quantum Field Theory

Dam T. Son[1] and Andrei O. Starinets[2]

[1] Institute for Nuclear Theory, University of Washington, Seattle, Washington 98195;
email: son@phys.washington.edu

[2] Perimeter Institute for Theoretical Physics, Waterloo, Ontario N2L 2Y5, Canada;
email: starina@perimeterinstitute.ca

Annu. Rev. Nucl. Part. Sci. 2007. 57:95–118

First published online as a Review in Advance on
April 20, 2007

The *Annual Review of Nuclear and Particle Science* is
online at http://nucl.annualreviews.org

This article's doi:
10.1146/annurev.nucl.57.090506.123120

Copyright © 2007 by Annual Reviews.
All rights reserved

0163-8998/07/1123-0095$20.00

Key Words

AdS/CFT correspondence, hydrodynamics

Abstract

We review recent progress in applying the AdS/CFT correspondence to finite-temperature field theory. In particular, we show how the hydrodynamic behavior of field theory is reflected in the low-momentum limit of correlation functions computed through a real-time AdS/CFT prescription, which we formulate. We also show how the hydrodynamic modes in field theory correspond to the low-lying quasi-normal modes of the AdS black p-brane metric. We provide proof of the universality of the viscosity/entropy ratio within a class of theories with gravity duals and formulate a viscosity bound conjecture. Possible implications for real systems are mentioned.

Contents

1. INTRODUCTION 96
2. HYDRODYNAMICS 97
   2.1. Kubo's Formula for Viscosity 99
   2.2. Hydrodynamic Modes 100
   2.3. Viscosity in Weakly Coupled Field Theories 101
3. AdS/CFT CORRESPONDENCE 102
   3.1. Review of AdS/CFT Correspondence at Zero Temperature 102
   3.2. Black Three-Brane Metric 106
4. REAL-TIME AdS/CFT 107
   4.1. Prescription for Retarded Two-Point Functions 109
   4.2. Calculating Hydrodynamic Quantities 110
5. THE MEMBRANE PARADIGM 112
6. THE VISCOSITY/ENTROPY RATIO 114
   6.1. Universality 114
   6.2. The Viscosity Bound Conjecture 115
7. CONCLUSION 116

1. INTRODUCTION

This review is about the recently emerging connection, through the gauge/gravity correspondence, between hydrodynamics and black hole physics. The study of quantum field theory at high temperature has a long history. It was first motivated by the Big Bang cosmology when it was hoped that early phase transitions might leave some imprints on the Universe (1). One of those phase transitions is the quantum chromodynamics (QCD) phase transition (which could actually be a crossover), which happened at a temperature around $T_c \sim 200$ MeV, when matter turned from a gas of quarks and gluons (the quark-gluon plasma, or QGP) into a gas of hadrons.

An experimental program was designed to create and study the QGP by colliding two heavy atomic nuclei. Most recent experiments were conducted at the Relativistic Heavy Ion Collider (RHIC) at Brookhaven National Laboratory. Although significant circumstantial evidence for the QGP was accumulated (2), a theoretical interpretation of most of the experimental data proved difficult because the QGP created at RHIC is far from being a weakly coupled gas of quarks and gluons. Indeed, the temperature of the plasma, as inferred from the spectrum of final particles, is only approximately 170 MeV, near the confinement scale of QCD. This is deep in the nonperturbative regime of QCD, where reliable theoretical tools are lacking. Most notably, the kinetic coefficients of the QGP, which enter the hydrodynamic equations (reviewed in Section 2), are not theoretically computable at these temperatures.

The paucity of information about the kinetic coefficients of the QGP in particular and of strongly coupled thermal quantum field theories in general is one of the main reasons for our interest in their computation in a class of strongly coupled field theories, even though this class does not include QCD. The necessary technological tool is the anti-de Sitter–conformal field theory (AdS/CFT) correspondence (3–5), discovered in the investigation of D-branes in string theory. This correspondence allows one to describe the thermal plasmas in these theories in terms of black holes in AdS space. The AdS/CFT correspondence is reviewed in Section 3.

The first calculation of this type, that of the shear viscosity in $\mathcal{N} = 4$ supersymmetric Yang-Mills (SYM) theory (6), is followed by the theoretical work to establish the rules of real-time finite-temperature AdS/CFT correspondence (7, 8). Applications of these rules to various special cases (9–12) clearly show that even very exotic field theories, when heated up to finite temperature, behave hydrodynamically at large distances and timescales (provided that the number of space-time dimensions is 2 + 1 or higher). This development is reviewed in Section 4. Moreover, the way AdS/CFT works reveals very deep connections to properties of black holes in classical gravity. For example, the hydrodynamic modes of a thermal medium are mapped, through the correspondence, to the low-lying quasi-normal modes of a black-brane metric. It seems that our understanding of the connection between hydrodynamics and black hole physics is still incomplete; we may understand more about gravity by studying thermal field theories. One idea along this direction is reviewed in Section 5.

From the point of view of heavy-ion (QGP) physics, a particularly interesting finding has been the formulation of a conjecture on the lowest possible value of the ratio of viscosity to volume density of entropy. This conjecture was motivated by the universality of this ratio in theories with gravity duals. This is reviewed in Section 6.

This review is written primarily for readers with a background in QCD and QGP physics who are interested in learning about AdS/CFT correspondence and its applications to finite-temperature field theory. Some parts of this review (for example, the section about hydrodynamics) should be useful for readers with a string theory or general relativity background who are interested in the connection between string theory, gravity, and hydrodynamics. The perspectives here are shaped by our personal taste and therefore may appear narrow, but the authors believe that this review may serve as the starting point to explore the much richer original literature. In this review we use the "mostly plus" metric signature $- + + +$.

2. HYDRODYNAMICS

From the modern perspective, hydrodynamics (13) is best thought of as an effective theory, describing the dynamics at large distances and timescales. Unlike the familiar effective field theories (for example, the chiral perturbation theory), it is normally formulated in the language of equations of motion instead of an action principle. The reason for this is the presence of dissipation in thermal media.

In the simplest case, the hydrodynamic equations are just the laws of conservation of energy and momentum,

$$\partial_\mu T^{\mu\nu} = 0. \qquad \qquad 1.$$

To close the system of equations, we must reduce the number of independent elements of $T^{\mu\nu}$. This is done through the assumption of local thermal equilibrium:

If perturbations have long wavelengths, the state of the system, at a given time, is determined by the temperature as a function of coordinates $T(\mathbf{x})$ and the local fluid velocity u^μ, which is also a function of coordinates $u^\mu(\mathbf{x})$. Because $u_\mu u^\mu = -1$, only three components of u^μ are independent. The number of hydrodynamic variables is four, equal to the number of equations.

In hydrodynamics we express $T^{\mu\nu}$ through $T(x)$ and $u^\mu(x)$ through the so-called constitutive equations. Following the standard procedure of effective field theories, we expand in powers of spatial derivatives. To zeroth order, $T^{\mu\nu}$ is given by the familiar formula for ideal fluids,

$$T^{\mu\nu} = (\varepsilon + P)u^\mu u^\nu + Pg^{\mu\nu},$$

2.

where ε is the energy density and P is the pressure. Normally one would stop at this leading order, but qualitatively new effects necessitate going to the next order. Indeed, from Equation 2 and the thermodynamic relations $d\varepsilon = TdS, dP = sdT$, and $\varepsilon + P = Ts$ (s is the entropy per unit volume), one finds that entropy is conserved (14):

$$\partial_\mu(s u^\mu) = 0.$$

3.

Thus, to have entropy production, one needs to go to the next order in the derivative expansion.

At the next order, we write

$$T^{\mu\nu} = (\varepsilon + P)u^\mu u^\nu + Pg^{\mu\nu} - \sigma^{\mu\nu},$$

4.

where $\sigma^{\mu\nu}$ is proportional to derivatives of $T(x)$ and $u^\mu(x)$ and is termed the dissipative part of $T^{\mu\nu}$. To write these terms, let us first fix a point x and go to the local rest frame where $u^i(x) = 0$. In this frame, in principle one can have dissipative corrections to the energy-momentum density $T^{0\mu}$. However, one recalls that the choice of T and u^μ is arbitrary, and thus one can always redefine them so that these corrections vanish, $\sigma^{00} = \sigma^{0i} = 0$, and so at a point x,

$$T^{00} = \varepsilon, \quad T^{0i} = 0.$$

5.

The only nonzero elements of the dissipative energy-momentum tensor are σ_{ij}. To the next-to-leading order there are extra contributions whose forms are dictated by rotational symmetry:

$$\sigma_{ij} = \eta\left(\partial_i u_j + \partial_j u_i - \frac{2}{3}\delta_{ij}\partial_k u^k\right) + \zeta\delta_{ij}\partial_k u^k.$$

6.

Going back to the general frame, we can now write the dissipative part of the energy-momentum tensor as

$$\sigma^{\mu\nu} = P^{\mu\alpha}P^{\nu\beta}\left[\eta\left(\partial_\alpha u_\beta + \partial_\beta u_\alpha - \frac{2}{3}g_{\alpha\beta}\partial_\lambda u^\lambda\right) + \zeta g_{\alpha\beta}\partial_\lambda u^\lambda\right],$$

7.

where $P^{\mu\nu} = g^{\mu\nu} + u^\mu u^\nu$ is the projection operator onto the directions perpendicular to u^μ.

If the system contains a conserved current, there is an additional hydrodynamic equation related to the current conservation,

$$\partial_\mu j^\mu = 0.$$

8.

The constitutive equation contains two terms:

$$j^\mu = \rho u^\mu - DP^{\mu\nu}\partial_\nu\alpha, \qquad\qquad 9.$$

where ρ is the charge density in the fluid rest frame and D is some constant. The first term corresponds to convection, the second one to diffusion. In the fluid rest frame, $\mathbf{j} = -D\nabla\rho$, which is Fick's law of diffusion, with D being the diffusion constant.

2.1. Kubo's Formula for Viscosity

As mentioned above, the hydrodynamic equations can be thought of as an effective theory describing the dynamics of the system at large lengths and timescales. Therefore one should be able to use these equations to extract information about the low-momentum behavior of Green's functions in the original theory.

For example, let us recall how the two-point correlation functions can be extracted. If we couple sources $J_a(\mathbf{x})$ to a set of (bosonic) operators $O_a(x)$, so that the new action is

$$S = S_0 + \int_x J_a(x)O_a(x), \qquad\qquad 10.$$

then the source will introduce a perturbation of the system. In particular, the average values of O_a will differ from the equilibrium values, which we assume to be zero. If J_a are small, the perturbations are given by the linear response theory as

$$\langle O_a(x) \rangle = - \int_y G^R_{ab}(x-y)J_b(y), \qquad\qquad 11.$$

where G^R_{ab} is the retarded Green's function

$$iG^R_{ab}(x-y) = \theta(x^0 - y^0)\langle [O_a(x), O_b(y)] \rangle. \qquad\qquad 12.$$

The fact that the linear response is determined by the retarded (and not by any other) Green's function is obvious from causality: The source can influence the system only after it has been turned on.

Thus, to determine the correlation functions of $T^{\mu\nu}$, we need to couple a weak source to $T^{\mu\nu}$ and determine the average value of $T^{\mu\nu}$ after this source is turned on. To find these correlators at low momenta, we can use the hydrodynamic theory. So far in our treatment of hydrodynamics we have included no source coupled to $T^{\mu\nu}$. This deficiency can be easily corrected, as the source of the energy-momentum tensor is the metric $g_{\mu\nu}$. One must generalize the hydrodynamic equations to curved space-time and from it determine the response of the thermal medium to a weak perturbation of the metric. This procedure is rather straightforward and in the interest of space is left as an exercise to the reader.

We concentrate on a particular case when the metric perturbation is homogeneous in space but time dependent:

$$g_{ij}(t, \mathbf{x}) = \delta_{ij} + h_{ij}(t), \quad h_{ij} \ll 1 \qquad\qquad 13.$$
$$g_{00}(t, \mathbf{x}) = -1, \quad g_{0i}(t, \mathbf{x}) = 0. \qquad\qquad 14.$$

Moreover, we assume the perturbation to be traceless, $h_{ii} = 0$. Because the perturbation is spatially homogeneous, if the fluid moves, it can only move uniformly:

$u^i = u^i(t)$. However, this possibility can be ruled out by parity, so the fluid must remain at rest all the time: $u^\mu = (1, 0, 0, 0)$. We now compute the dissipative part of the stress-energy tensor. The generalization of Equation 7 to curved space-time is

$$\sigma^{\mu\nu} = P^{\mu\alpha} P^{\nu\beta} \left[\eta (\nabla_\alpha u_\beta + \nabla_\beta u_\alpha) + \left(\zeta - \frac{2}{3} \eta \right) g_{\alpha\beta} \nabla \cdot u \right]. \qquad 15.$$

Substituting $u^\mu = (1, 0, 0, 0)$ and $g_{\mu\nu}$ from Equation 13, we find only contributions to the traceless spatial components, and these contributions come entirely from the Christoffel symbols in the covariant derivatives. For example,

$$\sigma_{xy} = 2\eta \Gamma^0_{xy} = \eta \partial_0 h_{xy}. \qquad 16.$$

By comparison with the expectation from the linear response theory, this equation means that we have found the zero spatial momentum, low-frequency limit of the retarded Green's function of T^{xy}:

$$G^R_{xy,xy}(\omega, \mathbf{0}) = \int dt\, d\mathbf{x}\, e^{i\omega t} \theta(t) \langle [T_{xy}(t, \mathbf{x}), T_{xy}(0, \mathbf{0})] \rangle = -i\eta\omega + O(\omega^2) \qquad 17.$$

(modulo contact terms). We have, in essence, derived the Kubo's formula relating the shear viscosity and a Green's function:

$$\eta = -\lim_{\omega \to 0} \frac{1}{\omega} \operatorname{Im} G^R_{xy,xy}(\omega, \mathbf{0}). \qquad 18.$$

There is a similar Kubo's relation for the charge diffusion constant D.

2.2. Hydrodynamic Modes

If one is interested only in the locations of the poles of the correlators, one can simply look for the normal modes of the linearized hydrodynamic equations, that is, solutions that behave as $e^{-i\omega t + i\mathbf{k} \cdot \mathbf{x}}$. Owing to dissipation, the frequency $\omega(\mathbf{k})$ is complex. For example, the equation of charge diffusion,

$$\partial_t \rho - D\nabla^2 \rho = 0, \qquad 19.$$

corresponds to a pole in the current-current correlator at $\omega = -i Dk^2$.

To find the poles in the correlators between elements of the stress-energy tensor, one can, without loss of generality, choose the coordinate system so that $\mathbf{k}$ is aligned along the x^3-axis: $\mathbf{k} = (0, 0, k)$. Then one can distinguish two types of normal modes:

1. Shear modes correspond to the fluctuations of pairs of components T^{0a} and T^{3a}, where $a = 1, 2$. The constitutive equation is

$$T^{3a} = -\eta \partial_3 u^a = -\frac{\eta}{\varepsilon + P} \partial_3 T^{0a}, \qquad 20.$$

and the equation for T^{0a} is

$$\partial_t T^{0a} - \frac{\eta}{\varepsilon + P} \partial_3^2 T^{0a} = 0. \qquad 21.$$

That is, it has the form of a diffusion equation for T^{0a}. Substituting $e^{-i\omega t + ikx^3}$ into the equation, one finds the dispersion law

$$\omega = -i\frac{\eta}{\varepsilon + P}k^2.$$ 22.

2. Sound modes are fluctuations of T^{00}, T^{03}, and T^{33}. There are now two conservation equations, and by diagonalizing them one finds the dispersion law

$$\omega = c_s k - \frac{i}{2}\left(\frac{4}{3}\eta + \zeta\right)\frac{k^2}{\varepsilon + P},$$ 23.

where $c_s = \sqrt{dP/d\varepsilon}$. This is simply the sound wave, which involves the fluctuation of the energy density. It propagates with velocity c_s, and its damping is related to a linear combination of shear and bulk viscosities.

In CFTs it is possible to use conformal Ward identities to show that the bulk viscosity vanishes: $\zeta = 0$. Hence, we shall concentrate our attention on the shear viscosity η.

2.3. Viscosity in Weakly Coupled Field Theories

We now briefly consider the behavior of the shear viscosity in weakly coupled field theories, with the $\lambda\phi^4$ theory as a concrete example. At weak coupling, there is a separation between two length scales: The mean free path of particles is much larger than the distance scales over which scatterings occur. Each scattering event takes a time of order T^{-1} (which can be thought of as the time required for final particles to become on shell). The mean free path ℓ_{mfp} can be estimated from the formula

$$\ell_{\mathrm{mfp}} \sim \frac{1}{n\sigma v},$$ 24.

where n is the density of particles, σ is the typical scattering cross section, and v is the typical particle velocity. Inserting the values for thermal $\lambda\phi^4$ theory, $n \sim T^3$, $\sigma \sim \lambda^2 T^{-2}$, and $v \sim 1$, one finds

$$\ell_{\mathrm{mfp}} \sim \frac{1}{\lambda^2 T} \gg \frac{1}{T}.$$ 25.

The viscosity can be estimated from kinetic theory to be

$$\eta \sim \varepsilon \ell_{\mathrm{mfp}},$$ 26.

where ε is the energy density. From $\varepsilon \sim T^4$ and the estimate of ℓ_{mft}, one finds

$$\eta \sim \frac{T^3}{\lambda^2}.$$ 27.

In particular, the weaker the coupling λ, the larger the viscosity η. This behavior is explained by the fact that the viscosity measures the rate of momentum diffusion. The smaller λ is, the longer a particle travels before colliding with another one, and the easier the momentum transfer.

It may appear counterintuitive that viscosity tends to infinity in the limit of zero coupling $\lambda \to 0$: At zero coupling there is no dissipation, so should the viscosity be zero? The confusion arises owing to the fact that the hydrodynamic theory, and hence the notion of viscosity, makes sense only on distances much larger than the mean free path of particles. If one takes $\lambda \to 0$, then to measure the viscosity one has to do the experiment at larger and larger length scales. If one fixes the size of the experiment and takes $\lambda \to 0$, dissipation disappears, but it does not tell us anything about the viscosity.

As will become apparent below, a particularly interesting ratio to consider is the ratio of shear viscosity to entropy density s. The latter is proportional to T^3; thus

$$\frac{\eta}{s} \sim \frac{1}{\lambda^2}. \qquad\qquad 28.$$

One has $\eta/s \gg 1$ for $\lambda \ll 1$. This is a common feature of weakly coupled field theories. Extrapolating to $\lambda \sim 1$, one finds $\eta/s \sim 1$. Theories with gravity duals are strongly coupled, and η/s is of order one. More surprisingly, this ratio is the same for all theories with gravity duals.

To compute rather than estimate the viscosity, one can use Kubo's formula. It turns out that one has to sum an infinite number of Feynman graphs to find the viscosity to leading order. Another way that leads to the same result is to first formulate a kinetic Boltzmann equation for the quasi-particles as an intermediate effective description, and then derive hydrodynamics by taking the limit of very long lengths and timescales in the kinetic equation. Interested readers should consult References 15 and 16 for more details.

3. AdS/CFT CORRESPONDENCE

3.1. Review of AdS/CFT Correspondence at Zero Temperature

This section briefly reviews the AdS/CFT correspondence at zero temperature. It contains only the minimal amount of materials required to understand the rest of the review. Further information can be found in existing reviews and lecture notes (17, 18).

The original example of AdS/CFT correspondence is between $\mathcal{N} = 4$ SYM theory and type IIB string theory on $\mathrm{AdS}_5 \times S^5$ space. Let us describe the two sides of the correspondence in some more detail.

The $\mathcal{N} = 4$ SYM theory is a gauge theory with a gauge field, four Weyl fermions, and six real scalars, all in the adjoint representation of the color group. Its Lagrangian can be written down explicitly, but is not very important for our purposes. It has a vanishing beta function and is a CFT (thus the CFT in AdS/CFT). In our further discussion, we frequently use the generic terms field theory or CFT for the $\mathcal{N} = 4$ SYM theory.

On the string theory side, we have type IIB string theory, which contains a finite number of massless fields, including the graviton, the dilaton Φ, some other fields (forms) and their fermionic superpartners, and an infinite number of massive string excitations. It has two parameters: the string length l_s (related to the slope parameter α' by $\alpha' = l_s^2$) and the string coupling g_s. In the long-wavelength limit, when all fields vary over length scales much larger than l_s, the massive modes decouple and one is left with type IIB supergravity in 10 dimensions, which can be described by an action (19)

$$S_{\mathrm{SUGRA}} = \frac{1}{2\kappa_{10}^2} \int d^{10}x \sqrt{-g}\, e^{-2\Phi} (\mathcal{R} + 4\partial^\mu \Phi \partial_\mu \Phi + \cdots), \qquad 29.$$

where κ_{10} is the 10-dimensional gravitational constant,

$$\kappa_{10} = \sqrt{8\pi G} = 8\pi^{7/2} g_s l_s^4, \qquad 30.$$

and … stay for the contributions from fields other than the metric and the dilaton. One of these fields is the five-form F_5, which is constrained to be self-dual. The type IIB string theory is based in a 10-dimensional space-time with the following metric:

$$ds^2 = \frac{r^2}{R^2}(-dt^2 + d\mathbf{x}^2) + \frac{R^2}{r^2}dr^2 + R^2 d\Omega_5^2. \qquad 31.$$

The metric is a direct product of a five-dimensional sphere ($d\Omega_5^2$) and another five-dimensional space-time spanned by t, $\mathbf{x}$, and r. An alternative form of the metric is obtained from Equation 31 by a change of variable $z = R^2/r$,

$$ds^2 = \frac{R^2}{z^2}(-dt^2 + d\mathbf{x}^2 + dz^2) + R^2 d\Omega_5^2. \qquad 32.$$

Both coordinates r and z are known as the radial coordinates. The limiting value $r = \infty$ (or $z = 0$) is known as the boundary of the AdS space.

It is a simple exercise to check that the $(t, \mathbf{x}, r)$ part of the metric is a space with constant negative curvature, or an AdS space. To support the metric 31 (i.e., to satisfy the Einstein equation), there must be some background matter field that gives a stress-energy tensor in the form of a negative cosmological constant in AdS$_5$ and a positive one in S^5. Such a field is the self-dual five-form field F_5 mentioned above.

Field theory has two parameters: the number of colors N and the gauge coupling g. When the number of colors is large, it is the 't Hooft coupling $\lambda = g^2 N$ that controls the perturbation theory. On the string theory side, the parameters are g_s, l_s, and radius R of the AdS space. String theory and field theory each have two dimensionless parameters that map to each other through the following relations:

$$g^2 = 4\pi g_s, \qquad 33.$$

$$g^2 N_c = \frac{R^4}{l_s^4}. \qquad 34.$$

Equation 33 tells us that, if one wants to keep string theory weakly interacting, then the gauge coupling in field theory must be small. Equation 34 is particularly interesting. It says that the large 't Hooft coupling limit in field theory corresponds to the limit when the curvature radius of space-time is much larger than the string length l_s. In this limit, one can reliably decouple the massive string modes and reduce string theory to supergravity. In the limit $g_s = 1$, $R \gg l_s$, one has classical supergravity instead of string theory. The practical utility of the AdS/CFT correspondence comes, in large part, from its ability to deal with the strong coupling limit in gauge theory.

One can perform a Kaluza-Klein reduction (20) by expanding all fields in S^5 harmonics. Keeping only the lowest harmonics, one finds a five-dimensional theory with the massless dilaton, SO(6) gauge bosons, and gravitons (21):

$$S_{5D} = \frac{N^2}{8\pi^2 R^3} \int d^5x \left(\mathcal{R}_{5D} - 2\Lambda - \frac{1}{2}\partial^\mu \Phi \partial_\mu \Phi - \frac{R^2}{8} F^a_{\mu\nu} F^{a\mu\nu} + \cdots \right). \qquad 35.$$

In AdS/CFT, an operator O of field theory is put in a correspondence with a field ϕ (bulk field) in supergravity. We elaborate on this correspondence below; here we keep the operator and the field unspecified. In the supergravity approximation, the mathematical statement of the correspondence is

$$Z_{4D}[J] = e^{iS[\phi_{cl}]}. \qquad 36.$$

On the left is the partition function of a field theory, where the source J coupled to the operator O is included:

$$Z_{4D}[J] = \int D\phi \exp\left(iS + i \int d^4x JO \right). \qquad 37.$$

On the right, $S[\phi_{cl}]$ is the classical action of the classical solution ϕ_{cl} to the field equation with the boundary condition

$$\lim_{z \to 0} \frac{\phi_{cl}(z, x)}{z^\Delta} = J(x). \qquad 38.$$

Here Δ is a constant that depends on the nature of the operator O (namely, on its spin and dimension). In the simplest case, $\Delta = 0$ and the boundary condition becomes $\phi_{cl}(z = 0) = J$. Differentiating Equation 36 with respect to J, one can find the correlation functions of O. For example, the two-point Green's function of O is obtained by differentiating $S_{cl}[\phi]$ twice with respect to the boundary value of ϕ,

$$G(x - y) = -i\langle TO(x)O(y)\rangle = -\frac{\delta^2 S[\phi_{cl}]}{\delta J(x)\delta J(y)}\bigg|_{\phi(z=0)=J}. \qquad 39.$$

The AdS/CFT correspondence thus maps the problem of finding quantum correlation functions in field theory to a classical problem in gravity. Moreover, to find two-point correlation functions in field theory, one can be limited to the quadratic part of the classical action on the gravity side.

The complete operator to field mapping can be found in References 5 and 17. For our purpose, the following is sufficient:

■ The dilaton Φ corresponds to $O = -L = \frac{1}{4}F^2_{\mu\nu} + \cdots$, where L is the Lagrangian density.

■ The gauge field A^a_μ corresponds to the conserved R-charge current $J^{a\mu}$ of field theory.

■ The metric tensor corresponds to the stress-energy tensor $T^{\mu\nu}$. More precisely, the partition function of a four-dimensional field theory in an external metric $g^0_{\mu\nu}$ is equal to

$$Z_{4D}\left[g^0_{\mu\nu}\right] = \exp(i S_{cl}[g_{\mu\nu}]), \qquad 40.$$

where a five-dimensional metric $g_{\mu\nu}$ satisfies the Einstein equations and has the following asymptotics at $z = 0$:

$$ds^2 = g_{\mu\nu}dx^\mu dx^\nu = \frac{R^2}{z^2}\left(dz^2 + g^0_{\mu\nu}dx^\mu dx^\nu\right). \tag{41}$$

From the point of view of hydrodynamics, the operator $F^2/4$ is not very interesting because its correlator does not have a hydrodynamic pole. In contrast, we find the correlators of the R-charge current and the stress-energy tensor to contain hydrodynamic information.

We simplify the graviton part of the action further. Our two-point functions are functions of the momentum $p = (\omega, \mathbf{k})$. We can choose spatial coordinates so that $\mathbf{k}$ points along the x^3-axis. This corresponds to perturbations that propagate along the x^3 direction: $h_{\mu\nu} = h_{\mu\nu}(t, r, x^3)$. These perturbations can be classified according to the representations of the $O(2)$ symmetry of the (x^1, x^2) plane. Owing to that symmetry, only certain components can mix; for example, h_{12} does not mix with any other components, whereas components h_{01} and h_{31} mix only with each other. We assume that only these three metric components are nonzero and introduce shorthand notations

$$\phi = h^1_2, \quad a_0 = h^1_0, \quad a_3 = h^1_3. \tag{42}$$

The quadratic part of the graviton action acquires a very simple form in terms of these fields:

$$S_{\text{quad}} = \frac{N^2}{8\pi^2 R^3} \int d^4x\, dr \sqrt{-g}\left(-\frac{1}{2}g^{\mu\nu}\partial_\mu\phi\partial_\nu\phi - \frac{1}{4g^2_{\text{eff}}}g^{\mu\alpha}g^{\nu\beta}f_{\mu\nu}f_{\alpha\beta}\right), \tag{43}$$

where $f_{\mu\nu} = \partial_\mu a_\nu - \partial_\nu a_\mu$ and $g^2_{\text{eff}} = g_{xx}$. In deriving Equation 43, our only assumption about the metric is that it has a diagonal form,

$$ds^2 = g_{tt}dt^2 + g_{rr}dr^2 + g_{xx}d\mathbf{x}^2, \tag{44}$$

so it can also be used below for the finite-temperature metric.

As a simple example, let us compute the two-point correlation function of T^{xy}, which corresponds to ϕ in gravity. The field equation for ϕ is

$$\partial_\mu(\sqrt{-g}\, g^{\mu\nu}\partial_\nu\phi) = 0. \tag{45}$$

The solution to this equation, with the boundary condition $\phi(p, z = 0) = \phi_0(p)$, can be written as

$$\phi(p, z) = f_p(z)\phi_0(p), \tag{46}$$

where the mode function $f_p(z)$ satisfies the equation

$$\left(\frac{f'_p}{z^3}\right)' - \frac{p^2}{z^3}f_p = 0, \tag{47}$$

with the boundary condition $f_p(0) = 1$. The mode equation (Equation 47) can be solved exactly. Assuming p is space like, $p^2 > 0$, the exact solution and its expansion around $z = 0$ is

$$f_p(z) = \frac{1}{2}(pz)^2 K_2(pz) = 1 - \frac{1}{4}(pz)^2 - \frac{1}{16}(pz)^4 \ln(pz) + O((pz)^4). \tag{48}$$

The second solution to Equation 47, $(pz)^2 I_2(pz)$, is ruled out because it blows up at $z \to \infty$.

We now substitute the solution into the quadratic action. Using the field equation, one can perform integration by parts and write the action as a boundary integral at $z = 0$. One finds

$$S = \frac{N^2}{16\pi^2} \int d^4x \, \frac{1}{z^3} \, \phi(x, z) \phi'(x, z)|_{z \to 0} = \int \frac{d^4p}{(2\pi)^4} \, \phi_0(-p) \, F(p, z) \, \phi_0(p)|_{z \to 0}, \qquad 49.$$

where

$$\mathcal{F}(p, z) = \frac{N^2}{16\pi^2} \frac{1}{z^3} f_{-p}(z) \partial_z f_p(z). \qquad 50.$$

Differentiating the action twice with respect to the boundary value ϕ_0, one finds

$$\langle T_{xy} T_{xy} \rangle_p = -2 \lim_{z \to 0} \mathcal{F}(p, z) = \frac{N^2}{64\pi^2} p^4 \ln(p^2). \qquad 51.$$

Note that we have dropped the term $\sim p^4 \ln z$, which, although singular in the limit $z \to 0$, is a contact term [i.e., a term proportional to a derivative of $\delta(x)$ after Fourier transform]. Removing such terms by adding local counter terms to the supergravity action is known as the holographic renormalization (22). It is, in a sense, a holographic counterpart to the standard renormalization procedure in quantum field theory, here applied to composite operators.

For time-like p, $p^2 < 0$, there are two solutions to Equation 47, which involve Hankel functions $H^{(1)}(z)$ and $H^{(2)}(z)$ instead of $K_2(z)$. Neither function blows up at $z \to \infty$, and it is not clear which should be picked. Here we encounter, for the first time, a subtlety of Minkowski-space AdS/CFT, which is discussed at great length in subsequent sections. At zero temperature this problem can be overcome by an analytic continuation from space-like p. However, this will not work at nonzero temperatures.

3.2. Black Three-Brane Metric

At nonzero temperatures, the metric dual to $\mathcal{N} = 4$ SYM theory is the black three-brane metric

$$ds^2 = \frac{r^2}{R^2}(-f dt^2 + d\mathbf{x}^2) + \frac{R^2}{r^2 f} dr^2 + R^2 d\Omega_5^2, \qquad 52.$$

with $f = 1 - r_0^4/r^4$. The event horizon is located at $r = r_0$, where $f = 0$. In contrast to the usual Schwarzschild black hole, the horizon has three flat directions $\mathbf{x}$. The metric 52 is thus called a black three-brane metric.

We frequently use an alternative radial coordinate u, defined as $u = r_0^2/r^2$. In terms of u, the boundary is at $u = 0$, the horizon at $u = 1$, and the metric is

$$ds^2 = \frac{(\pi T R)^2}{u^2}(-f(u) dt^2 + d\mathbf{x}^2) + \frac{R^2}{4u^2 f(u)} du^2 + R^2 d\Omega_5^2. \qquad 53.$$

The Hawking temperature is determined completely by the behavior of the metric near the horizon. Let us concentrate on the (t, r) part of the metric,

$$ds^2 = -\frac{4r_0}{R^2}(r - r_0)dt^2 + \frac{R^2}{4r_0(r - r_0)}dr^2. \qquad 54.$$

Change the radial variable from r to ρ,

$$r = r_0 + \frac{\rho^2}{r_0}, \qquad 55.$$

and the metric components become nonsingular:

$$ds^2 = \frac{R^2}{r_0^2}\left(d\rho^2 - \frac{4r_0^2}{R^2}\rho^2 dt^2\right). \qquad 56.$$

Note also that after a Wick rotation to Euclidean time τ, the metric has the form of the flat metric in cylindrical coordinates, $ds^2 \sim d\rho^2 + \rho^2 d\varphi^2$, where $\varphi = 2r_0 R^{-2}\tau$. To avoid a conical singularity at $\rho = 0$, φ must be a periodic variable with periodicity 2π. This fact matches with the periodicity of the Euclidean time in thermal field theory $\tau \sim \tau + 1/T$, from which one finds the Hawking temperature:

$$T_H = \frac{r_0}{\pi R^2}. \qquad 57.$$

One of the first finite-temperature predictions of AdS/CFT correspondence is that of the thermodynamic potentials of the $\mathcal{N} = 4$ SYM theory in the strong coupling regime. The entropy is given by the Bekenstein-Hawking formula $S = A/(4G)$, where A is the area of the horizon of the metric 52; the result can then be converted to parameters of the gauge theory using Equations 30, 33, and 34. One obtains

$$s = \frac{S}{V} = \frac{\pi^2}{2}N^2 T^3, \qquad 58.$$

which is three-quarters of the entropy density in $\mathcal{N} = 4$ SYM theory at zero 't Hooft coupling.

We now try to generalize the AdS/CFT prescription to finite temperature. In the Euclidean formulation of finite-temperature field theory, field theory lives in space-time with the Euclidean-time direction τ compactified. The metric is regular at $r = r_0$: If one views the (τ, r) space as a cigar-shaped surface, then the horizon $r = r_0$ is the tip of the cigar. Thus, r_0 is the minimal radius where the space ends, and there is no point in space with r less than r_0. The only boundary condition at $r = r_0$ is that fields are regular at the tip of the cigar, and the AdS/CFT correspondence is formulated as

$$Z_{4D}[J] = Z_{5D}[\phi]|_{\phi(z=0) \to J}. \qquad 59.$$

4. REAL-TIME AdS/CFT

In many cases we must find real-time correlation functions not given directly by the Euclidean path-integral formulation of thermal field theory. One example is the set of kinetic coefficients expressed, through Kubo's formulas, via a certain limit of real-time thermal Green's functions. Another related example appears if we want to directly find the position of the poles in the correlation functions that would correspond to the hydrodynamic modes.

In principle, some real-time Green's functions can be obtained by analytic continuation of the Euclidean ones. For example, an analytic continuation of a two-point Euclidean propagator gives a retarded or advanced Green's function, depending on the way one performs the continuation. However, it is often very difficult to directly compute a quantity of interest in that way. In particular, it is very difficult to get the information about the hydrodynamic (small ω, small $\mathbf{k}$) limit of real-time correlators from Euclidean propagators. The problem here is that we need to perform an analytic continuation from a discrete set of points in Euclidean frequencies (the Matsubara frequencies) $\omega = 2\pi i n$, where n is an integer, to the real values of ω. In the hydrodynamic limit, we are interested in real and small ω, whereas the smallest Matsubara frequency is already $2\pi T$.

Therefore, we need a real-time AdS/CFT prescription that would allow us to directly compute the real-time correlators. However, if one tries to naively generalize the AdS/CFT prescription, one immediately faces a problem. Namely, now $r = r_0$ is not the end of space but just the location of the horizon. Without specifying a boundary condition at $r = r_0$, there is an ambiguity in defining the solution to the field equations, even as the boundary condition at $r = \infty$ is set.

As an example, let us consider the equation of motion of a scalar field in the black hole background, $\partial_\mu(g^{\mu\nu}\partial_\nu\phi) = 0$. The solution to this equation with the boundary condition $\phi = \phi_0$ at $u = 0$ is $\phi(p, u) = f_p\phi_0(p)$, where $f_p(u)$ satisfies the following equation in the metric 53:

$$f_p'' - \frac{1 + u^2}{uf} f_p' + \frac{w^2}{uf^2} f_p - \frac{q^2}{uf} f_p = 0.$$

60.

Here the prime denotes differentiation with respect to u, and we have defined the dimensionless frequency and momentum:

$$w = \frac{\omega}{2\pi T}, \quad q = \frac{k}{2\pi T}.$$

61.

Near $u = 0$ the equation has two solutions, $f_1 \sim 1$ and $f_2 \sim u^2$. In the Euclidean version of thermal AdS/CFT, there is only one regular solution at the horizon $u = 1$, which corresponds to a particular linear combination of f_1 and f_2. However, in Minkowski space there are two solutions, and both are finite near the horizon. One solution termed f_p behaves as $(1 - u)^{-iw/2}$, and the other is its complex conjugate $f_p^* \sim (1 - u)^{iw/2}$. These two solutions oscillate rapidly as $u \to 1$, but the amplitude of the oscillations is constant. Thus, the requirement of finiteness of f_p allows for any linear combination of f_1 and f_2 near the boundary, which means that there is no unique solution to Equation 60.

4.1. Prescription for Retarded Two-Point Functions

Physically, the two solutions f_p and f_p^* have very different behavior. Restoring the $e^{-i\omega t}$ phase in the wave function, one can write

$$e^{-i\omega t} f_p \sim e^{-i\omega(t+r_*)}, \qquad\qquad 62.$$

$$e^{-i\omega t} f_p^* \sim e^{i\omega(t-r_*)}, \qquad\qquad 63.$$

where the coordinate

$$r_* = \frac{\ln(1-u)}{4\pi T} \qquad\qquad 64.$$

was introduced so that Equations 62 and 63 looked like plane waves. In fact, Equation 62 corresponds to a wave that moves toward the horizon (incoming wave) and Equation 63 to a wave that moves away from the horizon (outgoing wave).

The simplest idea, which is motivated by the fact that nothing should come out of a horizon, is to impose the incoming-wave boundary condition at $r = r_0$ and then proceed as instructed by the AdS/CFT correspondence. However, now we encounter another problem. If we write down the classical action for the bulk field, after integrating by parts we get contributions from both the boundary and the horizon:

$$S = \int \frac{d^4 p}{(2\pi)^4} \phi_0(-p)\mathcal{F}(p,z)\phi_0(p)|_{z=0}^{z=z_H}. \qquad\qquad 65.$$

If one tried to differentiate the action with respect to the boundary value ϕ_0, one would find

$$G(p) = \mathcal{F}(p,z)|_0^{z_H} + \mathcal{F}(-p,z)|_0^{z_H}. \qquad\qquad 66.$$

From the equation satisfied by f_p and from $f_p^* = f_{-p}$, it is easy to show that the imaginary part of $\mathcal{F}(p,z)$ does not depend on z; hence, the quantity $G(p)$ in Equation 66 is real. This is clearly not what we want, as the retarded Green's functions are, in general, complex. Simply throwing away the contribution from the horizon does not help because $\mathcal{F}(-p,z) = \mathcal{F}^*(p,z)$, owing to the reality of the equation satisfied by f_p.

A partial solution to this problem was suggested in Reference 7. The authors postulated that the retarded Green's function was related to the function $\mathcal{F}$ by the same formula that was found at zero temperature:

$$G^R(p) = -2 \lim_{z \to 0} \mathcal{F}(p,x). \qquad\qquad 67.$$

In particular, we throw away all contributions from the horizon. This prescription was established more rigorously in Reference 8 (following an earlier suggestion in Reference 23) as a particular case of a general real-time AdS/CFT formulation, which establishes the connection between the close-time-path formulation of real-time quantum field theory with the dynamics of fields in the whole Penrose diagram of the AdS black brane. Here we accept Equation 67 as a postulate and proceed to extract physical results from it. It is also easy to generalize this prescription to the case when we have more than one field. In that case, the quantity $\mathcal{F}$ becomes a matrix $\mathcal{F}_{ab}$, whose elements are proportional to the retarded Green's function G_{ab}.

4.2. Calculating Hydrodynamic Quantities

As an illustration of the real-time AdS/CFT correspondence, we compute the correlator of T_{xy}. First we write down the equation of motion for $\phi = b_y^x$:

$$\phi_p'' - \frac{1 + u^2}{uf}\phi_p' + \frac{w^2 - q^2 f}{uf^2}\phi_p = 0.\qquad 68.$$

In contrast to the zero-temperature equation, now ω and k enter the equation separately rather than through the combination $\omega^2 - k^2$. Thus, the Green's function will have no Lorentz invariance. The equation cannot be solved exactly for all ω and k. However, when ω and k are both much smaller than T, one can develop series expansion in powers of w and q. There are two solutions that are complex conjugates of each other. The solution that is an incoming wave at $u = 1$ and normalized to 1 at $u = 0$ is

$$f_p(z) = (1 - u^2)^{-iw/2} + O(w^2, q^2).\qquad 69.$$

The kinetic term in the action for ϕ is

$$S = -\frac{\pi^2 N^2 T^4}{8}\int du\,\frac{f}{u}\phi'^2.\qquad 70.$$

Applying the general formula of Equation 67, one finds the retarded Green's function of T_{xy},

$$G_{xy,xy}^R(\omega, k) = -\frac{\pi^2 N^2 T^4}{4}iw,\qquad 71.$$

and, using Kubo's formula for η, the viscosity,

$$\eta = \frac{\pi}{8}N^2 T^3.\qquad 72.$$

It is instructive to compute other correlators that have poles corresponding to hydrodynamic modes. As a warm-up, let us compute the two-point correlators of the R-charge currents, which should have a pole at $\omega = -i D k^2$, where D is the diffusion constant. We first write down Maxwell's equations for the bulk gauge field. Let the spatial momentum be aligned along the x^3-axis: $p = (\omega, 0, 0, k)$. Then the equations for A_0 and A_3 are coupled:

$$w A_0' + q f A_3' = 0,\qquad 73.$$

$$A_0'' - \frac{1}{uf}(q^2 A_0 + wq A_3) = 0,\qquad 74.$$

$$A_3'' + \frac{f'}{f}A_3' + \frac{1}{uf^2}(w^2 A_3 + wq A_0) = 0.\qquad 75.$$

One can eliminate A_3 and write down a third-order equation for A_0:

$$A_0''' + \frac{(uf)'}{uf}A_0'' + \frac{w^2 - q^2 f}{uf^2}A_0' = 0.\qquad 76.$$

Near $u = 1$ we find two independent solutions, $A_0' \sim (1 - u)^{\pm iw/2}$, and the incoming-wave boundary condition singles out $(1 - u)^{-iw/2}$. One can substitute $A_0' = (1 - u)^{-iw/2}F(u)$ into Equation 76. The resulting equation can be solved perturbatively in

w and q^2. We find

$$A_0' = C(1-u)^{-iw/2}\left(1 + \frac{iw}{2}\ln\frac{2u^2}{1+u} + q^2\ln\frac{1+u}{2u}\right). \qquad 77.$$

Using Equation 74, one can express C through the boundary values of A_0 and A_3 at $u = 0$:

$$C = \left.\frac{q^2 A_0 + wq A_3}{iw - q^2}\right|_{u=0}. \qquad 78.$$

Differentiating the action with respect to the boundary values, we find, in particular,

$$\langle J_0 J_0\rangle_p = \frac{N^2 T}{16\pi}\frac{k^2}{i\omega - Dk^2}, \qquad 79.$$

where

$$D = \frac{1}{2\pi T}. \qquad 80.$$

The correlator given by Equation 79 has the expected hydrodynamic diffusive pole, and D is the R-charge diffusion constant.

Similarly, one can observe the appearance of the shear mode in the correlators of the metric tensor. We note that the shear flow along the x^1 direction with velocity gradient along the x^3 direction involves T_{01} and T_{31}, hence the interesting metric components are $a_0 = b_0^1$ and $a_3 = b_3^1$. Two of the field equations are

$$a_0' - \frac{qf}{w}a_3' = 0, \qquad 81.$$

$$a_3'' - \frac{1+u^2}{uf}a_3' + \frac{1}{uf^2}(w^2 a_3 + wq a_0) = 0. \qquad 82.$$

They can be combined into a single equation:

$$a_0''' - \frac{2u}{f}a_0'' + \frac{2uf - q^2 f + w^2}{uf^2}a_0' = 0. \qquad 83.$$

Again, the solution can be found perturbatively in w and q:

$$a_0' = C(1-u)^{-iw/2}\left[u - iw\left(1 - u - \frac{u}{2}\ln\frac{1+u}{2}\right) + \frac{q^2}{2}(1-u)\right]. \qquad 84.$$

Applying the prescription, one finds the retarded Green's functions. For example,

$$G_{tx,tx}(\omega, k) = \frac{\xi k^2}{i\omega - Dk^2}, \qquad 85.$$

where

$$\xi = \frac{\pi}{8}N^2 T^3, \quad D = \frac{1}{4\pi T}. \qquad 86.$$

Thus, we found that the correlator contains a diffusive pole $\omega = -iDk^2$, just as anticipated from hydrodynamics. Furthermore, the magnitude of the momentum diffusion constant D also matched our expectation. Indeed, if one recalls the value of η from Equation 72 and the entropy density from Equation 58, one can check that

$$D = \frac{\eta}{\varepsilon + P}. \qquad 87.$$

5. THE MEMBRANE PARADIGM

Let us now look at the problem from a different perspective. The existence of hydro-dynamic modes in thermal field theory is reflected by the existence of the poles of the retarded correlators computed from gravity. Are there direct gravity counterparts of the hydrodynamic normal modes?

If the answer to this question is yes, then there must exist linear gravitational perturbations of the metric that have the dispersion relation identical to that of the shear hydrodynamic mode, $\omega \sim -iq^2$, and of the sound mode, $\omega = c_s q - i\gamma q^2$. It turns out that one can explicitly construct the gravitational counterpart of the shear mode. (It should be possible to perform the same construction for the sound mode, but this has not been noted in the literature.) Our discussion is physical but somewhat sketchy; for more details, see Reference 24.

First, let us construct a gravity perturbation that corresponds to a diffusion of a conserved charge (e.g., the R-charge in $\mathcal{N} = 4$ SYM theory). To keep the discussion general, we use the form of the metric 44, with the metric components unspecified. Our only assumptions are that the metric is diagonal and has a horizon at $r = r_0$, near which

$$g_{00} = -\gamma_0(r - r_0), \quad g_{rr} = \frac{\gamma_r}{r - r_0}. \qquad 88.$$

The Hawking temperature can be computed by the method used to arrive at Equation 57, and one finds $T = (4\pi)^{-1}(\gamma_0/\gamma_r)^{1/2}$.

We also assume that the action of the gauge field dual to the conserved current is

$$S_{\text{gauge}} = \int dx \sqrt{-g} \left(-\frac{1}{4g_{\text{eff}}^2} F^{\mu\nu} F_{\mu\nu} \right), \qquad 89.$$

where g_{eff} is an effective gauge coupling that can be a function of the radial coordinate r. For simplicity we set g_{eff} to a constant in our derivation of the formula for D; it can be restored by replacing $\sqrt{-g} \to \sqrt{-g}/g_{\text{eff}}^2$ in the final answer.

The field equations are

$$\partial_\mu \left(\frac{1}{g_{\text{eff}}^2} \sqrt{-g} F^{\mu\nu} \right) = 0. \qquad 90.$$

We search for a solution to this equation that vanishes at the boundary and satisfies the incoming-wave boundary condition at the horizon.

The first indication that one can have a hydrodynamic behavior on the gravity side is that Equation 90 implies a conservation law on a four-dimensional surface. We define the stretched horizon as a surface with constant r just outside the horizon,

$$r = r_h = r_0 + \varepsilon, \quad \varepsilon = r_0, \qquad 91.$$

and the normal vector n_μ directed along the r direction (i.e., perpendicularly to the stretched horizon). Then with any solution to Equation 90, one can associate a current on the stretched horizon:

$$j^\mu = n_\nu F^{\mu\nu}|_{r_h}. \qquad 92.$$

The antisymmetry of $F^{\mu\nu}$ implies that j^μ has no radial component, $j^r = 0$. The field equation (Equation 90) and the constancy of n_ν on the stretched horizon imply that this current is conserved: $\partial_\mu j^\mu = 0$. To establish the diffusive nature of the solution, we must show the validity of the constitutive equation $j^i = -D \partial_i j^0$.

Such a constitutive equation breaks time reversal and obviously must come from the absorptive boundary condition on the horizon. The situation is analogous to the propagation of plane waves to a nonreflecting surface in classical electrodynamics. In this case, we have the relation $\mathbf{B} = -\mathbf{n} \times \mathbf{E}$ between electric and magnetic fields. In our case, the corresponding relation is

$$F_{ir} = -\sqrt{\frac{\gamma_r}{\gamma_0}} \frac{F_{0i}}{r - r_0}, \qquad 93.$$

valid when r is close to r_0. This relates $j_i \sim F_{ir}$ to the parallel (to the horizon) component of the electric field F_{0i}, which is one of the main points of the membrane paradigm approach to black hole physics (25). We have yet to relate j_i to $j_0 \sim F_{0r}$, which is the component of the electric field normal to the horizon. To make the connection to F_{0r}, we use the radial gauge $A_r = 0$, in which

$$F_{0i} \approx -\partial_i A_0. \qquad 94.$$

Moreover, when k is small the fields change very slowly along the horizon. Therefore, at each point on the horizon the radial dependence of the scalar potential A_0 is determined by the Poisson equation,

$$\partial_r(\sqrt{-g}\, g^{rr}\, g^{00} \partial_r A_0) = 0, \qquad 95.$$

whose solution, which satisfies $A_0(r = \infty) = 0$, is

$$A_0(r) = C_0 \int_r^\infty dr' \frac{g_{00}(r')g_{rr}(r')}{\sqrt{-g(r')}}. \qquad 96.$$

This means that the ratio of the scalar potential A_0 to the electric field F_{0r} approaches a constant near the horizon:

$$\left. \frac{A_0}{F_{0r}} \right|_{r=r_0} = \frac{\sqrt{-g}}{g_{00}g_{rr}}(r_0) \int_{r_0}^\infty dr \frac{g_{00}g_{rr}}{\sqrt{-g}}(r). \qquad 97.$$

Combining the formulas $j^i \sim F_{0i} \sim \partial_i A_0$ and $A_0 \sim F_{0r} \sim j^0$, we find Fick's law $j^i = -D \partial_i j^0$, with the diffusion constant

$$D = \frac{\sqrt{-g}}{g_{xx} g_{\mathrm{eff}}^2 \sqrt{-g_{00}g_{rr}}}(r_0) \int_{r_0}^\infty dr \frac{-g_{00}g_{rr}g_{\mathrm{eff}}^2}{\sqrt{-g}}(r). \qquad 98.$$

Thus, we find that for a slowly varying solution to Maxwell's equations, the corresponding charge on the stretched horizon evolves according to the diffusion equation. Therefore, the gravity solution must be an overdamped one, with $\omega = -i D k^2$. This is an example of a quasi-normal mode. We also found the diffusion constant D directly in terms of the metric and the gauge coupling g_{eff}.

The reader may notice that our quasi-normal modes satisfy a vanishing Dirichlet condition at the boundary $r = \infty$. This is different from the boundary condition one uses to find the retarded propagators in AdS/CFT, so the relation of the quasi-normal modes to AdS/CFT correspondence may be not clear. It can be shown, however, that the quasi-normal frequencies coincide with the poles of the retarded correlators (A.O. Starinets, unpublished information).

We can now apply our general formulas to the case of $\mathcal{N} = 4$ SYM theory. The metric components are given by Equation 52. For the R-charge current $g_{\text{eff}} = $ constant, Equation 98 gives $D = 1/(2\pi T)$, in agreement with our AdS/CFT computation. For the shear mode of the stress-energy tensor, we have effectively $g_{\text{eff}}^2 = g_{xx}$, so $\mathcal{D} = 1/(4\pi T)$, which also coincides with our previous result. In both cases, the computation is much simpler than the AdS/CFT calculation.

6. THE VISCOSITY/ENTROPY RATIO

6.1. Universality

In all thermal field theories in the regime described by gravity duals, the ratio of shear viscosity η to (volume) density of entropy s is a universal constant equal to $1/(4\pi)$ [$\hbar/(4\pi k_B)$, if one restores $\hbar$, c, and the Boltzmann constant k_B].

One proof of the universality is based on the relationship between graviton's absorption cross section and the imaginary part of the retarded Green's function for T_{xy} (28). Another way to prove the universality (29) is via the direct AdS/CFT calculation of the correlation function in Kubo's formula 18. We, however, follow a different method. It is based on the formula for the viscosity derived from the membrane paradigm. A similar proof was given by Buchel & Liu (27).

The observation is that the shear gravitational perturbation with $k = 0$ can be found exactly by performing a Lorentz boost of the black-brane metric 52. Consider the coordinate transformations $r, t, x_i \to r', t', x'_i$ of the form

$$r = r',$$

$$t = \frac{t' + vy'}{\sqrt{1 - v^2}} \approx t' + vy',$$

$$y = \frac{y' + vt'}{\sqrt{1 - v^2}} \approx y' + vt',$$

$$x_i = x'_i,$$

99.

where $v < 1$ is a constant parameter and the expansion on the right corresponds to $v = 1$. In the new coordinates, the metric becomes

$$ds^2 = g_{00}dt'^2 + g_{rr}dr'^2 + g_{xx}(r)\sum_{i=1}^{p}(dx'^i)^2 + 2v(g_{00} + g_{xx})dt'dy'.$$

100.

This is simply a shear fluctuation at $k = 0$. In our language, the corresponding gauge potential is

$$a_0 = vg^{xx}(g_{00} + g_{xx}).$$

101.

This field satisfies the vanishing boundary condition $a_0(r = \infty) = 0$ owing to the restoration of Poincaré invariance at the boundary: $g_{00}/g_{xx} \to -1$ when $r \to \infty$. This clearly has a much simpler form than Equation 96 for the solution to the generic Poisson equation. The simple form of solution 101 is valid only for the specific case of the shear gravitational mode with $g_{\text{eff}}^2 = g_{xx}$. We have also implicitly used the fact that the metric satisfies the Einstein equations, with the stress-energy tensor on the right being invariant under a Lorentz boost.

Equation 97 now becomes

$$\frac{a_0}{f_{0r}}\Big|_{r\to r_0} = -\frac{1+g^{xx}g_{00}}{\partial_r(g^{xx}g_{00})}\Big|_{r\to r_0} = \frac{g_{xx}(r_0)}{\gamma_0}. \qquad 102.$$

The shear mode diffusion constant is

$$\mathcal{D} = \frac{a_0}{f_{0r}}\Big|_{r\to r_0}\frac{\sqrt{\gamma_0\gamma_r}}{g_{xx}(r_0)} = \sqrt{\frac{\gamma_r}{\gamma_0}} = \frac{1}{4\pi T}. \qquad 103.$$

Because $\mathcal{D} = \eta/(\varepsilon + P)$, and $\varepsilon + P = Ts$ in the absence of chemical potentials, we find that

$$\frac{\eta}{s} = \frac{1}{4\pi}. \qquad 104.$$

In fact, the constancy of this ratio has been checked directly for theories dual to Dp-brane (24), M-brane (11), Klebanov-Tseytlin and Maldacena-Nunez backgrounds (27), $\mathcal{N} = 2^*$ SYM theory (30), and others.

As remarked in Section 2, the ratio η/s is much larger than the one for weakly coupled theories. The fact that we found the ratio to be parametrically of order one implies that all theories with gravity duals are strongly coupled. In $\mathcal{N} = 4$ SYM theory, the ratio η/s has been computed to the next order in the inverse 't Hooft coupling expansion (31):

$$\frac{\eta}{s} = \frac{1}{4\pi}\left(1 + \frac{135\zeta(3)}{8(g^2 N)^{3/2}}\right). \qquad 105.$$

The sign of the correction can be guessed from the fact that in the limit of zero 't Hooft coupling $g^2 N \to 0$, the ratio diverges, $\eta/s \to \infty$.

6.2. The Viscosity Bound Conjecture

From our discussion above, one can argue that

$$\frac{\eta}{s} \geq \frac{\hbar}{4\pi} \qquad 106.$$

in all systems that can be obtained from a sensible relativistic quantum field theory by turning on temperatures and chemical potentials. The bound, if correct, implies that a liquid with a given volume density of entropy cannot be arbitrarily close to being a perfect fluid (which has zero viscosity). As such, it implies a lower bound on the viscosity of the QGP one may be creating at RHIC. Interestingly, some model calculations suggest that the viscosity at RHIC may be not too far away from the lower bound (32, 33).

A place where one may think that the bound should break down is within superfluids. The ability of a superfluid to flow without dissipation in a channel is sometimes described as zero viscosity. However, within the Landau's two-fluid model, any superfluid has a measurable shear viscosity (together with three bulk viscosities). For superfluid helium, the shear viscosity has been measured in a torsion-pendulum experiment by Andronikashvili (34). If one substitutes the experimental values, the ratio η/s for helium remains larger than $\hbar/4\pi k_B \approx 6.08 \times 10^{-13}$ Ks for all ranges of temperatures and pressures, by a factor of at least 8.8.

As discussed in Section 2.3, the ratio η/s is proportional to the ratio of the mean free path and the de Broglie wavelength of particles,

$$\frac{\eta}{s} \sim \frac{\ell_{\mathrm{mfp}}}{\lambda}. \qquad\qquad 107.$$

For the quasi-particle picture to be valid, the mean free path must be much larger than the de Broglie wavelength. Therefore, if the coupling is weak and the system can be described as a collection of quasi-particles, the ratio η/s is larger than 1. We have found that, within the $\mathcal{N} = 4$ SYM theory and, more generally, theories with gravity duals, even in the limit of infinite coupling, the ratio η/s cannot be made smaller than $1/(4\pi)$.

7. CONCLUSION

In this review, we cover only a small part of the applications of AdS/CFT correspondence to finite-temperature quantum field theory. Here we briefly mention further developments and refer the reader to the original literature for more details.

In addition to $\mathcal{N} = 4$ SYM theory, there exists a large number of other theories whose hydrodynamic behavior has been studied using the AdS/CFT correspondence, including the world-volume theories on M2- and M5-branes (11), theories on Dp-branes (24), and little string theory (35). In all examples, the ratio η/s is equal to $1/(4\pi)$, which is not surprising because the general proofs of Section 6 apply in these cases.

We have concentrated on the shear hydrodynamic mode, which has a diffusive pole ($\omega \sim -ik^2$). One can also compute correlators, which have a sound-wave pole from the AdS/CFT prescription (10). One such correlator is between the energy density T^{00} at two different points in space-time. The result confirms the existence of such a pole, with both the real part and imaginary part having exactly the values predicted by hydrodynamics (recall that in CFT the bulk viscosity is zero and the sound attenuation rate is determined completely by the shear viscosity).

Some of the theories listed above are CFTs, but many are not (e.g., the Dp-brane world-volume theories with $p \neq 3$). The fact that $\eta/s = 1/(4\pi)$ also in those theories implies that the constancy of this ratio is not a consequence of conformal symmetry. Theories with less than maximal number of supersymmetries have been found to have the universal value of η/s, for example, the $\mathcal{N} = 2^*$ theory (36), theories described by Klebanov-Tseytlin, and Maldacena-Nunez backgrounds (27). A common feature of these theories is that they all have a gravitational dual description. The bulk viscosity has been computed for some of these theories (35, 37).

Besides viscosity, one can also compute diffusion constants of conserved charges by using the AdS/CFT correspondence. Above we present the computation of the

R-charge diffusion constant in $\mathcal{N} = 4$ SYM theory; for similar calculations in some other theories, see References 11 and 24.

Recently, the AdS/CFT correspondence was used to compute the energy loss rate of a quark in the fundamental representation moving in a finite-temperature plasma (38–41). This quantity is of importance to the phenomenon of jet quenching in heavy-ion collisions.

So far, the only quantity that shows a universal behavior at the quantitative level, across all theories with gravitational duals, is the ratio of the shear viscosity and entropy density. Recently, many authors found that this ratio remains constant even at nonzero chemical potentials (42–46).

What have we learned from the application of AdS/CFT correspondence to thermal field theory? Although, at least at this moment, we cannot use the AdS/CFT approach to study QCD directly, we have found quite interesting facts about strongly coupled field theories. We have also learned new facts about quasi-normal modes of black branes. However, we have also found a set of puzzles: Why is the ratio of the viscosity and entropy density constant in a wide class of theories? Is there a lower bound on this ratio for all quantum field theories? Can this be understood without any reference to gravity duals? With these open questions, we conclude this review.

DISCLOSURE STATEMENT

The authors are not aware of any biases that might be perceived as affecting the objectivity of this review.

ACKNOWLEDGMENTS

The work of D.T.S. is supported in part by the U.S. Department of Energy under grant DE-FG02-00ER41132. Research at the Perimeter Institute is supported in part by the Government of Canada through NSERC and by the Province of Ontario through MEDT.

LITERATURE CITED

1. Kirzhnits DA, Linde AD. *Phys. Lett. B* 42:471 (1972)
2. Gyulassy M, McLerran L. *Nucl. Phys. A* 750:30 (2005)
3. Maldacena JM. *Adv. Theor. Math. Phys.* 2:231 (1998)
4. Gubser SS, Klebanov IR, Polyakov AM. *Phys. Lett. B* 428:105 (1998)
5. Witten E. *Adv. Theor. Math. Phys.* 2:253 (1998)
6. Policastro G, Son DT, Starinets AO. *Phys. Rev. Lett.* 87:081601 (2001)
7. Son DT, Starinets AO. *J. High Energy Phys.* 0209:042 (2002)
8. Herzog CP, Son DT. *J. High Energy Phys.* 0303:046 (2003)
9. Policastro G, Son DT, Starinets AO. *J. High Energy Phys.* 0209:042 (2002)
10. Policastro G, Son DT, Starinets AO. *J. High Energy Phys.* 0212:054 (2002)
11. Herzog CP. *J. High Energy Phys.* 0212:026 (2002)
12. Herzog CP. *Phys. Rev. D* 68:024013 (2003)

13. Forster D. *Hydrodynamic Fluctuations, Broken Symmetry, and Correlation Functions.* Reading, PA: Benjamin (1975)
14. Landau LD, Lifshitz EM. *Fluid Mechanics.* Oxford: Pergamon (1987)
15. Jeon S. *Phys. Rev. D* 52:3591 (1995)
16. Jeon S, Yaffe LG. *Phys. Rev. D* 53:5799 (1996)
17. Aharony O, et al. *Phys. Rep.* 323:183 (2000)
18. Klebanov IR. hep-th/0009139 (2000)
19. Polchinski J. *String Theory.* Cambridge: Cambridge Univ. Press (1998)
20. Appelquist T, Chodos A, Freund PG. *Modern Kaluza-Klein Theories.* Menlo Park, CA: Addison-Wesley (1987)
21. Kim HJ, Romans LJ, van Nieuwenhuizen P. *Phys. Rev. D* 32:389 (1985)
22. Bianchi M, Freedman DZ, Skenderis K. *Nucl. Phys. B* 631:159 (2002)
23. Maldacena JM. *J. High Energy Phys.* 0304:021 (2003)
24. Kovtun P, Son DT, Starinets AO. *J. High Energy Phys.* 0310:064 (2003)
25. Thorne KS, Price RH, MacDonald DA. *Black Hole: The Membrane Paradigm.* New Haven: Yale Univ. Press (1986)
26. Kovtun PK, Starinets AO. *Phys. Rev. D* 72:086009 (2005)
27. Buchel A, Liu JT. *Phys. Rev. Lett.* 93:090602 (2004)
28. Kovtun P, Son DT, Starinets AO. *Phys. Rev. Lett.* 94:111601 (2004)
29. Buchel A. *Phys. Lett. B* 609:392 (2005)
30. Buchel A. *Nucl. Phys. B* 708:451 (2005)
31. Buchel A, Liu JT, Starinets AO. *Nucl. Phys. B* 707:56 (2005)
32. Teaney D. *Phys. Rev. C* 68:034913 (2003)
33. Shuryak E. *Prog. Part. Nucl. Phys.* 53:273 (2004)
34. Andronikashvili E. *Zh. Eksp. Teor. Fiz.* 18:429 (1948)
35. Parnachev A, Starinets AO. *J. High Energy Phys.* 0510:027 (2005)
36. Buchel A, Liu JT. *J. High Energy Phys.* 0311:031 (2003)
37. Benincasa P, Buchel A, Starinets AO. *Nucl. Phys. B* 733:160 (2006)
38. Herzog CP, et al. *J. High Energy Phys.* 0607:013 (2006)
39. Liu H, Rajagopal K, Wiedemann UA. *Phys. Rev. Lett.* 97:182301 (2006)
40. Casalderrey-Solana J, Teaney D. *Phys. Rev. D* 74:085012 (2006)
41. Gubser SS. *Phys. Rev. D* 74:126005 (2006)
42. Son DT, Starinets AO. *J. High Energy Phys.* 0603:052 (2006)
43. Mas J. *J. High Energy Phys.* 0603:016 (2006)
44. Maeda K, Natsuume M, Okamura T. *Phys. Rev. D* 73:066013 (2006)
45. Saremi O. *J. High Energy Phys.* 0610:083 (2006)
46. Benincasa P, Buchel A, Naryshkin R. *Phys. Lett. B* 645:309 (2006)

Physics of String Flux Compactifications

Frederik Denef,[1] Michael R. Douglas,[2] and Shamit Kachru[3]

[1] Institute for Theoretical Physics, University of Leuven, B-3001 Heverlee, Belgium; email: frederik.denef@fys.kuleuven.be

[2] Department of Physics and Astronomy, Rutgers University, Piscataway, New Jersey 08855; email: mrd@physics.rutgers.edu

[3] Department of Physics and Stanford Linear Accelerator Center, Stanford University, Stanford, California 94305; email: skachru@leland.stanford.edu

Annu. Rev. Nucl. Part. Sci. 2007. 57:119–44

First published online as a Review in Advance on April 27, 2007

The *Annual Review of Nuclear and Particle Science* is online at http://nucl.annualreviews.org

This article's doi: 10.1146/annurev.nucl.57.090506.123042

Copyright © 2007 by Annual Reviews. All rights reserved

0163-8998/07/1123-0119$20.00

Key Words

string vacua, string phenomenology, moduli stabilization

Abstract

We provide a qualitative review of flux compactifications of string theory, focusing on broad physical implications and statistical methods of analysis.

Contents

1. INTRODUCTION 120
2. EXAMPLES OF FLUX VACUA 124
    2.1. A Toy Model 124
    2.2. IIa Flux Vacua 127
    2.3. More Realistic Models 129
3. STATISTICS OF VACUA 130
    3.1. The Bousso-Polchinski Model 130
    3.2. A Simple Toy Model 131
    3.3. Ensembles of Flux Vacua in String Theory 132
    3.4. Supersymmetry-Breaking Scale 135
4. IMPLICATIONS FOR THE TESTABILITY OF STRING
    THEORY 139

1. INTRODUCTION

String theory was first proposed as a candidate theory of quantum gravity in 1974 (1). Over the subsequent years, supersymmetric versions of the theory were developed and arguments were made that they were perturbatively finite (2). The discoveries of anomaly cancellation (3) and of quasi-realistic compactifications of the heterotic string (4) from 1984 to 1985, a period sometimes referred to as the first superstring revolution, led very rapidly to a broad consensus among particle physicists that superstring theory was a viable contender for a "theory of everything" that could describe all of fundamental physics.

At the time, there were several good reasons to be wary of this claim. There were many competing theories; besides the different varieties of superstrings, there was 11-dimensional supergravity. Another reason was the fact that string theory was defined strictly as a perturbative expansion, which could only be used directly at weak coupling. Besides the fact that the real world includes strongly coupled quantum chromodynamics, various arguments had been made that constructing a completely realistic model would require nonperturbative physics, most notably at the stage of supersymmetry breaking (5).

These doubts were addressed in a rather striking way in the second superstring revolution from 1994 to 1997, in which it was convincingly argued that all these theories are limits or aspects of one unified framework, now known as string theory or M theory, herewith referred to as string theory. The central idea, termed duality, is that the strong coupling limit of one of the various string theories can be equivalent (or dual) to another weakly coupled theory, one of the strings, or M theory—the 11D limit (6). Aside from providing a variety of nonperturbative definitions for string theory, these ideas also led to exact solutions for the effective Lagrangians of a variety of supersymmetric field theories, realizing phenomena such as confinement and chiral symmetry breaking (7). Thus, although there remained much to do, it now seemed reasonable to hope that with further theoretical progress, the remaining gaps such as supersymmetry breaking could be addressed.

Although the particle physics side of the story was moving along nicely, serious gaps remained on other fronts, especially cosmology. The well-known cosmological constant problem remained (8, 9), as well as the question of whether and how inflation could be described. On a different level, ever since the first studies of string compactification, it had appeared that making any concrete construction required making many arbitrary choices, such as choosing an extradimensional manifold, a gauge bundle, or a brane configuration, and so on. Other than the evident constraint that the correct choice should lead to physics that fits the observations, no principles had been proposed, even speculative ones, that would suggest that any of these choices were preferred. Thus, it was unclear how to obtain testable predictions from the theory.

A smaller-scale version of this problem is that the metric in the extra dimensions, as well as the other data of a solution of string theory, depends on continuous parameters, termed moduli. Thus, even having chosen a particular extradimensional manifold, one has many continuous parameters that enter into observable predictions. Again, no particular values of these parameters appeared to be preferred.

At first, one might compare this ambiguity with the choice of coupling constants in a renormalizable field theory. For example, the Standard Model has 19 parameters, and consistency conditions lead only to very weak constraints on these, such as the unitarity bound on the Higgs mass (10). However, the situation here is essentially different: String theory has no free parameters; rather each of the parameters of a solution corresponds to a scalar field in four dimensions. If this choice is unconstrained by the equations of motion, this implies that the scalar field is massless. Massless scalar fields typically (although not always) lead to modifications to the gravitational force law, which are not observed (11). Thus, this is a phenomenological problem known as the problem of moduli stabilization, which must be solved to obtain realistic models.

Having stated the problem, there is a simple argument for why it will generally solve itself in realistic models. The vacuum structure of a field theory is governed by the effective potential $V_{eff}(\phi)$, which incorporates all classical and quantum contributions to the potential energy, such as Casimir terms, loop corrections, instantons, and the like. These corrections can in principle be computed from the bare Lagrangian. In almost all cases where this has been done, the corrections take generic order-one values (in units of some fundamental scale) that respect the symmetries of the bare Lagrangian. In particular, this includes the masses of scalar fields. Thus, massless scalar fields never appear in practice, unless they are Goldstone bosons for a continuous symmetry or unless we can tune parameters (as is done in condensed matter systems to approach critical points).

A noteworthy exception is found in supersymmetric field theories, in which nonrenormalization theorems preclude corrections to the superpotential to all orders in perturbation theory. Although some theories admit nonperturbative corrections, others do not, and in those theories moduli are natural. However, a realistic model must break supersymmetry at some scale $M_{susy} \geq 1$ TeV, and below this scale the previous argument applies. Thus, the moduli will ultimately gain masses of order $m \sim c\, M_{susy}$, where c is often a small number of the order M_{susy}/M_p.

This solution turns out to be problematic (12, 13), as it leads to the so-called Polonyi problem (14), wherein the light moduli fields carry too much energy in the early Universe, leading to overclosure. Thus, one is led to look for other mechanisms that could give larger masses to moduli. A particularly simple mechanism, which we discuss in some detail in Section 2, is to postulate a background-generalized magnetic field in the extra dimensions, usually termed flux. The energy of such a field depends on the moduli and provides a new contribution to the effective potential. Its scale is set by the unit of quantization (a fundamental scale such as the string scale) and the inverse size of the extra dimensions. This is typically far higher than M_{susy} and solves the problem.

There is an unexpected side effect of this mechanism. On a qualitative level, it works for fairly generic nonzero choices of flux. Supposing that each flux can take of order 10 values, one finds of order 10^K distinct solutions, where K is the number of distinct topological types of flux (usually a Betti number) and the compactification manifolds used in string theory, say the Calabi-Yau manifolds, typically have $K \sim 30 - 200$. Thus, one finds a large vacuum multiplicity. Furthermore, physical predictions depend on this choice, both directly and because the values of the stabilized moduli depend on the flux.

As one might imagine and as we discuss below, this vacuum multiplicity makes it rather complicated to propose definitive tests of the theory. Philosophically, it seems very much at odds with the idea that a fundamental theory should be simple and unique. It is hard to know how much weight to put on such considerations, which do not in themselves bear on the truth or falsity of the theory as a description of nature, but certainly one should ask for more evidence before accepting this picture.

In a seemingly unrelated development beginning in the late 1990s, convincing evidence accumulated for a nonzero dark energy in our Universe, from the accelerated expansion as measured by observations of supernova, in precision measurements of the cosmic microwave background, and from other sources (15). Although not absolutely proven, the simplest model for this dark energy is a positive cosmological constant, of order $\Lambda = (0.71 \pm 0.01)\Omega$, where Ω is the critical density.

This observation brought the cosmological constant problem to the fore of fundamental physics. As long as the data were consistent with $\Lambda \sim 0$, the simplest hypothesis was that the correct theory would contain some mechanism that adjusted Λ to zero. Many proposals along these lines have been made, although none is generally accepted nor is there any proof that this is impossible. However, the existing proposals generally do not fit well within string theory, and it had been widely felt that a proper solution would require some essentially new idea.

The problem of fitting a specific nonzero Λ looks rather different; indeed the particular choice of value appears rather arbitrary. Its one evident property is that it is of the same order as the matter density, but only at the present epoch (the coincidence problem). This led to the study of proposals such as quintessence, in which the dark energy is time dependent; we only say here that this proposal has the advantage of being directly testable, and so far the evidence seems to be coming down on the other side, for a fixed cosmological constant.

There was one previous proposal for a mechanism that could explain a small nonzero cosmological constant, due to Banks (16), Linde (17), and particularly Weinberg (18). This was the idea that the underlying theory might manifest a large number of distinct vacua, which are stable at least on cosmological timescales, and realize different values of the cosmological term Λ. When Weinberg was writing, the direct observational bounds on Λ placed a bound on its value, which had just excluded the range of Λ s too large to allow successful galaxy formation. Weinberg advocated the point of view that, given that Λ is absurdly small compared to fundamental scales in nature (indeed, at the time he wrote, there was only a tiny upper bound on its value), one should postulate that the distribution of Λ in the different vacua is uniform in the observationally allowed range—that compatible with galaxy formation. It then follows statistically that most vacua compatible with observations would lie within a decade or two of the maximal allowed value. Thus, Weinberg predicted that the eventual observed Λ would be comparable to the maximal value compatible with galaxy formation. This has proved true.

An important ingredient in this argument is that the fundamental underlying theory should admit many vacua. Although from the beginning the choices present in string compactification made it clear that the theory has many distinct solutions, the number required by Weinberg's argument (roughly 10^{120}) was far larger than any reliable approximation of the number of metastable vacua. Although some estimates of very large numbers of constructions were made by counting distinct soluble points (lattice constructions or orbifold models) in continuous moduli spaces of vacua (19), it was widely believed and is still likely that these many constructions would collapse into a few after supersymmetry breaking because they are simply distinct points on the same supersymmetric moduli space. The number of discretely differing solutions seemed far smaller: For instance, although Calabi-Yau spaces are far from unique, the number of distinct choices after more than 20 years of systematically searching for constructions still numbers in the thousands. This fact, together with the inability to exhibit reliable metastable solutions after supersymmetry breaking, largely discouraged speculation on this topic.

The duality revolution, however, brought many new ingredients of string theory to the fore. One of these was the generalized gauge fields that couple to Dirichlet (or D) branes. Each of these gauge fields comes with an associated field strength, whose flux can thread the extra dimensions. Bousso & Polchinski (20) realized that in compact manifolds with sufficiently complicated topology, the number of flux choices allowed by the known constraints could easily exceed 10^{120}. This provided a large, discretely varying collection of models that could realize different values of the cosmological term. Although this model neglected many aspects of the physics, in particular simply freezing the moduli of the internal space by hand, it made it plausible that string theory could contain the large set of vacua required by Weinberg's argument.

In light of this combination of theoretical and observational considerations, finding concrete and computable models of string compactification in which the moduli problem is solved, and in which distinct choices of flux provide a large collection of vacua, has become very interesting. Starting around 2000, this problem received a great deal of attention (some representative works being References 21–26), with a milestone being the Kachru-Kallosh-Linde-Trivedi (KKLT) proposal (27) for a class of type IIB string theory compactifications in which all moduli are stabilized, by a combination of fluxes and nonperturbative effects, and supersymmetry is broken with a small positive cosmological constant. Subsequent work proposed more and more concrete realizations of this proposal (28–30), as well as alternate constructions using the other string theories or relying on different hypotheses for the scales and dominant terms in the effective potential. One can (heuristically) envision these potentials as taking values over a large configuration space, which has been termed the string landscape (31). A central question in recent years has been how to characterize the landscape and find dynamical mechanisms for populating and selecting among the vacua within it.

A full discussion of these constructions quickly becomes technical, and we refer the reader to reviews such as References 32–36 for the details. However, we outline one representative set of constructions in the next section to illustrate the ideas. The rest of the review addresses the questions of testability raised by the landscape. In Section 3, we discuss formal results for the distribution of vacua and their observable properties, and in Section 4, we survey the various approaches for getting concrete and testable predictions from the framework. Many related topics, such as inflation in string theory, had to be omitted for reasons of space. We were also limited to citing only major reviews and a few influential and/or particularly recent papers, and again direct readers to References 32–36 for a more complete bibliography.

2. EXAMPLES OF FLUX VACUA

This section describes in more detail the ingredients that enter in string flux compactifications. We first give a simple toy model that captures much of the relevant physics. In Section 2.2, we extend this to a full discussion of a simple class of flux vacua of IIa string theory. In Section 2.3, we briefly describe other ingredients that enter in making quasi-realistic models.

2.1. A Toy Model

The essential point is that fluxes and branes present in 10D string theory provide natural ingredients for stabilizing the moduli of a compactification. However, the basic ideas are independent of dimension, and the simplest illustration involves the 6D Einstein-Maxwell theory. We imagine, therefore, a 6D theory whose dynamical degrees of freedom include the metric g_{MN} and an Abelian Maxwell field F_{MN}. The Lagrangian takes the form

$$L = \int d^6 x \sqrt{-g} \left(M_6^4 \mathcal{R} - M_6^2 |F|^2 \right), \qquad\qquad 1.$$

where M_6 is a fundamental unit of mass (the 6D Planck scale).

To define a compactification of this 6D theory to four dimensions, one must choose a compact 2D internal space. The topologies of 2D manifolds without boundary are classified by the genus; that is, the possible spaces M_g are (hollow) donuts with $g = 0, 1, 2 \ldots$ holes. $g = 0$ is the sphere, $g = 1$ is the two-torus, and so forth.

Suppose we compactify the theory on the manifold M_g of genus g and total volume R^2. As an ansatz for the 6D metric, we could take

$$ds^2 = g_{\mu\nu}dx^\mu dx^\nu + R^2 \tilde{g}_{mn}dy^m dy^n, \qquad\qquad 2.$$

with $\tilde{g}$ a metric of unit volume on M_g. Our focus here is the dynamics of the modulus field $R(x)$, which enters into the ansatz (Equation 2).

We can expand the resulting 4D effective Lagrangian in powers of derivatives, as is standard in effective field theory. The leading terms arising from the gravitational sector are

$$M_6^4 R^2 \int d^4 x \sqrt{-g} \left(\left[\int d^2 y \sqrt{\tilde{g}} \mathcal{R}_2 \right] + R^2 \mathcal{R}_4 \right) + \cdots. \qquad\qquad 3.$$

Here, $\mathcal{R}_2$ is the curvature of $\tilde{g}$ and $\mathcal{R}_4$ is the curvature of the 4D space-time metric g. The ellipses include both gradients of R and terms involving the 4D gauge field.

There are two important points about the action as written above. First, defining $M_4^2 = M_6^4 R^2$ as the 4D Planck scale, we see the action is not in Einstein frame—the kinetic terms of R and the graviton are mixed. Second, the first term in brackets above is in fact a topological invariant, $\chi(M_g)$. For a surface of genus g, $\chi = 2 - 2g$. We therefore find a scalar potential for the modulus field unless $g = 1$.

To go to a 4D Einstein frame, we should redefine the 4D metric: $g \to h = R^2 g$. Then $\sqrt{h}\mathcal{R}_h = R^2 \sqrt{g}\mathcal{R}_g$, and we find a 4D Lagrangian:

$$M_4^2 \int d^4 x \sqrt{-h}(\mathcal{R}_h - V(R)). \qquad\qquad 4.$$

The potential $V(R)$ is given by

$$V(R) \sim (2g - 2)\frac{1}{R^4}. \qquad\qquad 5.$$

The R dependence follows from the Weyl rescaling above.

We have learned an interesting lesson. In the absence of background Maxwell fields, of the infinite family of compactification topologies parametrized by g, precisely one choice yields a static solution. The choice $g = 1$, the two-torus, yields a vanishing potential and a 4D flat-space solution. The positive-curvature $g = 0$ surface, the two-sphere S^2, naively has a negative potential $\sim -1/R^4$, yielding runaway to $R \to 0$ (where the regime of trustworthiness of our analysis breaks down). The negative curvature options, $g > 1$, instead yield positive potentials, which vanish as $R \to \infty$; they spontaneously decompactify back to the 6D flat-space solution.

All these features of our toy model have analogs in full-fledged string compactification. The T^2 here, which is Ricci flat, is analogous to the Calabi-Yau solutions, which string theorists have studied intensely. One difference is that whereas T^2 is a unique 2D manifold, there are at least thousands of topologically distinct Calabi-Yau threefolds, each leading to a class of compactifications. If our 6D theory were promoted to a supersymmetric Einstein-Maxwell theory, this 4D flat solution would have unbroken supersymmetry at the Kaluza-Klein (KK) scale (which partially explains the fascination that string theorists have had with the analog Calabi-Yau solutions). Note that the modulus field R survives unfixed in the 4D effective theory. It appears to 4D physicists as a massless scalar field. In fact, the T^2 has an additional modulus corresponding to its shape; the Calabi-Yau spaces also manifest both volume moduli and shape (complex structure) moduli.

The S^2 case, with positive curvature, mirrors the Einstein manifolds used in Freund-Rubin compactification (37). The negative curvature manifolds (here, the most common topologies) are just now being examined as candidate string compactifications (38).

2.1.1. Including fluxes and branes. Let us now turn on the Maxwell field. To find 4D vacuum solutions with maximal symmetry, we should allow only the gauge field configuration to be nontrivial on M_g. The most obvious possibility is to thread M_g with some number of units of magnetic flux of the gauge field

$$\int_{M_g} F = n. \qquad \qquad 6.$$

Like a magnetic flux threading a solenoid, this flux through M_g carries positive energy density. This makes an additional contribution to the 4D effective potential as a function of R. Naively, it scales like $1/R^2$—a factor of R^2 from the determinant of the metric and two factors of $1/R^2$ from the metric factors contracting the indices on F_{mn}. However, upon rescaling to reach the 4D Einstein frame, one multiplies the effective potential by an additional $1/R^4$, so the full potential takes the schematic form

$$V(R) \sim (2g - 2)\frac{1}{R^4} + \frac{n^2}{R^6}. \qquad \qquad 7.$$

In the presence of n units of magnetic flux, the status of the different compactification topologies changes.

Previously, the $g = 0$ S^2 compactification yielded solutions that run to (uncontrolled) small values of R. However, it is easy to see that with n units of flux, the positive flux energy can balance the negative curvature contribution at finite values of R, which grows with n. Therefore, for large n, one obtains reliable flux vacua from a large S^2, with n units of flux piercing it. These are the Freund-Rubin solutions.

For $g = 1$, where previously there was a moduli space of solutions, the flux causes a runaway to large R. Finally, for $g > 1$, because both terms in the potential are positive and vanish at large R, there is also a runaway.

The same behavior is obtained in the full string theory. The Freund-Rubin solutions also exist there and play a crucial role in, for example, the AdS/CFT correspondence. There are indeed no solutions of string theory that start from (supersymmetric) Calabi-Yau compactification and incorporate only flux; additional ingredients are needed (which we come to next). Finally, the negative curvature spaces also require additional ingredients to avoid runaway.

Next, we add branes: dynamical, fluctuating, lower-dimensional objects that can carry gauge and matter fields. A common example is the D-brane. These have positive tension and thus make positive contributions to $V(R)$. String theories also include fixed, nondynamical objects, the so-called orientifold planes (or O-planes), with negative tension.

We now modify our toy model to include some fixed $\mathcal{O}(1)$ number m of O3-planes in the background, at points on M_g. Taking into account Weyl rescaling, their contribution to the 4D effective potential is

$$\delta V_{O3} = -m\frac{1}{R^4}.\qquad\qquad 8.$$

We see that, after inclusion of these objects, the full effective potential of our toy model takes the general form

$$V(R) = (2g - 2)\frac{1}{R^4} - m\frac{1}{R^4} + n^2\frac{1}{R^6}.\qquad\qquad 9.$$

The physics of Equation 9 captures many of the qualitatively important facts in flux compactification (although the identical scaling of the O-plane and curvature contributions do not hold in the string analogs). Of course, it is an oversimplification: In real constructions there are many further constraints determining the number of branes, planes, and so forth. Including both branes and fluxes, if models with suitable m are available, (*a*) the modulus R of the T^2 compactification can be fixed by inclusion of O3-planes, and (*b*) the negative curvature models may now admit critical points of their potential at large R (for string theory examples, see Reference 39).

The vacua analogous to type *a* are our focus in much of this review. These are Calabi-Yau models with an underlying $\mathcal{N} = 2$ supersymmetry, broken to $\mathcal{N} = 1$ by the branes and O-planes. Other ingredients, such as nonperturbative physics, are required to stabilize all moduli and break supersymmetry, as described in Reference 32. In many of these models, the scale of supersymmetry breaking can be very low compared with the KK scale. Thus, these models are a logical place to search for string extensions of the Minimal Supersymmetric Standard Model (MSSM), which include its coupling to quantum gravity.

2.2. IIa Flux Vacua

The simplest full constructions of string vacua along these lines use IIa string theory (40–42). This theory contains p-form Ramond-Ramond (RR) field strengths F_p, with $p = 0, 2, 4, 6, 8$, and an Neveu-Schwarz-Neveu-Schwarz (NSNS) field strength H_3. Practically, this means that in addition to the possible fluxes threading nontrivial 0-, 2-, 3-, 4-, and 6-cycles of the 6D compactification manifold, the theory contains D0-, 2-, 4-, 6-, and 8-branes and NS5-branes. In addition to the D-branes, there can be O-planes of each even dimension <10. Any of these objects can form part of a solution with 4D rotational symmetry if they fill 4D space-time, and thus can wrap a $(p{-}3)$D cycle in the compactification manifold.

We can analyze the resulting effective potential in close analogy to the one for our toy model. Imagine compactification on a Ricci-flat 6D manifold M (say a Calabi-Yau) with $b_2 = b_4$ topologically distinct four-cycles, b_3 distinct three-cycles, no one-cycles or five-cycles, and volume R^6. Unlike our toy model, in string theory there is no predetermined dimensionless coupling; the 10D coupling g_s is determined by the dilaton field ϕ as $g_s = e^\phi$. A minimal discussion therefore focuses on the dynamics of the volume modulus R and ϕ, and we do that below, sketching the elaboration to include other moduli at the end. In a full discussion, all moduli must be fixed.

Because the manifold is Ricci flat, there is no contribution to the potential from the 10D curvature term. Thus, to prevent a runaway to large R, we must include O-planes. Given the assumed topology of M, these must be O6-planes wrapping three-cycles. We also need to know the g_s scaling of the various sources of energy in 10 dimensions. In 10D string frame, the Einstein term scales like $e^{-2\phi}$, whereas the D-brane/O-plane tensions scale like $e^{-\phi}$. In addition, the RR $|F|^2$ terms are independent of g_s (in the convention where the RR flux quantization is independent of g), whereas the NS $|H|^2$ arises at order $e^{-2\phi}$.

We can now write a schematic effective potential for R and e^ϕ, given any particular choice of ingredients. One natural class of compactifications involves N units of RR F_4 flux, $\mathcal{O}(1)$ units of F_0 flux and H_3 flux, and some fixed $\mathcal{O}(1)$ number of O6-planes. The resulting potential is

$$V = N^2 \frac{e^{4\phi}}{R^{14}} + \frac{e^{2\phi}}{R^{12}} - \frac{e^{3\phi}}{R^9} + \frac{e^{4\phi}}{R^6}. \qquad 10.$$

One sees that this potential admits vacua with $R \sim N^{1/4}$ and $g_s = e^\phi \sim N^{-3/4}$. By analogy with the Freund-Rubin toy model, this theory has vacua with large volume (and weak coupling) if one turns on a large number of flux quanta. An important distinction between this model and the Freund-Rubin vacua is that the size of the compact space M is parametrically smaller, in this leading approximation, than the curvature radius set by the leading estimate for the 4D cosmological term. Thus, these models are indeed models of 4D matter coupled to quantum gravity, over some range of scales.

This discussion has clearly oversimplified the complex problem of stabilizing Calabi-Yau moduli. However, the basic construction above does in fact yield models with all moduli stabilized in the full string theory. The simplest example (40) uses the Calabi-Yau orbifold T^6/Z_3^2. This space is a quotient of the six-torus by the action of two discrete symmetry groups. We start with the torus described by complex coordinates $z_i = x_i + iy_i$, where $i = 1, \ldots, 3$, subject to identifications

$$z_i \sim z_i + 1 \sim z_1 + \alpha, \qquad 11.$$

where $\alpha = e^{i\pi/3}$. This torus has a Z_3 symmetry T under which

$$T : (z_1, z_2, z_3) \to (\alpha^2 z_1, \alpha^2 z_2, \alpha^2 z_3). \qquad 12.$$

This action has a total of $3 \times 3 \times 3 = 27$ fixed points on the torus. Although these result in conical singularities, these are allowed in perturbative string theory (43). One can reduce this to nine singularities by quotienting by another freely acting Z_3 symmetry:

$$Q : (z_1, z_2, z_3) \to \left(\alpha^2 z_1 + \frac{1+\alpha}{3}, \alpha^4 z_2 + \frac{1+\alpha}{3}, z_3 + \frac{1+\alpha}{3} \right).$$

Orbifolds by discrete subgroups of $SU(3)$ give rise to (in general singular) Calabi-Yau manifolds, so our construction thus far gives an $\mathcal{N} = 2$ type IIa compactification on a Calabi-Yau space. To break the supersymmetry to $\mathcal{N} = 1$, we now orientifold by the simultaneous action of the Z_2 involution

$$\sigma : z_i \rightarrow -\bar{z}_i,\qquad\qquad 13.$$

with worldsheet parity reversal. The action of σ gives rise to a fixed three-plane $z_i = 0$, which is wrapped by a space-filling O6-plane.

The model we have described has 12 volume moduli, whose origin is as follows: Three of them simply rescale the volumes of the three two-tori involved in the construction. The other nine, one per singularity, are associated with exceptional cycles introduced by resolving the C^3/Z_3 fixed points. However, it has no complex structure moduli (it is rigid) because the Z_3 symmetries arise only for a unique shape of the covering torus, and because resolving the singularities introduces only volume moduli.

It is now straightforward to add ingredients that give rise to a potential analogous to that in Equation 10. We already have an O6-plane; one can thread the complementary three-cycle with H_3 flux and turn on F_0 flux and F_4 flux through the four-cycles transverse to the T^2 of the original $(T^2)^3$. The one novelty is the existence of the exceptional cycles, but these can be stabilized at large volume by four-form flux as well. The details can be found in Reference 40; the result is a model in which the leading effective potential stabilizes the moduli at weak string coupling and large volume, just as suggested by Equation 10. It should be admitted that this analysis is not an absolute proof of existence; for possible subtleties see Reference 44.

2.3. More Realistic Models

We have focused on the construction of models with computable moduli potentials. To make contact with observed physics, we also need a sector that gives rise to (a supersymmetric extension of) the Standard Model, and perhaps a sector responsible for supersymmetry breaking and its transmission to the Standard Model. There has been significant progress on constructing flux vacua, which incorporate all these elements.

One popular method for constructing Standard Model–like models in string theory is to use intersecting D-branes, for example, D6-branes in IIa theory. Stacks of N parallel branes manifest an $SU(N)$ gauge theory, and intersections of D6-branes at points in the extra dimensions localize chiral matter multiplets (45). These ingredients allow engineering the Standard Model and, indeed, fairly general $\mathcal{N} = 1$ gauge theories. For the state of the art, see Reference 46.

One can also use this to construct models of dynamical supersymmetry breaking. It recently became clear that even the simplest nonchiral gauge theories have supersymmetry-breaking vacua (47), and one can very simply engineer these constructions on D6-branes (48). Another option is to arrange for the flux potential itself to provide the supersymmetry breaking, as we discuss in the next section. Thus, it is quite plausible that the class of models described above includes the supersymmetric Standard Model and its various extensions, with supersymmetry breaking transmitted via gravity or gauge interactions.

3. STATISTICS OF VACUA

We now supply a brief introduction to and survey of the statistics of vacua and refer the reader to References 32 and 49 for more extensive reviews.

3.1. The Bousso-Polchinski Model

As noted in Reference 20, the freedom one has to turn on various independent flux quanta in string theory compactifications leads to ensembles of vacua with a variety or discretuum of low-energy effective parameters. This leads to a need for statistical analysis of the resulting vacuum distributions (50). This is true in particular for the cosmological constant, implying the natural existence of string vacua with exceedingly small effective 4D cosmological constants, such as our own, without the need to invoke any (so far elusive) dynamical mechanism to almost cancel the vacuum energy.

To see how this comes about, recall that the (classical) potential induced by a flux F characterized by flux quanta $N^i \in Z$, where $i = 1 \ldots K$, is given by

$$V_N(\phi) = V_0(\phi) + \int_X ||F||^2 = V_0(\phi) + \sum_{i,j} g_{ij}(\phi) N^i N^j, \qquad 14.$$

where ϕ denotes the moduli of the compactification manifold X and $g_{ij}(\phi)$ is some positive definite effective metric on the moduli space. The number of fluxes K is typically given by the number of homologically inequivalent closed cycles of some fixed dimension in X, which for the known examples of 6D Calabi-Yau manifolds is typically of order a few hundred. The bare potential $V_0(\phi)$ is taken to be negative. In string compactifications, it could, for example, come from O-plane contributions, and in this context V_0 (as well as g_{ij}) is of order of some fundamental scale such as the string or Kaluza-Klein scale.

Each vacuum of this model is characterized by a choice of flux vector N together with a minimum ϕ_* of $V_N(\phi)$. Finding these critical points explicitly is typically impossible, so to progress one must use indirect statistical methods, as we discuss below. However, before plunging in to this full, coupled problem, let us, following Reference 20, first simply freeze the moduli at some fixed value $\phi = \phi_0$ and ignore their dynamics altogether. In that case it is easy to compute the distribution of cosmological constant values: The number of vacua with cosmological constant $\Lambda = V_N(\phi_0)$ less than Λ_* is then given simply by the number of flux lattice points in a sphere of radius squared equal to $R^2 = |V_0| + \Lambda_*$, measured in the g_{ij} metric. When R is sufficiently large, this is well estimated by the volume of this ball, that is, $\mathrm{Vol}_K(R) = \frac{\pi^{K/2}}{\Gamma(1+K/2)} \frac{R^K}{\sqrt{\det g}}$, leading to a Λ distribution

$$d N_{\mathrm{vac}}(\Lambda) = \frac{\pi^{K/2}}{\Gamma(K/2)} \frac{(|V_0| + \Lambda)^{\frac{K}{2}-1}}{\sqrt{\det g}} d\Lambda \approx \left(\frac{2\pi e(|V_0| + \Lambda)}{\mu^4} \right)^{K/2} \frac{d\Lambda}{|V_0| + \Lambda}, \qquad 15.$$

where $\mu^4 := (\det g)^{1/K}$ can be interpreted as the mass scale of the flux part of the potential and where large K was assumed and Stirling's formula used to obtain the last approximate expression. Note that in particular at $\Lambda = 0$, for say $|V_0|/\mu^4 \sim \mathcal{O}(10)$, we get a vacuum density $d N_{\mathrm{vac}} \sim 10^{K/2} d\Lambda/|V_0|$. Hence, for K equal to a few hundred, there will be exponentially many vacua with Λ in the observed range $\Lambda \sim 10^{-120} M_p^4$, even if all fundamental scales setting the parameters of the potential are of order M_p^4.

In such a model, there is no need to postulate either anomalously large or small numbers, or an unknown dynamical mechanism, to obtain vacua with a small cosmological constant. Combined with a cosmological mechanism generating all possible vacua (such as eternal inflation) and Weinberg's argument, as discussed in the introduction, we have a candidate solution to the cosmological constant problem within string theory.

Of course, the Bousso-Polchinski model is only a crude approximation to actual flux vacua of string theory; in particular, freezing the moduli at arbitrary values is a major oversimplification. However, more general and more refined analyses of low-energy effective parameter distributions of actual string flux vacua that take moduli dynamics into account, initiated in Reference 50 and developed further in References 51–53, confirmed the general qualitative features of the model we just discussed. We now turn to an overview of these studies.

3.2. A Simple Toy Model

We demonstrate the basic ideas behind these techniques by considering the following ensemble of effective potentials:

$$V_{N,M}(\phi) = N\phi + M\frac{\phi^2}{2}, \quad N, \quad M \in Z, \quad N^2 + M^2 \leq L. \qquad 16.$$

Here, the (N, M) (crudely) models fluxes, while ϕ models a modulus field. To obtain a large number of vacua, we take L to be very large. Finding the critical points is trivial for this ensemble: They are given by $\phi_* = -N/M$ and are stable if $M > 0$. To find the distribution of vacua over ϕ space in the large L limit, we could in principle start from these explicit solutions. However, a more elegant and powerful method, which is also applicable to cases for which explicit solutions cannot be found, and to ensembles of actual string theory flux vacua, goes as follows.

The number of vacua ϕ_* in an interval I is given by

$$N_{\text{vac}}(I) = \sum_{N,M} \int_I d\phi \, \delta(V'(\phi)) \, | \, V''(\phi) \, | \, \theta(V''(\phi)), \qquad 17.$$

where $\theta(x) := 1$ if $x > 0$ and $\theta(x) := 0$ if $x < 0$. The integrand $\delta(V')|V''|$ gives a contribution of $+1$ for each critical point in I, while $\theta(V'')$ restricts to actual minima.

In the large L limit, we can approximate the sum over (N, M) by an integral and write

$$N_{\text{vac}}(I) \approx \int_I d\phi \, \rho(\phi), \, \rho(\phi) := \int dN\,dM\,\delta(V'(\phi))\,V''(\phi)\,\theta(V''(\phi)), \qquad 18.$$

where $\rho(\phi)$ can be interpreted as a vacuum number density on moduli space. To evaluate the integral over (N, M) at a given fixed ϕ, it is convenient to make the following linear change of variables $(N, M) \to (v', v'')$:

$$v' := V'(\phi) = N + M\phi, \quad v'' := V''(\phi) = M. \qquad 19.$$

This change of variables has Jacobian $= 1$, and the integration domain in the new variables is $L \geq N^2 + M^2 = (v' - v''\phi)^2 + (v'')^2$. The integral is now trivially evaluated, yielding the distribution

$$\rho(\phi) = \frac{L}{2}\frac{1}{1+\phi^2}.$$

20.

Note that this integrates to a total number of vacua $\mathcal{N}_{\mathrm{vac}}(\mathrm{IR}) \approx \pi L/2$. Indeed the total number of pairs (N, M) in the ensemble is approximated by the volume πL of the region $N^2 + M^2 \leq L$, each (N, M) leads to a unique critical point, and half of those are minima. The density (Equation 20) confirms the intuitive expectation that most vacua in this ensemble will be at order-one values of ϕ, but makes this far more precise.

Let us next consider a more general ensemble of the form

$$V_{N,M}(\phi) = Nf(\phi) + Mg(\phi),$$

21.

where f and g are arbitrary functions. Now it is no longer possible to proceed by finding explicit solutions—even finding a single explicit solution will typically be out of reach for simple choices of f and g. However, the previous computation is extended in a straightforward manner to this case, for general f and g, resulting in a vacuum number density

$$\rho(\phi)\,d\phi = \frac{L}{2}\frac{|f'g'' - f''g'|}{(f')^2 + (g')^2}d\phi = \frac{L}{2}\mathrm{sign}\left(\frac{g'}{f'}\right)' d\arctan\left(\frac{g'}{f'}\right).$$

22.

This illustrates the power of statistical methods over explicit constructions.

An interesting special case, which has an important counterpart in actual string flux vacua, is obtained by setting $f(\phi) = \phi$, $g(x) = \phi \log \phi - \phi$, and $\phi > 0$. Potentials with similar structure appear naturally in string theory, as we discuss in more detail below. The corresponding vacuum density is

$$\rho(\phi)\,d\phi = \frac{L}{2}\frac{1}{(1 + \log^2 \phi)}\frac{d\phi}{\phi} = \frac{L}{2}d\arctan\log\phi.$$

23.

Note that this distribution is approximately scale invariant, and thus naturally allows hierarchically small (and large) vacuum values of ϕ and, therefore, $V(\phi)$. For this particular ensemble, we can also see this directly, as the critical points can again be found explicitly; namely, $\phi_* = e^{-N/M}$.

3.3. Ensembles of Flux Vacua in String Theory

The most studied and best understood ensemble of flux vacua is the IIB ensemble, which arises by allowing two different kinds of fluxes (RR and NSNS) to be turned on through nontrivial three-cycles of a Calabi-Yau compactification space X. There are two kinds of geometric moduli, arising from complex structure (shape) and Kähler (size) deformations of X. Besides these, there is a universal modulus $\tau = C_0 + i/g_s$, where C_0 is the universal axionic scalar of the IIB theory and g_s is the string coupling constant. The complex structure moduli and τ appear nontrivially in the potential induced by the fluxes and as a result are generically stabilized at tree level. The Kähler moduli, however, do not appear nontrivially but can under certain conditions be stabilized by quantum effects (27). We assume this is the case and just freeze them in the following.

The potential for this ensemble has the standard $\mathcal{N} = 1$ supergravity form

$$V_N(z, \bar{z}) = e^K \left(g^{a\bar{b}} D_a W_N \overline{D_b W_N} - 3|W_N|^2 \right), \qquad 24.$$

where $g_{a\bar{b}} = \partial_a \bar{\partial}_{\bar{b}} K$, $D_a = \partial_a + \partial_a K$ and

$$W_N(z) = \sum_{i=1}^{K} N^i \Pi_i(z), \quad K(z, \bar{z}) = -\ln\left(Q^{ij} \Pi_i(z) \overline{\Pi_j(z)} \right). \qquad 25.$$

Here z^a denotes the complex structure coordinates together with τ, Q^{ij} is a known constant $K \times K$ matrix, and $\Pi_i(z)$ are certain complicated but in principle computable holomorphic functions (the periods) that depend on the Calabi-Yau X at hand. Note that the potential is quadratic in the N_i, as in the Bousso-Polchinski model. The number of flux quanta is $K = 2b_3(X)$, where $b_3(X)$ is the third Betti number of X, that is, the number of homologically independent three-cycles. For the quintic Calabi-Yau, which can be described as the zero locus of any homogeneous degree-five polynomial in $\mathbf{C}^5$ with points related by overall complex rescaling identified, one has $b_3 = 204$. Finally, there is a tadpole cancellation constraint on the fluxes, of the form

$$\frac{1}{2} Q^{ij} N_i N_j \leq L, \qquad 26.$$

where L depends again on the compactification data but typically ranges from $\mathcal{O}(10)$ to $\mathcal{O}(1000)$.

Supersymmetric vacua of this model are given by solutions of $D_a W_N(z) = 0$. Note that counting these amounts to a direct generalization of the counting problem of our toy model, with the periods $\Pi_i(z)$ generalizing $f(\phi)$ and $g(\phi)$. Indeed it turns out that approximate distributions can be derived at large L using essentially the same ideas (51, 52). The resulting distribution[1] of vacua over moduli space turns out to be surprisingly simple:

$$dN_{\text{vac}}(z) = \frac{(2\pi L)^{K/2}}{\left(\frac{K}{2}\right)! \pi^{K/4}} \det(R(z) + \omega(z)\mathbf{1}), \qquad 27.$$

where $R(z) = \frac{i}{2} R^a_{\ b c\bar{d}} dz^c \wedge d\bar{z}^{\bar{d}}$ is the curvature form of the metric $g_{a\bar{b}}$ and $\omega(z) = \frac{i}{2} g_{a\bar{b}} dz^a \wedge d\bar{z}^{\bar{b}}$ the Kähler form.

Note the close similarity of the z-independent prefactor with the prefactor of the Bousso-Polchinski distribution (Equation 15). Essentially, this arises because the constraint (Equation 26) combined with the condition of supersymmetry roughly restricts the fluxes to be contained in a sphere of radius proportional to $\sqrt{L}$. As a result, we get similarly huge numbers of actual IIB flux vacua in string theory. For

---

[1] To be more precise, this distribution is obtained by dropping the absolute value signs around the Jacobian determinant generalizing $|V''(\phi)|$ in Equation 18. As a result, some vacua are counted negatively. Hence, the density is an index density rather than an absolute density. This nevertheless gives a good estimate of the actual density, and in any case a lower bound. Absolute densities can be obtained as well, but are more complicated.

the three types of models analyzed in detail in Reference 28, we find N_{vac}(total) $\sim 10^{307}$, 10^{393}, and 10^{506}, respectively.

The last of these figures is the justification for the often quoted number 10^{500} as the total number of string vacua. However, as discussed in detail in Reference 32, there are many further uncertainties even in counting the known classes of vacua, so this famous number should not be taken too seriously. Furthermore, the number we are ultimately most interested in, namely, that of vacua similar to our own, is even less well understood at present; it could still be the case that only a few vacua (or even none) fit all the constraints implicit in the existing data. The point rather is that the problem of computing numbers of vacua with specific properties is both mathematically well posed (at least as much so as string theory itself is) and far easier than constructing the actual vacua. Thus, we can expect steady progress in this direction, leading to results of fairly direct phenomenological interest, as we explain shortly.

The distribution (Equation 27) has an interesting structure. It diverges (remaining integrable) when the curvature diverges, which happens near so-called conifold degenerations of the Calabi-Yau manifold X, corresponding to a three-sphere in X collapsing to zero size. Near this three-cycle, the Calabi-Yau can be described by an equation of the form $x_1^2 + x_2^2 + x_3^2 + x_4^2 = z$ in C^4, with the three-cycle being the real slice through it and $z = 0$ the conifold point. The vacuum density near this point is given by (52)

$$d N_{vac}(z) \sim \frac{d|z|}{|z| \log^2 |z|^{-2}} \sim d(\log |z|^{-1})^{-1}. \qquad 28.$$

Note this is approximately scale invariant, naturally allowing hierarchically small scales for z, similar to the toy model distribution in Equation 23. This is no coincidence: Near the conifold point, there is a pair of periods $\Pi(z) \sim (z, z \log z - z)$, just as in our toy model.

In the case at hand, small values of z give rise to large warped 5D anti–de Sitter space (AdS$_5$)-like throats in the compactification (25), which have a dual gauge theory description through the AdS-CFT correspondence (54). In terms of the dual gauge coupling g, the distribution is simply uniform $d N_{vac} \sim dg^2$. The enhancement of vacua at small z may be of more than academic interest: Such warped throats have many possible phenomenological uses. They can provide a natural embedding of the Randall-Sundrum scenario (55) in string theory, they give rise to natural models of warped (and hence exponentially low-scale) supersymmetry breaking (56), and some of the simplest inflationary scenarios in string theory make use of such throats (57).

From this one-parameter distribution, one can reasonably guess that the majority of IIB flux vacua will contain such warped throats, from the simple argument that the probability of having no such throats in an n-modulus case is expected to be roughly the n-th power of the probability of having no such throats in the one-modulus case, which becomes small at large n. Some more concrete evidence for this has been given in Reference 58.

Distributions of other quantities besides the moduli have also been obtained for IIB ensembles. For example, the distribution of cosmological constants of supersymmetric vacua near zero turns out to be uniform, $dN_{\text{vac}}(\Lambda) \sim \theta(-\Lambda)d\Lambda \sim d|W|^2$, and the same is true for the string coupling constant, $dN_{\text{vac}}(g_s) \sim dg_s$. If the Kähler moduli are stabilized according to the KKLT scenario, the distribution of KK scales M_{KK} is roughly given by $dN_{\text{vac}}(M_{KK}) \sim e^{-cM_{KK}^{-4}}d(M_{KK}^{-4})$, with c some constant, depending on the compactification manifold. This shows that low KK scales, (i.e., large extra dimensions) are statistically excluded in this scenario.

Distributions for other ensembles such as M theory or type IIa flux vacua have been worked out as well, and studies of distributions of discrete D-brane data such as gauge groups and matter content of intersecting brane models have been initiated. We refer the reader to Reference 32 and references therein for more details.

3.4. Supersymmetry-Breaking Scale

Perhaps most interesting from a phenomenological point of view is the scale of supersymmetry breaking. We would like to know, out of all the string theory vacua that agree with existing data, is it likely that we live in one with low-scale breaking, leading to the discovery of superpartners at the Large Hadron Collider (LHC)?

Of course, to answer this question conclusively, one would need a much better handle on many issues affecting both the distribution of vacua and also how cosmological selection mechanisms and the like influence actual probabilities on the string theory landscape. Because at present we know very little about the latter, we stick to the former and discuss the better defined number distributions of string flux vacua.

Assume that if many more vacua have property X than property not X, then X is favored, at least within the considerations we discuss. Another term for this type of consideration is stringy naturalness. Of course, if the probabilities of different vacua were roughly equal, this would lead directly to a statistical prediction. Even if this were not true, if the probabilities to obtain vacua were *uncorrelated* with the property of interest, one would also get a statistical prediction. However, a particular cosmological model might lead to probabilities that were actually *correlated* with the property of interest, requiring one to balance various competing effects. This is a very interesting possibility, but one requiring a discussion broader than what we can address here.

Thus, we would like to compute $dN_{\text{vac}}(M_{\text{susy}})$ given the observed values of $\Lambda \sim 0$ and $M_{EW} \sim 100\,\text{GeV}$, with M_{susy}^4 defined to equal the positive definite contribution to the supergravity potential, that is,

$$M_{\text{susy}}^4 \equiv \sum_i |F_i|^2 + \sum_\alpha D_\alpha^2, \qquad\qquad 29.$$

where $F_i = e^{K/2}D_i W$. Suppose this followed an approximate power law distribution, $dN_{\text{vac}}(M_{\text{susy}}) \sim M_{\text{susy}}^\alpha dM_{\text{susy}}$. Then for $\alpha < -1$, vacuum statistics would favor low-scale supersymmetry, whereas for $\alpha > -1$, it would not.

For comparison, let us begin with the standard prediction of field theoretic naturalness, implicit in the motivation often given for low-scale supersymmetry breaking based on the hierarchy problem. This is

$$dN_{\text{vac}}^{FT}(M_{\text{susy}}) \sim \left(\frac{M_{EW}^2 M_{\text{Pl}}^2}{M_{\text{susy}}^4} \right) \left(\frac{\Lambda}{M_{\text{susy}}^4} \right) f(M_{\text{susy}}), \qquad 30.$$

where the first factor represents the electroweak scale tuning and the second one the cosmological constant tuning, assuming a supersymmetric vacuum has cosmological constant zero (as is the case in rigid supersymmetric field theory). The factor $f(M)$ represents the a priori distribution coming out of the underlying theory, independent of these tuning requirements. If we grant that this is set by strong gauge dynamics, as in conventional field theory models of supersymmetry breaking, a reasonable ansatz may be $f(M) = dM/M$, analogous to Equation 28. This would lead to $\alpha = -9$ and a clear statistical prediction.

However, this leaves out all the fluxes and hidden sectors that were postulated in the Bousso-Polchinski model and that are generic in actual string theory compactifications. Because the expression in Equation 29 is a sum of squares, a simple model for their effect is that M_{susy} receives many independent positive contributions, about as many as there are fluxes or at least moduli, leading to prior distributions of the form (59, 60)

$$f(M_{\text{susy}}) \sim dM_{\text{susy}}^{4K}. \qquad 31.$$

This would lead to a large positive value for α and overwhelmingly prefer high-energy supersymmetry breaking.

Thus, depending on our microscopic picture, we arrive at very different conclusions. However, both of these simple considerations have flaws. The simplest problem with Equation 30 is the factor of $\Lambda/M_{\text{susy}}^4$. Instead, cosmological constant distributions of flux vacua generally go as Λ/M_F^4, with $M_F = M_{KK}$, $M_F = M_P$, or some other fundamental scale. In other words, the tuning problem of the cosmological constant is not helped by supersymmetry. Essentially, the reason is that in supergravity, the effective potential receives both positive and negative contributions, with negative contributions $\sim -|W|^2$ persisting for supersymmetric vacua. In the case of flux vacua, these negative contributions are distributed roughly uniformly up to some fundamental scale M_F^4, independent of the supersymmetry-breaking scale, as discussed at the end of Section 3.3. This leads to a tuning factor of Λ/M_F^4 instead of $\Lambda/M_{\text{susy}}^4$ in Equation 30, resulting in $\alpha = -5$, still favoring low-scale supersymmetry but rather less so.

There are many other implicit assumptions and even gaps in the arguments for Equation 30, many already recognized in the literature. Among those that implicitly favor low-scale breaking, the expression in Equation 30 assumes a generic solution to the μ problem, as well as to the other known problems of supersymmetric phenomenology such as flavor-changing neutral currents (FCNCs) and so on. As an example on the other side, there may exist some generic class of models in which the supersymmetric contributions to W are forced to be small, say by postulating an R symmetry, which is broken only along with supersymmetry breaking, restoring the $\Lambda/M_{\text{susy}}^4$ factor.

Let us however turn to the arguments leading to Equation 31, which if true would potentially outweigh all these other considerations. To examine this further, we need a microscopic model of supersymmetry breaking. In fact, one can expect a generic potential to contain many metastable supersymmetry-breaking minima, not because of any particular mechanism but simply because generic functions have many minima. Indeed, this was shown to be generic for IIB flux vacua in Reference 53, leading to the distribution

$$f(M_{\text{susy}}) \sim \left(\frac{M_{\text{susy}}}{M_F} \right)^{12} \qquad 32.$$

in the regime $M_{\text{susy}} \ll M_F$. This would still favor high supersymmetry-breaking scales in Equation 30, but much less so than in Equation 31. The flaw in the argument for the latter is that the different contributions to $F^2 = \sum_a |F_a|^2$ are *not* independent but correlated by the critical point condition $\partial_a V = 0$.

Although the specific power 12 may be surprising at first, it has a simple explanation (61, 62). Consider a generic flux vacuum with $M_{\text{susy}} \ll M_F$. Because one needs a goldstino for spontaneous supersymmetry breaking, at least one chiral superfield must have a low mass, call it ϕ. Generically, the flux potential gives order M_F masses to all the other chiral superfields, so they can be ignored and we can analyze the constraints in terms of an effective superpotential reduced to depend on the single field ϕ:

$$W(\phi) = W_0 + a\phi + b\phi^2 + c\phi^3 + \dots. \qquad 33.$$

The conditions for a metastable supersymmetric vacuum are then $|a| = M_{\text{susy}}^2$ and $|b| = 2|a|$ (this follows from the equation $V' = 0$), and finally $|c| \sim |a|$ (as explained in Reference 53 and many previous discussions). This is necessary so that $V'' > 0$. This also requires a lower bound on the curvature of the moduli space metric.

Furthermore, an analysis of flux superpotentials along the lines of our previous discussion bears out the expectation that the parameters (a, b, c) are independent and uniformly distributed complex parameters, up to the flux potential cutoff scale M_F. To get low-scale breaking, all three complex parameters must be tuned to be small in magnitude, leading directly to Equation 32.

The flaw in applying a standard naturalness argument here is very simple; one needs to tune several parameters in the microscopic theory to accomplish a single tuning at the low scale. Of course, if the underlying dynamics correlated these parameters, one could recover natural low-scale breaking. Thus, we have not replaced the paradigm of naturalness, but rather sharpened and extended it. Even if we granted that theories with such dynamics exist, the question becomes whether, among the many possibilities contained in the string theory landscape, the vacua realizing them are sufficiently numerous to dominate the simpler fine-tuning scenario. After all, the fine-tuning we are trying to explain is only of order $(100\,\text{GeV}/M_P)^2 \sim 10^{-34}$, and for all we know the fraction of string theory vacua that do this by low-scale supersymmetry breaking is even smaller. Although it will probably be some time before we can convincingly answer such questions, thinking about them has already shed new light on many old problems.

Having a precise microscopic picture from string theory becomes particularly important when one is evaluating the naturalness of discrete choices, or trying to weigh the importance of competing effects. For example, let us consider the possibility that some of the problems we discussed (such as large $|W|$ leading to Λ/M_F^4) could be solved by postulating a discrete R symmetry. Indeed, almost all existing proposals for natural models of low-scale breaking, such as those discussed in Reference 63, rely on this postulate.

Unfortunately for such proposals, there is a simple argument that discrete R symmetry is heavily disfavored in flux vacua (64). First, a discrete symmetry that acts on Calabi-Yau moduli space will have fixed points corresponding to particularly symmetric Calabi-Yau manifolds; at one of these points, it acts as a discrete symmetry of the Calabi-Yau. Such a symmetry of the Calabi-Yau will also act on the fluxes, trivially on some and nontrivially on others. To get a flux vacuum respecting the symmetry, one must turn on only invariant fluxes. Looking at examples, one finds that typically an order-one fraction of the fluxes transforms nontrivially; for definiteness, let us say half of them. Thus, applying Equation 27 and putting in some typical numbers for definiteness, we might estimate

$$\frac{N_{\text{vac symmetric}}}{N_{\text{vac all}}} \sim \frac{L^{K/2}}{L^K} \sim \frac{10^{100}}{10^{200}}.$$

Thus, discrete symmetries of this type come with a huge penalty. Although one can imagine discrete symmetries with other origins for which this argument may not apply, because W receives flux contributions, it clearly applies to the R symmetry desired in branch (3) and probably leads to suppressions far outweighing the potential gains. (Fortunately, this does not apply to R parity.)

This is in stark contrast to traditional naturalness considerations, in which all symmetries are natural. There are other examples of string vacuum distributions that come out differently from traditional expectations, or that are different among different classes of string theory vacua, as discussed in Reference 32 and 49 and references therein. Other distributions are simply not predicted at all by traditional naturalness arguments. A primary example is the distribution of gauge groups and charged matter content. Suppose we were to find evidence for a new strongly coupled gauge sector (perhaps responsible for supersymmetry breaking, perhaps not), but had very limited information about the matter spectrum, say a single resonance. What should we expect for the gauge group? Simple guesses may be $SU(2)$ (the smallest non-Abelian group) or $SU(4)$ (following the pattern $1 - 2 - 3 - \ldots$). Alternatively, a huge amount of theoretical work has been devoted to the proposition that $SU(N)$ gauge theory becomes simpler as N becomes large. Should we not give this intuition equal weight?

Although the considerations we just cited are unconvincing, string theory does predict a definite distribution of gauge theories and matter contents, which has been explored in numerous recent works, including References 65–70 and many more cited in the reviews. Besides bearing on the supersymmetry-breaking scale, such results could be useful in guiding searches for exotic matter, in motivating other proposals for dark matter, and so on.

4. IMPLICATIONS FOR THE TESTABILITY OF STRING THEORY

As of 2007, it seems fair to say that although string theory remains by far the best candidate we have for a complete theory of fundamental physics, there is still no compelling empirical evidence for or against the claim that the theory describes our Universe.

Many different approaches have been proposed to find such evidence. Perhaps the simplest, and certainly the best founded in the history of our subject, is to look for exotic or signature physics, which is easily modeled by string theory and not by other theories. There are many such phenomena, such as excitation states of the string (Regge recurrences). Each is associated with a new energy scale, for example, the string scale M_s for excited string states. As another example, in higher-dimensional theories, once one reaches energies of order of the higher-dimensional Planck scale M_P, one can have black hole creation. Presumably, Planck mass black holes decay very rapidly, and this process can be thought of as creating an unstable particle. However, one can argue that the resulting distribution of decay products will look very different from other interactions. In addition, the higher-dimensional Planck scale can in principle be far lower than the 4D Planck scale, perhaps low enough to make this observable (for a recent review, see Reference 71).

One can add the Kaluza-Klein scale M_{KK} to the list. Although, logically speaking, the observation of extra dimensions is not a direct test of string theory, clearly it would have an equally profound significance for fundamental physics. There are a number of other exotic phenomena of which one can say the same thing, such as the possibility of nontrivial 4D fixed-point field theories and phenomena related to warping. Again, it is not difficult to propose scenarios in which this new physics will not be seen until we reach a new energy scale M_{EX}. Thus, the larger problem is to decide whether any of these phenomena are relevant in our Universe.

The most direct approach to testing the theory is to find a way to probe these energy scales. At present, phenomenological constraints on all these new energy scales appear to be very weak, ranging from just above current collider bounds, around a few TeV, all the way up to the grand unified theory (GUT) and 4D Planck scales $M \geq 10^{16}$ GeV. In most of this parameter space, the exotic phenomena may well be inaccessible in terrestrial experiments and irrelevant in almost all astrophysical processes. Thus, although string theory can offer experimentalists many exciting possibilities, there is little in the way of guarantees, nor any clear way for such searches to falsify the theory.

Thus, a central task for string theorists is to better constrain these scales theoretically. Because parameters such as the size of the extra dimensions are moduli, the considerations discussed in our review are clearly very relevant for this, and there are already many interesting suggestions. For example, it appears that in KKLT IIB flux vacua with many moduli, large extra dimensions are disfavored. However, there is an alternate regime in IIB theory in which the structure of the effective potential requires large extra dimensions (72), and it appears that other constructions such as IIa and string theory flux vacua may statistically favor large extra dimensions (40, 73).

Thus, the picture at present is not very clear; furthermore it seems very likely a priori that considerations from early cosmology bear on this particular question. However, it is reasonable to expect significant theoretical progress on this question.

Again, this progress is likely to lead to statistical predictions in the sense that even if most vacua are shown to have some property (say for sake of discussion, large extra dimensions), there will be exceptions. To make a perhaps evident comment on the value of this, although in the hypothetical situation we discuss one would not be able to say that ruling out large extra dimensions would falsify string theory, one would at least know that one had drastically narrowed down the possibilities, allowing one to go on to determine the next most promising avenues for potential tests. In this way, getting a picture of the landscape is useful and perhaps even necessary in guiding the search for conclusive tests.

Let us now turn to the question of testability if we do not see exotic physics. To be more precise, suppose that all observed physics can be well described by some 4D effective field theory coupled to gravity. This is certainly true at present, and it may well turn out so for physics at 1 TeV and even 10 TeV as well. Although this would seem a frustrating possibility, one can certainly still hope to make contact with string theory from such data. After all, we believe that string theory has a finite number of vacua (74) and thus can lead to a finite number of 4D low-energy theories. Could we imagine showing that the data is fit by none of these theories, thus falsifying the theory?

Approaches of this type include the following:

1. One could try to find arguments that certain phenomena, which can be described by effective field theory, in fact cannot arise in string theory. For example, one can argue this for the time variation of the fine structure constant (75). One can also place bounds on gauge couplings (admittedly, far from the observed values) (76).

2. Similarly, we could try to use the phenomena to rule out competing theories, thereby getting circumstantial evidence for string theory. The basic example here is that, at present, there is no other generally accepted theory of 4D quantum gravity, and this is commonly taken as circumstantial evidence for string theory. Of course, one should not take this too seriously until it can be proven that alternatives do not exist. In our opinion, currently the most promising alternative to study is the idea that certain extended supergravity theories may provide finite theories of gravity (77, 78).

3. Perhaps physics at the few TeV scale of the LHC and the International Linear Collider (ILC) will show some remarkable simplicity that can easily be reproduced by string theory compactification. One idea is to focus on properties of the particular GUT theories obtained by heterotic string compactification. More recently, researchers have suggested that certain D-brane quiver gauge theories leading to Standard Model–like models are preferred (79, 80).

4. One could try to make statements about the likely distribution of predictions among all vacua of string theory. Then, to the extent that what we see is likely, we again obtain circumstantial evidence for string theory.

Again, approaches 1, 2, and 3 suffer from the general problem that the phenomena being discussed may not actually be properties of our Universe, whereas approach 4 suffers from the problem that we may just live in an unlikely Universe. Thus, one would probably need to combine information from all these approaches to make progress.

A more optimistic version of approach 4 holds that a better understanding of early cosmology and whatever mechanism populates the many vacua of the theory will lead to a strongly peaked probability distribution that selects one or a handful of the candidate vacua in a way amenable to calculation. The search for gauge-invariant and well-defined inflationary measures has been a 20-year struggle; recent progress is summarized in Reference 81. Note that should one find a natural and computable candidate measure, actually finding the preferred vacua may still be a daunting problem. It is not hard to imagine scenarios in which this is impossible, even in principle, because of fundamental limitations coming from the theory of computational complexity (82).

It is difficult at present to judge the prospects for any of these approaches. From thinking about historical analogies, a tentative conclusion for the problem at hand is that although a great deal can be learned on the theoretical side—perhaps eventually allowing us to propose a definite test—ultimately convincing evidence for string theory will probably have to come from observing some sort of exotic physics. A natural place to look for this is early cosmology, as the physics of inflation involves very high energies, with $V \sim M_{GUT}^4$ in many models. Several of the proposed models of inflation in string theory have characteristic signatures that (if sufficiently well measured) encode stringy physics. These include networks of cosmic D and F strings formed during the exit from brane inflation (83–85) and non-Gaussian signals in the cosmic microwave background radiation, which probe the specific nonlinearities of the Dirac-Born-Infeld (DBI) action (86–88).

Space prohibits a detailed discussion of these and many other interesting ideas. We conclude by noting that although the present situation is not very satisfactory, there is every reason to be optimistic. In string theory, we have a theoretical framework that on the one hand is grounded in precise mathematics (so that many, even most, theoretical suggestions can be falsified on internal grounds), yet on the other hand shows significant promise of making contact with observable physics. There are many well-motivated directions for improving the situation and good reasons to believe that substantial progress will be made in the future.

DISCLOSURE STATEMENT

The authors are not aware of any biases that might be perceived as affecting the objectivity of this review.

ACKNOWLEDGMENTS

The work of M.R.D. was supported by the Department of Energy grant number DE-FG02-96ER40959, and the work of S.K. was supported in part by a David and

Lucile Packard Foundation Fellowship, the Department of Energy under contract number DE-AC02-76SF00515, and the National Science Foundation under grant number 0244728. The work of F.D. was supported in part by the Belgian Federal Office for Scientific, Technical and Cultural Affairs through the Interuniversity Attraction Poles Programme P5/27, and by the European Community's Human Potential Programme under contract number MRTN-CT-2004-005104.

LITERATURE CITED

1. Scherk J, Schwarz JH. *Phys. Lett. B* 52:347 (1974)
2. Schwarz JH. *Phys. Rep.* 89:223 (1982)
3. Green MB, Schwarz JH. *Phys. Lett. B* 149:117 (1984)
4. Candelas P, Horowitz GT, Strominger A, Witten E. *Nucl. Phys. B* 258:46 (1985)
5. Dine M, Seiberg N. *Phys. Lett. B* 162:299 (1985)
6. Sen A. In *Fundamental Physics: Gravity, Gauge Theory And Strings*, ed. C Bachas, et al., p. 241. Berlin: Springer (2002)
7. Intriligator KA, Seiberg N. *Nucl. Phys. Proc. Suppl.* 45BC:1 (1996)
8. Weinberg S. *Rev. Mod. Phys.* 61:1 (1989)
9. Weinberg S. astro-ph/0005265 (2000)
10. Chanowitz MS. *Annu. Rev. Nucl. Part. Sci.* 38:323 (1988)
11. Adelberger EG, Heckel BR, Nelson AE. *Annu. Rev. Nucl. Part. Sci.* 53:77 (2003)
12. Banks T, Kaplan DB, Nelson AE. *Phys. Rev. D* 49:779 (1994)
13. de Carlos B, Casas JA, Quevedo F, Roulet E. *Phys. Lett. B* 318:447 (1993)
14. Coughlan GD, et al. *Phys. Lett. B* 131:59(1983)
15. Peebles PJE, Ratra B. *Rev. Mod. Phys.* 75:559 (2003)
16. Banks T. *Phys. Rev. Lett.* 52:1461 (1984)
17. Linde AD. *Rep. Prog. Phys.* 47:925 (1984)
18. Weinberg S. *Phys. Rev. Lett.* 59:2607 (1987)
19. Lerche W, Lust D, Schellekens AN. *Nucl. Phys. B* 287:477 (1987)
20. Bousso R, Polchinski J. *JHEP* 0006:006 (2000)
21. Gukov S, Vafa C, Witten E. *Nucl. Phys. B* 608:477 (2001)
22. Dasgupta K, Rajesh G, Sethi S. *JHEP* 9908:023 (1999)
23. Mayr P. *Nucl. Phys. B* 593:99 (2001)
24. Curio G, Klemm A, Lust D, Theisen S. *Nucl. Phys. B* 609:3 (2001)
25. Giddings SB, Kachru S, Polchinski J. *Phys. Rev. D* 66:106006 (2002)
26. Silverstein E. hep-th/0106209 (2001)
27. Kachru S, Kallosh R, Linde A, Trivedi SP. *Phys. Rev. D* 68:046005 (2003)
28. Denef F, Douglas MR, Florea B. *JHEP* 0406:034 (2004)
29. Denef F, et al. hep-th/0503124 (2005)
30. Lust D, et al. hep-th/0609013 (2006)
31. Susskind L. hep-th/0302219 (2003)
32. Douglas MR, Kachru S. hep-th/0610102 (2006)
33. Blumenhagen R, Kors B, Lust D, Stieberger S. hep-th/0610327 (2006)
34. Grana M. *Phys. Rep.* 423:91 (2006)
35. Silverstein E. hep-th/0405068 (2004)

36. Frey AR. hep-th/0308156 (2003)
37. Freund PGO, Rubin MA. *Phys. Lett. B* 97:233 (1980)
38. McGreevy J, Silverstein E, Starr D. hep-th/0612121 (2006)
39. Saltman A, Silverstein E. *JHEP* 0601:139 (2006)
40. DeWolfe O, Giryavets A, Kachru S, Taylor W. *JHEP* 0507:066 (2005)
41. Villadoro G, Zwirner F. *JHEP* 0506:047 (2005)
42. Camara PG, Font A, Ibanez LE. *JHEP* 0509:013 (2005)
43. Dixon LJ, Harvey JA, Vafa C, Witten E. *Nucl. Phys. B* 261:678 (1985)
44. Banks T, van den Broek K. hep-th/0611185 (2006)
45. Berkooz M, Douglas MR, Leigh RG. *Nucl. Phys. B* 480:265 (1996)
46. Blumenhagen R, Cvetic M, Langacker P, Shiu G. *Annu. Rev. Nucl. Part. Sci.* 55:71 (2005)
47. Intriligator K, Seiberg N, Shih D. *JHEP* 0604:021 (2006)
48. Ooguri H, Ookouchi Y. *Phys. Lett. B* 641:323 (2006)
49. Kumar J. *Int. J. Mod. Phys. A* 21:3441 (2006)
50. Douglas MR. *JHEP* 0305:046 (2003)
51. Ashok S, Douglas MR. *JHEP* 0401:060 (2004)
52. Denef F, Douglas MR. *JHEP* 0405:072 (2004)
53. Denef F, Douglas MR. *JHEP* 0503:061 (2005)
54. Klebanov IR, Strassler MJ. *JHEP* 0008:052 (2000)
55. Randall L, Sundrum R. *Phys. Rev. Lett.* 83:3370 (1999)
56. Kachru S, Pearson J, Verlinde HL. *JHEP* 0206:021 (2002)
57. Kachru S, et al. *JCAP* 0310:013 (2003)
58. Hebecker A, March-Russell J. hep-th/0607120 (2006)
59. Douglas MR. hep-th/0405279 (2004)
60. Susskind L. hep-th/0405189 (2004)
61. Dine M, O'Neil D, Sun Z. *JHEP* 0507:014 (2005)
62. Giudice GF, Rattazzi R. *Nucl. Phys. B* 757:19 (2006)
63. Dine M, Feng JL, Silverstein E. *Phys. Rev. D* 74:095012 (2006)
64. Dine M, Sun Z. *JHEP* 0601:129 (2006)
65. Blumenhagen R, et al. *Nucl. Phys. B* 713:83 (2005)
66. Dijkstra TPT, Huiszoon LR, Schellekens AN. *Nucl. Phys. B* 710:3 (2005)
67. Gmeiner F. *Fortsch. Phys.* 54:391 (2006)
68. Douglas MR, Taylor W. hep-th/0606109 (2006)
69. Kumar J, Wells JD. *JHEP* 0509:067 (2005)
70. Dienes KR. *Phys. Rev. D* 73:106010 (2006)
71. Landsberg G. *J. Phys. G* 32:R337 (2006)
72. Balasubramanian V, Berglund P, Conlon JP, Quevedo F. *JHEP* 0503:007 (2005)
73. Acharya BS, Denef F, Valandro R. *JHEP* 0506:056 (2005)
74. Acharya BS, Douglas MR. hep-th/0606212 (2006)
75. Banks T, Dine M, Douglas MR. *Phys. Rev. Lett.* 88:131301 (2002)
76. Arkani-Hamed N, Motl L, Nicolis A, Vafa C. hep-th/0601001 (2006)
77. Bern Z, Dixon LJ, Roiban R. hep-th/0611086 (2006)
78. Green MB, Russo JG, Vanhove P. hep-th/0611273 (2006)
79. Berenstein D, Pinansky S. hep-th/0610104 (2006)

80. Verlinde HL, Wijnholt M. hep-th/0508089 (2005)
81. Vilenkin A. hep-th/0609193 (2006)
82. Denef F, Douglas MR. hep-th/0602072 (2006)
83. Sarangi S, Tye SHH. *Phys. Lett. B* 536:185 (2002)
84. Copeland EJ, Myers RC, Polchinski J. *JHEP* 0406:013 (2004)
85. Dvali G, Vilenkin A. *JCAP* 0403:010 (2004)
86. Alishahiha M, Silverstein E, Tong D. *Phys. Rev. D* 70:123505 (2004)
87. Babich D, Creminelli P, Zaldarriaga M. *JCAP* 0408:009 (2004)
88. Chen X, Huang M, Kachru S, Shiu G. hep-th/0605045 (2006)

Systematic Errors

Joel Heinrich[1] and Louis Lyons[2]

[1] Department of Physics and Astronomy, University of Pennsylvania, Philadelphia, Pennsylvania 19104; email: heinrich@hep.upenn.edu

[2] Department of Physics, University of Oxford, Oxford OX1 3RH, United Kingdom; email: l.lyons@physics.ox.ac.uk

Annu. Rev. Nucl. Part. Sci. 2007. 57:145–69

First published online as a Review in Advance on May 2, 2007

The *Annual Review of Nuclear and Particle Science* is online at http://nucl.annualreviews.org

This article's doi:
10.1146/annurev.nucl.57.090506.123052

Copyright © 2007 by Annual Reviews.
All rights reserved

0163-8998/07/1123-0145$20.00

Key Words

systematics, nuisance parameters, confidence interval, *p*-values, Bayes, frequentism

Abstract

To introduce the ideas of statistical and systematic errors, this review first describes a simple pendulum experiment. We follow with a brief discussion of the Bayesian and frequentist approaches. Two widely used applications of statistical techniques in particle physics data include extracting ranges for parameters of interest (e.g., mass of the *W* boson, cross section for top production, neutrino mixing angles, etc.) and assessing the significance of possible signals (e.g., is there evidence for Higgs boson production?). These two topics are first discussed in the absence of systematics, and then methods of incorporating systematic effects are described. We give a detailed discussion of a Bayesian approach to setting upper limits on a Poisson process in the presence of background and/or acceptance uncertainties. The relevance of the choice of priors and how this affects the coverage properties of the method are described.

Contents

1. INTRODUCTION 146
    1.1. A Trivial Example: The Simple Pendulum 146
    1.2. Structure of this Review 148
2. BAYES AND FREQUENTISM.................................... 148
    2.1. Probability.. 148
    2.2. Bayes' Theorem... 149
    2.3. Bayesian Priors... 149
    2.4. Bayesian Intervals 150
    2.5. Frequentist Approach.................................... 150
    2.6. Comparison of Bayes and Frequentist Approaches............... 152
    2.7. Other Methods... 153
3. p-VALUES 153
    3.1. What p-Values Are 153
    3.2. Discrete Data 154
    3.3. What p-Values Are Not.................................. 155
4. METHODS OF INCLUDING SYSTEMATICS.................... 155
    4.1. Parameter Estimation 155
    4.2. p-Values 159
5. FULLY BAYESIAN CROSS-SECTION UPPER LIMITS........... 162
6. MISCELLANEOUS.. 164
    6.1. Blind Analyses 164
    6.2. Separating Signal from Background 164
    6.3. Simulation Issues.. 164
    6.4. Theory Uncertainties..................................... 165
    6.5. Fits with Correlated Systematics 166
    6.6. Asymmetric Errors 167
    6.7. The Best Linear Unbiased Estimate 167
7. CONCLUSIONS 167

שגיאות מי יבין מנסתרות נקני

Who can understand errors? Save me from hidden effects.
Psalm 19:13 $\pm$ 1

1. INTRODUCTION

1.1. A Trivial Example: The Simple Pendulum

The first experiment that many physics students perform at school is to measure g, the acceleration due to gravity, using a simple pendulum of length l. The time T for n swings is measured to determine the period $\tau = T/n$, and then g is $4\pi^2 l/\tau^2$. We use this example to introduce the basic ideas of statistical and, especially, systematic errors, the latter being the focus of this review. [See References 1–6 for previous reviews and notes on systematics, and the many research articles in the PHYSTAT series of conferences and workshops (7–12).]

Contributions to the statistical error on g arise from the limited accuracy with which we can measure T and l. These can be assessed from the measuring instruments, the quantities to be measured, and the experimentalists. Because the result will vary for repetitions of the measurement, the estimated statistical error can be checked with the standard deviation of a large set of independent measurements. Furthermore, because the statistical errors on a series of independent measurements are uncorrelated, their effect can be reduced by combining the results of such measurements. The usefulness of this reduction in statistical error may be limited by the fact that the systematic error does not automatically go down.

The measured quantities may also have systematic uncertainties, which could arise from uncertainties in the calibration of the clock and ruler used to measure T and l, respectively. These quantities can be determined by performing our own calibration in some subsidiary experiment. For example, we could use the ruler to estimate the length of an object, whose size is precisely known. In doing so we hope to have reduced the systematic error and replaced it by the smaller uncertainty of how well we know the new calibration of our ruler. We return to a more detailed consideration of subsidiary experiments in Sections 4.1.4, 4.1.5, and, especially, 5.

There are, however, other sources of possible systematics:

- The formula quoted for τ applies to *undamped* oscillations of a *small amplitude* of a simple pendulum, which consists of a *point mass* suspended by a *massless*, *extensionless string* from a *rigid support*. All italicized items are only approximated in a real experiment, and so corrections must be estimated for them or the result will be biased. The uncertainties in these corrections are systematics.

- We may be interested in the value of g at sea level, rather than at the elevation at which we performed the measurement. This requires a correction that depends not only on our elevation, but also on the material below us. We may have access to other people's estimate of this correction, but its uncertainty is a systematic. It is also possible that more than one such estimate exists, in which case we must decide how to combine them and how to deal with any discrepancies among the estimates.

The nature of systematic effects is such that they may not cause different answers when the experiment is repeated. Thus, a consistent set of results does not imply the absence of systematics. Furthermore, as is already apparent from the pendulum example, systematic effects occur not only on directly measured quantities. Thus, in general, the reliable assessment of systematics requires much more thought and work than for the corresponding statistical error.

Some errors are clearly statistical (e.g., those associated with the reading errors on T and l), and others are clearly systematic (e.g., the correction of the measured g to its sea level value). Others could be regarded as either statistical or systematic (e.g., the uncertainty in the recalibration of the ruler). Our attitude is that the type assigned to a particular error is not crucial. What is important is that possible correlations with other measurements are clearly understood.

The result of the experiment may be quoted as $g \pm \sigma_{stat} \pm \sigma_{syst}$, where the statistical and systematic errors are shown separately. If a single error is required, then typically σ_{stat} and σ_{syst} are combined in quadrature, on the grounds that they are uncorrelated. At the other extreme, some or all of the systematic errors can be shown individually. This would be useful in combining different measurements, for which some of the systematic effects may be correlated between the different measurements. Also, it is possible that in the future there may be an improvement in some relevant external information. (In the pendulum example, this could be the correction to sea level.) The way this reduces the systematic error can readily be assessed if this particular contribution is quoted separately.

The relevance of correlations can be seen by extending the above example to consider measuring the ratio of the g-values at two different locations. Then, for example, calibration errors in the T measurements at the two locations could cancel.

The examples discussed below are basically extensions of this simple case, but include the effects of non-Gaussian errors. Also the Poisson distribution, with its discrete observations, usually plays an important role in particle physics analyses.

1.2. Structure of this Review

The differences in the Bayesian and frequentist philosophies are described in Section 2, as is the construction of credible or confidence intervals for parameters. Section 3 deals with p-values for assessing the significance of any potential discoveries. These two sections largely avoid discussion of systematics. The various ways of including these effects for intervals and for p-values are considered in Section 4. A detailed example of calculating upper limits can be found in Section 5, while the last two sections contain a few miscellaneous items and the conclusions. The Supplemental Material (follow the Supplemental Material link from the Annual Reviews home page at **http://www.annualreviews.org**) contains a glossary.

2. BAYES AND FREQUENTISM

2.1. Probability

To a Bayesian, probability is interpreted as the degree of belief in a statement. It can vary from person to person because they can have different information about a situation. It is quantified by the fair bet: A Bayesian should be prepared to accept bets in either direction, with odds determined by the numerical value he assigns for the probability.

In contrast, frequentists define probability via a repeated series of almost identical trials; it is the limit of the fraction of successes as the number of trials tends to infinity Thus, frequentists will not assign a probability to a one-off event (e.g., will the first astronaut to Mars return to Earth alive?) or to the value of a physical constant (e.g., is the value of the strong coupling constant between 0.110 and 0.115?).

2.2. Bayes' Theorem

Bayes' theorem is derived from $P(A$ and $B)$, the probability that two events A and B both happen. Then

$$P(A \quad \text{and} \quad B) = P(A|B)P(B) = P(B|A)P(A), \qquad 1.$$

where $P(B)$ is the probability of B happening, while $P(A|B)$ is the conditional probability of A happening given that B has occurred. For example, A could be the probability that a high-energy proton-proton collision contains a top quark, and B the probability that it contains a W boson. Then Bayes' theorem

$$P(A|B) = P(B|A)P(A)/P(B) \qquad 2.$$

relates $P(A|B)$ to $P(B|A)$. Frequentists regard Bayes' theorem as completely noncontroversial, provided that the probabilities occurring in it are acceptable frequentist probabilities.

The dispute occurs when Bayesians choose B as data, and A as one or more parameter values. This then gives

$$p(\mu|x) \sim p(x|\mu)\pi(\mu), \qquad 3.$$

where the implied constant of proportionality is simply the normalization constant $1/\int p(x|\mu)\pi(\mu)\,d\mu$; the likelihood $p(x|\mu)$ is derived from the probability-density function (*pdf*) for obtaining measurements x, given the parameter μ; and the Bayes prior $\pi(\mu)$ specifies the assumed probability density for μ, before the experiment was performed. Here we use P to denote probabilities of discrete variables, and p or π for probability densities of continuous variables. In contrast $p(\mu|x)$ is the Bayesian posterior for μ and gives the probability density for μ after the data are obtained; it is determined by both the likelihood function and the prior. Bayes' theorem thus provides a way of using the data from our experiment to update our prior knowledge about a parameter.

All this is unacceptable for frequentists because they would object to assigning a probability distribution to a physical parameter μ. A Bayesian would counter this by explaining that the frequentist view of probability is too narrow and that Bayes' theorem should be interpreted in terms of degrees of belief.

2.3. Bayesian Priors

A practical problem arises in the Bayesian approach because of the need to choose the prior $\pi(\mu)$. If the value of the parameter had been well determined in some previous measurement as $a \pm b$, $\pi(\mu)$ could perhaps be a Gaussian distribution centered on a and with variance b^2. However, if we are looking for some previously unobserved process, it is more likely that very little is known prior to our experiment. Then we need to choose $\pi(\mu)$ to express our relative ignorance about μ.

An unreasonably popular choice is the uniform distribution

$$\pi(\mu) = \text{constant}, \qquad 4.$$

as it favors no particular value of μ. However, this prior implies that the range from 0 to 1 for μ is only as likely as that from 176,391.3 to 176,392.3. Another feature of a uniform prior over an infinite range is that it cannot be normalized.

Furthermore, although there does seem to be something natural in using a uniform distribution as an uninformative prior, the choice is in fact far from obvious. Priors uniform in μ^2, $\ln(\mu)$, or $1/\sqrt{\mu}$, and so on may be equally plausible and are different from our initial choice. There is thus arbitrariness in how ignorance is parametrized, and this will affect our posterior probability distribution.

The problem is exacerbated in several dimensions, with uniform priors having more undesirable properties. For example, in an analysis involving several different final states, each of which has its own acceptance A_i, the total acceptance $A_{tot} = \sum A_i$. Then priors that are uniform in each A_i (for positive A_i) correspond to a prior that grows with A_{tot}, and it is not obvious that this is what was intended (see Section 5).

The variety of Bayesians includes those who prefer subjective priors or objective ones. The former are happy that a different prior could be chosen by each scientist, as this encapsulates their varying degrees of knowledge and beliefs. Objective Bayesians try to find priors with desirable theoretical properties. Examples include the Jeffreys priors $1/\mu$ and $1/\sqrt{\mu}$ for specific cases.

A positive feature of the Bayesian approach is that it is simple to incorporate the fact that part of the infinite range for μ may be unphysical (e.g., a reaction rate or the mass of a particle should not be negative). Whatever functional form for $\pi(\mu)$ is used in the allowed region, it is set equal to zero where μ is unphysical. This ensures that any Bayesian interval for a parameter will always be completely physical.

2.4. Bayesian Intervals

The output of using Bayes' theorem is $p(\mu|x)$, the posterior probability distribution for the parameter of interest. This is supposed to contain a complete summary of what is known about μ. From it, various intervals at a given credible level α can be extracted. These could include, for example, upper or lower limits, central intervals [with probability $(1 - \alpha)/2$ on each side], or those defined by the largest probability density or some other ordering rule. Of all intervals at level α, those selected according to highest probability density have the shortest length in μ. However, they are not invariant with respect to nonlinear transformation of the parameter μ, for example, to μ^2 or to $1/\mu$. If there are several parameters of interest, the posterior $p(\mu_1, \mu_2, \ldots)$ can similarly be used to define a region in the space of the parameters at a credible level α.

2.5. Frequentist Approach

2.5.1. The Neyman construction.
The frequentist or classical approach to parameter determination is very different from the Bayesian one. It uses only $p(x|\mu)$ and never considers a prior $\pi(\mu)$ or posterior $p(\mu|x)$.

Figure 1 illustrates the construction of frequentist confidence intervals for a single non-negative parameter μ and a measurement x that depends on μ. For example, μ could be the temperature at the center of the Sun, and x the measured flux of solar

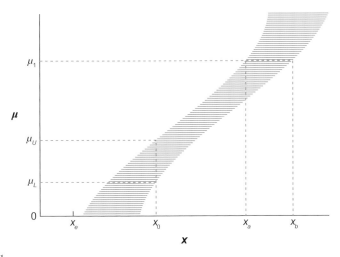

Figure 1

The Neyman construction for a confidence band at probability level α. It makes use of the *pdf* $p(x|\mu)$, which specifies the probability density for a measurement x for any value of the non-negative parameter μ. For the parameter value μ_1, there is a probability α of the measurement being in the region from x_a to x_b. This procedure is repeated for all possible μ, to build up the confidence belt (*shaded area*). For a particular measurement x_0, the values μ_L and μ_U at the edges of the confidence belt are found. The confidence interval from μ_L to μ_U thus specifies the values of μ for which our measurement x_0 is likely (at level α). For the shown confidence belt, a smaller measurement x_e would result in an empty confidence interval; there are no values of μ for which x_e is likely.

neutrinos, as determined by running a large neutrino detector for a month. We assume we know the *pdf* $p(x|\mu)$ for all physical μ. For a given μ, such as μ_1 in **Figure 1**, we can use the *pdf* to construct an interval from x_a to x_b such that the probability of x falling in this range is α, for example, 0.9. When x is continuous, it is possible to choose an x range such that the included probability is exactly α. This will not be so in general when x is discrete, in which case the conservative choice of having the probability at least equal to α is made.

The choice of a range in x is then repeated for every possible μ until the shaded area in **Figure 1** is built up. This shows likely measured results for any μ. Now assume we perform a measurement and obtain x_0. By finding where a vertical line at x_0 cuts the shaded region, limits μ_L and μ_U are established for μ. These then specify the range of values for μ for which x_0 is a likely result, as favored by the ordering rule (see Sections 2.5.2, 2.5.3, and 4.1.5).

Thus, the frequentist statement that $\mu_L \leq \mu \leq \mu_U$ at the 90% confidence level is a probability statement about μ_L and μ_U; if the measurement was repeated very many times and the range extracted for each of them, (at least) 90% of them should contain the true value of μ. This is known as coverage. This contrasts with the similar-looking statement by Bayesians. For them, μ_L and μ_U are determined in the particular experiment and are fixed. Then the statement specifies the percentage of the posterior probability distribution for μ within this range.

2.5.2. Feldman and Cousins' method.

The prescription for constructing a range in x for each μ is not unique in that there are infinitely many ranges that contain a fraction α of the *pdf*, and so a choice must be made among them; this is usually specified by an ordering rule. Thus, a central interval could have $(1 - \alpha)/2$ of the *pdf* on either side of the range, a range from x_L to $+\infty$ would be suitable for upper limits on μ, and so on. Feldman & Cousins (13) have exploited this freedom of choice, already noted by Neyman, to interesting effect.

In their unified approach to parameters with a physical bound (e.g., $\mu \geq 0$), Feldman and Cousins choose a likelihood-ratio ordering rule (see the Supplemental Material and Section 4.1.5), which results in favoring smaller values of the data x in the construction of the confidence interval. This has the consequence of greatly reducing the probability of having confidence intervals that contain only $\mu = 0$. Another feature of the Feldman-Cousins approach is that there is no longer the arbitrariness of choosing central intervals, upper limits, and so on (hence the name unified).

2.5.3. Several parameters.

In principle, the Neyman construction of **Figure 1** for one dimension of data and one physical parameter can be extended to cases where there are more data dimensions and/or parameters. For each possible combination of parameters, the region in data space at confidence level α is determined according to the chosen ordering rule. The actual data then serve to define a confidence region in the multiparameter space. This can then be projected down onto any given axis to produce a confidence interval for that parameter. This almost always results in overcoverage, which becomes more serious as the number of parameters increases. It thus becomes important to choose a good ordering rule so as to produce a region in parameter space that is narrow for the parameter of interest but elongated in the directions of the others. Because the ordering rule applies to the data at fixed parameter values, but the desired shape is in parameter space, this is nontrivial (14, 15). (See Section 4.1.5 for a fuller description.)

2.6. Comparison of Bayes and Frequentist Approaches

The main advantage of the frequentist method is that it does not require a prior and hence tends to be favored by particle physicists who want the data to speak for themselves and who prefer not to incorporate personal beliefs into their result. Another positive feature is that coverage is guaranteed. However, it becomes increasingly difficult computationally to perform the frequentist Neyman construction as the number of parameters and measurements increases.

Bayesians could argue that frequentists address the boring question about which values of the parameter are such that the data are likely, rather than the more interesting one about what parameter values we favor (after having performed the experiment and in light of all we previously knew about it). Bayesians also point out that frequentists need to define an ensemble of which their measurement is a member. Unfortunately, the choice of ensemble is often not unique (see the Supplemental Material and Reference 16).

Bayesian intervals are always physical, whereas frequentist ones can be empty. Frequentists have the freedom (or ambiguity) to choose the ordering rule, which results in different possible confidence ranges. This in some way mirrors the different credible intervals that can be extracted from a Bayesian posterior. Bayesian intervals can also be sensitive to the choice of the prior, especially for the case of upper limits.

2.7. Other Methods

Not all statistical procedures are Bayesian or frequentist. What distinguishes these two techniques is that they claim some self-consistent justification for their approach, whereas other methods are more ad hoc. Hence, the other possibilities do not usually achieve either coverage or Bayesian credibility.

Alternatives for parameter determination include the following:

- χ^2: This usually is used for a histogram of data. For large enough expected numbers of observations for each bin, the Gaussian approximation for the bin contents may be adequate. Then S, the weighted sum over all bins of squared deviations, behaves like χ^2 for the appropriate number of degrees of freedom. The best values of the parameters are determined by minimizing S, and regions in parameter space are defined by letting S increase from its minimum by the appropriate amount, depending on the effective number of free parameters and the required confidence level.
- Maximum likelihood: A confidence region for the parameters $p_1, p_2, \ldots, p_n$ involves determining a region in parameter space within a contour $\ln L(p_1, p_2, \ldots, p_n) = \ln L_{max}(p_1, p_2, \ldots, p_n) - k$, where k is an appropriately chosen constant. Alternatively, the likelihood could be profiled over several uninteresting parameters. Then if, for example, only one parameter p_1 remains, the profile likelihood is defined by $L_{prof}(p_1) = L(p_1, q_2, \ldots, q_n)$, where q_j are the values of p_j that maximize L for each particular p_1. Then the range corresponding to ± 1 error on p_1 is taken conventionally by finding where $\ln(L_{prof}(p_1))$ decreases by 0.5 compared with $\ln(L_{max})$. Adjusted likelihood methods (17) have been developed because the standard likelihood approach may not result in good coverage properties.
- Mixed methods: Because of the desire to use a frequentist method for the parameter of interest, and because of the practical difficulties of using a fully frequentist method for more than one or two nuisance parameters, Cousins & Highland (18) advocate using a mixed method, where a Bayesian approach is used for the systematics but the Neyman construction is used for the physics parameter. Their paper describes this technique for upper limits, but the mixed approach can be used in other contexts too. This is discussed further in Section 4.1.6.

3. p-VALUES

3.1. What p-Values Are

Both significance (e.g., claiming discovery of a new resonance) and goodness of fit (e.g., does the new resonance have a Breit-Wigner shape?) are traditionally quantified by p-values. A statistic t (a function of the observed data) is chosen, which becomes larger as the disagreement between the data and an assumed model increases. The p-value is the probability (when the assumed model is actually correct, the experiment is unbiased, etc.) that the value of t, over many repetitions of the experiment, will be greater than or equal to that observed in the data. (Because the calculation of a p-value involves an ensemble of experiments, the use of p-values is inherently non-Bayesian.) A small p-value is evidence that the assumed model does not match the data, casting doubt on any parameters derived from a fit using the model.

For example, if data x are expected to have a Gaussian distribution centered on zero and with unit variance, the p-value is simply the fractional area in the tail(s) beyond the measured x_0. (Whether the one-sided or two-sided tail area is considered depends on whether deviations in either direction are considered significant. Thus, in searching for oscillations of solar neutrinos, a reduced flux is expected, so only the lower tail is relevant.) The one-sided tail probability corresponding to $x_0 = 5.0$ is 3×10^{-7}. Often, in non-Gaussian situations, p-values are converted into the equivalent number of standard deviations for a Gaussian distribution, thereby providing a number that is easier to remember.

A second example is a data histogram with n_i events in the i-th bin, when μ_i is the expectation from our model. Pearson's χ^2 statistic

$$\chi^2_P(n_1, n_2 \ldots n_N) = \sum_{i=1}^{N} \frac{(n_i - \mu_i)^2}{\mu_i}$$

is often used to quantify goodness of fit. Here, μ_i is the variance of a Poisson distribution with mean μ_i. (Soon we are going to let μ_i have an uncertainty.) The p-value associated with the goodness of fit of the data to our model is then calculated using Poisson probabilities:

$$P = \sum_{K} \left[\prod_{i=1}^{N} \frac{e^{-\mu_i} \mu_i^{k_i}}{k_i!} \right],$$

where K is the set of all $(k_1, k_2 \ldots k_N)$ such that $\chi^2_P(k_1, k_2 \ldots k_N) \geq \chi^2_P(n_1, n_2 \ldots n_N)$. When μ_i are sufficiently large and assuming the model is correct, the distribution of χ^2_P is approximately independent of the actual values of μ_i, and the p-value can be approximated by the tail area beyond $t_0 = \chi^2_P(n_1, n_2 \ldots n_N)$ of a mathematical χ^2 distribution for N degrees of freedom (assuming no free parameters). A small p-value leads to rejection of the model. When the model under which the p-value is calculated has nuisance parameters (i.e., systematic uncertainties), the proper computation of the p-value is more complicated (see Section 4.2).

3.2. Discrete Data

When x is a continuous variable, the ideal p-value distribution is uniform between zero and unity. This implies that $\text{Prob}(p \leq p_0) = p_0$ for any p_0 between zero and unity. When x is discrete (e.g., for the data of a Poisson distribution), so are the possible p-values, and thus their distribution cannot be completely uniform. The discrete distribution is most uniform when the above equation is true for all achievable p-values.

3.3. What p-Values Are Not

It is vital to remember that a p-value is not the probability that the relevant hypothesis is true. Thus, statements such as "our data show that the probability that the Standard Model is true is below 1%" are incorrect interpretations of p-values. Similarly, a p-value of 10% or larger is not evidence that the null hypothesis is true; it is merely that the data are not inconsistent with it.

4. METHODS OF INCLUDING SYSTEMATICS

4.1. Parameter Estimation

We describe methods for incorporating systematics into measurements and limits.

4.1.1. Shift method. The shift method, based on linear propagation of errors, is simple but not always applicable. Given N nuisance parameters μ_i with uncorrelated Gaussian uncertainties σ_i, and an estimator of the parameter of interest $f(\mu_1, \mu_2, \ldots, \mu_N)$, the linear approximation yields

$$\sigma_f^2 \simeq \sum_{i=1}^{N} \left(\frac{\partial f}{\partial \mu_i} \right)^2 \sigma_i^2,$$

where σ_f is the combined systematic uncertainty of the measurement. If f is approximately linear over the region $\mu_i \pm \sigma_i$, the partial derivatives can be approximated by finite differences as

$$\frac{\partial f}{\partial \mu_i} \simeq \frac{f(\mu_1, \mu_2, \ldots, \mu_i + \sigma_i, \ldots, \mu_N) - f(\mu_1, \mu_2, \ldots, \mu_i, \ldots, \mu_N)}{\sigma_i} \equiv \frac{\Delta_i}{\sigma_i},$$

and one obtains $\sigma_f^2 \simeq \sum_{i=1}^{N} \Delta_i^2$. That is, one adds the 1σ shifts in quadrature. Given the full error matrix V_{ij} including correlations, this becomes

$$\sigma_f^2 \simeq \sum_{i,j} \Delta_i \Delta_j \left(\frac{V_{ij}}{\sigma_i \sigma_j} \right),$$

where the matrix in parentheses, the correlation matrix, has by construction ones along its main diagonal. (Contrast the multisim approach of Section 6.3.3.)

For example, m_H, the mass of the Higgs boson, could be determined from a fit to a mass spectrum, consisting of a peak above a background, some of which is caused by $t\bar{t}$ events. To estimate the contribution from the uncertainty of the mass of the top quark on a measurement of m_H, one simply reanalyzes the data with the top quark mass shifted by its 1σ uncertainty. Typically, this would also necessitate producing new Monte Carlo events at the shifted top mass, a straightforward, albeit time-consuming, task. The statistical error on the estimate of this contribution can be reduced by varying the nuisance parameter by more than one error. Then the resulting change is divided by the appropriate factor. This requires the dependence of the answer on the nuisance parameter to be linear over a wider range.

When one is setting a limit, rather than obtaining a point estimate with associated uncertainty, the shift method is not applicable. If the function f is significantly nonlinear in the nuisance parameters, the shift method is not reliable.

4.1.2. Likelihood method. Here we need the full likelihood as a function of both the parameters of interest and the nuisance parameters. (Although the likelihood also depends on the observed data, only its dependence on the parameters is explicit below.) Ideally, one uses the product of the actual likelihood functions from the main and subsidiary measurements. [Punzi (15) has discussed the case where only a range is defined for the nuisance parameter.] Otherwise, should only estimates $\hat{\mu}_i$ with an error matrix V_{ij} be available for a set of nuisance parameters μ_i, one typically approximates the likelihood associated with those nuisance parameters by a multidimensional Gaussian:

$$L \simeq \exp\left(-\frac{1}{2}\sum_{i,j}(\mu_i - \hat{\mu}_i)V_{ij}^{-1}(\mu_j - \hat{\mu}_j)\right).$$

We define

$$\ell(\vec{\mu}) = -2\sum_k \ln(L_k),$$

where L_k represent the component likelihoods (from independent parts of the overall *pdf*) for the parameters of interest and nuisance parameters, and the vector $\vec{\mu}$ contains both parameters of interest and nuisance parameters. [It is numerically more convenient to deal with -2 times the logarithm of the likelihood, as the value of the likelihood is often outside the range of floating point representation. The factor -2 is introduced by convention; with this factor, $\ell(\vec{\mu})$ is a χ^2 in the Gaussian case.] Given a starting point $\vec{\mu}_0$ close to the minimum of $\ell(\vec{\mu})$, the minimization package MINUIT (19) will numerically find the values $\hat{\vec{\mu}}$ that minimize $\ell(\vec{\mu})$, which represent maximum likelihood estimates for the parameters of interest and nuisance parameters. MINUIT (via its MIGRAD routine) will also produce a complete error matrix V_{ij}, which derives from the shape of the likelihood, for the parameters. By using $L = L_{main}L_{subsid}$, the statistical and systematic uncertainties are fully incorporated into the errors on the parameters of interest.

This method uses a numerically more precise approximation of the likelihood than the shift method. However, the method is not recommended for limits and assumes that the shape of the likelihood near its maximum can be approximated by a multidimensional Gaussian.

4.1.3. Profile likelihood. Here one obtains estimates of the parameters by maximizing the likelihood as above, but the uncertainties are handled differently. To define the profile likelihood, we divide the parameters into a single parameter of interest μ and the rest of the parameters $\vec{\phi}$, writing $L(\mu, \vec{\phi})$. The profile likelihood with respect to μ is then

$$L_P(\mu) = \text{maximum with respect to } \vec{\phi} \text{ of } L(\mu, \vec{\phi}).$$

Here also it is convenient to work with $\ell_P(\mu) = -2\ln(L_P(\mu))$. The uncertainties are given by

$$\ell_P(\hat{\mu} \pm \sigma_{\pm}) - \ell_P(\hat{\mu}) = 1,$$

where $\hat{\mu}$ is the value of μ that maximizes $L_P(\mu)$ and can be calculated using the MINUIT MINOS routine (see Reference 20 for further discussion). Because in general $\sigma_+ \neq \sigma_-$, these errors can be asymmetric. When $\mu + \sigma_+$ or $\mu - \sigma_-$ is outside the allowed region for μ, the method can be unreliable. Otherwise, the method does a better job of handling both the correlations between the parameters and any non-Gaussian behavior of the likelihood. Asymmetric errors give more information about the shape of the profile likelihood, but their proper interpretation is not always clear (21).

4.1.4. Fully Bayesian. The Bayesian approach requires a prior for the nuisance parameters. Because there may be correlations between the nuisance parameters, we write this as a joint nuisance prior $\pi(\vec{\phi})$. In cases where some groups of nuisance parameters are unrelated, the joint prior may be the product of several individual priors. Ideally, $\pi(\vec{\phi})$ would be derived from Bayesian posteriors provided by subsidiary measurements. Sometimes some portion of $\pi(\vec{\phi})$ is based, partially or wholly, on the physicist's judgment (or personal belief). It is important to state the source of the nuisance priors.

In some cases, the combined prior for μ and $\vec{\phi}$ is not factorizable into separate priors $\pi(\mu)$ and $\pi(\vec{\phi})$. For example, in a Poisson counting experiment with rate $s + b$, $1/\sqrt{s + b}$ is often suggested as a prior for the parameter of interest s, where the background rate b is the nuisance parameter. The posterior is then calculated as

$$p(\mu) = \frac{\int L(\mu, \vec{\phi}) \, \pi(\mu, \vec{\phi}) d\vec{\phi}}{\int\int L(\mu, \vec{\phi}) \, \pi(\mu, \vec{\phi}) \, d\vec{\phi} d\mu},$$

where $\pi(\mu, \vec{\phi})$ is the joint prior for all parameters.

When the prior can be factorized as $\pi(\mu, \vec{\phi}) = \pi(\mu)\pi(\vec{\phi})$, one may equivalently calculate the marginalized likelihood (or marginalized *pdf*)

$$L(\mu) = \int L(\mu, \vec{\phi}) \, \pi(\vec{\phi}) \, d\vec{\phi} \qquad\qquad 5.$$

first. Then one proceeds to treat the marginalized likelihood as one would a simple likelihood: multiply by the prior for μ to obtain the normalized posterior *pdf*:

$$p(\mu) = \frac{L(\mu)\pi(\mu)}{\int L(\mu)\pi(\mu) \, d\mu}.$$

As usual, the posterior is integrated to define an interval or limit. The Bayesian procedure works for all cases, for both one- and two-sided intervals, with no assumption of linearity or Gaussian shape. The marginalization integral may be difficult to calculate in some cases. One should check that the posterior is normalizable, in addition to checking for numerical accuracy, when integrating numerically.

The methods of Sections 4.1.1 through 4.1.3 are approximations of the fully Bayesian method, with a flat prior for μ. In the fully Bayesian approach, however, other priors are permitted, and the lack of a unique prior is viewed by frequentists as a drawback. This freedom of choice of prior extends to the nuisance parameters as well, as a nuisance prior for the main experiment is either based directly on personal belief or is the posterior of a subsidiary measurement, for which some choice of prior was also made. (A prior for the subsidiary measurement combined with the likelihood for the subsidiary measurement yields the subsidiary posterior; the subsidiary posterior becomes the nuisance prior in the main measurement.) Concerns about the answer's dependence on the choice of prior(s) may be alleviated by investigating the frequentist coverage of the method.

4.1.5. Fully frequentist. The fully frequentist method is based on the Neyman construction of confidence intervals. This requires the probability distribution of the data $\vec{x}$ given the parameter of interest μ and the nuisance parameters $\vec{\phi}$. In this case, $\vec{x}$ includes the data from the main measurement, and from the subsidiary measurements for the nuisance parameters.

To calculate intervals at confidence level α, one proceeds as follows: For each allowed value of $(\mu, \vec{\phi})$, one selects a region $R(\mu, \vec{\phi})$ in $\vec{x}$ space that would contain the result of the experiment (both the main and subsidiary experiments) with probability $= \alpha$ (discrete cases may require probability $\geq \alpha$). For observed data $\vec{x}$, the confidence interval is then the set $I(\vec{x})$ of all $(\mu, \vec{\phi})$ such that $R(\mu, \vec{\phi})$ contains $\vec{x}$.

For a given $\vec{x}$, the interval $I(\vec{x})$ will have some complicated shape in $(\mu, \vec{\phi})$ space. For any possible true value $(\mu, \vec{\phi})$, the distribution of resulting intervals $I(\vec{x})$ has the property that the true value $(\mu, \vec{\phi}) \in I(\vec{x})$ with probability α ($\geq \alpha$ for discrete cases). To produce an interval $[\mu_L, \mu_U]$ for μ alone (given $\vec{x}$), one takes the projection of $I(\vec{x})$ onto the μ-axis; $\mu_L \leq \mu$ for any $(\mu, \vec{\phi}) \in I(\vec{x})$, and $\mu_U \geq \mu$ for any $(\mu, \vec{\phi}) \in I(\vec{x})$. Then for any possible true value of $(\mu, \vec{\phi})$, the distribution of resulting intervals $[\mu_L, \mu_U]$ will also have the property that $\mu_L \leq \mu \leq \mu_U$ with probability $\geq \alpha$.

Projection induces overcoverage, especially with many nuisance parameters. Thus, one would like to define $R(\mu, \vec{\phi})$ so that the intervals $I(\vec{x})$ are already wide in the $\vec{\phi}$ variables. The regions $R(\mu, \vec{\phi})$ are often defined via an ordering rule

$$R(\mu, \vec{\phi}) = \{\vec{x} | \rho(\vec{x}, \mu, \vec{\phi}) \leq \rho_0(\alpha, \mu, \vec{\phi})\}.$$

Because of computational difficulty, there are few fully frequentist examples (14–16, 22, 23). Coverage is guaranteed by construction, but one should check for other pathologies, some of which are listed in Reference 24.

4.1.6. Mixed frequentist-Bayesian. Here one applies the Neyman construction to obtain an interval from $L(\mu)$ of Equation 5. Because only the parameter of interest remains after marginalization, the Neyman construction is not numerically difficult at this stage. The method is therefore easier to implement than the fully frequentist approach, while still avoiding the necessity of choosing a prior for the parameter of interest. References 18 and 25 describe the mixed approach.

Despite the Bayesian methodology for the nuisance parameters, the overall method may still have frequentist coverage. However, neither frequentist coverage nor Bayesian credibility for the parameter of interest is guaranteed by the method. Philosophically, the approach can be regarded as a black box whose properties must be determined. Investigation of coverage is therefore in order (if coverage is desired).

4.2. p-Values

Section 3 discussed p-values in the absence of systematic uncertainties. We now mention several ways of incorporating them into p-values. For each method described below, we assume that an appropriate statistic t, which may depend on data from both the main and subsidiary measurements, has been selected for the task at hand, and the p-values are the upper-tail probability of the distribution of the statistic t. Without any systematic uncertainties, the p-value corresponding to an observed value of the statistic t_0 is then

$$P(t_0) = \int_{t_0}^{\infty} f(t)\, dt,$$

where $f(t)$ is the probability density of the distribution of the statistic t.

Systematic uncertainties give the distribution of t a dependence on nuisance parameters $\vec{\phi}$: $f(t|\vec{\phi})$. In some cases, especially for goodness of fit, a parameter of interest can be a nuisance parameter for the purposes of calculating a p-value. (For example, in assessing whether a mass spectrum shows evidence for the production of the Higgs boson, the physically interesting mass of the Higgs is merely a nuisance parameter.) It is desirable to construct a goodness-of-fit statistic whose distribution is almost independent of unknown parameters, but exact independence is usually not possible. In the binned goodness-of-fit example, the Poisson likelihood ratio λ, reviewed in Reference 26, is less dependent than Pearson's χ^2 on the true values of the expected bin contents, but not completely independent. The distribution of either statistic may then depend on the parameter of interest, which becomes a nuisance parameter for the associated p-value.

The following sections describe methods for incorporating uncertainties on nuisance parameters into a p-value; they are discussed more fully in Reference 27. Inherent is the assumption that a small p-value leads to the rejection of the hypothesis in question, whereas a large p-value only means that the hypothesis in question is not rejected—the normal approach in questions of significance. A conservative approach is then to consider the largest possible p-value. In some goodness-of-fit applications, for example, performing checks that data match an expected distribution, the emphasis is somewhat different: A reasonably large p-value is interpreted to mean that no serious problems are present and it is safe to proceed with the analysis. In such applications, a method that tends to overestimate the p-value is not conservative; it tends to hide problems. Here, one may prefer a range, or distribution, of p-values that are likely to be correct, as a large goodness-of-fit p-value is intended to lend

Table 1 Errors in using *p*-values for goodness of fit and for significance

	Errors of first kind	**Errors of second kind**
Definition	Reject true H_0	Accept false H_0
p-value	Small	Larger
Goodness of fit	Loss of efficiency	Source of background
Significance	False discovery	Failure to discover
Effect of conservatism (larger *p*-value)	Fewer errors	More errors

p-values are used to quantify the degree of consistency between data and a null hypothesis H_0. In goodness of fit, we look for reasonable *p*-values (say $p > 1\%$) in order to not reject H_0 and, for example, to accept the values of some fitted parameters or to select a sample of a given type of events. For significance, we hope for very small *p*-values (say 10^{-7}) in order to reject H_0 and perhaps claim a discovery. The usual effect of nuisance parameters is an increase in *p*-values, which thus reduces the chance of false discovery but makes it more likely that H_0 may be incorrectly accepted.

confidence that the data are well modeled. For simplicity, the methods are illustrated for the significance case. How to properly incorporate uncertainties on nuisance parameters into a goodness-of-fit *p*-value is less well established. **Table 1** summarizes some differences in how systematics associated with *p*-values influence significance and goodness of fit.

4.2.1. Prior predictive. This is the method most commonly used in high-energy physics. One marginalizes the *pdf* over the priors for the nuisance parameters before calculating the *p*-value:

$$P(t_0) = \int_{t_0}^{\infty} \int f(t|\vec{\phi}) \, \pi(\vec{\phi}) \, d\vec{\phi} \, dt.$$

As with parameter determination, the prior $\pi(\vec{\phi})$ may be the posterior of a subsidiary measurement. This method may not be appropriate for goodness-of-fit *p*-values when the distribution of t depends strongly on the parameter(s) of interest, as the prior for the parameter of interest is typically chosen to be noninformative.

4.2.2. Posterior predictive. In a fully Bayesian method, the main experiment often has some information about the nuisance parameters. One can obtain the posterior $p(\vec{\phi})$ for the nuisance parameters by marginalizing the joint posterior over the parameter of interest μ,

$$p(\vec{\phi}) = \frac{\int L(\mu, \vec{\phi}) \, \pi(\mu, \vec{\phi}) \, d\mu}{\int \int L(\mu, \vec{\phi}) \, \pi(\mu, \vec{\phi}) \, d\mu \, d\vec{\phi}}.$$

The posterior-predictive *p*-value is then calculated as

$$P(t_0) = \int_{t_0}^{\infty} \int f(t|\vec{\phi}) p(\vec{\phi}) \, d\vec{\phi} \, dt.$$

Here also it is unclear that this method is appropriate for goodness-of-fit *p*-values when the distribution of t depends strongly on the parameter(s) of interest. The fact that the data are used twice, once to help determine $p(\vec{\phi})$ and again in the definition of $P(t_0)$, may represent a logical defect of the posterior-predictive method.

4.2.3. The plug-in method. The plug-in p-value is calculated as

$$P(t_0) = \int_{t_0}^{\infty} f(t|\vec{\phi}_0)\,dt,$$

where $\vec{\phi}_0$ is some estimate of the nuisance parameters. In some cases, as when $\vec{\phi}_0$ is estimated under the null hypothesis, the plug-in p-value is conservative. For example, with a null hypothesis of no signal, the background estimate must include all bins in a histogram, including those normally considered signal, to qualify as conservative. This to some extent compensates for the fact that the plug-in method does not account for the uncertainties associated with $\vec{\phi}$. Reference 28 shows how using a plug-in p-value for goodness of fit can lead to false confidence.

4.2.4. The supremum method. The supremum p-value is the largest possible p-value obtainable from any allowed values of the nuisance parameters:

$$P(t_0) = \text{maximum with respect to } \vec{\phi} \text{ of } \int_{t_0}^{\infty} f(t|\vec{\phi})\,dt.$$

It is useful when the statistic t can be chosen so that $f(t|\vec{\phi})$ is at least approximately independent of $\vec{\phi}$. Otherwise the supremum p-value is too conservative, destroying any effect.

As no prior is used, the supremum p-value is a completely frequentist construction. For example, suppose that there are n signal region events and m events in a background-only region, where, for simplicity, equal background contributions are expected in each region. Here the background-only region represents a subsidiary measurement, yielding the expected background in the signal region (with an uncertainty). As the statistic for a significance calculation, a frequentist approach may use the likelihood ratio (for equal expected rates in the two regions, compared with completely free rates), defined as

$$\lambda(n, m) = \frac{\text{max w.r.t. } \mu = v \text{ of } \exp(-\mu)\mu^n/n!\ \exp(-v)v^m/m!}{\text{max w.r.t. } \mu,\ v \text{ of } \exp(-\mu)\mu^n/n!\ \exp(-v)v^m/m!} = \frac{\left(\frac{n+m}{2}\right)^{n+m}}{n^n m^m}.$$

It is convenient to define the significance statistic as $t(n, m) = -2\ln[\lambda(n, m)]$. The p-value

$$P(n_0, m_0|v) = \sum_{t(n,m) \geq t(n_0, m_0)} \frac{e^{-v}v^n}{n!}\frac{e^{-v}v^m}{m!}$$

is evaluated under the no-signal hypothesis and depends on the value of the nuisance parameter v (the background rate in each bin). In evaluating the supremum p-value, we maximize $P(n_0, m_0|v)$ over the range $v = 0$ to ∞. This can be done numerically, and $P(n_0, m_0|v)$ does not depend strongly on v, so the method can be considered reasonable. For other choices of statistic, the p-value may be strongly dependent on the assumed value of a nuisance parameter. In such cases, the supremum p-value is useless.

The value of the nuisance parameter at which the maximum p-value occurs is not considered an estimate of that parameter; it is common for the p-value to have its maximum at a value of the nuisance parameter that is 10σ or more from its best estimate.

For goodness of fit, it may be more useful to quote both the minimum and maximum p-value, not just the maximum. For example, a range of 0.2–0.8 would be a clear indication of an acceptable fit, whereas a range of 10^{-4}–0.8 would be ambiguous.

4.2.5. Supremum over confidence interval.
In some cases it may be computationally difficult to maximize the p-value over an infinite range, or the p-value may be maximized at a value of a nuisance parameter excluded at very high confidence levels. A variation of the supremum method is to maximize the p-value over a confidence interval for $\vec{\phi}$. Suppose W is a region that contains the true value of $\vec{\phi}$ at confidence level $1 - \beta$. Then the p-value defined as

$$P(t_0) = \beta + \text{maximum with respect to } \vec{\phi} \in W \text{ of } \int_{t_0}^{\infty} f(t|\vec{\phi})\,dt$$

has valid frequentist coverage properties (29). No prior is necessary, but obtaining a confidence interval for the nuisance parameters requires that the *pdf* for the subsidiary measurements be available.

As defined, this p-value can never be less than β. One should choose a β much smaller than the smallest p-value to which one wishes to be sensitive. For example, to retain the possibility of obtaining p-values of order 10^{-10}, $\beta = 10^{-12}$ would be a reasonable choice. It is not valid to select β based on what is observed in the data; β must be chosen on other grounds.

The method allows one to ignore values of nuisance parameters rejectable at high confidence levels, at the cost of introducing an artificial floor below which the reported p-value may never descend. However, it is useful mainly in the same situation that the supremum p-value is useful: when $f(t|\vec{\phi})$ is at least approximately independent of $\vec{\phi}$.

5. FULLY BAYESIAN CROSS-SECTION UPPER LIMITS

Heinrich et al. (24) describe a fully Bayesian approach to upper limits on cross sections in the presence of a single nuisance parameter. In that example, the main measurement observes n events from a Poisson distribution with mean $s\varepsilon + b$, where the cross section s is the parameter of interest; the acceptance ε is a nuisance parameter; and, to keep the example simple, the background b is taken to be a known constant.

A subsidiary measurement to determine ε is specified in which m events are observed from a Poisson process with mean $\kappa\varepsilon$, where κ is a known constant. In the subsidiary measurement, a prior for ε of the form ε^{q-1} yields a posterior for ε that becomes the prior:

$$\pi(\varepsilon) = \frac{(\kappa\varepsilon)^{\mu} e^{-\kappa\varepsilon}}{\varepsilon\Gamma(\mu)}$$

for ε in the main measurement, where $\mu = m+q$. This is a gamma distribution, with mean μ/κ and variance μ/κ^2.

A prior for the cross section s, which in Reference 24 is chosen to have the form s^{r-1}, is also necessary. The posterior for the cross section s, which in this simple case can be obtained analytically, is then

$$p(s) = \frac{\Gamma(\mu+n)}{\Gamma(\mu-r)\Gamma(r+n)} \frac{s^{r+n-1}\kappa^{\mu-r}}{(s+\kappa)^{\mu+n}} \frac{M(-n, 1-n-\mu, b(s+\kappa)/s)}{M(-n, 1-n-r, b)}.$$

The confluent hypergeometric function (30) $M(-n, a, x)$ with integer $n \geq 0$ is a polynomial of degree n in x. As $M(-n, 1-n-r, b)$ is a polynomial in b whose highest-order term contains the factor b^n/r, with $b > 0$, the posterior approaches a δ function at $s = 0$ in the limit as $r \to 0$. In the presence of background, a $1/s$ prior has too much weight at $s = 0$.

Other popular choices of priors for s are $r = 1$ (a flat prior) and $r = 1/2$ (a $1/\sqrt{s}$ prior). A flat prior for s is conservative, as it results in mild overcoverage for upper limits on s. Coverage drops as r decreases. Consequently, for priors of the form s^{r-1} in this problem, optimal coverage properties for upper limits are achieved by an r somewhere in the range of $0 < r \leq 1$.

Reference 31 advocates the use of coverage as a diagnostic for the selection of priors—one approach within a larger objective Bayesian methodology—and seems a practical way to reduce the ambiguity associated with the choice of prior. Objective Bayesians attempt to minimize the influence of priors on the posterior, rather than derive them from personal belief, as in the subjective Bayesian approach.

When not derived directly from the posterior of a subsidiary measurement, it is common to assume a Gaussian form for the prior for ε (set to zero for $\varepsilon < 0$). When combined with a flat prior for s, this leads to a posterior density for s that is not normalizable, behaving like $1/s$ at large s. Some blame the Gaussian ε prior, others the flat prior for s, but certainly the combination of the two is pathological. (Even here, the flat prior is still conservative; upper limits become infinite.)

However, in this example it is only for s that flat is conservative. A flat subsidiary prior for ε is less conservative than a $1/\sqrt{\varepsilon}$ prior. An ε^{q-1} subsidiary prior yields larger upper limits for s as q decreases because underestimating the acceptance means overestimating the cross section. This reversal becomes important in the multichannel generalization of the problem; Reference 32 shows that a flat prior for ε can lead to significant undercoverage when there are more than two channels with Poisson subsidiary measurements. For N channels with acceptances ε_i, assigning a flat prior to each ε_i results in an effective prior for the total acceptance $\varepsilon = \sum \varepsilon_i$ proportional to ε^{N-1}. For large N and for small numbers of events in the subsidiary experiments, this can lead to significant undercoverage.

6. MISCELLANEOUS

In this section, we discuss several different issues of relevance to systematics. As with many statistical situations but especially with systematics, although textbook problems have unique solutions, in real life the issues are such that decisions often involving personal judgment must be made.

6.1. Blind Analyses

Corrections for potential systematic biases can be significant in magnitude and, as already mentioned, can involve personal choices. There is thus the opportunity/danger that a physicist can decide which corrections to apply in order, subconsciously, to adjust the corrections until a desired result is obtained. (Similar remarks apply to the choice of statistical technique.) This can be avoided if a blind analysis is employed. Various techniques for blind analyses are described in References 33–36.

6.2. Separating Signal from Background

Almost every analysis in particle physics involves the separation of the wanted signal from various sources of background. This is often achieved by using a classifier that has been trained on simulated samples of signal and backgrounds, which have been checked to describe the actual data with reasonable accuracy.

The traditional way of handling systematic effects is to train the classifier with simulated samples for which the nuisance parameters are fixed at their optimal values, and then to investigate how the classifier's performance is affected when these nuisance parameters are changed. Better performance of the classifier may be achieved if the generated training samples already contain the systematic uncertainties. That is, the nuisance parameters should be varied randomly event by event over their expected probability distributions (37, 38).

6.3. Simulation Issues

6.3.1. Reweighting. One method for estimating a contribution to systematics is to determine δa, the change in the answer, when the nuisance parameter v is changed by its uncertainty σ_v (see Section 4.1.1). This usually involves generating Monte Carlo samples with v equal to its best value v_0 and to $v_0 + \sigma_v$ (and maybe $v_0 - \sigma_v$ too).

If the Monte Carlo samples at v_0 and at $v_0 + \sigma_v$ are generated independently, the uncertainty on δa due to the limited Monte Carlo statistics will be $\sqrt{2}\sigma_{MC}$, where σ_{MC} is the statistical error on each. If instead the sample at $v_0 + \sigma_v$ is obtained by a suitable reweighting of the events at v_0, the error on the difference can be even smaller than σ_{MC} because of the correlations between the samples, provided the events' reweighting factors are not too different from each other. Thus, for example, a set of events generated from an exponential decay distribution with a lifetime $\tau = \tau_1$ can be converted to a distribution with lifetime $\tau = \tau_2$ by reweighting each event with decay time t by a factor of $\tau_1/\tau_2 \times \exp(-[1/\tau_2 - 1/\tau_1]t)$.

It is better to vary v in the direction such that the parts of the distribution with smaller numbers of events are reweighted downward rather than upward. For example, it is better to reweight a Gaussian to a narrower width rather than to a larger one.

6.3.2. Allowance for statistical errors of simulation.

With a contribution to the systematic quantities estimated as $c_i \pm d_i$ (where d_i is the statistical error on c_i arising from the limited simulation statistics), the question is, what should be used for the actual contribution to our error estimate? Suggestions have included c_i, $c_i + d_i$, $\sqrt{c_i^2 + d_i^2}$, $\sqrt{c_i^2 - d_i^2}$, and c_i if c_i is larger than d_i but otherwise zero, and so on. We suggest an aggressive approach in which the contribution to the variance is taken as $c_i^2 - d_i^2$, even if this is negative. Assuming that the separate contributions are independent, the total variance for the systematic is then $\sum(c_i^2 - d_i^2)$, except that if this sum were negative, we would set it to zero. Our logic is that, if a potential systematic were in fact negligible, we would expect c_i to be Gaussian distributed about zero with variance d_i^2. If each contribution to the variance were taken as the larger of $c_i^2 - d_i^2$ and zero, we would overestimate our systematic. A problem this suggestion shares with most others is that, if one of the contributions to the systematics is poorly determined, then so is the total systematic.

6.3.3. Unisim or multisim?

The contribution from a possible systematic can be estimated by seeing the change in the answer a when the nuisance parameter is varied by its uncertainty. The various contributions from each systematic are calculated separately and then combined (see Section 4.1.1). It essentially makes use of the first-order Taylor expansion

$$a = a_0 + \sum \frac{\partial a}{\partial v_i} \delta v_i. \qquad\qquad 6.$$

The simulations to estimate the derivatives are termed unisims by the MiniBooNE Collaboration.

An alternative, even for just one source of systematic, is to generate separate simulated samples, in each of which the values for each systematic have been chosen randomly from their expected distributions and then an answer a is extracted for each sample. The width of the distribution in a values then is used to estimate the total effect of all the systematics. The advantage is that it can handle non-Gaussian distributions and also nonlinearities in the dependence of a on the nuisance parameters. In some cases there can also be savings in the total amount of simulation needed by this multisim approach, to achieve the same accuracy as for the corresponding unisim. Roe (39) has compared the relative accuracies of the two methods.

6.4. Theory Uncertainties

In some cases, the extraction of a physical result requires some theoretical formulation, and there can be more than one variant of how this is implemented. (For example, several different sets of parton distribution functions exist.) The question arises of how the different results contribute to the systematic error. Again, this is a situation without a unique answer.

If there are only two possible versions of the theory, it is probably best to quote separately the results for each. This deals most simply with the situation in which one version is subsequently ruled out by other data. However, if there are multiple choices, quoting the result for each may not be practical. It can also depend on whether the different theoretical variants are regarded as equally plausible. If so, it is common to quote the result as the average, and in the case of two possibilities, half the difference is usually taken as the error. If the different results are regarded as samples of possible results (which from a frequentist viewpoint is hard to maintain), it may not be unreasonable to calculate their standard deviation; for two values, this would give 0.7 times their difference.

A similar situation arises when different functional forms are used to parametrize a distribution. For example, a peak may have more events in its tails than is appropriate for the Gaussian distribution used to fit the data. This is simpler to deal with if the alternative can be formulated in such a way that the original distribution is nested within it. Thus, a sum of Gaussians or a Student's t distribution with N degrees of freedom may be used to describe the heavier tails. Such considerations can be especially important for nuisance parameters, where the approximation of (truncated) Gaussian distributions may be overly optimistic, especially in the tails. This can cause problems when high significances (e.g., larger than 5σ) are demanded and have been calculated using an incorrect probability density distribution.

6.5. Fits with Correlated Systematics

In comparing a theoretical distribution with a data histogram where systematic effects are involved, the relevant χ^2 expression is

$$S = \sum \left(y_i^{\text{data}} - y_i^{\text{pred}} \right) H_{ij} \left(y_j^{\text{data}} - y_j^{\text{pred}} \right), \qquad 7.$$

where H_{ij} is the inverse error matrix on the data y_i^{data}, and the correlations are likely due to the systematics. Demortier (40) showed that this is equivalent to minimizing

$$S' = \sum \frac{\left(y_i^{\text{data}} - z_i^{\text{pred}}(c_1, c_2, \ldots) \right)^2}{(\sigma_{\text{stat}})_i^2} + \sum c_s^2, \qquad 8.$$

where $z_i^{\text{pred}}(c_1, c_2, \ldots)$ is the sum of the predictions y_i^{pred} in the absence of systematics plus the possible systematic effects, and c_s are the coefficients of the different systematics. Each systematic is assumed to have a defined magnitude in each of the n bins of the experimental distribution. With enough data, more than n different coefficients c_i for the magnitudes of the systematic sources can be extracted from the data. For low statistics, Equation 8 can be modified by using the likelihood based on the Poisson distribution of the small number of events in each bin of a histogram.

6.6. Asymmetric Errors

Sometimes separate values are quoted for positive and for negative errors, for example, $1.0^{+0.4}_{-0.2}$ ps for a lifetime. These can arise from statistical errors where the likelihood function has a non-Gaussian shape, or for systematic errors when there are asymmetric changes as a nuisance parameter is changed upward and downward by one error from its central value.

When various contributions with uncorrelated but asymmetric errors are combined, a popular method is to combine the upper errors in quadrature, and similarly for the lower errors, but there is no basis for this. The quadrature rule is applicable when the separate errors can be in the same direction or can tend to cancel. But upper errors are always in the same sense, so the justification disappears. Barlow (21) has discussed ways of dealing with asymmetric errors for calculating χ^2 or for combining different measurements of the same quantity.

6.7. The Best Linear Unbiased Estimate

A well-known technique for combining several measurements $a_i \pm \sigma_i$ of a single quantity a is to use the best linear unbiased estimate (41). This is

$$\hat{a} = \sum \beta_i a_i, \qquad\qquad 9.$$

where the coefficients satisfy $\sum \beta_i = 1$ and are chosen such that the error on $\hat{a}$ is minimized. For uncorrelated errors, this is equivalent to minimizing $\sum (a_i - \hat{a})^2 / \sigma_i^2$. When each error is expressed as $\pm(\sigma_{\text{stat}})_i \pm(\sigma_{\text{syst}})_i$, the method can still be used, with the possibility of including any correlations between the various errors. The method is not recommended for highly correlated results, where $\hat{a}$ can be outside the range of the individual a_i, and its value depends sensitively on the exact values of the correlation coefficients. An advantage of extracting the coefficients β_i is that this enables the separate contributions of the statistical and systematic errors to $\hat{a}$ to be calculated.

7. CONCLUSIONS

Various methods for incorporating systematic effects in parameter estimation have been discussed. These range from the fully Bayesian approach to a complete frequentist Neyman construction. There are also many other recipes that are neither fully Bayesian nor fully frequentist, for example, profile likelihood, mixed Bayes-frequentist, and so on. Although many physicists would regard the fully frequentist method as desirable (in that it avoids the necessity to choose priors and is guaranteed not to undercover), practicalities generally require the use of some other technique. The Bayesian calculation of upper limits on a Poisson production process was discussed in Section 5. It resulted in a usable algorithm, with reasonable coverage properties, provided priors were appropriately chosen.

In searching for new physics, possible discrepancies between the data and the null hypothesis (currently the Standard Model for particle physics) are expressed as *p*-values. Possible ways of incorporating nuisance parameters were listed in Section 4.2. Even though it is possible to specify the properties of various procedures for incorporating nuisance parameters in calculations of intervals or *p*-values, by their very nature systematic errors are such that their evaluation will often depend on the experience and judgment of the experimentalist.

DISCLOSURE STATEMENT

The authors are not aware of any biases that might be perceived as affecting the objectivity of this review.

ACKNOWLEDGMENTS

We wish to thank our experimental colleagues, too numerous to list individually, for many stimulating and vigorous discussions over the years, and especially the CDF Statistics Committee, who also provided valuable feedback on this review. We are also grateful to those statisticians who have participated in the PHYSTAT series of conferences and had the patience to explain relevant aspects of statistics to us. The Leverhulme Foundation kindly provided L.L. with a grant, which partially supported this work.

LITERATURE CITED

1. Barlow R. See Ref. 9, p. 134 (2002)
2. Sinervo P. See Ref. 10, p. 122 (2003)
3. Cousins RD. See Ref. 11, p. 75 (2006)
4. Punzi G. CDF Note 7975. **http://www-cdf.fnal.gov/physics/statistics/notes/punzi-systdef.ps** (2001)
5. Blocker C. CDF Note 6506. **http://www-cdf.fnal.gov/publications/cdf6506_systematics.ps** (2001)
6. Barlow R, et al. (BaBar Stat. Work. Group) *Chapter 7.* **http://www.slac.stanford.edu/BFROOT/www/Statistics/** (2002)
7. James F, Lyons L, Perrin Y, eds. *CERN Yellow Report.* CERN 2000–005. Geneva, Switz.: CERN (2000)
8. Lyons L, et al. *Proc. Fermilab Confid. Limits Workshop*, Batavia, Illinois, 2000. **http://conferences.fnal.gov/cl2k/** (2000)
9. Whalley MR, Lyons L, eds. *Proc. Conf. Adv. Stat. Tech. Part. Phys.* IPPP/02/39. Durham, UK: Univ. Durham. 333 pp. (2002)
10. Lyons L, Mount R, Reitmeyer R, eds. *Proc. PHYSTAT 2003 Stat. Probl. Part. Phys. Astrophys. Cosmol.* SLAC-R-703. Menlo Park, CA: Stanford Linear Accel. Cent. 334 pp. (2003)
11. Lyons L, Karagoz Unel M, eds. *Proc. PHYSTAT05 Stat. Probl. Part. Phys. Astrophys. Cosmol.* London: Imp. Coll. Press. 310 pp. (2006)
12. Linnemann J, Lyons L, Reid N. *BIRS PHYSTAT Workshop Stat. Inference Probl. High Energy Phys. Astron.*, Banff, 2006. **http://www.pims.math.ca/birs/birspages.php?task=displayevent&event_id=06w5054** (2006)

13. Feldman GJ, Cousins RD. *Phys. Rev. D* 57:3873 (1998)
14. Cousins RD. *Nucl. Instrum. Methods A* 417:391 (1998)
15. Punzi G. See Ref. 11, p. 88 (2006)
16. Demortier L. physics/0312100 v2 (2003)
17. Reid N. See Ref. 10, p. 265 (2003)
18. Cousins RD, Highland VL. *Nucl. Instrum. Methods A* 320:331 (1992)
19. James F. Function minimization and error analysis reference manual. *CERN Program Libr. Long Writeup D506.* Geneva, Switz.: CERN (1998)
20. Rolke WA, Lopez AM, Conrad J. See Ref. 11, p. 97 (2006)
21. Barlow R. See Ref. 10, p. 250 (2003)
22. Nicolo D, Signorelli G. See Ref. 9, p. 152 (2002)
23. Sen B, Walker M, Woodroofe M. *Statistica Sin.* 17:In press
24. Heinrich J, et al. physics/0409129 (2004)
25. Conrad J, Tegenfeldt F. See Ref. 11, p. 93 (2006)
26. Baker S, Cousins R. *Nucl. Instrum. Methods A* 221:437 (1984)
27. Demortier L. CDF Note 8662. **http://www-cdf.fnal.gov/publications/cdf8662_p_values_in_detail.pdf** (2007); Cranmer K. See Ref. 11, p. 112 (2006); Linnemann J. See Ref. 10, p. 35 (2003)
28. Heinrich J. See Ref. 10, p. 52 (2003)
29. Berger RL, Boos DD. *J. Am. Stat. Assoc.* 89:1012 (1994)
30. Slater LJ. In *Handbook of Mathematical Functions*, ed. M Abramowitz, IA Stegun, p. 503. New York: Dover (1968)
31. Bayarri MJ, Berger JO. *Stat. Sci.* 19:58 (2004)
32. Heinrich J. See Ref. 11, p. 98 (2006)
33. Harrison P. See Ref. 9, p. 278 (2002)
34. Roodman A. See Ref. 10, p. 166 (2003)
35. Heinrich J. CDF Note 6576. **http://www-cdf.fnal.gov/publications/cdf6576_blind.pdf** (2003)
36. Lyons L. *Physics perspective.* Presented at Stat. Chall. Mod. Astron., 4th, State College, 2006. CDF Note 8514. **http://www-cdf.fnal.gov/publications/cdf8514_HEP_Astro_Stats.ps** (2006)
37. Lyons L. *Nucl. Instrum. Methods A* 324:565 (1993)
38. Neal R. BIRS PHYSTAT Workshop Stat. Inference Probl. High Energy Phys. Astron., Banff, 2006. **http://www.pims.math.ca/birs/birspages.php?task=displayevent&event_id=06w5054** (2006)
39. Roe B. *Nucl. Instrum. Methods A* 570:159 (2007)
40. Demortier L. CDF Note 8661. **http://www.cdf.fnal.gov/publications/cdf8661_chi2fit_w_corr_syst.pdf** (1999)
41. Lyons L, Gibaut D, Clifford P. *Nucl. Instrum. Methods A* 270:110 (1988)

Two-Photon Physics
in Hadronic Processes

Carl E. Carlson[1] and Marc Vanderhaeghen[1,2]

[1] Department of Physics, College and William and Mary, Williamsburg, Virginia 23187; email: carlson@physics.wm.edu

[2] Thomas Jefferson National Accelerator Facility, Newport News, Virginia 23606; email: marcvdh@jlab.org

Annu. Rev. Nucl. Part. Sci. 2007. 57:171–204

First published online as a Review in Advance on May 4, 2007

The *Annual Review of Nuclear and Particle Science* is online at http://nucl.annualreviews.org

This article's doi: 10.1146/annurev.nucl.57.090506.123116

Copyright © 2007 by Annual Reviews. All rights reserved

0163-8998/07/1123-0171$20.00

Key Words

electron scattering, form factors, two-photon exchange processes

Abstract

Two-photon exchange contributions to elastic electron scattering are reviewed. The apparent discrepancy between unpolarized Rosenbluth and polarization transfer experiments in the extraction of elastic nucleon form factors is discussed, as well as the understanding of this puzzle in terms of two-photon exchange corrections. Calculations of such corrections within both partonic and hadronic frameworks are reviewed. In view of recent spin-dependent electron scattering data, the relation of the two-photon exchange process to the hyperfine splitting in hydrogen is critically examined. The imaginary part of the two-photon exchange amplitude as can be accessed from the beam-normal spin asymmetry in elastic electron-nucleon scattering is reviewed. Further extensions and open issues in this field are outlined.

Contents

1. INTRODUCTION 172
2. TWO-PHOTON EXCHANGE IN ELASTIC
   ELECTRON-NUCLEON SCATTERING 176
   2.1. Hard Two-Photon Exchange 176
   2.2. Elastic Electron-Nucleon Scattering Observables 177
   2.3. Partonic Calculation of the Two-Photon Exchange 179
   2.4. Results 180
   2.5. Remarks on Related Topics 183
3. POLARIZABILITY CORRECTIONS TO HYDROGEN
   HYPERFINE SPLITTING 184
   3.1. Two-Photon Exchange and Atomic Structure 184
   3.2. Hyperfine Splitting Calculations 185
   3.3. Numerics, Especially for Δ_{pol} 188
   3.4. Remarks and Prospects 190
4. BEAM-NORMAL AND TARGET-NORMAL SPIN
   ASYMMETRIES 190
   4.1. The Imaginary (Absorptive) Part of the Two-Photon Exchange
        Amplitude 191
   4.2. Results and Discussion 194
5. CONCLUSIONS 198

I like a thing simple but it must be simple through complication. Everything must come into your scheme, otherwise you cannot achieve real simplicity.

Gertrude Stein: *What Are Masterpieces* (1)

1. INTRODUCTION

Elastic electron-nucleon (eN) scattering in the one-photon exchange approximation is a time-honored tool for accessing information on the structure of hadrons. Experiments with increasing precision have become possible in recent years, triggered mainly by new techniques to perform polarization experiments at electron scattering facilities. This opened a new frontier in the measurement of hadron structure quantities, such as its electroweak form factors (FFs), parity-violating (PV) effects, nucleon polarizabilities, transition FFs, or the measurement of spin-dependent structure functions, to name a few. For example, experiments using polarized electron beams and measuring the ratio of the recoil nucleon in-plane polarization components have profoundly extended our understanding of the nucleon electromagnetic FFs; for recent reviews on nucleon FFs, see, for example, References 2–4.

For the proton, such polarization experiments access the ratio G_E/G_M of the proton's electric (G_E) to magnetic (G_M) FFs directly from the ratio of the transverse and longitudinal polarizations in elastic eN scattering as (5)

$$\frac{P_s}{P_l} = -\sqrt{\frac{2\varepsilon}{\tau(1+\varepsilon)}}\frac{G_E(Q^2)}{G_M(Q^2)}. \qquad 1.$$

Here,

$$\tau \equiv \frac{Q^2}{4M^2}, \quad \frac{1}{\varepsilon} \equiv 1 + 2(1+\tau)\tan^2\frac{\theta}{2}, \qquad 2.$$

where $Q^2 = -q^2$ is the momentum transfer squared, θ is the laboratory scattering angle, and $0 \leq \varepsilon \leq 1$. Recently, this ratio was measured at the Jefferson Laboratory (JLab) out to a space-like momentum transfer Q^2 of 5.6 GeV^2 (6–8). Surprisingly, these experiments extracted a ratio of G_E/G_M, which is clearly at variance with un-polarized measurements (9–11) that use the Rosenbluth separation technique, which measures the angular dependence of the differential cross section for elastic eN scattering at fixed Q^2:

$$\frac{d\sigma}{d\Omega_{Lab}} \propto G_M^2 + \frac{\varepsilon}{\tau}G_E^2, \qquad 3.$$

where the proportionality factor is well known and isolates the ε-dependent term. In each case, the quoted formulas assume single-photon exchange between electron and nucleon.

To explain the discrepancy between the two experimental techniques, suspicion falls on the Rosenbluth measurements, but not because of experimental problems per se. The Rosenbluth formula (Equation 3) at high Q^2 has a numerically big term, G_M^2, and a small term. The results for G_E come from the small term. Any omitted ε-dependent corrections to the large term can thus have a strikingly large effect on G_E.

Two-photon exchange (**Figure 1**) is one possible culprit. The subject has a long history. In fact, since the late 1950s and during the 1960s, when eN scattering was measured systematically at the Stanford Linear Accelerator Center (SLAC) to access nucleon electromagnetic FFs, the validity of the one-photon exchange approximation has been discussed both theoretically and experimentally. Because the one-photon exchange cross section depends quadratically on the lepton charge, the difference be-tween electron-nucleon and positron-nucleon cross sections is a test for two- or mul-tiphoton exchange processes. Early comparisons of electron- and positron-nucleon scattering cross sections were consistent with equal cross sections (12, 13), but the precision achieved in those early investigations could not exclude two-photon ex-change effects at the few-percent level of the cross section. Contributions of this size

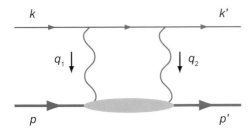

Figure 1

Two-photon exchange in the lepton-nucleon scattering process $l(k) + N(p) \rightarrow l(k') + N(p')$, where k, k' and p, p' are the four-momenta of leptons and nucleons, respectively.

can be expected owing to the additional electromagnetic coupling in the two-photon exchange diagram, which brings in the suppression factor $\alpha = e^2/(4\pi) \simeq 1/137$.

On the theory side, the corrections to elastic eN scattering of order e^2 relative to the Born approximation, also known as radiative corrections, have a long history (see, e.g., References 14 and 15), which were heavily applied in the analysis of early eN scattering experiments. The infrared divergences that were associated with one photon in **Figure 1** being soft—that is, having a vanishingly small four-momentum—had been extracted; they cancelled against infrared divergences from soft bremsstrahlung. However, the contributions when both photons are hard (i.e., when the momentum transfer of both individual photons is large) were not calculated in those works because of insufficient knowledge of the intermediate hadron state. The early calculators were well aware of the omission and explicitly expressed hope that the missing contributions would be small.

In the late 1950s, Drell and collaborators (16, 17) attempted some early estimates of the two-photon exchange contribution of **Figure 1** with two hard photons. Those works constructed a nonrelativistic model for the blob in **Figure 1**, including the nucleon and lowest-nucleon-resonance contribution, the $\Delta(1232)$. The calculation found that the resonance contribution to the two-photon exchange diagram affects the cross sections at the $\sim 1\%$ level. Owing to the nonrelativistic nature of the calculation, the result was limited to approximately 1 GeV electron beam energy. In subsequent works (e.g., References 18 and 19), two-photon exchange effects were approximately calculated to higher energies. In particular, Greenhut (19) evaluated the contribution of higher nucleon-resonance intermediate states, with masses up to 1.7 GeV, when evaluating the blob in **Figure 1**. They found that the dispersive (real) part of the two-photon amplitude yields an electron-to-positron cross-section ratio that deviates from unity at the 1%–2% level in the few-GeV region. The relative smallness of the resonance contribution originates partly because the real parts of the resonance amplitudes change sign in the integrand that enters the evaluation of the box diagram.

Triggered by the experimental discrepancy between polarization transfer and Rosenbluth measurements of the proton FF ratio G_E/G_M at larger Q^2 in recent years, the field has seen a new life. In 2003, Guichon & Vanderhaeghen (20) noticed that the general form of the two-photon exchange graphs could be expressed in an effective current-current form, but with an extra structure beyond those that gave G_M and G_E. Furthermore, if this extra term had the size one might estimate from perturbation theory, then its interference with the one-photon exchange amplitude could be comparable in size to the $(G_E)^2$ term in the Rosenbluth cross section at large Q^2. In addition, there could be ε-dependent modifications to the $(G_M)^2$ term. Hence, there was motivation for a precise evaluation. Realistic calculations of elastic eN scattering beyond the Born approximation are required to demonstrate in a quantitative way that 2γ exchange effects can indeed resolve this discrepancy at larger Q^2. We review several such attempts and describe the present status of this field. On the experimental side, in recent years there have been first attempts to extract the effect of two-photon exchange contributions in a quantitative way from the electron-proton scattering data (21).

Besides offering a way to explain the glaring discrepancy between two methods of measuring the proton electric FF, the study of two-photon processes also contain opportunities to access nucleon structure physics, which surpasses the information contained in nucleon FFs. The possibility arises because a successful two-photon calculation involving a hadronic system requires knowledge of hadronic structure, of a sort that has been available only recently. For example, one line of investigation arises when the virtuality of one or both of the photons in the two-photon process is large compared to a nucleon mass scale. In that case, the hard scale allows one to access the Compton scattering subprocess on a quark within the nucleon. The new (nonperturbative) pieces of information that one then accesses from such a process are the quark correlation functions within the nucleon, also known as generalized parton distributions (GPDs); for recent reviews, see References 22–26. We review how these GPDs can in turn be used to estimate the two-photon exchange diagram of **Figure 1** at large Q^2.

Another line of calculations involves the hadronic corrections to ultrahigh-precision atomic physics experiments, such as the hydrogen hyperfine splitting (hfs). Theoretical predictions for quantities such as the Lamb shift in hydrogen or the hydrogen hfs can currently be performed within quantum electrodynamics to such accuracy that the leading theoretical uncertainties are related to the nuclear size or the nuclear excitation spectrum; for reviews, see, for example, References 27 and 28. For the hfs in hydrogen, which is known to 13 significant figures, current theoretical understanding is at the parts-per-million level. The leading theoretical uncertainty involves the calculation of the two-photon exchange graph of **Figure 1** for zero momentum transfer between the bound electron and the proton, allowing for all nucleon-excited states in the blob. The current theoretical understanding of these two-photon exchange corrections is reviewed in this work.

There are several physical problems in which a one-photon exchange potential is not sufficiently accurate. Besides the description of simple atoms, such as hydrogen and helium, for a precise description of positronium, one must also include two- and multiphoton exchange effects. In particular, in the interaction of electrically neutral systems, such as neutral atoms and molecules, the effect of two-photon exchange gives the dominant contribution to the forces between such systems. For a theoretical review of such dispersion forces, see, for example, Reference 29.

To push the precision frontier further in electron scattering as well as in the hadronic corrections to atomic physics quantities, one needs a good control of 2γ exchange mechanisms and needs to understand how they may or may not affect different observables. This justifies a systematic study of such 2γ exchange effects, both theoretically and experimentally. Besides the real (dispersive) part of the 2γ exchange amplitude, which can be accessed by reversing the sign of the lepton charge, precise measurements of the imaginary part of the 2γ exchange amplitude also became possible in very recent years. The imaginary (absorptive) part of the 2γ exchange amplitude can be accessed through a single spin asymmetry (SSA) in elastic eN scattering, when either the target or beam spin is polarized normal to the scattering plane, as was discussed some time ago (30, 31). As time-reversal invariance forces this SSA to vanish for one-photon exchange, it is of order $\alpha = e^2/(4\pi) \simeq 1/137$. Furthermore, to polarize an ultrarelativistic particle in the direction normal to its momentum involves a suppression factor m/E (with m the mass and E the energy of the particle), which typically is of order $10^{-4}–10^{-3}$ for the electron when the electron beam energy is in the 1 GeV range. Therefore, the resulting target-normal SSA can be expected to be of order 10^{-2}, whereas the beam-normal SSA is of order $10^{-6}–10^{-5}$. Measurements of small asymmetries of the order of parts per million are quite demanding experimentally, but have been performed in very recent years and are also reviewed in this work.

The outline of the present work is as follows. In Section 2, we review the elastic eN scattering beyond the Born approximation and highlight the discrepancy in the extraction of G_E/G_M using polarization transfer and unpolarized (Rosenbluth) measurements. We give a brief review of the different attempts that have recently been made to explain this difference in terms of two-photon exchange corrections, and present in more detail a partonic description at larger momentum transfers. In Section 3, we describe the hadronic corrections of the hydrogen hfs, based on the latest evaluation of the forward polarized structure functions that enter in the calculation of the two-photon exchange diagram. In Section 4, we review the beam and target single spin asymmetries, which measure the imaginary part of the two-photon exchange amplitude. In particular, we give an overview of the recent high-precision measurements in case of a polarized beam with normal beam spin polarization. We conclude in Section 5 and spell out a few open issues in this field.

2. TWO-PHOTON EXCHANGE IN ELASTIC ELECTRON-NUCLEON SCATTERING

2.1. Hard Two-Photon Exchange

In this section, we study attempts to completely evaluate the two-photon exchange contributions to elastic electron-proton scattering, specifically including the exchange of two hard photons, which can probe well inside the proton and which require detailed knowledge of proton structure to evaluate. The immediate motivation, already given in the introduction, is the conflict between the Rosenbluth, with pre-2003 radiative corrections, and the polarization measurements of G_E/G_M for the proton. The difference between the two techniques is a factor of four at $Q^2 = 5.6$ GeV^2. (The data appear on plots later in this section, when we discuss how well the proposed resolutions of the conundrum are working.)

Modern quantitative calculations either treat the hadronic intermediate state in **Figure 1** as a proton plus a set of resonances, or else treat it in a constituent picture using GPDs. In the earliest of the modern calculations, Blunden et al. (32) evaluated the two-photon exchange amplitude, keeping just the elastic nucleon intermediate state. They found that the two-photon exchange correction with an intermediate nucleon has the proper sign and magnitude to partially resolve the discrepancy between the two experimental techniques. Later, the same group, joined by Kondratyuk (34), included contributions of the $\Delta(1232)$ in the intermediate state, which partly canceled the elastic terms. Most recently, Kondratyuk & Blunden (35) included five more baryon resonances in the intermediate state. Although the overall contribution of the additional resonances was not large, the totality of their corrections with their choices for the γ-nucleon-resonance vertices led to good agreement with the Rosenbluth data using the FFs obtained from polarization data.

Borisyuk & Kobushkin (36) also considered two-photon corrections with elastic nucleon intermediate states and used dispersive techniques to reduce the necessary integrals to ones involving the vertex FFs with only space-like momentum transfers. They reduced it further to a single numerical integral for sufficiently low Q^2. (They do not show any Rosenbluth-type plots, but their plotted results for (say) $\delta G_M / G_M$—notation defined below—are in line with results known from what is described in the rest of this section, despite the rather different methodology.)

A group (37, 38), including the present authors, calculated the hard two-photon elastic eN scattering amplitude at large momentum transfers by relating the required virtual Compton process on the nucleon to GPDs, which also enter in other wide-angle scattering processes. This approach effectively sums all possible excitations of inelastic nucleon intermediate states. They found that the two-photon corrections to the Rosenbluth process indeed can substantially reconcile the two ways of measuring G_E / G_M.

Rosenbluth data are also available where the recoiling proton, rather than the electron, is detected (11). These data appear to match the data where the scattered electron was detected. The two-photon exchange contributions are the same regardless of what particle is detected. However, the bremsstrahlung corrections are different. We defer detailed discussion of the proton-detected data pending reassessment of the original (15, 39) and the new (11, 40) proton-observed bremsstrahlung calculations.

2.2. Elastic Electron-Nucleon Scattering Observables

To describe elastic eN scattering,

$$l(k, b) + N(p, \lambda_N) \rightarrow l(k', b') + N(p', \lambda'_N), \qquad 4.$$

where b, b', λ_N, and λ'_N are helicities, we adopt the definitions

$$P = \frac{p + p'}{2}, \quad K = \frac{k + k'}{2}, \quad q = k - k' = p' - p, \qquad 5.$$

define the Mandelstam variables

$$s = (p + k)^2, \quad t = q^2 = -Q^2, \quad u = (p - k')^2, \qquad 6.$$

let $v \equiv K \cdot P$, and let M be the nucleon mass.

The T-matrix helicity amplitudes are given by

$$T^{b'\,b}_{\lambda'_N, \lambda_N} \equiv \langle k', b'; p', \lambda'_N | T | k, b; p, \lambda_N \rangle. \qquad 7.$$

Parity invariance reduces the number of independent helicity amplitudes from 16 to 8. Time-reversal invariance further reduces the number to 6 (41). Furthermore, in a gauge theory, lepton helicity is conserved to all orders in perturbation theory when the lepton mass is zero. We neglect the lepton mass. This finally reduces the number of independent helicity amplitudes to 3, which one may, for example, choose as

$$T_{+,+}^{+,+}; \quad T_{-,-}^{+,+}; \quad T_{-,+}^{+,+} = T_{+,-}^{+,+}. \qquad 8.$$

(The phase in the last equality is for particle momenta in the x-z plane.)

Alternatively, one can expand in terms of a set of three independent Lorentz structures, multiplied by three generalized FFs. One such T-matrix expansion is

$$
\begin{aligned}
T_{b,\lambda'_N \lambda_N} &= \frac{e^2}{Q^2} \bar{u}(k', b) \gamma_\mu u(k, b) \\
&\times \bar{u}(p', \lambda'_N) \left(\tilde{G}_M \gamma^\mu - \tilde{F}_2 \frac{P^\mu}{M} + \tilde{F}_3 \frac{\gamma \cdot K P^\mu}{M^2} \right) u(p, \lambda_N).
\end{aligned} \qquad 9.
$$

The expansion is general. The overall factors and the notations $\tilde{G}_M$ and $\tilde{F}_2$ were chosen (20) to have a straightforward connection to the standard FFs in the one-photon exchange limit.

If desired, one may replace the $\tilde{F}_3$ term by an axial-like term, using the identity

$$
\begin{aligned}
\bar{u}(k')\gamma \cdot P u(k) \times \bar{u}(p')\gamma \cdot K u(p) &= \frac{s-u}{4} \bar{u}(k')\gamma_\mu u(k) \times \bar{u}(p')\gamma^\mu u(p) \\
&+ \frac{t}{4} \bar{u}(k')\gamma_\mu \gamma_5 u(k) \times \bar{u}(p')\gamma^\mu \gamma^5 u(p),
\end{aligned} \qquad 10.
$$

which is valid for massless leptons and any nucleon mass. We continue, however, with the T-matrix in the form shown in Equation 9. An equivalent expansion has also been studied in Reference 42.

The scalar quantities $\tilde{G}_M$, $\tilde{F}_2$, and $\tilde{F}_3$ are complex functions of two variables, say ν and Q^2. We also use

$$\tilde{G}_E \equiv \tilde{G}_M - (1 + \tau)\tilde{F}_2. \qquad 11.$$

To separately identify the one- and two-photon exchange contributions, we use the notations $\tilde{G}_M = G_M + \delta\tilde{G}_M$ and $\tilde{G}_E = G_E + \delta\tilde{G}_E$, where G_M and G_E are the usual proton magnetic and electric FFs, which are functions of Q^2 only and are defined from matrix elements of the electromagnetic current. The amplitudes $\tilde{F}_3 = \delta\tilde{F}_3$, $\delta\tilde{G}_M$, and $\delta\tilde{G}_E$ originate from processes involving the exchange of at least two photons and are of order e^2 (relative to the factor e^2 in Equation 9).

The unpolarized cross section is

$$\frac{d\sigma}{d\Omega_{Lab}} = \frac{\tau \sigma_R}{\varepsilon(1 + \tau)} \frac{d\sigma_{NS}}{d\Omega_{Lab}}, \qquad 12.$$

where the no-structure cross section is

$$\frac{d\sigma_{NS}}{d\Omega_{Lab}} = \frac{4\alpha^2 \cos^2(\theta/2)}{Q^4} \frac{E'^3}{E}, \qquad 13.$$

where E and E' are the incoming and outgoing electron Lab energies. Other quantities are defined after Equation 1. The reduced cross section including the two-photon exchange correction becomes (20)

$$\sigma_R = G_M^2 + \frac{\varepsilon}{\tau}G_E^2 + 2G_M\mathcal{R}\left(\delta\tilde{G}_M + \varepsilon\frac{\nu}{M^2}\tilde{F}_3\right) + 2\frac{\varepsilon}{\tau}G_E\mathcal{R}\left(\delta\tilde{G}_E + \frac{\nu}{M^2}\tilde{F}_3\right) + \mathcal{O}(e^4),$$
14.

where $\mathcal{R}$ stands for the real part.

The general expressions for the double polarization observables, including two-photon exchange, are (20)

$$P_t = A_t = -\sqrt{\frac{2\varepsilon(1-\varepsilon)}{\tau}}\frac{(2b)}{\sigma_R}$$

$$\times \left\{G_E G_M + G_E\mathcal{R}(\delta\tilde{G}_M) + G_M\mathcal{R}\left(\delta\tilde{G}_E + \frac{\nu}{M^2}\tilde{F}_3\right) + \mathcal{O}(e^4)\right\},$$

$$P_l = -A_l = \sqrt{1-\varepsilon^2}\frac{(2b)}{\sigma_R}\left\{G_M^2 + 2G_M\mathcal{R}\left(\delta\tilde{G}_M + \frac{\varepsilon}{1+\varepsilon}\frac{\nu}{M^2}\tilde{F}_3\right) + \mathcal{O}(e^4)\right\},$$
15.

where $b = \pm 1/2$ is the helicity of the electron and where we assumed $m_e = 0$. The polarizations are related to the analyzing powers A_t or A_l by time-reversal invariance. That the polarization P_l is unity in the backward direction $\varepsilon = 0$ follows generally from lepton-helicity conservation and angular-momentum conservation.

2.3. Partonic Calculation of the Two-Photon Exchange

To estimate the two-photon exchange contribution to $\tilde{G}_M$, $\tilde{F}_2$, and $\tilde{F}_3$ at large momentum transfers, a partonic calculation was performed in References 37 and 38, as illustrated in **Figure 2**. In such a calculation, one starts from the subprocess on a quark, denoted by the scattering amplitude H in **Figure 2**. This hard scattering amplitude consists of the two-photon exchange direct and crossed box diagrams of **Figure 3**. The two-photon exchange contribution to the elastic electron scattering off spin 1/2 Dirac particles was first calculated in Reference 43. The equivalent of the amplitude $\tilde{G}_M$ for a quark target, which contains an IR divergence, is separated into a soft and hard part. The soft part corresponds to the situation where one of the photons in **Figure 3** carries zero four-momentum, and is obtained by replacing the

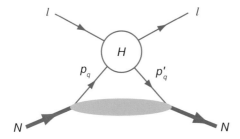

Figure 2

Handbag approximation for the elastic lepton-nucleon scattering at large momentum transfers. In the partonic scattering process (indicated by H), the lepton scatters from quarks in the nucleon, with momenta p_q and p_q'. The blob represents the generalized parton distributions of the nucleon.

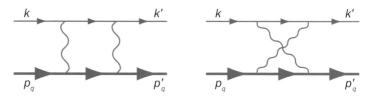

Figure 3

Direct and crossed box diagrams that describe the two-photon exchange contribution to the lepton-quark scattering process, corresponding with the circle denoted by H in **Figure 2.**

other photon's four-momentum by q in both the numerator and denominator of the loop integral (44); for details, see References 37 and 38.

To obtain the amplitude on the nucleon, the quarks are embedded in the proton as described through the nucleon's GPDs. In the resulting calculation of the handbag diagrams, both photons are coupled to the same quark. However, there are also contributions from processes in which the photons interact with different quarks. One can show that the IR contributions from these processes, which are proportional to the products of the charges of the interacting quarks, added to the soft contributions from the handbag diagrams give the same result as the soft contributions calculated with just a nucleon intermediate state (45). Thus, the low-energy theorem for Compton scattering is satisfied. In the handbag approximation, the hard parts that appear when the photons couple to different quarks, the so-called cat's ears diagrams, are neglected. For the real parts, the IR divergence arising from the direct and crossed box diagrams, at the nucleon level, is cancelled when adding the bremsstrahlung contribution from the interference of diagrams where a soft photon is emitted from the electron and from the proton. The remaining finite soft part was calculated in Reference 46, which we verified explicitly, and applies when the outgoing electron is detected.

The resulting finite hard parts of the handbag diagram are then evaluated using a model for the GPDs. The GPDs can be measured in deeply virtual or wide-angle Compton scattering and have been reviewed in References 22–25 and elsewhere. The GPD models used in the following numerical estimates are detailed in References 38 and 47; see also Reference 48.

2.4. Results

2.4.1. Cross section and polarization transfer. **Figure 4** shows the reduced differential cross section for electron-proton scattering σ_R, for two values of Q^2. There are three items on each graph: One is the data; the next is the straight line, which is the result of the 1γ exchange calculation using G_E/G_M, taken from the polarization data (7), with a reasonable and commonly used choice for G_M (49). The slope of this straight line is too flat to fit the data, reflecting the conflict between the polarization measurements and the Rosenbluth measurements with the hard 2γ corrections. Third are the slightly curved lines, showing the results of the 2γ corrections while still using the G_E/G_M ratio from the polarization data. Results are shown for two different GPD models, described in References 37, 38, and 47; they do not differ

Cross section for e-p elastic scattering

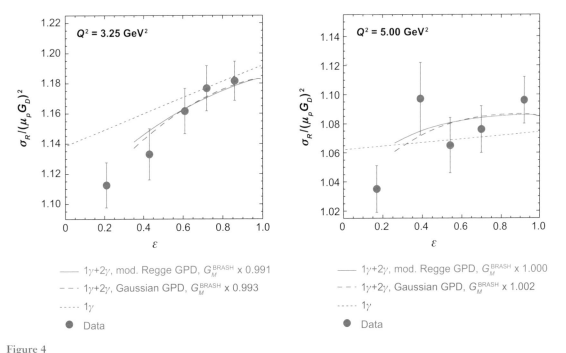

Figure 4

Rosenbluth plots for elastic electron-proton (e-p) scattering: σ_R divided by $(\mu_p G_D)^2$, with $G_D = (1 + Q^2/0.71)^{-2}$. Dotted blue curves represent the Born approximation, using G_E/G_M from polarization data (6, 7). Solid green curves represent the full calculation, using the modified Regge generalized parton distribution (GPD), for the kinematical range $-u > M^2$. Dashed purple curves are the same as the solid curves, but use the Gaussian GPD. The data are taken from Reference 9; the figure and calculation are taken from Reference 38.

greatly. (The renormalization of G_M that we have allowed does not affect the slope.) One sees that the hard 2γ corrections steepen the average slope and improve agreement with the data. Note also the nonlinearity in the Rosenbluth plot, which can be checked with a more precise experiment.

Figure 5 presents the 2γ results in a different way. The plot shows the extracted G_E/G_M versus Q^2. One set of data, falling linearly with Q^2, is from the polarization experiments. Another set of data, roughly constant in Q^2 and plotted with inverted triangles, is from Rosenbluth data analyzed using only the Mo-Tsai (14) radiative corrections. The solid squares show the best-fit G_E/G_M from data from Reference 9 when analyzed including the hard 2γ corrections. One sees that for Q^2 in the 2–3 GeV^2 range, the G_E/G_M extracted using the Rosenbluth method and including the 2γ corrections agree well with the polarization transfer results. At higher Q^2, there is at least partial reconciliation between the two methods. There is one more point on **Figure 5** that shows the result of also including some hard bremsstrahlung corrections, which will be discussed below.

Figure 5

Determinations of the proton G_E/G_M ratio. The polarization data are from Gayou et al. (7) and Punjabi et al. (8), and the Rosenbluth data are from Andivahis et al. (9), which include only the well-known Mo-Tsai corrections. Our Rosenbluth G_E/G_M include the two-photon corrections, and for one point, also a hard bremsstrahlung correction, still using Andivahis et al.'s data. Some of our points for the Rosenbluth results are slightly offset horizontally for clarity.

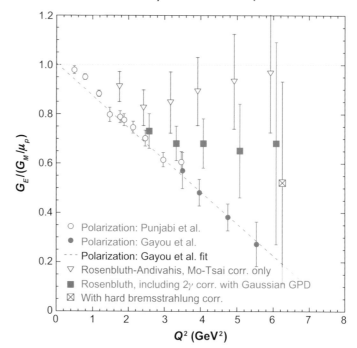

Rosenbluth with 2γ corrections vs. polarization data

○ Polarization: Punjabi et al.
● Polarization: Gayou et al.
--- Polarization: Gayou et al. fit
▽ Rosenbluth-Andivahis, Mo-Tsai corr. only
■ Rosenbluth, including 2γ corr. with Gaussian GPD
⊠ With hard bremsstrahlung corr.

Using the model calculation of References 38 and 47, workers also checked that the 2γ corrections do not impact the polarization measurements as strongly as the Rosenbluth measurements. Most of the effect is on P_s, at smaller ε, which is changed by a few percent relative to the 1γ result and which could be seen in a precise experiment (50). The effect on P_l was negligibly small.

2.4.2. Positron-proton versus electron-proton scattering. Positron-proton and electron-proton scattering have the opposite sign for the two-photon corrections relative to the one-photon terms. Hence, one expects elastic e^+p and e^-p scattering to differ by a few percent. **Figure 6** shows our results for three different Q^2 values. These curves are obtained by adding our two-photon box calculation, minus the corresponding part of the soft-only calculation in Reference 14, to the one-photon calculations; hence, they are meant to be compared with data in which the corrections given in Reference 14 have already been made. Each curve is based on the Gaussian GPD and is cut off at low ε when $-u = M^2$. Early data from SLAC are available (12); more precise data are anticipated from JLab (51). [Reference 12 used the Meister-Yennie (15) soft corrections rather than those of Mo and Tsai. We have checked that for these kinematics, the difference between them is smaller than 0.1%, which is negligible compared to the size of the error bars.]

2.4.3. Results from single-baryon intermediate states. We have focused on a partonic view of two-photon physics. The results when viewing the hadronic

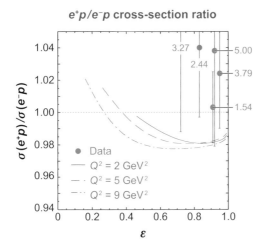

e^+p/e^-p cross-section ratio

Figure 6

Ratio of elastic e^+/e^- cross sections on the proton. The generalized parton distribution calculations of the 2γ correction are for Q^2 of 2, 5, and 9 GeV^2, for the kinematical range with $-u > M^2$. Also shown are all known data, from Reference 12, with $Q^2 > 1.5$ GeV^2 (the missing central value is at 1.111). The numbers near the data give Q^2 for that point in GeV^2.

intermediate state as a proton, or a proton plus a set of resonances, are similar (32–36). The effect, in a calculation with just a proton intermediate state, of the extra 2γ corrections upon extracting G_E/G_M from a Rosenbluth experiment is shown in **Figure 7**. Furthermore, corrections to the polarization experiments are just a few percent, but in the opposite direction from the partonic evaluation, with nearly all the effect coming upon P_t and not upon P_l (33).

2.5. Remarks on Related Topics

Bystritskiy et al. (52) have suggested that hard bremsstrahlung may cause the difference between the Rosenbluth and polarization results. Bremsstrahlung is a process where a real photon is emitted. If the photon energy is sufficiently low, the experimenters will fail to see it and will count the reaction as elastic. Usual bremsstrahlung

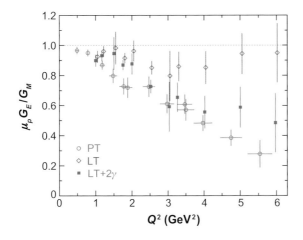

Figure 7

Extracting G_E/G_M with 2γ corrections calculated using a single proton as the hadronic intermediate state. PT is G_E/G_M obtained from the polarization transfer experiments, LT is G_E/G_M obtained from a Rosenbluth experiment using only the Mo-Tsai radiative corrections, and LT+2γ includes the extra 2γ corrections done this way. Figure adapted from Reference 33, and we thank the authors for providing it.

calculations are for soft bremsstrahlung, where the emitted photon energy is kept only to linear order in denominators and entirely omitted in numerators. Soft bremsstrahlung multiplies all amplitudes by the same factor. It changes the slope on a Rosenbluth plot by a well-known amount. If one makes no approximations in the photon energy, there can be different effects on different spin amplitudes. Thus, the claim is that emitted photons energetic enough to affect the spin structure of the calculation but small enough to escape detection give rise to the difference between the two methods of measuring G_E/G_M. A contrasting numerical claim is that hard bremsstrahlung effects are noticeable and helpful in reconciling the Rosenbluth and polarization experiments, but are not decisive. Along these lines is a result from Afanasev (53) for hard bremsstrahlung at $Q^2 = 6$ GeV^2, which has been added to the 2γ results and is included in **Figure 5**. These contrasting claims clearly need adjudication, but an independent reexamination is not available as of this writing.

Electron or muon scattering off deuterons or larger nuclei has not been within the scope of the present review. Larger nuclei have a factor Z advantage in the relative size of the 2γ and 1γ effects, although breakup effects vitiate this advantage for elastic scattering except at low energy. One can examine some of the work that sought evidence of 2γ effects in larger nuclei in References 54–59.

Two-photon exchange effects also affect PV elastic electron-proton scattering via their interference with the lowest-order Z-exchange diagram. Afanasev & Carlson (60) pointed this out and found that the 2γ exchange also led to extra terms with different τ and ε dependences than those known from analyses using only Born diagrams. The calculated size of the effects, using the partonic model at Q^2 of several GeV^2, was $\mathcal{O}(1\%)$. This is below present experimental uncertainties, but PV experiments with $\mathcal{O}(1/2\%)$ uncertainties are planned.

Arrington & Sick (61) considered the effects that the most recent and precise low-Q^2 determinations of G_E and G_M would have upon PV elastic electron-proton scattering and included the two-photon correction terms pointed out in Reference 60. The actual two-photon calculations at their Q^2 were done using single-hadron intermediate states (33, 62).

3. POLARIZABILITY CORRECTIONS TO HYDROGEN HYPERFINE SPLITTING

3.1. Two-Photon Exchange and Atomic Structure

We begin with some explanation of how this piece of atomic physics fits properly into a review of two-photon physics. Hydrogen hfs in the ground state is known to 13 significant figures in frequency units (63),

$$E_{\mathrm{hfs}}(e^- p) = 1\ 420.405\ 751\ 766\ 7(9)\ \mathrm{MHz}. \qquad 16.$$

This accuracy is remarkable to theorists, who currently hope to obtain a calculation accurate to the parts per million level. To reach this goal, some improvement is still needed, and the current best calculations are a few parts per million away from the data.

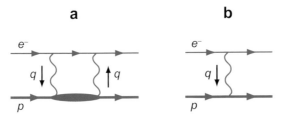

a   b

Figure 8

(a) Generic two-photon
exchange diagram, giving
proton-structure
corrections to hyperfine
splitting. (b) One-photon
exchange diagram.

The main uncertainty in calculating the hfs in hydrogen comes from the
hadronic, or proton structure, corrections. The generic process that contributes is two-photon exchange, shown in **Figure 8a**, which involves the proton
structure because individually each of the exchanged photons could be quite
energetic.

One-photon exchange, **Figure 8b**, does not involve proton structure, at the accuracy needed for the present purpose. The characteristic momentum of the electron
in a hydrogen atom is of $\mathcal{O}(\alpha m_e)$, which is very low on a nuclear physics scale. Hence,
the q^2 of an exchanged single photon is very low, and the variation of the proton
FF from its $q^2 = 0$ value is minimal. One can show that keeping the electron momentum gives corrections of $\mathcal{O}(\alpha m_e/M)$ smaller than what comes from two-photon
exchange. Hence, one sets the momenta of the electrons to zero. (For information,
in the one-photon exchange hfs calculation, there arises a q^2 factor in the numerator
that cancels the $1/q^2$ from the photon propagator; then the neglect of the electron
momenta is safe.)

3.2. Hyperfine Splitting Calculations

The calculated hfs in hydrogen is (63–65)

$$E_{\text{hfs}}(e^- p) = \left(1 + \Delta_{\text{QED}} + \Delta^p_{\text{weak}} + \Delta_{\text{str}}\right) E^p_F; \qquad 17.$$

the two-photon exchange lies in the structure-dependent term Δ_{str}. The Fermi energy
is

$$E^p_F = \frac{8\alpha^3 m_r^3}{3\pi} \mu_B \mu_p = \frac{16\alpha^2}{3} \frac{\mu_p}{\mu_B} \frac{R_\infty}{(1 + m_e/M)^3}, \qquad 18.$$

where $m_r = m_e M/(M + m_e)$ is the reduced mass. By convention, in E^p_F, one uses the
actual magnetic moment for the proton and the Bohr magneton for the electron (note
that μ_B can be used to replace the electron mass), and R_∞ is the Rydberg constant
in frequency units. The second form allows optimal accuracy in evaluating E^p_F. The
least accurately known quantity is the ratio μ_p/μ_B, which is known to eight significant
figures. Hence, to the parts-per-million level, E^p_F is known more than sufficiently well.
The QED (27, 28) and weak interaction corrections (66) are well known and are not
discussed, except to mention that the QED corrections could be obtained, for the
present purposes, without calculation. They are the same as for muonium, so it is
possible to obtain them to more than adequate accuracy using muonium hfs data and
a judicious scaling (67, 68).

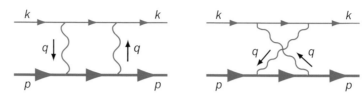

Figure 9

Two-photon exchange diagrams for the elastic proton-structure corrections to hyperfine splitting.

The structure-dependent corrections are, in a standard treatment, divided into Zemach, recoil, and polarizability corrections:

$$\Delta_{\text{str}} = \Delta_Z + \Delta_R^p + \Delta_{\text{pol}}. \qquad 19.$$

The first two terms arise when the blob in **Figure 8a** is just a proton, as in **Figure 9**, and together they are known as the elastic corrections. The electron-photon vertex is well known, and the proton-photon vertex is given by (69, 70)

$$\Gamma_\mu = \gamma_\mu F_1(Q^2) + \frac{i}{2M}\sigma_{\mu\nu}q^\nu F_2(Q^2) \qquad 20.$$

(for the photon with incoming momentum q), if the intermediate proton is on-shell. Of course, in general it is not. However, one can show that the imaginary part of the diagram does come only from kinematics where the intermediate electron and proton are on-shell. Hence, one can correctly use the above vertex to calculate the imaginary part of the diagrams and then obtain the whole of the diagram using a dispersion relation (71).

In the nonrelativistic limit, the recoil terms are zero and the Zemach term is not. (The nonrelativistic limit is $M \to \infty$, with m_e and the proton size fixed. Proton size information is embedded in the FFs F_1 and F_2.) The Zemach term (70) was calculated long ago, and in modern form is

$$\Delta_Z = \frac{8\alpha m_r}{\pi} \int_0^\infty \frac{dQ}{Q^2}\left[G_E(Q^2)\frac{G_M(Q^2)}{1+\kappa_p} - 1\right] = -2\alpha m_r r_Z, \qquad 21.$$

where $\kappa_p \simeq 1.79$ is the proton anomalous magnetic moment and where the last equality defines the Zemach radius r_Z. The charge and magnetic FFs are linear combinations of F_1 and F_2:

$$\begin{aligned} G_M &= F_1 + F_2, \\ G_E &= F_1 - \frac{Q^2}{4M^2}F_2. \end{aligned} \qquad 22.$$

The recoil corrections are not displayed in this review because they are somewhat long (although not awful; see References 69 and 70). An important point is that they do depend on the FFs and, hence, upon the proton structure. However, their numerical value is fairly steady by present standards of accuracy when they are evaluated using different up-to-date analytic FFs based on fits to the scattering data.

One obtains the polarizability corrections by summing all contributions where the intermediate hadronic state, the blob in **Figure 8a**, is not a single photon. Paralleling the elastic case, one can show that the imaginary part of this diagram comes only from configurations where the intermediate electron plus hadronic state is kinematically on-shell, that is, physically realizable. Hence, one can calculate the imaginary part of the box diagram if one has data on inelastic electron-proton scattering. Then, further paralleling the elastic case, one obtains the full box diagram via a dispersion relation (71).

To see some detail, the lower half of **Figure 8a** is the same as forward Compton scattering of off-shell photons from protons, which is given in terms of the matrix element

$$T_{\mu\nu}(q, p, S) = \frac{i}{2\pi M} \int d^4\xi \, e^{iq\cdot\xi} \, \langle pS|T\{j_\mu(\xi), \, j_\nu(0)\}|pS\rangle, \qquad 23.$$

where j_μ is the electromagnetic current and the states are proton states of momentum p and spin-4 vector S. The spin dependence is in the antisymmetric part:

$$T^A_{\mu\nu} = \frac{i}{M\nu}\varepsilon_{\mu\nu\alpha\beta}q^\alpha \left[(H_1(\nu, q^2) + H_2(\nu, q^2))S^\beta - H_2(\nu, q^2)\frac{S\cdot q p^\beta}{p\cdot q}\right]. \qquad 24.$$

The two structure functions H_1 and H_2 depend on q^2 and on the lab frame photon energy ν, defined by $M\nu = p\cdot q$.

The optical theorem relates the imaginary part of the forward Compton amplitude to the inelastic scattering cross section for off-shell photons on protons. The relations are precisely

$$\text{Im } H_1(\nu, q^2) = \frac{1}{\nu}g_1(\nu, Q^2) \quad \text{and} \quad \text{Im } H_2(\nu, q^2) = \frac{M}{\nu^2}g_2(\nu, Q^2), \qquad 25.$$

where g_1 and g_2 are functions appearing in the cross section and are measured (73–77) at SLAC, HERMES, JLab, and elsewhere.

Using the Compton amplitude to evaluate the inelastic part of the two-photon loops leads to

$$\Delta_{\text{pol}} = \frac{E_{2\gamma}}{E_F}\bigg|_{\text{inel}} = \frac{2\alpha m_e}{(1+\kappa_p)\pi^3 M}\int \frac{d^4Q}{(Q^4 + 4m_e^2 Q_0^2)\, Q^2}$$
$$\times \left\{(2Q^2 + Q_0^2)\, H_1^{\text{inel}}(iQ_0, -Q^2) - 3Q^2 Q_0^2 H_2^{\text{inel}}(iQ_0, -Q^2)\right\}, \qquad 26.$$

where we have Wick rotated the integral so that $Q_0 = -i\nu$, $\vec{Q} = \vec{q}$, and $Q^2 \equiv Q_0^2 + \vec{Q}^2$. The dispersion relation that gives H_1 is, assuming no subtraction,

$$H_1^{\text{inel}}(\nu_1, q^2) = \frac{1}{\pi}\int_{\nu_{th}^2}^\infty d\nu \frac{\text{Im}\, H_1(\nu, q^2)}{\nu^2 - \nu_1^2}, \qquad 27.$$

where the integral is only over the inelastic region ($\nu > \nu_{th}$, the inelastic threshold), and a similar relation holds for H_2. (The no-subtraction assumption is discussed in Reference 78.)

Integrating what can be integrated, and neglecting m_e inside the integral, yields the expression (79–84)

$$\Delta_{\text{pol}} = \frac{\alpha m_e}{2(1 + \kappa_p)\pi M}(\Delta_1 + \Delta_2), \tag{28.}$$

where, with $\tau = \nu^2/Q^2$,

$$\Delta_1 = \frac{9}{4}\int_0^\infty \frac{dQ^2}{Q^2}\left\{F_2^2(Q^2) + 4M\int_{\nu_{th}}^\infty \frac{d\nu}{\nu^2}\beta(\tau)g_1(\nu, Q^2)\right\},$$

$$\Delta_2 = -12M\int_0^\infty \frac{dQ^2}{Q^2}\int_{\nu_{th}}^\infty \frac{d\nu}{\nu^2}\beta_2(\tau)g_2(\nu, Q^2). \tag{29.}$$

The auxiliary functions are

$$\beta(\tau) = \frac{4}{9}[-3\tau + 2\tau^2 + 2(2 - \tau)\sqrt{\tau(\tau + 1)}],$$

$$\beta_2(\tau) = 1 + 2\tau - 2\sqrt{\tau(\tau + 1)}. \tag{30.}$$

The integral for Δ_1 actually requires further comment. Only the second term comes from the procedure just outlined; historically, it was thought convenient to add the first term and then subtract the same term from the recoil corrections. This stratagem allows the electron mass to be taken to zero in Δ_1. The individual terms in Δ_1 diverge (they would not had the electron mass been kept), but the whole is finite because of the Gerasimov-Drell-Hearn (85, 86) sum rule,

$$4M\int_{\nu_{th}}^\infty \frac{d\nu}{\nu^2}g_1(\nu, 0) = -\kappa_p^2, \tag{31.}$$

coupled with the observation that the auxiliary function $\beta(\tau)$ becomes unity as we approach the real-photon point.

3.3. Numerics, Especially for Δ_{pol}

The polarizability corrections have some history. Considerations of Δ_{pol} were begun by Iddings in 1965 (79), improved by Drell & Sullivan in 1966 (80), and given the present notation by De Rafael in 1971 (81). However, no sufficiently good spin-dependent data existed, so it was several decades before the formula could be evaluated to a result incompatible with zero. In 2002, Faustov & Martynenko (83) became the first to use $g_{1,2}$ data to obtain results inconsistent with zero. They obtained

$$\Delta_{\text{pol}}(\text{F \& M 2002}) = (1.4 \pm 0.6) \text{ ppm}. \tag{32.}$$

However, only data from SLAC were available. None of the SLAC data had Q^2 below 0.30 GeV^2; Δ_1 and Δ_2 are sensitive to the behavior of the structure functions at low Q^2. Also in 2002 there appeared analytic expressions for $g_{1,2}$ fit to data by Simula et al. (87), which included JLab as well as SLAC data. Simula et al. did not integrate their results to obtain Δ_{pol}, but had they done so, they would have obtained $\Delta_{\text{pol}} = (0.4 \pm 0.6)$ ppm (84).

More recently, Faustov et al. (88) reanalyzed their result for Δ_{pol} and obtained a somewhat larger value,

$$\Delta_{\text{pol}} (\text{FGM 2006}) = (2.2 \pm 0.8) \text{ ppm.} \qquad 33.$$

The data underlying this result, however, still do not include the lower Q^2 data from JLab, which are noted immediately below.

Data for $g_1(v, q^2)$ are improved thanks to the EG1 experiment at JLab, which had its first data run in 2000–2001. A preliminary analysis of these data became available in 2005 (77); final data are anticipated soon. The Q^2 range measured in this experiment went down to 0.05 GeV^2. Using analytic forms checked against the preliminary data, Nazaryan et al. (84) evaluated Δ_{pol} and obtained

$$\Delta_{\text{pol}} (\text{NCG 2006}) = (1.3 \pm 0.3) \text{ ppm.} \qquad 34.$$

This is similar to the 2002 Faustov-Martynenko result, but with a claim that the newer data allow a smaller uncertainty limit.

Table 1 lists the numerical values of the corrections compared with the experimental value of the hfs. For the polarizability corrections, we used the value from Reference 84 on the grounds that it was based on the most complete inelastic electron-proton scattering data. For the Zemach term, we used the value (90) based on Sick's (91) FF fits, because those fits emphasized the low-Q^2 elastic scattering data that dominate the Zemach integral. The values for the recoil terms and weak interaction corrections have lower uncertainty limits. We took the former from Reference 67, and they are also discussed in Reference 64; the latter may be found in References 64 and 66.

Thus, the sum of what one may argue are the best calculated corrections falls short of the data by approximately 2 ppm, or approximately 2.8 standard deviations. Of course, some judgment has entered the choice of numbers. Other FF fits to ostensibly

Table 1   Up-to-date corrections to hydrogenic hyperfine structure

Quantity	Value (ppm)	Uncertainty (ppm)
$(E_{\text{hfs}}(e^- p)/E_F^p) - 1$	1103.48	0.01
Δ_{QED}	1136.19	0.00
Δ_Z (using Reference 90)	−41.59	0.46
Δ_R^p	5.84	0.15
Δ_{pol} (from Reference 84)	1.30	0.30
Δ_{weak}^p	0.06	–
Total	1101.80	0.60
Deficit	1.68	0.60

The first row with numbers gives the target value based on the experimental data and the best evaluation of the Fermi energy (known to eight figures), derived from known physical constants. The corrections are listed next. [The Zemach term includes a 1.53% correction from higher-order electronic contributions (89), as well as a +0.07 ppm correction from muonic vacuum polarization and a +0.01 ppm correction from hadronic vacuum polarization (64).] The total of all corrections is 1.68 ± 0.60 ppm short of the experimental value.

the same data give different values for Δ_Z; for example, Kelly's (91) FFs lead to $\Delta_Z = -40.93 \pm 49$ ppm. Note that using the Kelly-based value of Δ_Z and the latest value of Δ_{pol} from Faustov et al. leads to excellent agreement of the measured hfs.

3.4. Remarks and Prospects

Thus, the calculated hfs in atomic hydrogen is an example of two-photon physics and requires proton-structure information measured at nuclear and particle physics laboratories. Until 2006, the largest uncertainty was in the proton polarizability corrections, which are related to data from polarized inelastic electron-proton scattering. The numerical value of the polarizability contributions to hydrogen hyperfine structure, based on the most recent proton-structure function data is $\Delta_{\mathrm{pol}} = (1.3 \pm 0.3)$ ppm. This is quite similar to the Faustov-Martyenko 2002 result, which we think is remarkable given the improvement in the data upon which it is based. Most of the calculated Δ_{pol} comes from integration regions where the photon four-momentum squared is small, $Q^2 < 1$ GeV^2.

There is still a modest discrepancy between what we think are the best hydrogen hfs calculations and experiment, on the order of 2 ppm. Where might the problem lie? It may be in the use of the unsubtracted dispersion relation; or it may be in the value of the Zemach radius, which taken at face value now contributes the largest single uncertainty among the hadronic corrections to hfs; or perhaps it is a low-Q^2 surprise in g_1 or g_2. It is, at any rate, not a statistical fluctuation in the hfs data.

An interplay between the fields of atomic and nuclear or particle physics may be relevant to sorting out the problem. For example, the best values of the proton charge radius currently come from small corrections accurately measured in the atomic Lamb shift (91). The precision of the atomic measurement of the proton charge radius can increase markedly if the Lamb shift is measured in muonic hydrogen (92), and data may be taken in 2007 at the Paul Scherrer Institute. In the present context, the charge radius is noticed by its effect on determinations of the Zemach radius.

We close this section by noting that the final EG1 data analysis from JLab/CLAS should be released soon, and this may shift the value of Δ_{pol} somewhat. We also note that one can keep the lepton masses so as to calculate muonic hydrogen hfs, and such calculations have already appeared (88, 93).

4. BEAM-NORMAL AND TARGET-NORMAL SPIN ASYMMETRIES

In this section we discuss the imaginary part of the two-photon exchange amplitude. It can be accessed in the beam- or target-normal spin asymmetries in elastic eN scattering, and it measures the nonforward structure functions of the nucleon. After briefly reviewing the theoretical formalism, we discuss calculations in the threshold region, in the resonance region, and in the diffractive region—corresponding to high energy and forward angles—as well as in the hard scattering region.

The imaginary (absorptive) part of the 2γ exchange amplitude can be accessed through an SSA in elastic eN scattering, when either the target or beam spin is polarized normal to the scattering plane, as was discussed some time ago (30, 31, 94). As time-reversal invariance forces this SSA to vanish for one-photon exchange, it is of order $\alpha = e^2/(4\pi) \simeq 1/137$. Furthermore, to polarize an ultrarelativistic particle in the direction normal to its momentum involves a suppression factor m/E (with m the mass and E the energy of the particle), which for the electron is of order 10^{-4}–10^{-3} when the electron beam energy is in the 1-GeV range. Therefore, the resulting target-normal SSA is expected to be of order 10^{-2}, whereas the beam-normal SSA is of order 10^{-6}–10^{-5}. A measurement of such small asymmetries is quite demanding experimentally. However, in the case of a polarized lepton beam, asymmetries of the order of parts per million are currently accessible in PV elastic eN scattering experiments. The PV asymmetry involves a beam spin polarized along its momentum. However, the SSA for an electron beam spin normal to the scattering plane can also be measured using the same experimental setups. First measurements of this beam-normal SSA at beam energies up to 1 GeV have yielded values near -10 ppm (95–97) in the forward angular range and up to an order of magnitude larger in the backward angular range (98). At higher beam energies, first results for the beam-normal SSA in elastic eN scattering experiments were also reported recently (97, 99, 100).

First estimates of the target-normal SSA in elastic eN scattering were performed in References 31 and 94. In those works, the 2γ exchange with a nucleon intermediate state (so-called elastic or nucleon pole contribution) was calculated, and the inelastic contribution was estimated in a very forward angle approximation. Estimates within this approximation have also been reported for the beam-normal SSA (101). The general formalism for elastic eN scattering with lepton-helicity flip, which is needed to describe the beam-normal SSA, was developed in Reference 102. Furthermore, the beam-normal SSA was also estimated at large momentum transfers Q^2 in Reference 102 using a parton model, which was crucial (37) for interpreting the results from unpolarized elastic eN scattering, as discussed in Section 2. In the handbag model of References 37, 38, and 102, the corresponding 2γ exchange amplitude was expressed in terms of GPDs, and the real and imaginary parts of the 2γ exchange amplitude were related through a dispersion relation. Hence, in the partonic regime, a direct comparison of the imaginary part with experiment can provide a very valuable cross-check on the calculated result for the real part.

To extend such a dispersion-relation formalism to lower momentum transfers, as was discussed some time ago in the literature (56, 57), the necessary first step is a precise knowledge of the imaginary part of the two-photon exchange amplitude, which enters in both the beam- and target-normal SSAs. Using unitarity, one can relate the imaginary part of the 2γ amplitude to the electro-absorption amplitudes on a nucleon (see **Figure 10**).

4.1. The Imaginary (Absorptive) Part of the Two-Photon Exchange Amplitude

An observable that is directly proportional to the imaginary part of the two- (or multi-) photon exchange is given by the elastic scattering of an unpolarized electron

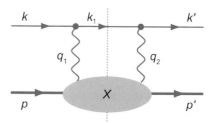

Figure 10

The two-photon exchange diagram. The blob represents the response of the nucleon to the scattering of the virtual photon. In the imaginary part of the two-photon amplitude, the intermediate state indicated by the vertical dashed line is on-shell.

on a proton target polarized normal to the scattering plane (or the recoil polarization normal to the scattering plane, which is exactly the same, assuming time-reversal invariance). For a target polarized perpendicular to the scattering plane, the corresponding SSA, which we refer to as the target-normal spin asymmetry (A_n), is defined by

$$A_n = \frac{\sigma_\uparrow - \sigma_\downarrow}{\sigma_\uparrow + \sigma_\downarrow}, \qquad 35.$$

where $\sigma_\uparrow$ ($\sigma_\downarrow$) denotes the cross section for an unpolarized beam and for a nucleon spin parallel (antiparallel) to the normal polarization vector, defined as

$$S^\mu = (0, \vec{S}_n), \quad \vec{S}_n \equiv (\vec{k} \times \vec{k}')/|\vec{k} \times \vec{k}'|. \qquad 36.$$

As De Rujula et al. (31) have shown, the target (or recoil) normal spin asymmetry is related to the absorptive part of the elastic eN scattering amplitude as

$$A_n = \frac{2 \, \mathrm{Im} \left(\sum_{\mathrm{spins}} T_{1\gamma}^* \cdot \mathrm{Abs} T_{2\gamma} \right)}{\sum_{\mathrm{spins}} |T_{1\gamma}|^2}, \qquad 37.$$

where $T_{1\gamma}$ denotes the one-photon exchange amplitude. Because the one-photon exchange amplitude is purely real, the leading contribution to A_n is of order $O(e^2)$, and is due to an interference between one- and two-photon exchange amplitudes. When neglecting terms that correspond with electron-helicity flip (i.e., setting $m_e = 0$), A_n can be expressed in terms of the invariants for elastic eN scattering, defined in Equation 9 as (37)

$$A_n = \sqrt{\frac{2\,\varepsilon\,(1+\varepsilon)}{\tau}} \left(G_M^2 + \frac{\varepsilon}{\tau} G_E^2 \right)^{-1} \times \left\{ - G_M \mathcal{I} \left(\delta\tilde{G}_E + \frac{\nu}{M^2} \tilde{F}_3 \right). \right.$$
$$\left. + G_E \mathcal{I} \left(\delta\tilde{G}_M + \left(\frac{2\varepsilon}{1+\varepsilon} \right) \frac{\nu}{M^2} \tilde{F}_3 \right) \right\} + \mathcal{O}(e^4), \qquad 38.$$

where $\mathcal{I}$ denotes the imaginary part.

For a beam polarized perpendicular to the scattering plane, one can also define an SSA analogously, as in Equation 35 and as noted in Reference 101, where $\sigma_\uparrow$ ($\sigma_\downarrow$) now denotes the cross section for an unpolarized target and for an electron beam spin parallel (antiparallel) to the normal polarization vector, given by Equation 36.

We refer to this asymmetry as the beam-normal spin asymmetry (B_n). It vanishes explicitly when $m_e = 0$, as it involves an electron-helicity flip. The general eN scattering amplitude including lepton-helicity flip involves six invariant amplitudes and has been worked out in Reference 102, where the expression for B_n can also be found. As for A_n, B_n also vanishes in the Born approximation and is therefore of order e^2. The imaginary part of the two-photon exchange amplitude, which enters the beam- or target-normal SSA, is related to the absorptive part of the doubly virtual Compton scattering tensor on the nucleon, as shown in **Figure 10**. The discontinuity of the two-photon exchange amplitude, shown in **Figure 10**, can be expressed as

$$\text{Abs}\, T_{2\gamma} = e^4 \int \frac{d^3\vec{k}_1}{(2\pi)^3 2E_{k_1}} \bar{u}(k', h')\gamma_\mu(\gamma \cdot k_1 + m_e)\gamma_\nu u(k, h)\frac{1}{Q_1^2 Q_2^2} W^{\mu\nu}(p', \lambda'_N; p, \lambda_N),$$

39.

where the momenta are defined as indicated on **Figure 10**, with $q_1 \equiv k - k_1$, $q_2 \equiv k' - k_1$, and $q_1 - q_2 = q$. Here $h(h')$ denote the helicities of the initial (final) electrons, and $\lambda_N(\lambda'_N)$ denote the helicities of the initial (final) nucleons. In Equation 39, the hadronic tensor $W^{\mu\nu}(p', \lambda'_N; p, \lambda_N)$ corresponds with the absorptive part of the doubly virtual Compton scattering tensor with two space-like photons:

$$W^{\mu\nu}(p', \lambda'_N; p, \lambda_N) = \sum_X (2\pi)^4 \delta^4(p + q_1 - p_X)\langle p', \lambda'_N | J^{\dagger\mu}(0)|X\rangle$$
$$\times \langle X|J^\nu(0)|p, \lambda_N\rangle,$$

40.

where the sum goes over all possible on-shell intermediate hadronic states X (denoting $p_X^2 \equiv W^2$). Note that in the limit $p' = p$, Equation 40 reduces to the forward tensor for inclusive eN scattering and can be parametrized by the usual four-nucleon forward structure functions. In the nonforward case, however, the absorptive part of the doubly virtual Compton scattering tensor of Equation 40, which enters in the evaluation of target- and beam-normal spin asymmetries, depends upon 18 invariant amplitudes (103). Although this may seem to be a forbiddingly large number of new functions, we may use the unitarity relation to express the full nonforward tensor in terms of electroproduction amplitudes $\gamma^* N \to X$. The number of intermediate states X that one considers in the calculation will then limit how high in energy one can reliably calculate the hadronic tensor in Equation 40. In the following section, the tensor $W^{\mu\nu}$ is discussed for the elastic contribution ($X = N$) in (a) the resonance region as a sum over all πN intermediate states (i.e., $X = \pi N$), using a phenomenological state-of-the-art calculation for the $\gamma^* N \to \pi X$ amplitudes, (b) the diffractive region (corresponding with high-energy, forward scattering) where it can be related to the total photo-absorption cross section on a proton, and (c) the hard scattering region where it can be related to nucleon GPDs.

There are special regions in the phase-space integral of Equation 39, corresponding with near singularities, which may give important contributions (logarithmic enhancements) under some kinematical conditions. When one of the photons becomes soft and the second carries the full momentum transfer, one speaks of the quasi-virtual Compton scattering, whereas when the intermediate electron line is soft, both photons have near-zero virtuality, corresponding with quasi-real Compton scattering.

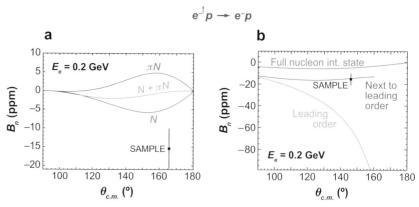

Figure 11

Beam-normal spin asymmetry for $e^{-\uparrow} p \rightarrow e^- p$ at a beam energy $E_e = 0.2$ GeV as a function of the *c.m.* scattering angle. (*a*) Calculation from Reference 105 for different hadronic intermediate states (X) in the blob of **Figure 10**: N (*red curve*), πN (*blue curve*), sum of N and πN (*gray curve*). (*b*) Calculation from Reference 104, where the nucleon intermediate state is expanded to leading order (*orange curve*) and next to leading order (*purple curve*) in E_e/M. For comparison, the full nucleon intermediate state result (*red curve*) is also shown. The data point is from the SAMPLE Collaboration (95).

4.2. Results and Discussion

4.2.1. Threshold region. In Reference 104, the beam-normal spin asymmetry was studied at low energies in an effective theory of electrons, protons, and photons. This calculation, in which pions are integrated out, effectively corresponds with the nucleon intermediate state contribution only, expanded to second order in E_e/M. To this order, the calculation includes the recoil corrections to the scattering from a point charge, the nucleon charge radius, and the nucleon isovector magnetic moment. **Figure 11** (right panel) shows that the theory expanded up to second order in E_e/M (indicated by the full results) can give a good account of the SAMPLE data point at the low-energy $E_e = 0.2$ GeV.

However, when doing the full calculation for the nucleon intermediate state, which is model independent (as it only involves on-shell $\gamma^* NN$ matrix elements), the result is further reduced, as seen in **Figure 11** (left panel). Inclusion of threshold pion electroproduction contributions, arising from the πN intermediate states, partly cancels the elastic contributions. Because in this low-energy region the matrix elements are rather well known, it is not clear at present how to get a better agreement with the rather large asymmetry measured by SAMPLE (95).

4.2.2. Resonance region. When measuring the imaginary part of the elastic eN amplitude through a normal SSA at sufficiently low energies—below or around two-pion production threshold—using pion electroproduction experiments as input, one is in a regime where these electroproduction amplitudes are relatively well known. As both photons in the 2γ exchange process are virtual and integrated over, an observable such as the beam- or target-normal SSA is sensitive to the electroproduction amplitudes on the nucleon for a range of photon virtualities. This may provide information on resonance transition FFs complementary to the information obtained from current pion electroproduction experiments.

In Reference 105, the imaginary part of the two-photon exchange amplitude was calculated by relating it through unitarity to the contribution of $X = N$ and $X = \pi N$ intermediate state contributions. For the πN intermediate state contribution, the corresponding pion electroproduction amplitudes were taken from the phenomenological MAID analysis (106), which contains both resonant and nonresonant pion production mechanisms. The calculation in Reference 105 shows that at forward angles, the quasi-real Compton scattering at the endpoint $W = W_{\mathrm{max}}$ yields only a very small contribution, which grows larger when going to backward angles. This quasi-RCS contribution is of opposite sign as the remainder of the integrand and therefore determines the position of the maximum (absolute) value of B_n when going to backward angles.

In **Figure 12**, the results for B_n are shown at different beam energies below $E_e = 1$ GeV. Clearly, at energies $E_e = 0.3$ GeV and higher, the nucleon intermediate state (elastic part) yields only a very small relative contribution. Therefore, B_n is a direct measure of the inelastic part that gives rise to sizably large asymmetries, of the order of several tens of parts per million in the backward angular range, driven mainly by the quasi-RCS near singularity. First results from the A4 Collaboration for B_n at backward angles (for E_e around 0.3 GeV) indeed point toward a large B_n value of order -100 ppm for $\theta_{c.m.}$ around 150° (98). At forward angles, the sizes of the predicted asymmetries are compatible with the first high-precision measurements performed by the A4 Collaboration (96), although the model slightly overpredicts (in absolute value) B_n at $E_e = 0.570$ GeV and 0.855 GeV.

4.2.3. High-energy, forward scattering (diffractive) region. References 107 and 108 showed that at very high energies and forward scattering angles (so-called diffractive limit), B_n is dominated by the quasi-real Compton singularity. In this (extreme forward limit) case, the hadronic tensor can be expressed in terms of the total photoabsorption cross section on the proton $\sigma_{\mathrm{tot}}^{\gamma p}$, allowing us to express B_n through the simple analytic expression

$$B_n = -\frac{m_e \sqrt{Q^2} \sigma_{\mathrm{tot}}^{\gamma p}}{8\pi^2} \frac{G_E}{\tau G_M^2 + \varepsilon G_E^2} \left[\log \frac{Q^2}{m_e^2} - 2 \right]. \qquad 41.$$

Note that the quasi-real Compton singularity gives rise to a (single) logarithmic enhancement factor that is at the origin of the relatively large value of B_n.

In **Figure 13**, the estimate from Reference 107, which is based on Equation 41, is shown for different parameterizations of the total photo-absorption cross section. The beam-normal spin asymmetry has been measured at SLAC (E-158) at $E_e = 46$ GeV ($\sqrt{s} \simeq 9$ GeV) and very forward angle ($Q^2 \simeq 0.05$ GeV²). First results (99) indicate a value of $B_n \simeq -3.5 \rightarrow -2.5$ ppm, confirming the estimate shown in **Figure 13**.

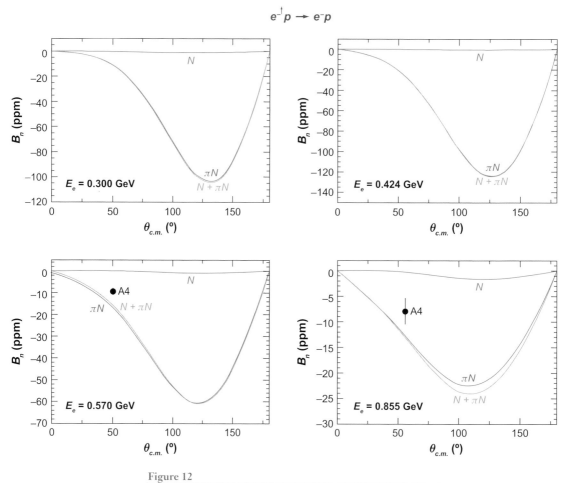

Figure 12

B_n for $e^{-\uparrow} p \to e^- p$ as a function of the *c.m.* scattering angle at different beam energies, as indicated in the figure. The calculations are for different hadronic intermediate states (X) in the blob of **Figure 10**: N (*red curves*), πN (*blue curves*), sum of N and πN (*gray curves*). The data points are from the A4 Collaboration (MAMI) (96). Calculations and figure adapted from Reference 105.

At intermediate energies, around $E_e \simeq 3$ GeV, and forward angles, B_n has also been measured by the HAPPEX and G0 Collaborations. The simple diffractive formula of Equation 41 does not apply rigorously any more, and one must calculate corrections due to the deviation from forward scattering. Such calculations were recently performed in References 107 and 109 in different model approaches, where the calculation from Reference 109 includes subleading terms in Q^2. The predicted asymmetries are in basic agreement with first results reported by HAPPEX (97) and

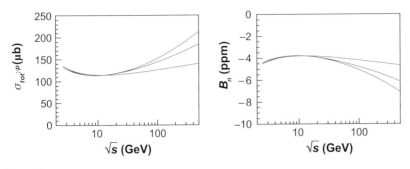

Figure 13

Energy dependence of B_n for $e^{-\uparrow} p \rightarrow e^- p$ (*right*) at $Q^2 = 0.05$ GeV^2 (corresponding with very forward scattering angles), using Equation 41 for different parameterizations of the total photo-absorption cross section on the proton. (*Left*) Block-Halzen $\log$ fit (*purple curves*), Block-Halzen $\log^2$ fit (*red curves*), and Donnachie-Landshoff fit (*blue curves*). Calculations taken from Reference 107.

G0 (100). In **Table 2**, the present status of measurements of beam-normal spin asymmetries, most of them in the forward angular range, are shown.

4.2.4. Hard scattering region. In the hard scattering region, the beam- and target-normal spin asymmetries were estimated in References 38 and 102 through the scattering off a parton, which is embedded in the nucleon through a GPD. Using the same phenomenological parametrizations for the GPDs, workers found that B_n yielded values around +1 ppm to +1.5 ppm in the few-GeV beam energy range (see **Figure 14**). In particular, the forward angular range for $e^{-\uparrow} p \rightarrow e^- p$ scattering was a favorable region to get information on the inelastic part of B_n. Because in the handbag calculation real and imaginary parts are linked, a direct measurement of B_n may yield a valuable cross-check for the real part, which was crucial in understanding the unpolarized cross-section data for $e^- p \rightarrow e^- p$ at large momentum transfer. A measurement of A_n at intermediate Q^2 values on a neutron target is also planned (110).

Table 2   Summary of measurements of the beam-normal spin asymmetry in elastic electron-proton scattering

Collaboration	E_e (GeV)	Q^2 (GeV^2)	B_n (ppm)
SAMPLE (95)	0.192	0.10	-16.4 ± 5.9
A4 (96)	0.570	0.11	-8.59 ± 0.89
A4 (96)	0.855	0.23	-8.52 ± 2.31
HAPPEX (97)	3.0	0.11	-6.7 ± 1.5
G0 (100)	3.0	0.15	-4.06 ± 1.62
G0 (100)	3.0	0.25	-4.82 ± 2.85
E-158 (99)	46	0.06	$-3.5 \rightarrow -2.5$

Figure 14

B_n for elastic $e^- p$ scattering as function of ε at different values of Q^2, as indicated in the figure. The thick upper curves ($B_n > 0$) are the generalized parton distribution calculations for the kinematical range where $s, -u > M^2$. For comparison, the elastic contribution is also displayed (*thin lower curves*; $B_n < 0$). Calculations and figure adapted from Reference 102.

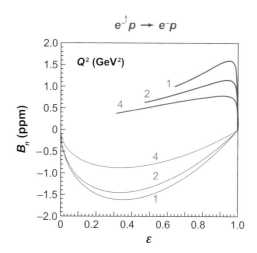

5. CONCLUSIONS

The striking difference between the unpolarized (Rosenbluth) and polarization transfer measurements of the proton G_E/G_M FF ratio has triggered a renewed interest in the field of two-photon exchange in eN scattering experiments. Theoretical calculations within both a hadronic and partonic framework, which were reviewed here, made it very likely that hard two-photon exchange corrections are the main culprit in the difference between both experimental techniques. Despite the long history of two-photon exchange corrections, it is interesting to note that concepts developed over the past decade, such as GPDs that describe two-photon processes with one or two large photon virtualities, enter when quantifying two-photon exchange corrections at larger Q^2.

The model-independent finding is that the hard two-photon corrections hardly affect polarization transfer results, but they do correct the slope of the Rosenbluth plots at larger Q^2 in an important way: toward reconciling the Rosenbluth data with the G_E/G_M ratio, which decreases linearly in Q^2, from the polarization data. This being said, it is fair to state at this point that neither of the current calculations convincingly quantifies the effect, owing to either uncontrolled approximations (such as using ad hoc assumptions for off-shell vertices in hadronic calculations) or extrapolations beyond their region of validity (such as applying partonic calculations in the low momentum transfer regime). The present and forthcoming high-precision electron scattering data, which aim at testing two-photon exchange effects, have shifted the emphasis, however, from the qualitative to the quantitative realm.

Besides entering as corrections to elastic eN scattering data, two-photon exchange processes are also the leading corrections that improve our quantitative understanding of the hfs in hydrogen. We reviewed in this work that our best estimate, based on the recent data for polarized nucleon structure functions that enter the polarizability correction to the hydrogen hfs, yields a result $\Delta_{\text{pol}} = (1.3 \pm 0.3)$ ppm. Combined with recent estimates of the Zemach radius, the correction to the hydrogen hfs falls short of the data by approximately 2 ppm, or approximately 2.8 standard deviations.

We have also reviewed the recent measurements of large beam-normal single spin asymmetries, of order -10 to -100 ppm and measured to parts-per-million-level accuracy, which arose over the past few years as an interesting spin-off of the high-precision PV experiments in eN scattering. We discussed how such asymmetries, which measure the imaginary part of the two-photon exchange amplitude, can be expressed through nonforward nucleon structure functions and can provide a new tool to access hadron structure information.

FUTURE ISSUES

We end this review by spelling out a few open issues and challenges (both theoretical and experimental) in this field:

1. Experimental measurements of the two-photon exchange processes: To use electron scattering as a precision tool, it is clearly worthwhile to arrive at a quantitative understanding of two-photon exchange processes. This calls for detailed experimental studies, and several new experiments are already planned. A first type of experiment is to perform high-precision unpolarized experiments and look for nonlinearities in the Rosenbluth plots. A recent study (111) found that the nonlinearities in ε in current Rosenbluth data remain small over a fairly large Q^2 range. Such experiments, however, cannot separate the part of the two-photon corrections that is linear in ε from that which appears to be the dominant source of corrections in theoretical calculations. The difference between elastic e^- and e^+ scattering on a proton target directly accesses the real part of the two-photon exchange amplitude and the linear part in ε. At intermediate Q^2 values, some forthcoming e^+/e^- experiments at JLab (51) and VEPP-3 (112) will allow one to systematically study and quantify two-photon exchange effects. Also forthcoming is an experimental check of the predicted small effect of two-photon processes on the polarization data by measuring the ε dependence in polarization transfer experiments (50).

2. Further refinements on the theoretical side: The calculations of the real part of the two-photon amplitude for elastic eN scattering described in this review are performed either within a hadronic framework at lower Q^2 values or within a partonic picture at larger Q^2 values. The hadronic calculation assumes some off-shell vertex functions and performs a four-dimensional integral of the box diagram. It would be desirable to have a dispersion-relation framework where the real part is explicitly calculated as an integral over the imaginary part, which can be related to observables (i.e., on-shell amplitudes) using unitarity. In the forward limit, such dispersion relations are underlying, for example, the evaluation of the hfs in hydrogen. In the nonforward case, the convergence of such dispersion relations would need further study. Within the partonic framework, a full calculation of the processes with two active quarks (the so-called cat's ears diagrams) remains to be performed.

3. Systematic exploration of the imaginary part of the two-photon exchange amplitude: The ongoing experiments for both the beam- and target-normal SSA in elastic eN scattering will trigger further theoretical work, and a cross-fertilization between theory and experiment can be expected in this field.

4. Two-photon exchange effects in resonance production processes: De facto, all present measurements of resonance transition FFs use the double polarization technique. However, the sensitivity of unpolarized measurements of G_E/G_M to two-photon exchange effects also signals that such potential effects can be expected when measuring resonance transition FFs out to larger Q^2 values, as is performed at, for example, JLab. In particular, a first theoretical study of the two-photon exchange contribution to the $eN \rightarrow e\Delta(1232) \rightarrow e\pi N$ process with the aim of a precision study of the ratios of electric quadrupole (E2) and Coulomb quadrupole (C2) to the magnetic dipole (M1) $\gamma^* N\Delta$ transitions was performed recently (113). The two-photon exchange amplitude has been related to the $N \rightarrow \Delta$ GPDs. It was found that the C2/M1 ratio at larger Q^2 values depends strongly on whether this quantity is obtained from an interference cross section or from the Rosenbluth-type cross sections, similar to the elastic $eN \rightarrow eN$ process. It will be interesting to confront these results with upcoming new Rosenbluth separation data at intermediate Q^2 values to arrive at a precision extraction of the large Q^2 behavior of the R_{EM} and R_{SM} ratios.

5. Two-photon exchange effects in deep-inelastic scattering processes: Two-photon exchange effects were also studied in the past in deep-inelastic scattering processes (see, e.g., References 114–116). They can be expected to affect the extraction of the longitudinal structure function from L/T Rosenbluth-type experiments in the deep-inelastic scattering region, which remains to be quantified.

6. Two-photon exchange effects in PV elastic eN scattering and related study of γZ box-diagram effects: In recent years, an unprecedented precision has been achieved in PV electron scattering experiments. Two-photon exchange processes may be relevant in the interpretation of the current generation of high-precision parity-violation experiments, in particular for using those data to determine the strange-quark content of the proton; for a first study, see Reference 60.

   A related issue in the interpretation of PV electron scattering experiments are the γZ box-diagram contributions. Such calculations have been performed for atomic PV (117) that correspond with zero momentum transfer, as well as in the deep-inelastic scattering region (118), by calculating the γZ exchange between an electron and a quark. A full calculation in the small and intermediate Q^2 regime, where many current PV experiments are performed, is definitely a worthwhile topic for further research.

7. Two-photon exchange effects in hydrogen hfs: More accurate data, and data at lower Q^2, are forthcoming on polarized inelastic electron-proton scattering. These data will allow a more precise evaluation of the polarizability corrections to hydrogen hfs. In addition, a measurement of hfs in muonic hydrogen may be possible (92). The calculated corrections for muonic hydrogen have different weightings because of the muon mass, and with calculations such as those in References 84, 88, and 93 updated with new scattering data, the result of the measurement could be presented as an independent accurate determination of the proton's Zemach radius.

DISCLOSURE STATEMENT

The authors are not aware of any biases that might be perceived as affecting the objectivity of this review.

ACKNOWLEDGMENTS

We thank A. Afanasev, S. Brodsky, Y.C. Chen, M. Gorchtein, K. Griffioen, P. Guichon, V. Nazaryan, V. Pascalutsa, and B. Pasquini for collaborations on the subjects reviewed in this work. Furthermore, we thank A. Afanasev, D. Armstrong, J. Arrington, T. Averett, L. Capozza, K. de Jager, M. Gorchtein, F. Maas, Ch. Perdrisat, and M. Ramsey-Musolf for discussions and correspondence during the course of this work. The work of C.E.C. is supported by the National Science Foundation under grant PHY-0555600. The work of M.V. is supported in part by the Department of Energy grant DE-FG02-04ER41302 and contract DE-AC05-06OR23177, under which Jefferson Science Associates, LLC, operates the Jefferson Laboratory.

LITERATURE CITED

1. Stein G. In *What are Masterpieces?* New York: Pitman. 104 pp. (1970)
2. Hyde-Wright CE, de Jager K. *Annu. Rev. Nucl. Part. Sci.* 54:217 (2004)
3. Arrington J, Roberts CD, Zanotti JM. nucl-th/0611050 (2006)
4. Perdrisat CF, Punjabi V, Vanderhaeghen M. hep-ph/0612014 (2006)
5. Akhiezer AI, Rosentsweig LN, Shmushkevich IM. *Sov. Phys. JETP* 6:588 (1958); Scofield J. *Phys. Rev.* 113:1599 (1959); Scofield J. *Phys. Rev.* 141:1352 (1966); Dombey N. *Rev. Mod. Phys.* 41:236 (1969); Akhiezer AI, Rekalo MP. *Sov. J. Part. Nucl.* 4:277 (1974); Arnold RG, Carlson CE, Gross F. *Phys. Rev.* C 23:363 (1981)
6. Jones MK, et al. (Jefferson Lab Hall A Collab.) *Phys. Rev. Lett.* 84:1398 (2000)
7. Gayou O, et al. (Jefferson Lab Hall A Collab.) *Phys. Rev. Lett.* 88:092301 (2002)
8. Punjabi V, et al. *Phys. Rev.* C 71:055202 (2005). Erratum. *Phys. Rev.* C 71:069902 (2005)
9. Andivahis L, et al. *Phys. Rev.* D 50:5491 (1994)

10. Christy ME, et al. (E94110 Collab.) *Phys. Rev. C* 70:015206 (2004)

11. Qattan IA, et al. *Phys. Rev. Lett.* 94:142301 (2005)

12. Mar J, et al. *Phys. Rev. Lett.* 21:482 (1968)

13. Hartwig S, et al. *Lett. Nuovo Cim.* 12:30 (1975)

14. Mo LW, Tsai YS. *Rev. Mod. Phys.* 41:205 (1969)

15. Meister N, Yennie DR. *Phys. Rev.* 130:1210 (1963)

16. Drell SD, Ruderman MA. *Phys. Rev.* 106:561 (1957)

17. Drell SD, Fubini S. *Phys. Rev.* 113:741 (1959)

18. Werthamer NR, Ruderman MA. *Phys. Rev.* 123:1005 (1961)

19. Greenhut GK. *Phys. Rev.* 184:1860 (1969)

20. Guichon PAM, Vanderhaeghen M. *Phys. Rev. Lett.* 91:142303 (2003)

21. Arrington J. *Phys. Rev. C* 69:032201 (2004); *Phys. Rev. C* 71:015202 (2005)

22. Ji XD. *J. Phys. G* 24:1181 (1998)

23. Goeke K, Polyakov MV, Vanderhaeghen M. *Prog. Part. Nucl. Phys.* 47:401 (2001)

24. Diehl M. *Phys. Rep.* 388:41 (2003)

25. Ji XD. *Annu. Rev. Nucl. Part. Sci.* 54:413 (2004)

26. Belitsky AV, Radyushkin AV. *Phys. Rep.* 418:1 (2005)

27. Eides MI, Grotch H, Shelyuto VA. *Phys. Rep.* 342:63 (2001)

28. Karshenboim SG. *Phys. Rep.* 422:1 (2005)

29. Feinberg G, Sucher J, Au CK. *Phys. Rep.* 180:83 (1989)

30. Barut AO, Fronsdal C. *Phys. Rev.* 120:1871 (1960)

31. De Rujula A, Kaplan JM, De Rafael E. *Nucl. Phys. B* 35:365 (1971)

32. Blunden PG, Melnitchouk W, Tjon JA. *Phys. Rev. Lett.* 91:142304 (2003)

33. Blunden PG, Melnitchouk W, Tjon JA. *Phys. Rev. C* 72:034612 (2005)

34. Kondratyuk S, Blunden PG, Melnitchouk W, Tjon JA. *Phys. Rev. Lett.* 95:172503 (2005)

35. Kondratyuk S, Blunden PG. *Phys. Rev. C* 75:038201 (2007)

36. Borisyuk D, Kobushkin A. *Phys. Rev. C* 74:065203 (2006); Borisyuk D, Kobushkin A. nucl-th/0612104 (2004)

37. Chen YC, et al. *Phys. Rev. Lett.* 93:122301 (2004)

38. Afanasev AV, et al. *Phys. Rev. D* 72:013008 (2005)

39. Krass AS. *Phys. Rev.* 125:2172 (1962)

40. Ent R, et al. *Phys. Rev. C* 64:054610 (2001)

41. Goldberger ML, Nambu Y, Oehme R. *Ann. Phys.* 2:226 (1957)

42. Rekalo MP, Tomasi-Gustafsson E. *Eur. Phys. J. A* 22:331 (2004)

43. Van Nieuwenhuizen P. *Nucl. Phys. B* 28:429 (1971)

44. Grammer GJ, Yennie DR. *Phys. Rev. D* 8:4332 (1973)

45. Brodsky SJ, Primack JR. *Ann. Phys.* 52:315 (1969)

46. Maximon LC, Tjon JA. *Phys. Rev. C* 62:054320 (2000)

47. Guidal M, Polyakov MV, Radyushkin AV, Vanderhaeghen M. *Phys. Rev. D* 72:054013 (2005)

48. Diehl M, Feldmann T, Jakob R, Kroll P. *Eur. Phys. J. C* 39:1 (2005)

49. Brash EJ, Kozlov A, Li S, Huber GM. *Phys. Rev. C* 65:051001 (2002)

50. Gilman R, Pentchev L, Perdrisat C, Suleiman R. *JLab Exp. Propos. E-04–019*, Jefferson Lab., Newport News, Va. (2004)

51. Brooks W. *JLab Exp. Propos. E-04-116*, Jefferson Lab., Newport News, Va. (2004)
52. Bystritskiy YM, Kuraev EA, Tomasi-Gustafsson E. hep-ph/0603132 (2006)
53. Afanasev A. *Radiative corrections to parity-violating electron scattering experiments*. Presented at Workshop Precis. ElectroWeak Interact., Williamsburg, Va. (2005); Afanasev A, Akushevich I, Merenkov N. *Phys. Rev. D* 64:113009 (2001)
54. Dong YB, Kao CW, Yang SN, Chen YC. *Phys. Rev. C* 74:064006 (2006)
55. Rekalo MP, Tomasi-Gustafsson E, Prout D. *Phys. Rev. C* 60:042202 (1999)
56. Bordes J, Penarrocha JA, Bernabeu J. *Phys. Rev. D* 35:3310 (1987)
57. Penarrocha JA, Bernabeu J. *Ann. Phys.* 135:321 (1981)
58. Bernabeu J, Penarrocha JA. *Phys. Rev. D* 22:1082 (1980)
59. Offermann EAJM, et al. *Phys. Rev. C* 44:1096 (1991)
60. Afanasev AV, Carlson CE. *Phys. Rev. Lett.* 94:212301 (2005)
61. Arrington J, Sick I. nucl-th/0612079 (2006)
62. Blunden PG, Sick I. *Phys. Rev. C* 72:057601 (2005)
63. Karshenboim SG. *Can. J. Phys.* 77:241 (1999)
64. Volotka AV, Shabaev VM, Plunien G, Soff G. *Eur. Phys. J. D* 33:23 (2005)
65. Dupays A, et al. *Phys. Rev. A* 68:052503 (2003)
66. Eides MI. *Phys. Rev. A* 53:2953 (1996)
67. Brodsky SJ, Carlson CE, Hiller JR, Hwang DS. *Phys. Rev. Lett.* 94:022001 (2005); Erratum. *Phys. Rev. Lett.* 94:169902 (2005)
68. Friar JL, Sick I. *Phys. Rev. Lett.* 95:049101 (2005); Brodsky SJ, Carlson CE, Hiller JR, Hwang DS. *Phys. Rev. Lett.* 95:049102 (2005)
69. Bodwin GT, Yennie DR. *Phys. Rev. D* 37:498 (1988)
70. Martynenko AP. *Phys. Rev. A* 71:022506 (2005)
71. Drechsel D, Pasquini P, Vanderhaeghen M. *Phys. Rep.* 378:99 (2003)
72. Zemach AC. *Phys. Rev.* 104:1771 (1956)
73. Anthony PL, et al. (E155 Collab.) *Phys. Lett. B* 493:19 (2000)
74. Anthony PL, et al. (E155 Collab.) *Phys. Lett. B* 553:18 (2003)
75. Fatemi R, et al. (CLAS Collab.) *Phys. Rev. Lett.* 91:222002 (2003)
76. Yun J, et al. *Phys. Rev. C* 67:055204 (2003)
77. Deur A. nucl-ex/0507022 (2005)
78. Carlson CE. hep-ph/0611206 (2006); Carlson CE. physics/0610289 (2006)
79. Iddings CK. *Phys. Rev.* 138:B446 (1965)
80. Drell SD, Sullivan JD. *Phys. Rev.* 154:1477 (1967)
81. De Rafael E. *Phys. Lett. B* 37:201 (1971)
82. Gnädig P, Kuti J. *Phys. Lett. B* 42:241 (1972)
83. Faustov RN, Martynenko AP. *Eur. Phys. J. C* 24:281 (2002); Faustov RN, Martynenko AP. *Phys. Atom. Nucl.* 65:265 (2002) [*Yad. Fiz.* 65:291 (2002)]
84. Nazaryan V, Carlson CE, Griffioen KA. *Phys. Rev. Lett.* 96:163001 (2006)
85. Gerasimov SB. *Sov. J. Nucl. Phys.* 2:430 (1966) [*Yad. Fiz.* 2:598 (1966)]
86. Drell SD, Hearn AC. *Phys. Rev. Lett.* 16:908 (1966)
87. Simula S, Osipenko M, Ricco G, Taiuti M. *Phys. Rev. D* 65:034017 (2002)
88. Faustov RN, Gorbacheva IV, Martynenko AP. hep-ph/0610332 (2006)

89. Karshenboim SG. *Phys. Lett.* 225A:97 (1997)

90. Friar JL, Sick I. *Phys. Lett. B* 579:285 (2004)

91. Mohr PJ, Taylor BN. *Rev. Mod. Phys.* 72:351 (2000); Mohr PJ, Taylor BN. *Rev. Mod. Phys.* 77:1 (2005); Sick I. *Phys. Lett. B* 576:62 (2003); Kelly JJ. *Phys. Rev. C* 70:068202 (2004)

92. Antognini A, et al. *AIP Conf. Proc.* 796:253 (2005)

93. Faustov RN, Cherednikova EV, Martynenko AP. *Nucl. Phys. A* 703:365 (2002)

94. De Rujula A, Kaplan JM, De Rafael E. *Nucl. Phys. B* 53:545 (1973)

95. Wells SP, et al. (SAMPLE Collab.) *Phys. Rev. C* 63:064001 (2001)

96. Maas FE, et al. (A4 Collab.) *Phys. Rev. Lett.* 94:082001 (2005)

97. Kaufmann L. (Happex Collab.) Proc. Parit. Viol. Hadron. Struct. (PAVI06), Milos, Greece. *Eur. Phys. J. A.* In press

98. Capozza L. (A4 Collab.) Proc. Parit. Viol. Hadron. Struct. (PAVI06), Milos, Greece. *Eur. Phys. J. A.* In press

99. Kumar K (E-158 Collab.) Proc. Parit. Viol. Hadron. Struct. (PAVI06), Milos, Greece. *Eur. Phys. J. A.* In press

100. Phillips SK, King P (G0 Collab.) *Transverse beam spin asymmetries from the G0 forward-angle measurement.* Presented at Div. Nucl. Phys. Meet., Nashville, Tenn. (2006)

101. Afanasev A, Akusevich I, Merenkov NP. hep-ph/0208260 (2002)

102. Gorchtein M, Guichon PAM, Vanderhaeghen M. *Nucl. Phys. A* 741:234 (2004)

103. Tarrach R. *Nuovo Cim. A* 28:409 (1975)

104. Diaconescu L, Ramsey-Musolf MJ. *Phys. Rev. C* 70:054003 (2004)

105. Pasquini B, Vanderhaeghen M. *Phys. Rev. C* 70:045206 (2004)

106. Drechsel D, Hanstein O, Kamalov S, Tiator L. *Nucl. Phys. A* 645:145 (1999)

107. Afanasev AV, Merenkov NP. *Phys. Lett. B* 599:48 (2004); Afanasev AV, Merenkov NP. hep-ph/0407167 (2004)

108. Gorchtein M. *Phys. Rev. C* 73:035213 (2006); *Phys. Rev. C* 73:055201 (2006)

109. Gorchtein M. *Phys. Lett. B* 644:322 (2007)

110. Averett T, Chen JP, Jiang X. *JLab Exp. Propos. E-05-015*, Jefferson Lab., Newport News, Va. (2005)

111. Tvaskis V, et al. *Phys. Rev. C* 73:025206 (2006)

112. Arrington J, et al. nucl-ex/0408020 (2004)

113. Pascalutsa V, Carlson CE, Vanderhaeghen M. *Phys. Rev. Lett.* 96:012301 (2006)

114. Bartels J. *Nucl. Phys. B* 82:172 (1974)

115. Kingsley RL. *Nucl. Phys. B* 46:615 (1972); Fishbane PM, Kingsley RL. *Phys. Rev. D* 8:3074 (1973)

116. Bodwin GT, Stockham CD. *Phys. Rev. D* 11:3324 (1975)

117. Marciano WJ, Sirlin A. *Phys. Rev. D* 27:552 (1983); Marciano WJ, Sirlin A. *Phys. Rev. D* 29:75 (1984). Erratum. *Phys. Rev. D* 31:213 (1985)

118. Böhm M, H Spiesberger H. *Nucl. Phys. B* 294:1081 (1987)

Glauber Modeling in High-Energy Nuclear Collisions

Michael L. Miller,[1] Klaus Reygers,[2]
Stephen J. Sanders,[3] and Peter Steinberg[4]

[1] Laboratory for Nuclear Science, Massachusetts Institute of Technology, Cambridge, Massachusetts 02139; email: mlmiller@mit.edu

[2] Institute for Nuclear Physics, University of Münster, D-48149 Münster, Germany; email: reygers@ikp.uni-muenster.de

[3] Department of Physics and Astronomy, University of Kansas, Lawrence, Kansas, 66045; email: ssanders@ku.edu

[4] Physics Department, Brookhaven National Laboratory, Upton, New York 11973; email: peter.steinberg@bnl.gov

Annu. Rev. Nucl. Part. Sci. 2007. 57:205–43

First published online as a Review in Advance on May 9, 2007

The *Annual Review of Nuclear and Particle Science* is online at http://nucl.annualreviews.org

This article's doi:
10.1146/annurev.nucl.57.090506.123020

Copyright © 2007 by Annual Reviews.
All rights reserved

0163-8998/07/1123-0205$20.00

Key Words

heavy ion physics, number of participating nucleons, number of binary collisions, impact parameter, eccentricity

Abstract

We review the theoretical background, experimental techniques, and phenomenology of what is known in relativistic heavy ion physics as the Glauber model, which is used to calculate geometric quantities. A brief history of the original Glauber model is presented, with emphasis on its development into the purely classical, geometric picture used for present-day data analyses. Distinctions are made between the optical limit and Monte Carlo approaches, which are often used interchangeably but have some essential differences in particular contexts. The methods used by the four RHIC experiments are compared and contrasted, although the end results are reassuringly similar for the various geometric observables. Finally, several important RHIC measurements are highlighted that rely on geometric quantities, estimated from Glauber calculations, to draw insight from experimental observables. The status and future of Glauber modeling in the next generation of heavy ion physics studies is briefly discussed.

Contents

1. INTRODUCTION 206
2. THEORETICAL FOUNDATIONS OF GLAUBER MODELING.. 207
  2.1. A Brief History of the Glauber Model 207
  2.2. Inputs to Glauber Calculations 208
  2.3. Optical-Limit Approximation 210
  2.4. The Glauber Monte Carlo Approach 212
  2.5. Differences between Optical and Monte Carlo Approaches 213
  2.6. Glauber Model Systematics 215
3. RELATING THE GLAUBER MODEL TO EXPERIMENTAL
  DATA 216
  3.1. Methodology 216
  3.2. Experimental Details 220
  3.3. Acceptance Biases 226
  3.4. Estimating Geometric Quantities 226
4. GEOMETRIC ASPECTS OF P+A AND A+A PHENOMENA 231
  4.1. Inclusive Charged-Particle Yields (Total and Midrapidity) 231
  4.2. Hard Scattering: T_{AB} Scaling 232
  4.3. Eccentricity and Relation to Elliptic Flow 235
  4.4. Eccentricity Fluctuations 237
  4.5. J/ψ Absorption in Normal Nuclear Matter 237
5. DISCUSSION AND THE FUTURE 240

1. INTRODUCTION

Ultrarelativistic collisions of nuclei produce the highest multiplicities of outgoing particles of all subatomic systems known in the laboratory. Thousands of particles are created when two nuclei collide head-on, generating event images of dramatic complexity compared with proton-proton collisions. The latter is a natural point of comparison, as nuclei are themselves made up of nucleons, that is, protons and neutrons. Thus, it is natural to ask just how many nucleons are involved in any particular collision, or, more reasonably, in a sample of selected collisions. It is also interesting to consider other ways to characterize the overlapping nuclei, for example, their shape.

Although this problem may seem intractable—with the femtoscopic length scales involved precluding direct observation of the impact parameter (b), number of participating nucleons (N_{part}), or binary nucleon-nucleon collisions (N_{coll})—theoretical techniques have been developed to allow estimation of these quantities from experimental data. These techniques, which consider the multiple scattering of nucleons in nuclear targets, are generally referred to as Glauber models, after Roy Glauber. Glauber pioneered the use of quantum-mechanical scattering theory for composite systems, describing nontrivial effects discovered in cross sections for proton-nucleus (p+A) and nucleus-nucleus (A+B) collisions at low energies.

Over the years, workers developed a variety of methods for calculating geometric quantities relevant for p+A and A+B collisions (1–3). Moreover, a wide variety of experimental methods were established to connect distributions of measured quantities to the distributions of geometric quantities. This review is an attempt to explain how the RHIC experiments grappled with this problem, and largely succeeded, because of both good-sense experimental and theoretical thinking.

Heavy ion physics entered a new era with the turn on of the RHIC collider in 2000 (4). Previous generations of heavy ion experiments searching for the quark-gluon plasma (QGP) focused on particular signatures suggested by theory. From this strategy, experiments generally measured observables in different regions of phase space or focused on particular particle types. The RHIC experiments were designed in a comprehensive way, with certain regions of phase space covered by multiple experiments. This allowed systematic cross checks of various measurements between experiments, increasing the credibility of the separate results (5–8).

Among the most fundamental observables shared between the experiments were those relating to the geometry of the collision. Identical zero-degree calorimeters (ZDCs) (9) were installed in all experiments to estimate the number of spectator nucleons, and all experiments had coverage for multiplicities and energy measurements over a wide angular range. This allowed a set of systematic studies of centrality in d+Au and A+A collisions, providing one of the first truly extensive data sets, all of which were characterized by geometric quantities.

This review covers the basic information a newcomer to the field should know to understand how centrality is estimated by the experiments, and how this is related to A+B collisions. Section 2 discusses the history of the Glauber model by reference to the theoretical description of A+B collisions. Section 3 addresses how experiments measure centrality by a variety of methods and relates it by a simple procedure to geometrical quantities. Section 4 illustrates the relevance of various geometrical quantities by reference to actual RHIC data. These examples show how a precise quantitative grasp of the geometry allows qualitatively new insights into the dynamics of A+B collisions. Finally, Section 5 assesses the current state of knowledge and points to new directions in our understanding of nuclear geometry and its impact on present and future measurements.

2. THEORETICAL FOUNDATIONS OF GLAUBER MODELING

2.1. A Brief History of the Glauber Model

The Glauber model was developed to address the problem of high-energy scattering with composite particles. This was of great interest to both nuclear and particle physicists, both of whom have benefited from the insights of Glauber in the 1950s. In his 1958 lectures, Glauber presented his first collection of various papers and unpublished work from the 1950s (2). Some of this was updated work started by Moliere in the 1940s, but much of it was in direct response to the new experiments involving protons, deuterons, and light nuclei. Until then, there had been no systematic calculations treating the many-body nuclear system as either a projectile or target. Glauber's work put the quantum theory of collisions of composite objects on a firm basis and provided a consistent description of experimental data for protons incident on deuterons and larger nuclei (10, 11). Most striking were the observed dips in the elastic peaks, whose position and magnitude were predicted by Glauber's theory by Czyż and Lesniak in 1967 (12).

In the 1970s, high-energy beams of hadrons and nuclei were systematically scattered off of nuclear targets. Glauber's work was found to have utility in describing total cross sections, for which factorization relationships were found (e.g., $\sigma_{AB}^2 \sim \sigma_{AA}\sigma_{BB}$) (13, 14). In 1969 Czyż & Maximon (1) applied the theory in its most complete form to p+A and A+B collisions, focusing mainly on elastic collisions. Finally, Bialas et al. (3, 15) applied Glauber's approach to inelastic nuclear collisions, after they had already applied their wounded-nucleon model to hadron-nucleus collisions. This is essentially the bare framework of the traditional Glauber model, with all of the quantum mechanics reduced to its simplest form (16). The main remaining feature of the original Glauber calculations is the optical limit, used to make the full multiple-scattering integral numerically tractable.

Bialas et al.'s (15) approach introduced the basic functions used in more modern language, including the thickness function and a prototype of the nuclear overlap function T_{AB}. This paper also introduced the convention of using the optical limit for analytical and numerical calculations, despite full knowledge that the real Glauber calculation is a 2(A+B+1)-dimensional integral over the impact parameter and each of the nucleon positions. A similar convention exists in most theoretical calculations of geometrical parameters to this day.

With the advent of desktop computers, the Glauber Monte Carlo (GMC) approach emerged as a natural framework for use by more realistic particle-production codes (17, 18). The idea was to model the nucleus in the simplest way, as uncorrelated nucleons sampled from measured density distributions. Two nuclei could be arranged with a random impact parameter b and projected onto the x-y plane. Then, interaction probabilities could be applied by using the relative distance between two nucleon centroids as a stand-in for the measured inelastic nucleon-nucleon cross section.

The GMC approach was first applied to high-energy heavy ion collisions in the HIJET model (17) and has shown up in practically all A+A simulation codes. This includes HIJING (19), VENUS (20), RQMD(21), and all models that require specific production points for the different subprocesses in an A+B collision, rather than just aggregate values.

2.2. Inputs to Glauber Calculations

In all calculations of geometric parameters using a Glauber approach, some experimental data must be given as model inputs. The two most important are the nuclear charge densities measured in low-energy electron scattering experiments and the energy dependence of the inelastic nucleon-nucleon cross section.

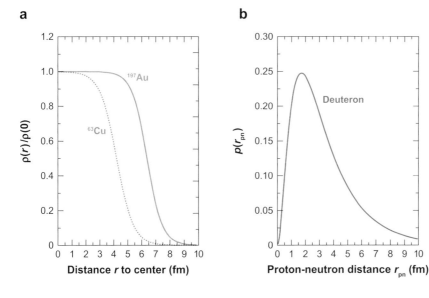

a

Figure 1

(*a*) Density distributions for
nuclei used at RHIC.
(*b*) Distribution of the
proton-neutron distance in
the deuteron as given by the
Hulthén wave function.

b

2.2.1. Nuclear charge densities. The nucleon density is usually parameterized by a Fermi distribution with three parameters:

$$\rho(r) = \rho_0 \frac{1 + w(r/R)^2}{1 + \exp\left(\frac{r-R}{a}\right)}, \qquad 1.$$

where ρ_0 corresponds to the nucleon density in the center of the nucleus, R corresponds to the nuclear radius, a to the skin depth, and w characterizes deviations from a spherical shape. For ^{197}Au ($R = 6.38$ fm, $a = 0.535$ fm, $w = 0$) and ^{63}Cu ($R = 4.20641$ fm, $a = 0.5977$ fm, $w = 0$), the nuclei so far employed at RHIC, $\rho(r)/\rho_0$ is shown in **Figure 1a** with the Fermi distribution parameters as given in References 22 and 23. In the Monte Carlo procedure, the radius of a nucleon is drawn randomly from the distribution $4\pi r^2 \rho(r)$ (where the absolute normalization is of course irrelevant).

At RHIC, effects of cold nuclear matter have been studied with the aid of d+Au collisions. In the Monte Carlo calculations, the deuteron wave function was represented by the Hulthén form (24, 25)

$$\phi(r_{\text{pn}}) = \frac{1}{\sqrt{2\pi}} \frac{\sqrt{ab(a+b)}}{b-a} \left(\frac{e^{-a r_{\text{pn}}} - e^{-b r_{\text{pn}}}}{r_{\text{pn}}} \right), \qquad 2.$$

with parameters $a = 0.228$ fm^{-1} and $b = 1.18$ fm^{-1} (26). The variable r_{pn} in Equation 2 denotes the distance between the proton and the neutron. Accordingly, r_{pn} was drawn from the distribution $p(r_{\text{pn}}) = 4\pi r_{\text{pn}}^2 \phi^2(r_{\text{pn}})$, which is shown in **Figure 1b**.

2.2.2. Inelastic nucleon-nucleon cross section. In the context of high-energy nuclear collisions, we are typically interested in multiparticle nucleon-nucleon

Figure 2

The inelastic nucleon-nucleon cross section σ_{inel}^{NN} as parameterized by PYTHIA (*solid line*) (34, 35), in addition to data on total and elastic nucleon-nucleon cross sections as a function of $\sqrt{s}$ (36). The stars indicate the nucleon-nucleon cross section used for Glauber Monte Carlo calculations at RHIC ($\sigma_{inel}^{NN} = 32.3$, 35.6, 40, and 42 mb at $\sqrt{s_{NN}} = 19.6$, 62.4, 130, and 200 GeV, respectively).

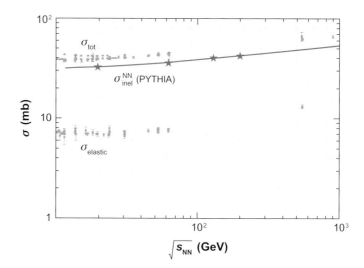

processes. As the cross section involves processes with low momentum transfer, it is impossible to calculate this using perturbative quantum chromodynamics (QCD). Thus, the measured inelastic nucleon-nucleon cross section (σ_{inel}^{NN}) is used as an input and provides the only nontrivial beam-energy dependence for Glauber calculations. From $\sqrt{s_{NN}} \sim 20$ GeV (CERN SPS) to $\sqrt{s_{NN}} = 200$ GeV (RHIC), σ_{inel}^{NN} increases from ~ 32 mb to ~ 42 mb, as shown in **Figure 2**. Diffractive and elastic processes, which are typically ignored in high-energy multiparticle nuclear collisions, are generally active out to large impact parameters, and thus require full quantum-mechanical wave functions.

2.3. Optical-Limit Approximation

The Glauber model views the collision of two nuclei in terms of the individual interactions of the constituent nucleons (see, e.g., Reference 27). In the optical limit, the overall phase shift of the incoming wave is taken as a sum over all possible two-nucleon (complex) phase shifts, with the imaginary part of the phase shifts related to the nucleon-nucleon scattering cross section through the optical theorem (28, 29). The model assumes that at sufficiently high energies, these nucleons will carry sufficient momentum that they will be essentially undeflected as the nuclei pass through each other. Workers also assumed that the nucleons move independently in the nucleus and that the size of the nucleus is large compared to the extent of the nucleon-nucleon force. The hypothesis of independent linear trajectories of the constituent nucleons makes it possible to develop simple analytic expressions for the A+B interaction cross section and for the number of interacting nucleons and the number of nucleon-nucleon collisions in terms of the basic nucleon-nucleon cross section.

In **Figure 3**, two heavy ions, target A and projectile B, are shown colliding at relativistic speeds with impact parameter **b** (for colliding beam experiments, the

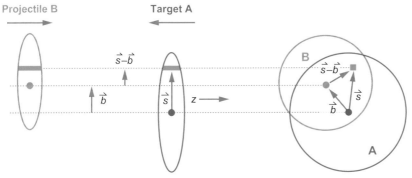

a

Side view

Projectile B   Target A

$\vec{s}-\vec{b}$

$\vec{b}$ $\vec{s}$ $z \longrightarrow$

b

Beam-line view

B $\vec{s}-\vec{b}$ $\vec{s}$

$\vec{b}$

A

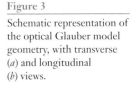

Figure 3

Schematic representation of the optical Glauber model geometry, with transverse (*a*) and longitudinal (*b*) views.

distinction between the target and projectile nuclei is a matter of convenience). We focus on the two flux tubes located at a displacement **s** with respect to the center of the target nucleus and a distance **s** − **b** from the center of the projectile. During the collision these tubes overlap. The probability per unit transverse area of a given nucleon being located in the target flux tube is $\hat{T}_A(\mathbf{s}) = \int \hat{\rho}_A(\mathbf{s}, z_A) dz_A$, where $\hat{\rho}_A(\mathbf{s}, z_A)$ is the probability per unit volume, normalized to unity, for finding the nucleon at location $(\mathbf{s}, z_A)$. A similar expression follows for the projectile nucleon. The product $\hat{T}_A(\mathbf{s})\hat{T}_B(\mathbf{s} - \mathbf{b})\, d^2s$ then gives the joint probability per unit area of nucleons being located in the respective overlapping target and projectile flux tubes of differential area d^2s. Integrating this product over all values of **s** defines the thickness function $\hat{T}(\mathbf{b})$, with

$$\hat{T}_{AB}(\mathbf{b}) = \int \hat{T}_A(\mathbf{s})\, \hat{T}_B(\mathbf{s} - \mathbf{b})\, d^2s. \qquad 3.$$

Notice that $\hat{T}(\mathbf{b})$ has the unit of inverse area. We can interpret this as the effective overlap area for which a specific nucleon in A can interact with a given nucleon in B. The probability of an interaction occurring is then $\hat{T}(\mathbf{b})\, \sigma_{\text{inel}}^{\text{NN}}$, where $\sigma_{\text{inel}}^{\text{NN}}$ is the inelastic nucleon-nucleon cross section. Elastic processes lead to very little energy loss and are consequently not considered in the Glauber model calculations. Once the probability of a given nucleon-nucleon interaction has been found, the probability of having n such interactions between nuclei A (with A nucleons) and B (with B nucleons) is given as a binomial distribution:

$$P(n, \mathbf{b}) = \binom{AB}{n} \left[\hat{T}_{AB}(\mathbf{b})\, \sigma_{\text{inel}}^{\text{NN}}\right]^n \left[1 - \hat{T}_{AB}(\mathbf{b})\, \sigma_{\text{inel}}^{\text{NN}}\right]^{AB-n}, \qquad 4.$$

where the first term is the number of combinations for finding n collisions out of AB possible nucleon-nucleon interactions, the second term the probability for having exactly n collisions, and the last term the probability of exactly $AB - n$ misses.

On the basis of this probability distribution, a number of useful reaction quantities can be found. The total probability of an interaction between A and B is

$$\frac{d^2\sigma_{\text{inel}}^{AB}}{db^2} \equiv p_{\text{inel}}^{AB}(b) = \sum_{n=1}^{AB} P(n, \vec{b}) = 1 - \left[1 - \hat{T}_{AB}(\mathbf{b})\,\sigma_{\text{inel}}^{NN}\right]^{AB}. \qquad 5.$$

The vector impact parameter can be replaced by a scalar distance if the nuclei are not polarized. In this case, the total cross section can be found as

$$\sigma_{\text{inel}}^{AB} = \int_0^\infty 2\pi b\,db \left\{1 - \left[1 - \hat{T}_{AB}(b)\,\sigma_{\text{inel}}^{NN}\right]^{AB}\right\}. \qquad 6.$$

The total number of nucleon-nucleon collisions is

$$N_{\text{coll}}(b) = \sum_{n=1}^{AB} n\,P(n, b) = AB\,\hat{T}_{AB}(b)\,\sigma_{\text{inel}}^{NN}, \qquad 7.$$

using the result for the mean of a binomial distribution. The number of nucleons in the target and projectile nuclei that interact is known as either the number of participants or the number of wounded nucleons. The number of participants (or wounded nucleons) at impact parameter b is given by (15, 30)

$$\begin{aligned}
N_{\text{part}}(\mathbf{b}) &= A \int \hat{T}_A(\mathbf{s})\left\{1 - \left[1 - \hat{T}_B(\mathbf{s} - \mathbf{b})\,\sigma_{\text{inel}}^{NN}\right]^{B}\right\} d^2s \\
&+ B \int \hat{T}_B(\mathbf{s} - \mathbf{b})\left\{1 - \left[1 - \hat{T}_A(\mathbf{s})\,\sigma_{\text{inel}}^{NN}\right]^{A}\right\} d^2s, \qquad 8.
\end{aligned}$$

where the integral over the bracketed terms gives the respective inelastic cross sections for nucleon-nucleus collisions:

$$\sigma_{\text{inel}}^{A(B)} = \int d^2s \left\{1 - \left[1 - \sigma_{\text{inel}}^{NN}\,\hat{T}_{A(B)}(\mathbf{s})\right]^{A(B)}\right\}. \qquad 9.$$

The optical form of the Glauber theory is based on continuous nucleon density distributions. The theory does not locate nucleons at specific spatial coordinates, as is the case for the Monte Carlo formulation discussed in the next section. This difference between the optical and Monte Carlo approaches can lead to subtle differences in calculated results, as discussed below.

2.4. The Glauber Monte Carlo Approach

The virtue of the Monte Carlo approach for the calculation of geometry-related quantities such as the average number of participants $\langle N_{\text{part}} \rangle$ and nucleon-nucleon collisions $\langle N_{\text{coll}} \rangle$ is its simplicity. Moreover, it is possible to simulate experimentally observable quantities such as the charged-particle multiplicity and to apply similar centrality cuts, as in the analysis of real data. In the Monte Carlo ansatz, the two colliding nuclei are assembled in the computer by distributing the A nucleons of nucleus A and the B nucleons of nucleus B in a three-dimensional coordinate system according to the respective nuclear density distribution. A random impact parameter b is then drawn from the distribution $d\sigma/db = 2\pi b$. An A+B collision is treated as

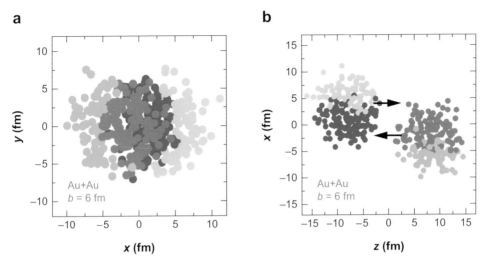

a

b

Figure 4

A Glauber Monte Carlo event (Au+Au at $\sqrt{s_{NN}} = 200$ GeV with impact parameter $b = 6$ fm) viewed (*a*) in the transverse plane and (*b*) along the beam axis. The nucleons are drawn with radius $\sqrt{\sigma_{inel}^{NN}/\pi}/2$. Darker circles represent participating nucleons.

a sequence of independent binary nucleon-nucleon collisions. That is, the nucleons travel on straight-line trajectories, and the inelastic nucleon-nucleon cross section is assumed to be independent of the number of collisions a nucleon underwent before. In the simplest version of the Monte Carlo approach, a nucleon-nucleon collision takes place if the nucleons' distance d in the plane orthogonal to the beam axis satisfies

$$d \leq \sqrt{\sigma_{inel}^{NN}/\pi}, \qquad \qquad 10.$$

where σ_{inel}^{NN} is the total inelastic nucleon-nucleon cross section. As an alternative to the black-disk nucleon-nucleon overlap function, for example, a Gaussian overlap function can be used (31). An illustration of a GMC event for a Au+Au collision with impact parameter $b = 6$ fm is shown in **Figure 4**. $\langle N_{part} \rangle$ and $\langle N_{coll} \rangle$ and other quantities are then determined by simulating many A+B collisions.

2.5. Differences between Optical and Monte Carlo Approaches

It is often overlooked that the various integrals used to calculate physical observables in the Glauber model are predicated on a particular approximation known as the optical limit. This limit assumes that scattering amplitudes can be described by an eikonal approach, where the incoming nucleons see the target as a smooth density. This approach captures many features of the collision process, but does not completely capture the physics of the total cross section. Thus, it tends to lead to distortions in the estimation of N_{part} and N_{coll} compared to similar estimations using the GMC approach.

This can be seen by simply looking at the relevant integrals. The full (2A+2B)-dimensional integral to calculate the total cross section is (15)

$$\sigma_{\text{inel}}^{\text{AB}} = \int d^2 b \int d^2 s_1^A \cdots d^2 s_A^A d^2 s_1^B \cdots d^2 s_B^B \times \hat{T}_A\left(\mathbf{s_1^A}\right) \cdots \hat{T}_A\left(\mathbf{s_A^A}\right) \hat{T}_B\left(\mathbf{s_1^B}\right) \cdots \hat{T}_B\left(\mathbf{s_B^B}\right)$$

$$\times \left\{ 1 - \prod_{j=1}^{B} \prod_{i=1}^{A} \left[1 - \hat{\sigma}\left(\mathbf{b} - \mathbf{s_i^A} + \mathbf{s_j^B}\right) \right] \right\},$$ 11.

where $\hat{\sigma}(\mathbf{s})$ is normalized such that $\int d^2 s\, \hat{\sigma}(\mathbf{s}) = \sigma_{\text{inel}}^{\text{NN}}$, whereas the optical-limit version of the same calculation is (see Equation 6)

$$\sigma_{\text{AB}} = \int d^2 b \left\{ 1 - \left[1 - \sigma_{\text{inel}}^{\text{NN}} \hat{T}_{\text{AB}}(b) \right]^{AB} \right\}.$$ 12.

These expressions are generally expected to be the same for large A (and B) and/or when $\sigma_{\text{inel}}^{\text{NN}}$ is sufficiently small (15). The main difference between the two is that many terms in the full calculation are missing in the optical-limit calculation. These are the terms that describe local density fluctuations event by event. Thus, in the optical limit, each nucleon in the projectile sees the oncoming target as a smooth density.

One can test this interpretation to first order by plotting the total cross section as a function of $\sigma_{\text{inel}}^{\text{NN}}$ for an optical-limit calculation as well as for a GMC calculation with the same parameters, as shown in the left panel of **Figure 5**. As the nucleon-nucleon cross section becomes more point like, the optical and GMC cross sections converge.

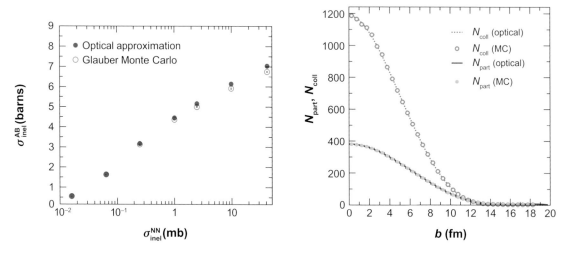

Figure 5

(*Left*) The total cross section, calculated in the optical approximation and with a Glauber Monte Carlo (MC)—both with identical nuclear parameters—as a function of $\sigma_{\text{inel}}^{\text{NN}}$, the inelastic nucleon-nucleon cross section. (*Right*) N_{coll} and N_{part} as a function of impact parameter, calculated in the optical approximation (*lines*) and with a Glauber Monte Carlo (*symbols*). The two are essentially identical out to $b = 2R_A$. Figure courtesy of M.D. Baker, M.P. Decowski, and P. Steinberg.

This confirms the general suspicion that it is the ability of GMC calculations to introduce shadowing corrections that reduces the cross section relative to the optical calculations (11).

However, when calculating simple geometric quantities such as N_{part} and N_{coll} as a function of impact parameter, there is little difference between the two calculations, as shown in the right panel of **Figure 5**. The only substantial difference comes at the highest impact parameters, something discussed in Section 3.4.1. Fluctuations are also sensitive to this difference, but there is insufficient room in this review to discuss more recent developments (32).

2.6. Glauber Model Systematics

As discussed above, the Glauber model depends on the inelastic nucleon-nucleon cross section σ_{inel}^{NN} and the geometry of the interacting nuclei. In turn, σ_{inel}^{NN} depends on the energy of the reaction, as shown in **Figure 2**, and the geometry depends on the number of nucleons in the target and projectile.

Figure 6a shows the effect of changing σ_{inel}^{NN} on the calculated values of N_{part} and N_{coll} for a ^{197}Au+^{197}Au reaction over the range of σ_{inel}^{NN} values relevant at RHIC. The secondary axis shows the values of $\sqrt{s_{NN}}$ corresponding to the σ_{inel}^{NN} values. The values are shown for central events, with impact parameter $b < 2$ fm, and for a minimum-bias (MB) throw of events. The error bars, which extend only beyond the symbol size for the N_{coll} ($b < 2$ fm) results, are based on an assumed uncertainty of σ_{inel}^{NN} at a given energy of 3 mb. In general, the Glauber calculations show only a weak energy dependence over the energy range covered by the RHIC accelerator.

Figure 6b shows the dependence of N_{part} and N_{coll} on the system size for central and MB events, with values calculated for identical particle collisions of the indicated systems at a fixed value of $\sigma_{inel}^{NN} = 42$ mb (corresponding to $\sqrt{s_{NN}} = 200$ GeV). Because the Glauber model is largely dependent on the geometry of the colliding nuclei, some simple scalings can be expected for N_{part} and N_{coll}.

a

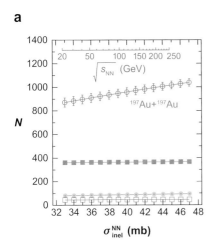

b

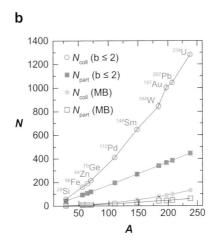

Figure 6

(*a*) Dependence of N_{part} and N_{coll} on σ_{inel}^{NN} for central ($b \leq 2$ fm) and minimum-bias (MB) events for a ^{197}Au+^{197}Au reaction. (*b*) Dependence of these values on the system size, with $\sigma_{inel}^{NN} = 42$ mb.

Figure 7

Geometric scaling behavior
of N_{coll} as discussed in the
text. The calculations are
done with $\sigma_{inel}^{NN} = 42$ mb.

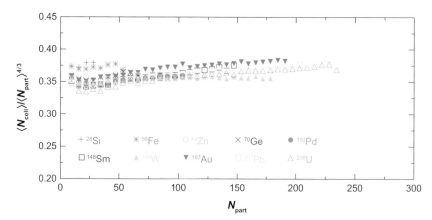

N_{part} should scale with the volume of the interaction region. In **Figure 6b** this is seen by the linear dependence of N_{part} for central collisions on A, where the common volume V of the largely overlapping nuclei in central collisions is proportional to A for a saturated nuclear density. In a collision of two equal nuclei (A+A), the average number of collisions per participant scales as the length $l_z \propto N_{part}^{1/3}$ of the interaction volume along the beam direction so that the number of collisions roughly follows

$$N_{coll} \propto N_{part}^{4/3}, \qquad\qquad 13.$$

independent of the size of the nuclei. This scaling relationship is demonstrated in **Figure 7**, where $N_{coll}/N_{part}^{4/3}$ is plotted as a function of N_{part} for the systems shown in **Figure 6b**. The result is confirmed experimentally in Reference 33, where the Cu+Cu and Au+Au systems are compared. The geometric nature of the Glauber model is evident.

3. RELATING THE GLAUBER MODEL TO EXPERIMENTAL DATA

Unfortunately, neither N_{part} nor N_{coll} can be measured directly in a RHIC experiment. Mean values of such quantities can be extracted for classes of measured events (N_{evt}) via a mapping procedure. Typically, a measured distribution (e.g., dN_{evt}/dN_{ch}) is mapped to the corresponding distribution obtained from phenomenological Glauber calculations. This is done by defining centrality classes in both the measured and calculated distributions and then connecting the mean values from the same centrality class in the two distributions. The specifics of this mapping procedure differ both between experiments as well as between collision systems within a given experiment. Herein we briefly summarize the principles and various implementations of centrality definition.

3.1. Methodology

The basic assumption underlying centrality classes is that the impact parameter b is monotonically related to particle multiplicity, both at midrapidity and forward

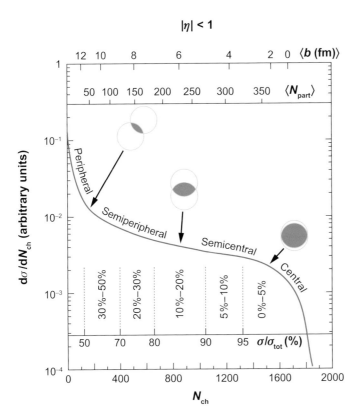

Figure 8

An illustrated example of the correlation of the final-state-observable total inclusive charged-particle multiplicity N_{ch} with Glauber-calculated quantities (b, N_{part}). The plotted distribution and various values are illustrative and not actual measurements (T. Ullrich, private communication).

rapidity. For large b events ("peripheral") we expect low multiplicity at midrapidity and a large number of spectator nucleons at beam rapidity, whereas for small b events ("central") we expect large multiplicity at midrapidity and a small number of spectator nucleons at beam rapidity (**Figure 8**). In the simplest case, one measures the per-event charged-particle multiplicity (dN_{evt}/dN_{ch}) for an ensemble of events. Once the total integral of the distribution is known, centrality classes are defined by binning the distribution on the basis of the fraction of the total integral. The dashed vertical lines in **Figure 8** represent a typical binning. The same procedure is then applied to a calculated distribution, often derived from a large number of Monte Carlo trials. For each centrality class, the mean value of Glauber quantities (e.g., $\langle N_{part} \rangle$) for the Monte Carlo events within the bin (e.g., 5%–10%) is calculated. Potential complications to this straightforward procedure arise from various sources: event selection, uncertainty in the total measured cross section, fluctuations in both the measured and calculated distributions, and finite kinematic acceptance.

3.1.1. Event selection. All four RHIC experiments share a common detector to select MB heavy ion events. The ZDCs are small acceptance hadronic calorimeters with an angular coverage of $\theta \leq 2$ mrad with respect to the beam axis (9). Situated behind the charged-particle steering DX magnets of RHIC, the ZDCs are primarily

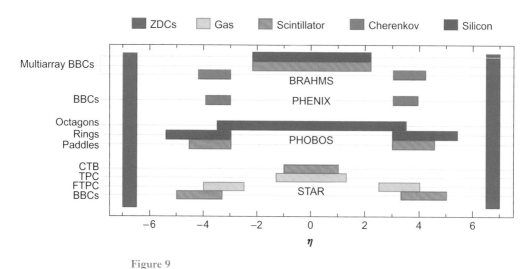

Figure 9

Pseudorapidity coverage of the centrality selection detectors of the four RHIC experiments
(37–40). BBCs, beam-beam counter arrays; ZDCs, zero-degree calorimeters.

sensitive to neutral spectators. For Au+Au collisions at $\sqrt{s_{NN}} = 130$ GeV and above,
the ZDCs are ∼100% efficient for inelastic collisions, thus providing an excellent
MB trigger. The RHIC experiments often apply an online timing cut to select events
within a given primary vertex interval ($\sim|z_{\text{vertex}}| < 30$ cm). Further coincidence with
fast detectors near midrapidity is often also used to suppress background events such
as beam-gas collisions. Experiment-specific event selection is described in detail in
Section 3.2. **Figure 9** displays the pseudorapidity coverage of the suite of subsystems
used to define centrality (both offline and at the trigger level) (37–40). With the
exception of the STAR TPC and FTPCs, all other subsystems are intrinsically fast
and available for event triggering.

3.1.2. Centrality observables. In MB p+p and p+$\bar{\text{p}}$ collisions at high energy, the
charged-particle multiplicity $dN_{\text{evt}}/dN_{\text{ch}}$ has been measured over a wide range of rapidity and is well described by a negative binomial distribution (NBD) (41). However,
the multiplicity is also known to scale with the hardness (q^2) of the collision—the multiplicity for hard jet events is significantly higher than that of MB collisions. In heavy
ion collisions, we manipulate the fact that the majority of the initial-state nucleon-
nucleon collisions will be analogous to MB p+p collisions, with a small perturbation
from much rarer hard interactions. The final integrated multiplicity of heavy ion
events is then roughly described as a superposition of many NBDs, which quickly
approaches the Gaussian limit.

The total inclusive charged-particle multiplicity N_{ch} can be measured offline by
counting charged tracks (e.g., STAR TPC) or estimated online from the total energy
deposited in a detector divided by the typical energy deposition per charged particle
(e.g., PHOBOS Paddle scintillation counters). As shown in **Figure 9**, there is a wide
variation in the acceptance of various centrality detectors at RHIC. PHOBOS, with

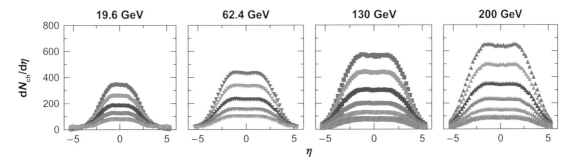

Figure 10

Charged-particle multiplicity (PHOBOS) in Au+Au collisions at various center-of-mass energies for $|\eta| < 5$ (57, 58). The different colors represent different centrality selections.

the largest acceptance in η, is well suited to measure $\frac{d^2 N_{evt}}{dN_{ch} d\eta}$ (**Figure 10**). These data illustrate two key features of particle production in A+B collisions. At a fixed beam energy, there is no dramatic change in shape as the centrality changes. However, reducing the beam energy does change the shape substantially, as the maximum rapidity varies as $\ln(\sqrt{s_{NN}})$. Thus, the same trigger detector may have a very different overall efficiency at different beam energies.

N_{ch} can be simulated via various methods, but all require the coupling of a Glauber calculation to a model of charged-particle production, either dynamic [e.g., HIJING (19)] or static (randomly sampled from a Gaussian, Poisson, or negative binomial). Most follow the general prescription that the multiplicity scales approximately with N_{part}. For an optical Glauber calculation, the simulated multiplicity (N_{ch}^{sim}) can be calculated semi-analytically, assuming that each participant contributes a given value of N_{ch}, which is typically drawn from one of the aforementioned static probability distributions (42). The same can be done for a GMC simulation with the added advantage that the detector response to such events can be simulated, thus enabling an equivalent comparison of simulated and measured N_{ch} distributions. Various dynamical models of heavy ion collisions exist and can also be coupled to detector simulations. In all cases, the exact value of N_{part}, N_{coll}, b, and N_{ch}^{sim} are known for each event.

3.1.3. Dividing by percentile of total inelastic cross section. With a measured and simulated dN_{evt}/dN_{ch} distribution in hand, one can then perform the mapping procedure to extract mean values. Suppose that the measured and simulated distributions are both one-dimensional histograms. For each histogram, the total integral is calculated and centrality classes are defined in terms of a fraction of the total integral. Typically, the integration is performed from large to small values of N_{ch}. For example, the 10%–20% most central class is defined by the boundaries n_{10} and n_{20}, which satisfy

$$\frac{\int_{\infty}^{n_{10}} \frac{dN_{evt}}{dN_{ch}} \, dN_{ch}}{\int_{\infty}^{0} \frac{dN_{evt}}{dN_{ch}} \, dN_{ch}} = 0.1 \quad \text{and} \quad \frac{\int_{\infty}^{n_{20}} \frac{dN_{evt}}{dN_{ch}} \, dN_{ch}}{\int_{\infty}^{0} \frac{dN_{evt}}{dN_{ch}} \, dN_{ch}} = 0.2. \qquad 14.$$

Figure 11

Spectator-energy
deposition in the
zero-degree calorimeters
(ZDCs) as a function of
charged-particle
multiplicity in the
beam-beam counter arrays
(BBCs) (PHENIX) (49).

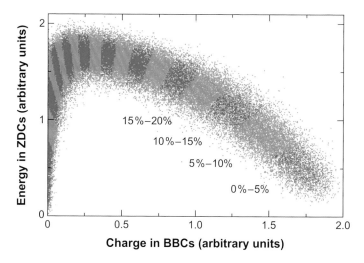

See, for example, **Figure 8**. The same procedure is performed on both the measured and simulated distribution. Note that n_i^{measured} need not equal $n_i^{\mathrm{simulated}}$. This nontrivial fact implies that the mapping procedure is robust to an overall scaling of the simulated N_{ch} distribution compared to the measured distribution. One can extend the centrality classification beyond one dimension by studying the correlation of beam-rapidity spectator multiplicity with midrapidity particle production (**Figure 11**). Although the distribution is somewhat asymmetric, the naive expectations of Section 3.1 are clearly upheld and the mapping procedure proceeds as in the one-dimensional case.

Once a centrality class is defined in simulation, the mean values of quantities such as N_{part} can be calculated for events that fall in that centrality bin. Systematic uncertainty in the total measured cross section propagates into a leading systematic uncertainty on the Glauber quantities extracted via the mapping process. This uncertainty can be directly propagated by varying the value of the denominator in Equation 14 accordingly and recalculating $\langle N_{\mathrm{part}} \rangle$ and so on. For example, for Au+Au collisions at $\sqrt{s_{\mathrm{NN}}} = 200$ GeV in the 10%–20% (60%–80%) most central bin, the STAR Collaboration quotes values of $\langle N_{\mathrm{part}} \rangle \approx 234$ (21), with an uncertainty of $\sim$6 (5) from the 5% uncertainty on the total cross section alone (43). Clearly the uncertainty in the total cross section becomes increasingly important as one approaches the most peripheral collisions.

3.2. Experimental Details

3.2.1. BRAHMS.
The BRAHMS experiment uses the charged-particle multiplicity observed in a pseudorapidity range of $-2.2 \leq \eta \leq 2.2$ to determine reaction centrality (44–46). The multiplicities are measured in a multiplicity array consisting of an inner barrel of Si strip detectors (SiMA) and an outer barrel of plastic scintillator tile detectors (TMA) (37, 47) for collisions within 36 cm of the nominal vertex location. Both arrays cover the same pseudorapidity range for collisions at the nominal vertex. Beam-beam counter arrays (BBCs) located on either side of the nominal interaction

point at a distance of 2.2 m of the nominal vertex extend the pseudorapidity coverage for measuring charged-particle pseudorapidity densities. These arrays consist of Cherenkov UV-transmitting plastic radiators coupled to photomultiplier tubes. **Figure 12** shows the normalized multiplicity distribution measured in the multiplicity array for the ^{197}Au+^{197}Au reaction at $\sqrt{s_{NN}} = 130$ GeV.

The BRAHMS reference multiplicity distribution requires coincident signals in the experiment's two ZDC detectors, an interaction vertex located within 30 cm of

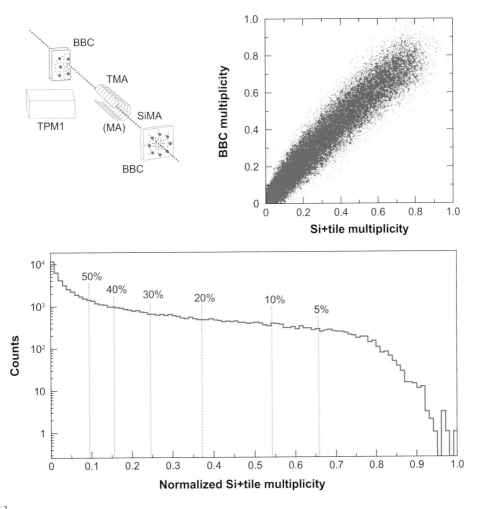

Figure 12

Normalized multiplicity distribution for the ^{197}Au+^{197}Au reaction at $\sqrt{s_{NN}} = 130$ GeV as measured in the BRAHMS multiplicity array (MA) (44). The inset shows the correlation pattern of multiplicities measured in the beam-beam counter array (BBC) and the MA. The vertical lines indicate multiplicity values associated with the indicated centrality cuts. The schematic shows the relative locations of the Si (SiMA) and tile (TMA) elements of the MA, the BBCs, and the front time-projection chamber (TPM1) of the midrapidity spectrometer.

the nominal vertex location, and that there be at least four hits in the TMA. This additional requirement largely removes background contributions from beam-residual gas interactions and from very peripheral collisions involving only electromagnetic processes. The collision vertex can be determined by either the BBCs, the ZDC counters, or a time-projection chamber that is part of a midrapidity spectrometer arm (TPM1 in **Figure 12**).

A simulation of the experimental response based on realistic GEANT3 simulations (48) and using the HIJING Monte Carlo event generator (19) for input was used to estimate the fraction of the inelastic scattering yield that was missed in the experiment's MB event selection. Multiplicity spectra using the simulated events are compared with the experimental spectra. The shapes of the spectra are found to agree very well for an extended range of multiplicities in the TMA array above the threshold multiplicity set for the event selection. The simulated events are then used to extrapolate the experimental spectrum below threshold. Using this procedure, it is estimated that the MB event selection criteria selects $(93 \pm 3)\%$ of the total nuclear cross section for Au+Au collisions at $\sqrt{s_{NN}} = 200$ GeV, down to $(87 \pm 7)\%$ for Cu+Cu collisions at $\sqrt{s_{NN}} = 62.4$ GeV.

3.2.2. PHENIX. We consider two examples: Au+Au at $\sqrt{s_{NN}} = 200$ GeV and Cu+Cu at $\sqrt{s_{NN}} = 22.4$ GeV. The MB trigger condition for Au+Au collisions at $\sqrt{s_{NN}} = 200$ GeV was based on BBCs (38). The two BBCs ($3.1 \leq |\eta| \leq 3.9$) each consist of 64 photomultipliers that detect Cherenkov light produced by charged particles traversing quartz radiators. On the hardware level, an MB event was required to have at least two photomultiplier hits in each BBC. Some analyses only used events with an additional hardware coincidence of the two ZDCs. Moreover, it was required that the interaction vertex along the beam axis (z-axis) reconstructed on the basis of the arrival time difference in the two BBCs lie within ± 30 cm of the nominal vertex.

The efficiency of accepting inelastic Au+Au collisions under the condition of having at least two photomultiplier hits in each BBC ($N_{PMT}^{BBC} \geq 2$) was determined with the aid of HIJING Monte Carlo events (19) and a detailed simulation of the BBC response (48). With an offline vertex cut of ± 30 cm, these simulations yield an efficiency of $(92.3 \pm 2)\%$ for Au+Au at $\sqrt{s_{NN}} = 200$ GeV (49). The systematic uncertainties reflect uncertainties of (*a*) dN_{ch}/dy in HIJING, (*b*) the shape of the z-vertex distribution, and (*c*) the stability of the photomultiplier gains.

The additional requirement of a ZDC coincidence removes remaining background events from beam-gas interactions, but possibly also leads to a small inefficiency for peripheral collisions. The efficiency of accepting real Au+Au collisions with $N_{PMT}^{BBC} \geq 2$ under the condition of a coincidence of the ZDCs was estimated to be $99_{-1.5}^{+1.0}\%$. Combining the efficiencies of the BBC and ZDC requirement as well as the offline vertex cut, PHENIX finds that its sample of MB events in Au+Au collisions at $\sqrt{s_{NN}} = 200$ GeV corresponds to $91.4_{-3.0}^{+2.5}\%$ of the total inelastic cross section. Centrality classes were then defined by cuts on the two-dimensional distribution of the ZDC energy as a function of the BBC signal, as shown in **Figure 11**.

As a second example, we consider the centrality selection in Cu+Cu collisions at $\sqrt{s_{NN}} = 22.4$ GeV. At this energy, the beam rapidity $y_{beam} \approx 3.2$ lies within the

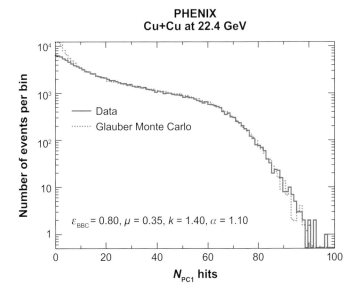

PHENIX
Cu+Cu at 22.4 GeV

Figure 13

Measured and simulated distribution of the Pad Chamber 1 (PC1) hit multiplicity used as a centrality variable in Cu+Cu collisions at $\sqrt{s_{\mathrm{NN}}} = 22.4$ GeV (PHENIX). BBC, beam-beam counter array.

pseudorapidity range of the BBCs. The BBCs were still used as MB trigger detectors ($N_{\mathrm{PMT}}^{\mathrm{BBC}} \geq 1$ in both BBCs). However, a monotonic relation between the BBC signal and the impact parameter was no longer obvious. Thus, the hit multiplicity (N_{PC1}) measured with a Pad Chamber 1 (PC1) detector (50) at midrapidity ($|\eta| \leq 0.35$) was used as a centrality variable.

The PC1 multiplicity distribution was simulated on the basis of a convolution of the N_{part} distribution from a GMC and an NBD. A nonlinear scaling of the average particle multiplicity with N_{part} was allowed: Workers assumed that the number of independently decaying precursor particles (ancestors, N_{ancestor}) is given by $N_{\mathrm{ancestor}} = N_{\mathrm{part}}^{\alpha}$. The number of measured PC1 hits per precursor particle was assumed to follow an NBD:

$$P_{\mu,k}(n) = \frac{\Gamma(n+k)}{\Gamma(n+1)\Gamma(k)} \cdot \frac{(\mu/k)^n}{(\mu/k+1)^{n+k}}. \qquad 15.$$

In a GMC event, the NBD was sampled N_{ancestor} times to obtain the simulated PC1 multiplicity for this event. The PC1 multiplicity distribution was simulated for a grid of values for μ, k, and α to find optimal values. **Figure 13** shows the measured and simulated PC1 distribution, in addition to the best estimate of the BBC trigger efficiency ($\varepsilon_{\mathrm{BBC}} \approx 0.8$) corresponding to the difference at small N_{PC1} (see Reference 49 for a similar study in Au+Au collisions at $\sqrt{s_{\mathrm{NN}}} = 19.6$ GeV). Given the good agreement between the measured and simulated distribution in **Figure 13**, centrality classes for Cu+Cu collisions at $\sqrt{s_{\mathrm{NN}}} = 22.4$ GeV were defined by identical cuts on the measured and simulated N_{PC1}.

3.2.3. PHOBOS. As discussed above, PHOBOS measures centrality with two sets of 16 scintillator paddle counters covering $3 < |\eta| < 4.5$ (39). Good events are defined by having a time difference of less than 4 ns between the first hit impinging

Figure 14

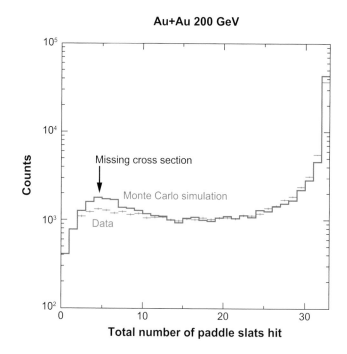

Figure 14

PHOBOS data (39) showing the number of paddle slats, out of 32 total, hit by charged particles in triggered events. The data are compared with a full Monte Carlo simulation of unbiased HIJING events. The data are normalized around N(*slats*) ~16 in order to compare data and the full simulations and thus to estimate the fraction of events lost to the trigger conditions. Figure adapted with permission from Reference 39.

on each paddle counter (limiting the vertex range) and either a coincidence between the PHOBOS ZDCs or a high-summed energy signal in the paddles (to avoid the slight inefficiency in the ZDCs at small impact parameter at low energies).

PHOBOS estimates the observed fraction of the total cross section by measuring the distribution of the total number of paddle slats, as shown in **Figure 14**. Most of the variation in this quantity essentially measures the very-low-multiplicity part of the multiplicity distribution, as the bulk of the events have sufficient multiplicity to fire all of the paddles simultaneously. The inefficiency is determined by matching the plateau structure in the data to that in HIJING and measuring how many of the low-multiplicity events are missed in the data relative to the Monte Carlo calculation. This accounts for a variety of instrumental effects in an aggregate way, and the difference between the estimated value and 100% sets the scale for the systematic uncertainty.

The total detection efficiency for 200 GeV Au+Au collisions is 97%, and 88% when requiring that more than two slats hit on each set of 16 paddle counters. This last requirement dramatically reduces background events taken to tape, and the relative efficiency is straightforward to measure with the events triggered on a coincidence of one or more hits in each set of counters.

For events with much lower multiplicities and/or lower energies, both the multiplicity and rapidity reach are substantially smaller. This strongly impacts the efficiency of the paddle counters and thus potentially distorts the centrality estimation. For these, PHOBOS uses the full distribution of multiplicities measured in several η regions of the Octagon and Ring multiplicity counters and matches them to the distributions measured in a Monte Carlo simulation, typically HIJING. Once the overall multiplicity scale is fit, the difference in the integrals between the data and Monte Carlo simulation gives a reasonable estimate of the fraction of the observed total cross section.

3.2.4. STAR. STAR defines centrality classes for Cu+Cu and Au+Au (d+Au) using charged-particle tracks reconstructed in the TPC (FTPC) over full azimuth and $|\eta| < 0.5$ $(2.5 < \eta < 4)$. Background events are removed by requiring the reconstruction of a primary vertex in addition to either a coincident ZDC (130/200 GeV Au+Au/Cu+Cu) or BBC (62.4 GeV Au+Au/Cu+Cu) signal. Vertex-reconstruction inefficiency in low-multiplicity events reduces the fraction of the total measured cross section to, for example, $(95 \pm 5)\%$ for 130 GeV Au+Au. For MB events, centrality is defined offline by binning the measured dN_{evt}/dN_{ch} distribution by a fraction of the total cross section. Glauber calculations are performed using a Monte Carlo calculation. STAR enhances central events via an online trigger using a coincidence between the MB ZDC condition and the large energy deposition in the Central Trigger Barrel—a set of 240 scintillating slats covering full azimuth and $-1 < \eta < 1$. After offline cuts, the central trigger corresponds to the 0%–5% most central class of events. STAR has several methods of extracting mean values of Glauber quantities.

1. STAR reports little dependence on the mean values of N_{part} and N_{coll} extracted via the aforementioned mapping procedure when vastly different models of particle production were used to simulate the charged-particle multiplicity. Thus, for many analyses (62.4/130/200 GeV Au+Au), STAR bypasses simulation of the multiplicity distribution and instead defines centrality bins from the Monte Carlo–calculated $d\sigma/dN_{part}$ and $d\sigma/dN_{coll}$ distributions themselves (43, 51, 52). Mean values of Glauber quantities are extracted by binning the calculated distribution (e.g., $d\sigma/dN_{part}$) analogously to the measured dN_{evt}/dN_{ch}. Potential biases due to lack of fluctuations in simulated particle production were evaluated and found to be negligible for all but the most peripheral events. Further uncertainties in the extraction of $\langle N_{part} \rangle$ and $\langle N_{coll} \rangle$ are detailed in References 43 and 51 and are dominated by uncertainty in σ_{inel}^{NN}, the Woods-Saxon parameters of Au and Cu, and the 5% uncertainty in the measured cross section.
2. For Cu+Cu and recent studies of elliptic-flow fluctuations in Au+Au, STAR has invoked a full simulation of the TPC multiplicity distribution (53), analogous to that performed for previous d+Au studies described below.
3. For d+Au events, centrality was defined by both measuring and simulating the charged-particle multiplicity in the FTPC in the direction of the initial Au beam. The simulated distribution was constructed using a GMC model coupled to a random sampling of an NBD. The NBD parameters were taken directly from measurements from the UA5 Collaboration at the same rapidity and energy (41). For each Monte Carlo event, the NBD was randomly sampled N_{part} times. After accounting for tracking efficiency, the simulated N_{ch} distribution was found to be in good agreement with the data (54). The mean values of various Glauber quantities were then extracted, as described in Section 3.1.3. A second class of events was also used, where a single neutron was tagged in the ZDC in the direction of the initial d beam. These single-neutron events are

essentially peripheral p+Au collisions, and the corresponding FTPC multiplicity is again well described by the GMC-based simulation (54).

3.3. Acceptance Biases

Because centrality is estimated using quantities that vary monotonically with particle multiplicity, one must be careful to avoid associating fluctuations of an observable with fluctuations in the geometric quantities themselves. This is especially true when estimating the yield per participant pair, when one is estimating the participants from the yield itself. Of course, in heavy ion collisions, the extraordinarily high multiplicities reduce the effect of autocorrelation bias, as was estimated by STAR (52). However, the RHIC experiments have found that lower multiplicities and lower energies are quite challenging. Estimating the number of participants in d+Au proved particularly delicate owing to autocorrelations, which were reduced (in HIJING simulations) by using large regions in pseudorapidity positioned far forward and backward of midrapidity (7).

3.4. Estimating Geometric Quantities

3.4.1. Total geometric cross section.
The total geometric cross section for the collision of two nuclei A and B, that is, the integral of the distributions $d\sigma/db$ shown in **Figure 15**, is a basic quantity that can be easily calculated in the GMC approach. It corresponds to all GMC events with at least one inelastic nucleon-nucleon collision. In ultrarelativistic A+B collisions, the de Broglie wavelength of the nucleons is small compared to their transverse extent so that quantum-mechanical effects are negligible. Hence, the total geometric cross section is expected to be a good approximation of the total inelastic cross section. For the reaction systems in **Figure 15** (d+Au, Cu+Cu, and Au+Au at $\sqrt{s_{NN}} = 200$ GeV), the Monte Carlo calculations yield

Figure 15

Total geometrical cross section from Glauber Monte Carlo calculations (d+Au, Cu+Cu, and Au+Au at $\sqrt{s_{NN}} = 200$ GeV). The green line represents an optical-limit calculation for Au+Au.

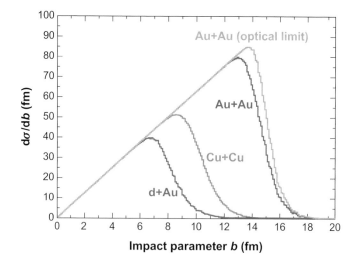

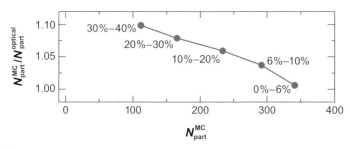

Figure 16

The ratio of N_{part} calculated with a Glauber Monte Carlo approach ($N_{\mathrm{part}}^{\mathrm{MC}}$) to that calculated with an optical approximation ($N_{\mathrm{part}}^{\mathrm{optical}}$) for the same fraction of the total inelastic Au+Au cross section ($\sqrt{s_{\mathrm{NN}}} = 130$ GeV), plotted as a function of $N_{\mathrm{part}}^{\mathrm{MC}}$ (60). Figure adapted with permission from Reference 60.

$\sigma_{\mathrm{geo}}^{\mathrm{dAu}} \approx 2180$ mb, $\sigma_{\mathrm{geo}}^{\mathrm{CuCu}} \approx 3420$ mb, and $\sigma_{\mathrm{geo}}^{\mathrm{AuAu}} \approx 6840$ mb, respectively. The systematic uncertainties are of the order of 10% and are dominated by the uncertainties of the nuclear density profile.

Also shown in **Figure 15** is a comparison with an optical calculation of $\mathrm{d}\sigma/\mathrm{d}b$, which shows the effect described in Section 2.3. Optical-limit calculations do not naturally contain the terms in the multiple-scattering integral, where nucleons hide behind each other. This leads to a slightly larger cross section ($\sigma_{\mathrm{geo,\,optical}}^{\mathrm{AuAu}} \approx 7280$ mb). Although this seems like a small perturbation on $\sigma_{\mathrm{geo}}^{\mathrm{AB}}$, it has a surprisingly large effect on the extraction of N_{part} and N_{coll}. This does not come from any fundamental change in the mapping of impact parameter onto those variables. **Figure 5** shows the mean value of N_{coll} (upper curve) and N_{part} (lower curve) as a function of b, where the two track each other very precisely over a large range in impact parameter, well within the range usually measured by the RHIC experiments. The problem comes in when dividing a sample up into bins in a fractional cross section. Although it is straightforward to estimate the most central bins, one finds a systematic difference of N_{part} between the two calculations as the geometry gets more peripheral. This is shown in **Figure 16** for $\sqrt{s_{\mathrm{NN}}} = 130$ GeV by comparing the Monte Carlo calculation in HIJING with the optical-limit calculation in Reference 42.

3.4.2. Participants (N_{part}) and binary collisions (N_{coll}). As described in Section 2, the Glauber model is a multiple collision model that treats an A+B collision as an independent sequence of nucleon-nucleon collisions. A participating nucleon or wounded nucleon is defined as a nucleon that undergoes at least one inelastic nucleon-nucleon collision. The centrality of an A+B collision can be characterized both by the number of participating nucleons N_{part} and by the number of binary nucleon-nucleon collisions N_{coll}. **Figure 17** shows the average number of participants $\langle N_{\mathrm{part}} \rangle$ and nucleon-nucleon collisions $\langle N_{\mathrm{coll}} \rangle$ as a function of the impact parameter b for Au+Au and Cu+Cu collisions at $\sqrt{s_{\mathrm{NN}}} = 200$ GeV. The event-by-event fluctuations of these quantities for a fixed impact parameter are illustrated by the scatter plots.

Figure 17

Average number of
participants $\langle N_{\text{part}} \rangle$ and
binary nucleon-nucleon
collisions $\langle N_{\text{coll}} \rangle$, in
addition to event-by-event
fluctuations of these
quantities in the Glauber
Monte Carlo calculation
as a function of the impact
parameter b.

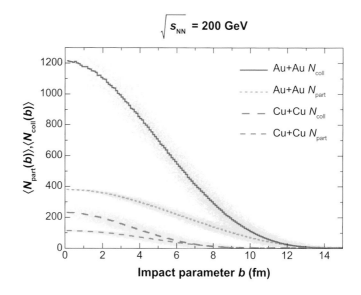

The shapes of the N_{part} and N_{coll} distributions shown in **Figure 18** for Au+Au
collisions reflect the fact that peripheral A+B collisions are more likely than central
collisions. $\langle N_{\text{part}} \rangle$ and $\langle N_{\text{coll}} \rangle$ for a given experimental centrality class, for example,
the 10% most central collisions, depend on the fluctuations of the centrality variable,
which is closely related to the geometrical acceptance of the respective detector. By
simulating the fluctuations of the experimental centrality variable and applying similar
centrality cuts as in the analysis of real data, one obtains N_{part} and N_{coll} distributions for

Figure 18

N_{part} and N_{coll} distributions
in Au+Au collisions at
$\sqrt{s_{\text{NN}}} = 200$ GeV from a
Glauber Monte Carlo
calculation. By applying cuts
on simulated centrality
variables, in this case the
beam-beam counter array
and zero-degree calorimeter
signal as measured by
PHENIX, one obtains N_{part}
and N_{coll} distributions for
the respective centrality
class.

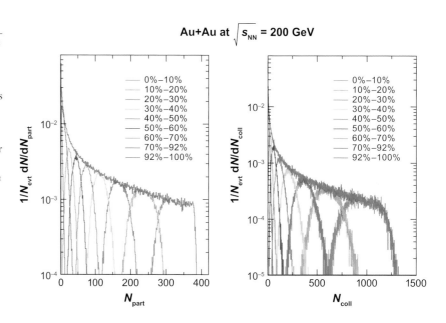

each centrality class. For peripheral classes, the bias introduced by the inefficiency of the experimental MB trigger must be taken into account by applying a corresponding trigger threshold on the GMC events. Experimental observables such as particle multiplicities can then be plotted as a function of the mean value of N_{part} and N_{coll} distributions.

The systematic uncertainties of N_{part}, N_{coll}, and other calculated quantities are estimated by varying various model parameters. **Figure 19** shows such a study for Au+Au collisions at $\sqrt{s_{NN}} = 200$ GeV (PHENIX). The following effects were considered:

- The default value of the nucleon-nucleon cross section of $\sigma_{NN} = 42$ mb was changed to 39 mb and 45 mb.
- Woods-Saxon parameters were varied to determine uncertainties related to the nuclear density profile.
- Effects of a nucleon hard core were studied by requiring a minimum distance of 0.8 fm between two nucleons of the same nucleus without distorting the radial density profile.
- Parameters of the BBC and ZDC simulation (e.g., parameters describing the finite resolution of these detectors) were varied.
- The black-disk nucleon-nucleon overlap function was replaced by a gray disk and a Gaussian overlap function (31) without changing the total inelastic nucleon-nucleon cross section.
- The origin of the centrality cuts applied in the scatter plot of ZDC versus BBC space was modified in the Glauber calculation.
- The uncertainty of the efficiency of the MB trigger leads to uncertainties as to which percentile of the total inelastic cross section is actually selected with certain centrality cuts. The centrality cuts applied on the centrality observable simulated with the GMC were varied accordingly to study the influence on $\langle N_{part} \rangle$ and $\langle N_{coll} \rangle$.
- Even if the MB trigger efficiency were precisely known, potential instabilities of the centrality detectors could lead to uncertainties as to which percentile of the total cross section is selected. This has been studied by comparing the number of events in each experimental centrality class for different run periods. The effect on $\langle N_{part} \rangle$ and $\langle N_{coll} \rangle$ was again studied by varying the cuts on the simulated centrality variable accordingly.

The total systematic uncertainty indicated by the shaded boxes in **Figure 19** was obtained by adding the deviations from the default result for each item in the above list in quadrature. The uncertainty of N_{part} decreases from $\sim$20% in peripheral collisions to $\sim$3% in central Au+Au collisions. N_{coll} has similar uncertainties as N_{part} for peripheral Au+Au collisions. For $N_{part} > 100$ (or $N_{coll} > 200$), the relative systematic uncertainty of N_{coll} remains constant at $\sim$10%. Similar estimates for the systematic uncertainties of N_{part} and N_{coll} at the CERN SPS energy of $\sqrt{s_{NN}} = 17.2$ GeV were reported in Reference 55.

For the comparison of observables related to hard processes in A+A and p+p collisions, it is advantageous to introduce the nuclear overlap function $\langle T_{AB} \rangle_f$ for a

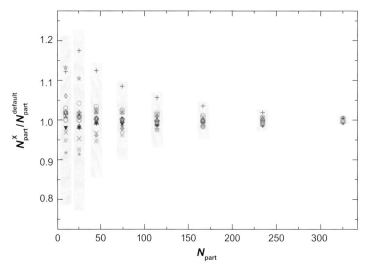

- ▼ $\sigma_{NN}^{inel} = 39$ mb
- ▲ $\sigma_{NN}^{inel} = 45$ mb
- ○ Woods-Saxon: $R = 6.65$ fm, $d = 0.55$ fm
- □ Woods-Saxon: $R = 6.25$ fm, $d = 0.53$ fm
- ⬚ Nucleons with hard core ($r_{hard} = 0.4$ fm)
- △ Different simulation of ZDC energy
- ◇ Different BBC fluctuations

- ⊕ Gray disk NN overlap function
- ✳ Gaussian NN overlap function
- ✴ Variation in 2D BBC-ZDC centrality selection
- + BBC trigger eff. (91.4+2.5)% (more central)
- ✕ BBC trigger eff. (91.4−3.0)% (less central)
- ○ Sys. error of exp. centr. sel. (more central)
- ✕ Sys. error of exp. centr. sel. (less central)

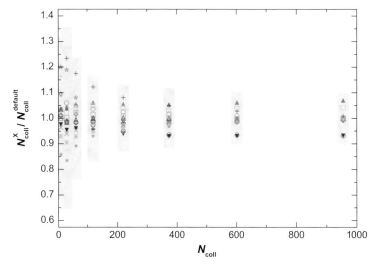

Figure 19

Effect of various parameters in the Glauber Monte Carlo calculation on N_{part} and N_{coll} for Au+Au collisions at $\sqrt{s_{NN}} = 200$ GeV. BBC, beam-beam counter array; ZDC, zero-degree calorimeter.

certain centrality class f (see Section 4.2) that is calculated in the GMC approach as

$$\langle T_{AB} \rangle_f = \langle N_{coll} \rangle_f / \sigma_{inel}^{NN}. \qquad 16.$$

The uncertainty of the inelastic nucleon-nucleon cross section σ_{inel}^{NN} does not contribute to the systematic uncertainty. Apart from this, $\langle T_{AB} \rangle_f$ has the same systematic uncertainties as $\langle N_{coll} \rangle_f$.

3.4.3. Eccentricity. One of the surprising features of the RHIC data was the strong event-by-event asymmetries in the azimuthal distributions. This has been attributed to the phenomenon of elliptic flow—the transformation of spatial asymmetries into momentum asymmetries by means of hydrodynamic evolution. For any hydrodynamic model to be appropriate, the system must be sufficiently opaque (where opacity is the product of density times interaction cross section) such that the system equilibrates locally at early times. This suggests that the relevant geometric quantity for controlling elliptic flow is the shape of the overlap region, which sets the scale for the gradients that develop.

The typical variable used to quantify this shape is the eccentricity, defined as

$$\varepsilon = \frac{\langle y^2 - x^2 \rangle}{\langle y^2 + x^2 \rangle}. \qquad 17.$$

Just as with other variables discussed here, there are two ways to calculate this. In the optical limit, one performs the averages at a fixed impact parameter, weighting either by the local participant or the binary collision density. In the Monte Carlo approach, one simply calculates the moments of the participants themselves. Furthermore, one can calculate these moments with the x-axis oriented in two natural frames. The first is along the nominal reaction plane (estimated using spectator nucleons). The second is along the short principal axis of the participant distribution itself (56). The only mathematical difference between the two definitions involves the incorporation of the correlation coefficient $\langle xy \rangle$:

$$\epsilon_{part} = \frac{\sqrt{(\sigma_x^2 - \sigma_y^2)^2 + 4(\sigma_{xy})^2}}{\sigma_x^2 + \sigma_y^2}. \qquad 18.$$

A comparison of the two definitions is shown in **Figure 20**. One sees very different limiting behavior at very large and small impact parameters. At large impact parameter, fluctuations due to small numbers of participants drive $\epsilon_{std} \to 0$, but $\epsilon_{part} \to 1$. As $b = 0$, ϵ_{std} also goes to zero as the system becomes radially symmetric, whereas ϵ_{part} now picks up the fluctuations and remains finite. The relevance of these two quantities to actual data is discussed in Sections 4.3 and 4.4.

4. GEOMETRIC ASPECTS OF P+A AND A+A PHENOMENA

4.1. Inclusive Charged-Particle Yields (Total and Midrapidity)

The total multiplicity in hadronic reactions is a measure of the degrees of freedom released in the collision. In the 1970s, researchers discovered that the total number of

Figure 20

PHOBOS calculation of eccentricity in Au+Au and Cu+Cu collisions as a function of N_{part} (74). Both standard and participant eccentricities are shown. Figure adapted with permission from Reference 74.

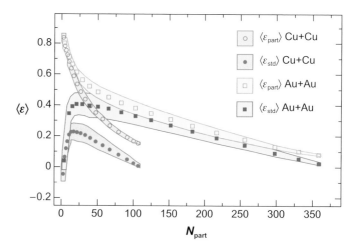

particles produced in p+A collisions was proportional to the number of participants; that is, $N_{ch} \propto N_{part} = \nu + 1$, where ν is defined as the number of struck nucleons in the nucleus (61). This experimental fact was instrumental in establishing N_{part} as a fundamental physical variable. The situation became more interesting when the total multiplicity was measured in Au+Au at four RHIC energies, spanning an order of magnitude in $\sqrt{s_{NN}}$, and was approximately proportional to N_{part} there as well. This is shown in **Figure 21** with PHOBOS data from References 58 and 59, and is striking if one considers the variety of physics processes that should contribute to the total multiplicity.

By contrast, the inclusive charged-particle density near midrapidity $[dN/d\eta(|\eta| < 1)]$ does not scale linearly with $N_{part}/2$. This is shown in **Figure 22** with STAR data, with N_{part} estimated both from an optical calculation as well as a GMC calculation (62). The comparison shows why care must be taken in the estimation of N_{part}, as using one or the other gives better agreement with very different models. Eskola et al.'s (63) saturation model shows scaling with N_{part} and agrees with the data if N_{part} is estimated with an optical calculation. However, it disagrees with the data when the GMC approach is used. Better agreement with the data can be found with a so-called two-component model. For example, in Reference 42

$$\frac{dN}{d\eta} = n_{pp}\left[(1-x)N_{part} + xN_{coll}\right]. \qquad 19.$$

This model can be fit to the data by a judicious choice of x, the parameter that controls the admixture of hard particle production. However, there is no evidence of any energy dependence to this parameter (64, 65) from $\sqrt{s_{NN}} = 19.6$ to 200 GeV, suggesting that the source of this dependence has little to nothing to do with hard or semihard processes.

4.2. Hard Scattering: T_{AB} Scaling

The number of hard processes between point-like constituents of the nucleons in an A+B collision is proportional to the nuclear overlap function $T_{AB}(b)$ (66–71). This

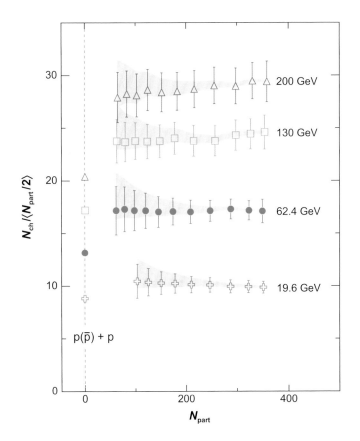

Figure 21

Total inclusive
charged-particle
multiplicity (N_{ch}) divided by
the number of participating
nucleon pairs ($N_{part}/2$) from
PHOBOS data at four
RHIC energies (58, 59).
The data are compared with
p+p data or interpolations
to unmeasured energies at
$N_{part} = 2$. Figure adapted
with permission from
References 58 and 59.

follows directly from the factorization theorem in the theoretical description of hard interactions within perturbative QCD (72). In detail, the average yield for a hard process with cross section σ_{hard}^{pp} in p+p collisions per encounter of two nuclei A and B with impact parameter b is given by

$$N_{hard}^{AB,enc}(b) = T_{AB}(b)\sigma_{hard}^{pp}.$$
20.

Here, $T_{AB}(b)$ is normalized so that $\int T_{AB}(b) \, d^2b = AB$. σ_{hard}^{pp} can, for example, represent the cross section for the production of charm-anticharm quark pairs ($\bar{c}c$) or high-transverse-momentum (p_T) direct photons in proton-proton collisions.

At large impact parameter, not every encounter of two nuclei A and B leads to an inelastic collision. Hence, the average number of hard processes per inelastic A+B collision is given by

$$N_{hard}^{AB}(b) = \frac{T_{AB}(b)}{p_{inel}^{AB}(b)}\sigma_{hard},$$
21.

where $p_{inel}^{AB}(b)$ is the probability of an inelastic A+B collision. In the optical limit, $p_{inel}^{AB}(b)$ is given by (see Equation 5)

$$p_{inel}^{AB}(b) = 1 - \left(1 - \sigma_{inel}^{NN}\frac{T_{AB}(b)}{AB}\right)^{AB}.$$
22.

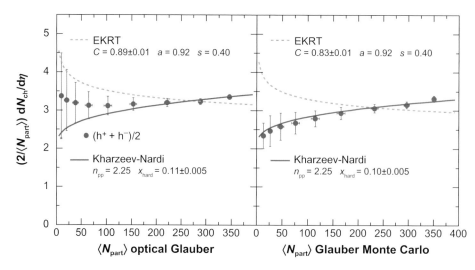

Figure 22

Inclusive charged-particle multiplicity near midrapidity [$dN/d\eta(|\eta| < 1)$] divided by the number of participating nucleon pairs ($N_{\mathrm{part}}/2$) estimated using the optical approximation (*left*) and a Glauber Monte Carlo calculation (*right*) from STAR. The data are compared with two-component fits and a calculation based on parton saturation [EKRT (63)]. Figure adapted with permission from Reference 63.

Particle yields at RHIC are usually measured as a function of p_T. If an invariant cross section $d\sigma^{pp}/dp_T$ for a hard scattering process leading to the production of a certain particle x has been measured in p+p collisions, then according to Equation 21 the invariant multiplicity of x per inelastic A+B collision with impact parameter b is given by

$$\frac{1}{N_{\mathrm{inel}}^{AB}} \frac{dN_x^{AB}}{dp_T} = \frac{T_{AB}(b)}{p_{\mathrm{inel}}^{AB}(b)} \frac{d\sigma_x^{pp}}{dp_T}. \qquad 23.$$

This baseline expectation is based purely on nuclear geometry and assumes the absence of any nuclear effects. In reality, one must average over a certain impact parameter distribution. As an example, we consider a centrality class f, which corresponds to a fixed impact parameter range $b_1 \le b \le b_2$. Taking the average in this range over the impact parameter distribution [weighting factor $d\sigma^{AB}/db = 2\pi b\, p_{\mathrm{inel}}^{AB}(b)$] leads to

$$\frac{1}{N_{\mathrm{inel}}^{AB}} \frac{dN_x^{AB}}{dp_T}\bigg|_{\mathrm{f}} = \langle T_{AB}\rangle_{\mathrm{f}} \frac{d\sigma_x^{pp}}{dp_T}, \qquad 24.$$

with

$$\langle T_{AB}\rangle_{\mathrm{f}} = \frac{\int d^2b\, T_{AB}(b)}{\int d^2b\, p_{\mathrm{inel}}^{AB}(b)} = \frac{\int_{b_1}^{b_2} db\, 2\pi b\, T_{AB}(b)}{\int_{b_1}^{b_2} db\, 2\pi b\, p_{\mathrm{inel}}^{AB}(b)}. \qquad 25.$$

Averaging over the full impact parameter range ($b_1 = 0$, $b_2 = \infty$) yields $\langle T_{AB}\rangle_{\mathrm{f}} = AB/\sigma_{\mathrm{geo}}^{AB}$. In the GMC approach, $\langle T_{AB}\rangle_{\mathrm{f}}$ for a certain centrality class is calculated as

$$\langle T_{AB}\rangle_{\mathrm{f}} = \langle N_{\mathrm{coll}}\rangle_{\mathrm{f}}/\sigma_{\mathrm{inel}}^{NN}, \qquad 26.$$

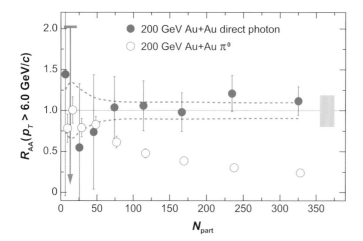

Figure 23

Nuclear modification factor R_{AA} in Au+Au collisions at $\sqrt{s_{NN}} = 200$ GeV for direct-photon and neutral-pion yields integrated above $p_T = 6$ GeV/c as a function of N_{part} (73). The dashed lines indicate the systematic uncertainties of $\langle T_{AB}\rangle_f$ used in the calculation of R_{AA}. Figure adapted with permission from Reference 73.

where the averaging is done for all A+B collisions with at least one inelastic nucleon-nucleon collision and whose simulated centrality variable belongs to centrality class f.

To quantify nuclear effects on particle production in hard scattering processes, the nuclear modification factor $R_{AB}(p_T)$ is defined as

$$R_{AB}(p_T) = \frac{\left(N_{inel}^{AB}\right)^{-1} dN_x^{AB}/dp_T}{\langle T_{AB}\rangle_f \, d\sigma_x^{pp}/dp_T}. \qquad 27.$$

At high p_T ($p_T \geq 2-3$ GeV/c for hadrons and $p_T \geq 4-6$ GeV/c for direct photons), particle production is expected to be dominated by hard processes such that, in the absence of nuclear effects, R_{AB} should be unity. Owing to their electromagnetic nature, high-p_T direct photons are essentially unaffected by the hot and dense medium produced in an A+B collision so that they should exhibit T_{AB} scaling (**Figure 23**). Direct photons indeed follow T_{AB} scaling over the entire centrality range, whereas neutral pions are strongly suppressed in central collisions. This is one of the major discoveries at RHIC. The direct-photon measurement is an experimental proof of T_{AB} scaling of hard processes in A+B collisions. With this observation, the most natural explanation for the suppression of high-p_T neutral pions is energy loss of partons from hard scattering in a QGP (jet quenching) (5–8).

4.3. Eccentricity and Relation to Elliptic Flow

Hydrodynamic calculations suggest that spatial asymmetries in the initial state are mapped directly into in the final-state momentum distribution. At midrapidity, these anisotropies are manifest in the azimuthal (ϕ) distributions of inclusive and identified charged particles, with the modulation of $dN/d\phi \sim 1 + 2v_2 \cos[2(\phi - \Psi_R)]$ characterized by the second Fourier coefficient v_2, where Ψ_R defines the angle of the reaction plane for a given event. It is generally assumed that v_2 is proportional to the event eccentricity ϵ, which was introduced above. Glauber calculations are used to estimate the eccentricity, for either an ensemble of events or on an event-by-event basis.

For much of the RHIC program, both calculations were typically carried out in the standard reference frame, with the x-axis oriented along the reaction plane. Using this calculation method was apparently sufficient to compare hydrodynamic calculations with Au+Au data. However, it was always noticed that the most central events, which should trend to $\epsilon = 0$, tended to have a significant v_2 value. This led to the study of the participant eccentricity, calculated with the x-axis oriented along the short principal axis of the approximately elliptical distribution of participants in a Monte Carlo approach (56), described above.

Figure 24, taken from Reference 74, shows v_2—the second Fourier coefficient $[\langle\cos(2[\phi - \Psi_R])\rangle]$ of the inclusive particle yield relative to the estimated reaction

Figure 24

(*a*) PHOBOS measurements (74) of v_2 for 62.4 and 200 GeV Cu+Cu and Au+Au collisions (four system-energy combinations in all). v_2 divided by standard eccentricity (*b*) and v_2 divided by participant eccentricity (*c*), showing an approximate scaling for all the system-energy combinations. Figure adapted with permission from Reference 74.

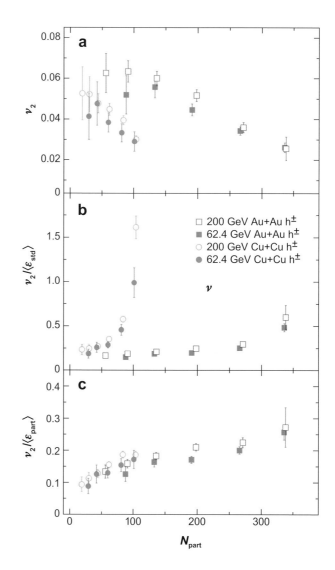

plane angle—as a function of N_{part}. In hydrodynamic models, v_2 is proportional to the eccentricity, suggesting that v_2/ϵ should be a scaling variable. Although the raw values of v_2 as a function of N_{part} peak at similar levels in Au+Au and Cu+Cu, dividing by the standard eccentricity makes the two data sets diverge. However, dividing by ϵ_{part} shows that the two systems have similar $v_2/\epsilon_{\text{part}}$ at the same N_{part}. This shows that the participant eccentricity, a quantity calculated in a simple GMC approach, drives the hydrodynamic evolution of the system for very different energies and geometries.

4.4. Eccentricity Fluctuations

As described above, one of the most spectacular measurements at RHIC was the large value of elliptic flow in Au+Au collisions, suggestive of a Perfect Liquid. After the initial measurement (75), much attention was given to potential biases that could artificially inflate the extraction of v_2 from the data, such as nonflow effects (e.g., correlations from jet fragmentation, resonance decay) and event-by-event fluctuations in v_2. Reference 32 was one of the first analyses to study the effects of fluctuations on extraction of v_2. Using the assumption $v_2 \sim \epsilon_{\text{std}}$, fluctuations in ϵ_{std} were studied using a GMC calculation and comparing $\langle \epsilon_{\text{std}}^n \rangle$ to $\langle \epsilon_{\text{std}} \rangle^n$. Fluctuations were found to play a significant role, where different methods of extraction (e.g., two-particle versus higher-order cumulants) gave results differing by as much as a factor of two, with the most significant differences found for the most central (0%–5%) and most peripheral (60%–80%) event classes.

Recently, in References 53 and 76, the STAR and PHOBOS collaborations have reported measurements of not only $\langle v_2 \rangle$, but also the rms width σ_{v_2}. **Figure 25** shows the distribution of $\sigma_{v_2}/\langle v_2 \rangle$ versus $\langle b \rangle$ from Au+Au data. The measurements are compared with calculations from various dynamical models as well as GMC calculations using both $\sigma_{\epsilon_{\text{std}}}/\langle \epsilon_{\text{std}} \rangle$ and $\sigma_{\epsilon_{\text{part}}}/\langle \epsilon_{\text{part}} \rangle$. Clearly, the ϵ_{std} description is ruled out, whereas the ϵ_{part} description is in good agreement within the measured uncertainties, implying that the measurements are sensitive to the initial conditions. The agreement with the ϵ_{part} Glauber calculation further implies that the measured v_2 fluctuations are fully accounted for by the fluctuations in the initial geometry, leaving little room for other sources (e.g., Color Glass Condensate). We note that these analyses are new and the physics conclusions are far from final. However, this is another excellent example in which Glauber calculations are critical for interpreting RHIC data.

4.5. J/ψ Absorption in Normal Nuclear Matter

Owing to the large mass of the charm quark, $c\bar{c}$ pairs are expected to be produced only in hard processes in the initial phase of an A+B collision. The production rate for $c\bar{c}$ pairs is thus calculable within perturbative QCD, which makes them a calibrated probe of later stages of a heavy ion collision. In particular, workers suggested that free color charges in a QGP could prevent the formation of a J/ψ from the initially produced $c\bar{c}$ pairs so that J/ψ suppression was initially considered a key signature of a QGP formation (77). However, a suppression of J/ψ s relative to the expected

a

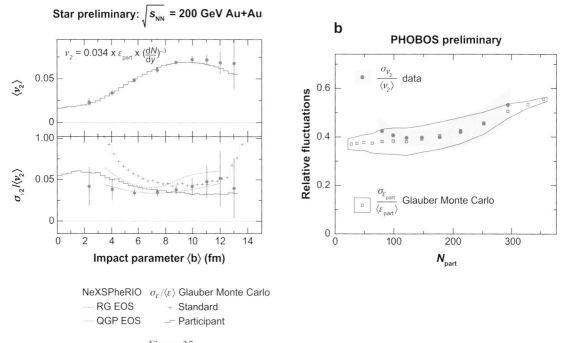

Star preliminary: $\sqrt{s_{NN}}$ = 200 GeV Au+Au

$v_2 = 0.034 \times \varepsilon_{part} \times (\frac{dN}{dy})^{-3}$

PHOBOS preliminary

$\frac{\sigma_{v_2}}{\langle v_2 \rangle}$ data

$\frac{\sigma_{\varepsilon_{part}}}{\langle \varepsilon_{part} \rangle}$ Glauber Monte Carlo

NeXSPheRIO $\sigma_\varepsilon / \langle \varepsilon \rangle$ Glauber Monte Carlo
...... RG EOS   + Standard
—— QGP EOS   ⌐ Participant

Figure 25

(*a*) Elliptic-flow mean value (*top*) and rms width scaled by the mean (*bottom*), as measured by STAR, compared with various dynamical models and a Glauber Monte Carlo calculation (53). (*b*) Gaussian width of v_2 divided by mean v_2 from PHOBOS (76). Figure adapted with permission from References 53 and 76.

production rate for hard processes was already seen in p+A collisions. Thus, it became clear that the conventional J/ψ suppression in p+A collisions needed to be quantified and extrapolated to A+B collisions before any conclusions could be drawn about a possible QGP formation in A+B collisions. Cold nuclear matter effects, which affect J/ψ production, include the modification of the parton distribution in the nucleus (shadowing) and the absorption of preresonant $c\bar{c}$ pairs (78). For the extrapolation of the latter effect from p+A to A+A collisions, the Glauber model is frequently used (79), as described in the following.

We first concentrate on p+A collisions (**Figure 26**). Conventional J/ψ suppression is thought to be related to the path along the z-axis in normal nuclear matter that a preresonant $c\bar{c}$ pair created at point (b, z_A) needs to travel. Because the $c\bar{c}$ pair is created in a hard process, the location of its production is indeed rather well defined. Usually the effects of processes that inhibit the formation of a J/ψ from the preresonant $c\bar{c}$ pair are parameterized with a constant absorption cross section σ_{abs}. Furthermore, formation-time effects are often neglected. Under these assumptions, the probability for the breakup of a preresonant $c\bar{c}$ state created at (**b**, z_A) in a collision

p+A

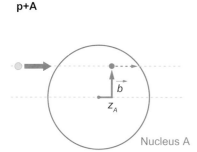

A+B

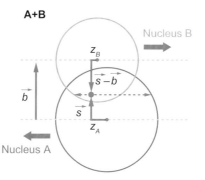

Nucleus B

Nucleus A

Nucleus A

Figure 26

Sketch illustrating the calculation of the J/ψ absorption in normal nuclear matter in proton-nucleus (*left*) and nucleus-nucleus (*right*) collisions. The dashed arrows indicate the paths of a $c\bar{c}$ pair created in a hard process.

with a certain nucleon j of nucleus A then reads (27, 30)

$$p_{\mathrm{abs}}(\mathrm{b}, z_{\mathrm{A}}) = \sigma_{\mathrm{abs}}\hat{T}_{\mathrm{A}>}(\mathrm{b}, z_{\mathrm{A}}), \quad \text{with} \quad \hat{T}_{\mathrm{A}>}(\mathrm{b}, z_{\mathrm{A}}) = \int_{z_{\mathrm{A}}}^{\infty} \hat{\rho}_{\mathrm{A}}(\mathrm{b}, z)\,\mathrm{d}z. \qquad 28.$$

Here, $\hat{\rho}_{\mathrm{A}}$ is the density profile of nucleus A, normalized so that integration over full space yields unity. Hence, the total survival probability for the $c\bar{c}$ pair is

$$p_{\mathrm{surv}}^{\mathrm{A}}(\mathrm{b}, z_{\mathrm{A}}) = (1 - \sigma_{\mathrm{abs}}\hat{T}_{\mathrm{A}>}(\mathrm{b}, z_{\mathrm{A}}))^{A-1} \approx \exp(-(A-1)\sigma_{\mathrm{abs}}\hat{T}_{\mathrm{A}>}(\mathrm{b}, z_{\mathrm{A}})), \qquad 29.$$

where the approximation holds for large nuclei $A \gg 1$. The term $A - 1$ reflects the fact that the $c\bar{c}$-producing nucleon does not contribute to the absorption. The spatial distribution of the produced $c\bar{c}$ pairs follows $\hat{\rho}_{\mathrm{A}}(\mathrm{b}, z)$. Thus, for impact-parameter-averaged p+A collisions (MB), the expression for the J/ψ absorption in the Glauber model reads

$$S_{\mathrm{pA}} = \frac{\sigma(p + A \to J/\psi)}{A \cdot \sigma(p + p \to J/\psi)} = \int \mathrm{d}^2 b \; \mathrm{d}z_{\mathrm{A}}\hat{\rho}_{\mathrm{A}}(\mathrm{b}, z_{\mathrm{A}})p_{\mathrm{surv}}^{\mathrm{A}}(\mathrm{b}, z_{\mathrm{A}}). \qquad 30.$$

Fitting the Glauber model expectation to p+A data yields absorption cross sections of the order of a few millibarns at CERN SPS energies (80).

As illustrated by the dashed arrow in the right panel of **Figure 26**, a $c\bar{c}$ pair created in a collision of two nuclei A and B must pass through both nuclei. Analogous to Equations 28 and 29, the J/ψ survival probability for the path in nucleus B can be written as

$$p_{\mathrm{surv}}^{\mathrm{B}}(\mathrm{s}-\mathrm{b}, z_{\mathrm{B}}) = (1 - \sigma_{\mathrm{abs}}\hat{T}_{B<}(\mathrm{s}-\mathrm{b}, z_{\mathrm{B}}))^{B-1}, \quad \hat{T}_{B<}(\mathrm{s}-\mathrm{b}, z_{\mathrm{B}}) = \int_{-\infty}^{z_{\mathrm{B}}} \hat{\rho}_{\mathrm{B}}(\mathrm{s}-\mathrm{b}, z)\,\mathrm{d}z. \quad 31.$$

The spatial $c\bar{c}$ production probability density follows $\hat{\rho}_{\mathrm{A}}(s, z_{\mathrm{A}}) \times \hat{\rho}_{\mathrm{B}}(\mathrm{s} - \mathrm{b}, z_{\mathrm{B}})$, so that the normal J/ψ suppression as estimated with the Glauber model for A+B collisions with fixed impact parameter b is given by

$$\frac{\mathrm{d}S_{\mathrm{AB}}}{\mathrm{d}^2 b}(b) = \frac{1}{AB\sigma(p + p \to J/\psi)}\frac{\mathrm{d}\sigma(AB \to J/\psi)}{\mathrm{d}^2 b}$$

$$= \int \mathrm{d}^2 s \; \mathrm{d}z_{\mathrm{A}} \; \mathrm{d}z_{\mathrm{B}} \; \hat{\rho}_{\mathrm{A}}(s, z_{\mathrm{A}})\hat{\rho}_{\mathrm{B}}(\mathrm{s} - \mathrm{b}, z_{\mathrm{B}})p_{\mathrm{surv}}^{\mathrm{A}}(s, z_{\mathrm{A}})p_{\mathrm{surv}}^{\mathrm{B}}(\mathrm{s} - \mathrm{b}, z_{\mathrm{B}}). \qquad 32.$$

Here, $\mathrm{d}S_{\mathrm{AB}}/\mathrm{d}^2 b$ is normalized so that $\int \mathrm{d}^2 b \; \mathrm{d}S_{\mathrm{AB}}(b)/\mathrm{d}^2 b = 1$ for $\sigma_{\mathrm{abs}} = 0$. Sometimes the J/ψ suppression observed in p+A collisions is extrapolated to A+B collisions within the Glauber framework by calculating an effective path length $L = L_A + L_B$ so that the expected normal J/ψ suppression can be written as (79–81)

$$S_{AB} = \exp(-L\rho_0\sigma_{abs}),\qquad\qquad 33.$$

where $\rho_0 = 0.17$ fm^{-1} is the nucleon density in the center of heavy nuclei. At the CERN SPS, a suppression stronger than expected from the absorption in cold nuclear matter has been observed in central Pb+Pb collisions (81). This so-called anomalous J/ψ suppression has been discussed as a potential signal for a QGP formation at the CERN SPS energy.

5. DISCUSSION AND THE FUTURE

The Glauber model as used in ultrarelativistic heavy ion physics is based purely on nuclear geometry. What is left from its origin as a quantum-mechanical multiple-scattering theory is the assumption that A+B collisions can be viewed as sequence of nucleon-nucleon collisions and that individual nucleons travel on straight-line trajectories. With the number of participants N_{part} and the number of binary nucleon-nucleon collisions N_{coll}, the Glauber model introduces quantities that are essentially not measurable. Only the forward energy in fixed-target experiments has a rather direct relation to N_{part}.

The motivation for using these rather theoretical quantities is manifold. One main reason for using geometry-related quantities such as N_{part} calculated with the Glauber model is the possibility of comparing centrality-dependent observables measured in different experiments. Moreover, the comparison of a different reaction system as a function of geometric quantities often leads to new insights. Basically all experiments calculate N_{part} and N_{coll} in a similar way using a Monte Carlo implementation of the Glauber model so that the theoretical bias introduced in the comparisons is typically small. Thus, the Glauber model provides a crucial interface between theory and experiment.

The widespread use of the Glauber model is related to the fact that indeed many aspects of ultrarelativistic A+B collisions can be understood purely based on geometry. A good example is the total charged-particle multiplicity that scales as N_{part} over a wide centrality and center-of-mass energy range. Another example is the anisotropic momentum distribution of low-p_T ($p_T < \sim 2$ GeV/c) particles with respect to the reaction plane. This so-called elliptic flow has its origin in the spatial anisotropy of the initial overlap volume in noncentral A+B collisions. It is a success of the Glauber model that event-by-event fluctuations of the spatial anisotropy of the overlap zone as calculated in the Monte Carlo approach appear to be relevant for understanding the measured elliptic flow. In this way, a precise understanding of the Glauber picture has been of central concern for understanding the matter produced at RHIC as a near perfect fluid.

The study of particle production in hard scattering processes is another important field of application for the Glauber model. According to the QCD factorization theorem, the only difference between p+p and A+A collisions in the perturbative QCD description in the absence of nuclear effects is the increased parton flux. This corresponds to a scaling of the invariant particle yields with N_{coll} as calculated with the Glauber model. The scaling of hard processes with N_{coll} or T_{AB} was confirmed by the measurement of high-p_T direct photons in Au+Au collisions at RHIC. This supported the interpretation of a deviation from T_{AB} scaling for neutral pions and other hadrons (high-p_T hadron suppression) as a result of parton energy loss in a QGP.

Future heavy ion experiments, both at RHIC and at the LHC, will further push our understanding of nuclear geometry. As the RHIC experiments study more complex multiparticle observables, the understanding of fluctuations and correlations even in something as apparently simple as the GMC will become a limiting factor in interpreting data. And as the study of high-p_T phenomena involving light and heavy flavors becomes prominent in the RHIC II era, the understanding of nuclear geometry, both experimentally and theoretically, will limit the experimental systematic errors. At the LHC, the precision of the geometric calculations will be limited by the knowledge of σ_{inel}^{NN}, which should be measured in the first several years of the p+p program. After that, Glauber calculations will be a central part of understanding the baseline physics of heavy ions at the LHC in terms of nuclear geometry. We hope this review will prepare the next generation of relativistic heavy ion physicists for tackling these issues.

DISCLOSURE STATEMENT

The authors are not aware of any biases that might be perceived as affecting the objectivity of this review.

ACKNOWLEDGMENTS

The authors would like to thank our colleagues for illuminating discussions, especially Mark Baker, Andrzej Bialas, Wit Busza, Patrick Decowski, Jamie Dunlop, Roy Glauber, Ulrich Heinz, Constantin Loizides, Steve Manly, Alexander Milov, Dave Morrison, Jamie Nagle, Mike Tannenbaum, and Thomas Ullrich. We would like to thank the editorial staff of Annual Reviews for their advice and patience. M.L.M. acknowledges the support of the MIT Pappalardo Fellowship in Physics. This work was supported in part by the Office of Nuclear Physics of the U.S. Department of Energy under contracts DE-AC02-98CH10886, DE-FG03-96ER40981, and DE-FG02-94ER40818.

LITERATURE CITED

1. Czyż W, Maximon LC. *Ann. Phys.* 52:59 (1969)
2. Glauber RJ. In *Lectures in Theoretical Physics*, ed. WE Brittin, LG Dunham, 1:315. New York: Interscience (1959)
3. Bialas A, Bleszyński M, Czyż W. *Acta Physiol. Pol. B* 8:389 (1977)
4. Back BB, et al. *Phys. Rev. Lett.* 85:3100 (2000)
5. Arsene I, et al. *Nucl. Phys. A* 757:1 (2005)

6. Adcox K, et al. *Nucl. Phys. A* 757:184 (2005)

7. Back BB, et al. *Nucl. Phys. A* 757:28 (2005)

8. Adams J, et al. *Nucl. Phys. A* 757:102 (2005)

9. Adler C, et al. *Nucl. Instrum. Methods A* 470:488 (2001)

10. Franco V, Glauber RJ. *Phys. Rev.* 142:1195 (1966)

11. Glauber RJ. *Phys. Rev.* 100:242 (1955)

12. Czyż W, Lesniak L. *Phys. Lett. B* 24:227 (1967)

13. Fishbane PM, Trefil JS. *Phys. Rev. Lett.* 32:396 (1974)

14. Franco V. *Phys. Rev. Lett.* 32:911 (1974)

15. Bialas A, Bleszyński M, Czyż W. *Nucl. Phys. B* 111:461 (1976)

16. Glauber RJ. *Nucl. Phys. A* 774:3 (2006)

17. Ludlam TW, Pfoh A, Shor A. HIJET. A Monte Carlo Event Generator for P Nucleus and Nucleus Nucleus Collisions. *Brookhaven Natl. Lab. Tech. Rep. BNL-51921*, Brookhaven Natl. Lab., Upton, N.Y. (1985)

18. Shor A, Longacre RS. *Phys. Lett. B* 218:100 (1989)

19. Wang XN, Gyulassy M. *Phys. Rev. D* 44:3501; code HIJING 1.383 (1991)

20. Werner K. *Phys. Lett. B* 208:520 (1988)

21. Sorge H, Stoecker H, Greiner W. *Nucl. Phys. A* 498:567C (1989)

22. Collard HR, Elton LRB, Hofstadter R. In *Landolt-Börnstein, Numerical Data and Functional Relationships in Science and Technology*, Vol. 2: *Nuclear Radii*, ed. H Schopper, p. 1. Berlin: Springer-Verlag. 54 pp. (1967)

23. De Jager CW, De Vries H, De Vries C. *Atom. Data Nucl. Data Tabl.* 36:495 (1987)

24. Hulthén L, Sagawara M. *Handb. Phys.* 39:1 (1957)

25. Hodgson PE. In *Nuclear Reactions and Nuclear Structure*, p. 453. Oxford: Clarendon. 661 pp. (1971)

26. Adler SS, et al. *Phys. Rev. Lett.* 91:072303 (2003)

27. Wong CY. In *Introduction to High-Energy Heavy-Ion Collisions*, p. 251. Singapore: World Sci. 516 pp. (1994)

28. Chauvin J, Bebrun D, Lounis A, Buenerd M. *Phys. Rev. C.* 83:1970 (1983)

29. Wibig T, Sobczynska D. *J. Phys. G* 24:2037 (1998)

30. Kharzeev D, Lourenco C, Nardi M, Satz H. *Z. Phys. C* 74:307 (1997)

31. Pi H. *Comput. Phys. Commun.* 71:173 (1992)

32. Miller M, Snellings R. nucl-ex/0312008 (2003)

33. Alver B, et al. *Phys. Rev. Lett.* 96:212301 (2006)

34. Guillaud JP, Sobol A. Simulation of diffractive and non-diffractive processes at the LHC energy with the PYTHIA and PHOJET MC event generators. *CNRS Tech. Rep. LAPP-EXP 2004–06*, CNRS, Paris, France (2004)

35. Sjostrand T, Mrenna S, Skands P. *JHEP* 0605:026 (2006)

36. Yao YM, et al. *J. Phys. G* 33:1 (2006)

37. Adamczyk M, et al. *Nucl. Instrum. Methods A* 499:437 (2003)

38. Allen M, et al. *Nucl. Instrum. Methods A* 499:549 (2003)

39. Back BB, et al. *Nucl. Instrum. Methods A* 499:603 (2003)

40. Braem A, et al. *Nucl. Instrum. Methods A* 499:720 (2003)

41. Ansorge RE, et al. *Z. Phys. C* 43:357 (1989)

42. Kharzeev D, Nardi M. *Phys. Lett. B* 507:121 (2001)

43. Adler C, et al. *Phys. Rev. Lett.* 89:202301 (2002)
44. Bearden IG, et al. *Phys. Lett. B* 523:227 (2001)
45. Bearden IG, et al. *Phys. Rev. Lett.* 88:202301 (2002)
46. Arsene I, et al. *Phys. Rev. Lett.* 94:032301 (2005)
47. Lee YK, et al. *Nucl. Instrum. Methods A* 516:281 (2004)
48. Appl. Softw. Group. *GEANT detector description and simulation tool. w5013 edition*, **http://wwwasdoc.web.cern.ch/wwwasdoc/geant_html3/geantall.html**. Geneva: CERN (1993)
49. Adler SS, et al. *Phys. Rev. C* 71:034908 (2005). Erratum. *Phys. Rev. C* 71:049901 (2005)
50. Aronson SH, et al. *Nucl. Instrum. Methods A* 499:480 (2003)
51. Adams J, et al. *Phys. Rev. C* 70:044901 (2004)
52. Adams J, et al. *Phys. Rev. Lett.* 91:172302 (2003)
53. Sorensen P. nucl-ex/0612021 (2006)
54. Adams J, et al. *Phys. Rev. Lett.* 91:072304 (2003)
55. Aggarwal MM, et al. *Eur. Phys. J. C* 18:651 (2001)
56. Manly S, et al. *Nucl. Phys. A* 774:523 (2006)
57. Back BB, et al. *Phys. Rev. Lett.* 91:052303 (2003)
58. Back BB, et al. *Phys. Rev. C* 74:021901 (2006)
59. Back BB, et al. *Phys. Rev. C* 74:021902 (2006)
60. Back BB, et al. *Phys. Rev. C* 65:031901 (2002)
61. Elias JE, et al. *Phys. Rev. Lett.* 41:285 (1978)
62. Adams J, et al. nucl-ex/0311017 (2003)
63. Eskola KJ, Kajantie K, Ruuskanen PV, Tuominen K. *Nucl. Phys. B* 570:379 (2000)
64. Back BB, et al. *Phys. Rev. C.* 70:021902 (2004)
65. Back BB, et al. *Phys. Rev. C* 65:061901 (2002)
66. Eskola KJ, Kajantie K, Lindfors J. *Nucl. Phys. B* 323:37 (1989)
67. Eskola KJ, Vogt R, Wang XN. *Int. J. Mod. Phys. A* 10:3087 (1995)
68. Vogt R. *Heavy Ion Phys.* 9:339 (1999)
69. Arleo F, et al. hep-ph/0311131 (2003)
70. Jacobs P, Wang XN. *Prog. Part. Nucl. Phys.* 54:443 (2005)
71. Tannenbaum MJ. nucl-ex/0611008 (2006)
72. Owens JF. *Rev. Mod. Phys.* 59:465 (1987)
73. Adler SS, et al. *Phys. Rev. Lett.* 94:232301 (2005)
74. Alver B, et al. nucl-ex/0610037 (2006)
75. Ackermann KH, et al. *Phys. Rev. Lett.* 86:402 (2001)
76. Alver B, et al. nucl-ex/0702036 (2006)
77. Matsui T, Satz H. *Phys. Lett. B* 178:416 (1986)
78. Vogt R. *Phys. Rep.* 310:197 (1999)
79. Gerschel C, Hufner J. *Annu. Rev. Nucl. Part. Sci.* 49:255 (1999)
80. Alessandro B, et al. *Eur. Phys. J. C* 33:31 (2004)
81. Alessandro B, et al. *Eur. Phys. J. C* 39:335 (2005)

The Cosmic Microwave Background for Pedestrians: A Review for Particle and Nuclear Physicists

Dorothea Samtleben,[1] Suzanne Staggs,[2] and Bruce Winstein[3]

[1] Max-Planck-Institut für Radioastronomie, D-53121 Bonn, Germany;
email: dsamtleb@mpifr-bonn.mpg.de

[2] Department of Physics, Princeton University, Princeton, New Jersey 08544;
email: staggs@princeton.edu

[3] Department of Physics, Enrico Fermi Institute, and Kavli Institute for Cosmological Physics, University of Chicago, Chicago, Illinois 60637;
email: bruce@kicp.uchicago.edu

Annu. Rev. Nucl. Part. Sci. 2007. 57:245–83

The *Annual Review of Nuclear and Particle Science* is online at http://nucl.annualreviews.org

This article's doi:
10.1146/annurev.nucl.54.070103.181232

Copyright © 2007 by Annual Reviews.
All rights reserved

0163-8998/07/1123-0245$20.00

Key Words

CMB, particle physics, cosmology, inflation, polarization, receivers

Abstract

We intend to show how fundamental science is drawn from the patterns in the temperature and polarization fields of the cosmic microwave background (CMB) radiation, and thus to motivate the field of CMB research. We discuss the field's history, potential science and current status, contaminating foregrounds, detection and analysis techniques, and future prospects. Throughout the review we draw comparisons to particle physics, a field that has many of the same goals and that has gone through many of the same stages.

Contents

1. INTRODUCTION 246
   1.1. The Standard Paradigm 247
   1.2. Fundamental Physics in the CMB 248
   1.3. History 250
   1.4. Introduction to the Angular Power Spectrum 251
   1.5. Current Understanding of the Temperature Field 252
   1.6. Acoustic Oscillations 252
   1.7. How Spatial Modes Look Like Angular Anisotropies 254
   1.8. CMB Polarization 255
   1.9. Processes after Decoupling: Secondary Anisotropies 257
   1.10. What We Learn from the CMB Power Spectrum 259
   1.11. Discussion of Cosmological Parameters 262
2. FOREGROUNDS 264
   2.1. Overview 264
   2.2. Foreground Removal 268
3. METHODS OF DETECTION 268
   3.1. The CMB Experiment Basics 268
   3.2. The Detection Techniques 269
   3.3. Observing Strategies 275
   3.4. Techniques of Data Analysis 277
4. FUTURE PROSPECTS 279
   4.1. The Next Satellite Experiment 281
5. CONCLUDING REMARKS 281

1. INTRODUCTION

What is all the fuss about "noise"? In this review we endeavor to convey the excitement and promise of studies of the cosmic microwave background (CMB) radiation to scientists not engaged in these studies, particularly to particle and nuclear physicists. Although the techniques for both detection and data processing are quite far apart from those familiar to our intended audience, the science goals are aligned. We do not emphasize mathematical rigor,[1] but rather attempt to provide insight into (*a*) the processes that allow extraction of fundamental physics from the observed radiation patterns and (*b*) some of the most fruitful methods of detection.

In Section 1, we begin with a broad outline of the most relevant physics that can be addressed with the CMB and its polarization. We then treat the early history of the field, how the CMB and its polarization are described, the physics behind the acoustic peaks, and the cosmological physics that comes from CMB studies. Section 2 presents the important foreground problem: primarily galactic sources of microwave

---

[1]This review complements the one by Kamionkowski & Kosowsky (1).

radiation. The third section treats detection techniques used to study these extremely faint signals. The promise (and challenges) of future studies is presented in the last two sections. In keeping with our purposes, we do not cite an exhaustive list of the ever expanding literature on the subject, but rather indicate several particularly pedagogical works.

1.1. The Standard Paradigm

Here we briefly review the now standard framework in which cosmologists work and for which there is abundant evidence. We recommend readers to the excellent book *Modern Cosmology* by Dodelson (2). Early in its history (picoseconds after the Big Bang), the energy density of the Universe was divided among matter, radiation, and dark energy. The matter sector consisted of all known elementary particles and included a dominant component of dark matter, stable particles with negligible electromagnetic interactions. Photons and neutrinos (together with the kinetic energies of particles) comprised the radiation energy density, and the dark energy component—some sort of fluid with a negative pressure—appears to have had no importance in the early Universe, although it is responsible for its acceleration today.

Matter and radiation were in thermal equilibrium, and their combined energy density drove the expansion of space, as described by general relativity. As the Universe expanded, wavelengths were stretched so that particle energies (and hence the temperature of the Universe) decreased: $T(z) = T(0)(1 + z)$, where z is the redshift and $T(0)$ is the temperature at $z = 0$, or today. There were slight overdensities in the initial conditions that, throughout the expansion, grew through gravitational instability, eventually forming the structure we observe in today's Universe: myriad stars, galaxies, and clusters of galaxies.

The Universe was initially radiation dominated. Most of its energy density was in photons, neutrinos, and kinetic motion. After the Universe cooled to the point at which the energy in rest mass equaled that in kinetic motion (matter-radiation equality), the expansion rate slowed and the Universe became matter dominated, with most of its energy tied up in the masses of slowly moving, relatively heavy stable particles: the proton and deuteron from the baryon sector and the dark matter particle(s). The next important era is termed either decoupling, recombination, or last scattering. When the temperature reached roughly 1 eV, atoms (mostly H) formed and the radiation cooled too much to ionize. The Universe became transparent, and it was during this era that the CMB we see today was emitted, when physical separations were 1000 times smaller than today ($z \approx 1000$). At this point, less than one million years into the expansion, when electromagnetic radiation ceased playing an important dynamical role, baryonic matter began to collapse and cool, eventually forming the first stars and galaxies. Later, the first generation of stars and possibly supernova explosions seem to have provided enough radiation to completely reionize the Universe (at $z \approx 10$). Throughout these stages, the expansion was decelerated by the gravitational force on the expanding matter. Now we are in an era of cosmic acceleration ($z \leq 2$), where we find that approximately 70% of the energy density is in the fluid that causes the acceleration, 25% is in dark matter, and just 5% is in baryons, with a negligible amount in radiation.

1.2. Fundamental Physics in the CMB

The CMB is a record of the state of the Universe at a fraction of a million years after the Big Bang, after a quite turbulent beginning, so it is not immediately obvious that any important information survives. Certainly the fundamental information available in the collisions of elementary particles is best unraveled by observations within nanoseconds of the collision. Yet even in this remnant radiation lies the imprint of fundamental features of the Universe at its earliest moments.

1.2.1. CMB features: evidence for inflation.
One of the most important features of the CMB is its Planck spectrum. It follows the blackbody curve to extremely high precision, over a factor of approximately 1000 in frequency (see **Figure 1**). This implies that the Universe was in thermal equilibrium when the radiation was released (actually long before, as we see below), which was at a temperature of approximately 3000 K. Today it is near 3 K.

An even more important feature is that, to better than a part in 10^4, this temperature is the same over the entire sky. This is surprising because it strongly implies that everything in the observable Universe was in thermal equilibrium at one time in its evolution. Yet at any time and place in the expansion history of the Universe,

Figure 1

Measurements of the CMB flux versus frequency, together with a fit to the data. Superposed are the expected blackbody curves for $T = 2$ K and $T = 40$ K.

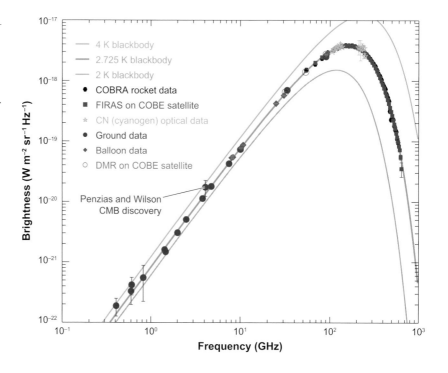

there is a causal horizon defined by the distance light (or gravity) has traveled since the Big Bang; at the decoupling era, this horizon corresponded to an angular scale of approximately 1°, as observed today. The uniformity of the CMB on scales well above 1° is termed the horizon problem.

The most important feature is that there are differences in the CMB temperature from place to place, at the level of 10^{-5}, and that these fluctuations have coherence beyond the horizon at the time of last scattering. The most viable notion put forth to address these observations is the inflationary paradigm, which postulates a very early period of extremely rapid expansion of the Universe. Its scale factor increased by approximately 21 orders of magnitude in only approximately 10^{-35} s. Before inflation, the small patch that evolves into our observable Universe was likely no larger across than the Planck length, its contents in causal contact and local thermodynamic equilibrium. The process of superluminal inflation disconnects regions formally in causal contact. When the expansion slowed, these regions came back into the horizon and their initial coherence became manifest.

The expansion turns quantum fluctuations into (nearly) scale-invariant CMB inhomogeneities, meaning that the fluctuation power is nearly the same for all three-dimensional Fourier modes. So far, observations agree with the paradigm, and scientists in the field use it to organize all the measurements. Nevertheless, we are far from understanding the microphysics driving inflation. The number of models and their associated parameter spaces greatly exceed the number of relevant observables. New observations, particularly of the CMB polarization, promise a more direct look at inflationary physics, moving our understanding from essentially kinematical to dynamical.

1.2.2. Probing the Universe when $T = 10^{16}$ GeV. For particle physicists, probing microphysics at energy scales beyond accelerators using cosmological observations is attractive. The physics of inflation may be associated with the grand unification scale, and if so, there could be an observable signature in the CMB: gravity waves. Metric perturbations, or gravity waves (also termed tensor modes), would have been created during inflation, in addition to the density perturbations (scalar modes) that give rise to the structure in the Universe today.

In the simplest of inflationary models, there is a direct relation between the energy scale of inflation and the strength of these gravity waves. The notion is that the Universe initially had all its energy in a scalar field Φ displaced from the minimum of its potential V. $V(\Phi)$ is suitably constructed so that Φ slowly rolls down its potential, beginning the inflationary era of the Universe, which terminates only when Φ approaches its minimum.

Inflation does not predict the level of the tensor (or even scalar) modes. The parameter $r = T/S$ is the tensor-to-scalar ratio for fluctuation power; it depends on the energy scale at which inflation began. Specifically, the initial height of the potential V_i depends on r, as $V_i = r(0.003M_{pl})^4$. A value of $r = 0.001$, perhaps the smallest detectable level, corresponds to $V_i^{0.25} = 6.5 \times 10^{15}$ GeV.

The tensor modes leave distinct patterns on the polarization of the CMB, which may be detectable. This is now the most important target for future experiments.

They also have effects on the temperature anisotropies, which currently limit r to less than approximately 0.3.

1.2.3. How neutrino masses affect the CMB. It is a remarkable fact that even a slight neutrino mass affects the expansion of the Universe. When the dominant dark matter clusters, it provides the environment for baryonic matter to collapse, cool, and form galaxies. As described above, the growth of these structures becomes more rapid in the matter-dominated era. If a significant fraction of the dark matter were in the form of neutrinos with electron-volt-scale masses (nonrelativistic today), these would have been relativistic late enough in the expansion history that they could have moved away from overdense regions and suppressed structure growth. Such suppression alters the CMB patterns and provides some sensitivity to the sum of the neutrino masses. Note also that gravitational effects on the CMB in its passage from the epoch (or surface) of last scattering to the present leave signatures of that structure and give an additional (and potentially more sensitive) handle on the neutrino masses (see Section 1.9.2).

1.2.4. Dark energy. We know from the CMB that the geometry of the Universe is consistent with being flat. That is, its density is consistent with the critical density. However, the overall density of matter and radiation discerned today (the latter from the CMB directly) falls short of accounting for the critical density by approximately a factor of three, with little uncertainty. Thus, the CMB provides indirect evidence for dark energy, corroborating supernova studies that indicate a new era of acceleration. Because the presence and possible evolution of a dark energy component alter the expansion history of the Universe, there is the promise of learning more about this mysterious component.

1.3. History

In 1965, Penzias & Wilson (3), in trying to understand a nasty noise source in their experiment to study galactic radio emission, discovered the CMB—arguably the most important discovery in all the physical sciences in the twentieth century. Shortly thereafter, scientists showed that the radiation was not from radio galaxies or reemission of starlight as thermal radiation. This first measurement was made at a central wavelength of 7.35 cm, far from the blackbody peak. The reported temperature was $T = 3.5 \pm 1$ K. However, for a blackbody, the absolute flux at any known frequency determines its temperature. **Figure 1** shows the spectrum of detected radiation for different temperatures. There is a linear increase in the peak position and in the flux at low frequencies (the Rayleigh-Jeans part of the spectrum) as temperature increases.

Multiple efforts were soon mounted to confirm the blackbody nature of the CMB and to search for its anisotropies. Partridge (4) gives a very valuable account of the early history of the field. However, there were false observations, which was not surprising given the low ratio of signal to noise. Measurements of the absolute CMB temperature are at milli-Kelvin levels, whereas relative measurements between two places on the sky are at micro-Kelvin levels. By 1967, Partridge and Wilkinson had shown, over large regions of the sky, that $\frac{\Delta T}{T} \leq (1 - 3) \times 10^{-3}$, leading to the conclusion that the Universe was in thermal equilibrium at the time of decoupling (4). However, nonthermal injections of energy even at much earlier times, for example, from the decays of long-lived relic particles, would distort the spectrum. It is remarkable that current precise measurements of the blackbody spectrum can push back the time of significant injections of energy to when the Universe was barely a month old (5). Thus, recent models that attribute the dark matter to gravitinos as decay products of long-lived supersymmetric weakly interacting massive particles (SUSY WIMPs) (6) can only tolerate lifetimes of less than approximately one month.

The solar system moves with velocity $\beta \approx 3 \times 10^{-3}$, causing a dipole anisotropy of a few milli-Kelvins, first detected in the 1980s. (Note that the direction of our motion was not the one initially hypothesized from motions of our local group of galaxies.) The first detection of primordial anisotropy came from the COBE satellite (7) in 1992, at the level of 10^{-5} (30 μK), on scales of approximately 10° and larger. The impact of this detection matched that of the initial discovery. It supported the idea that structure in the Universe came from gravitational instability to overdensities. The observed anisotropies are a combination of the original ones at the time of decoupling and the subsequent gravitational red- or blueshifting as photons leave over- or underdense regions.

1.4. Introduction to the Angular Power Spectrum

Here we describe the usual techniques for characterizing the temperature field. First, we define the normalized temperature Θ in direction $\hat{\mathbf{n}}$ on the celestial sphere by the deviation ΔT from the average: $\Theta(\hat{\mathbf{n}}) = \frac{\Delta T}{\langle T \rangle}$. Next, we consider the multipole decomposition of this temperature field in terms of spherical harmonics Y_{lm}:

$$\Theta_{lm} = \int \Theta(\hat{\mathbf{n}}) Y_{lm}^*(\hat{\mathbf{n}}) \, d\Omega, \qquad 1.$$

where the integral is over the entire sphere.

If the sky temperature field arises from Gaussian random fluctuations, then the field is fully characterized by its power spectrum $\Theta_{lm}^* \Theta_{l'm'}$. The order m describes the angular orientation of a fluctuation mode, but the degree (or multipole) l describes its characteristic angular size. Thus, in a Universe with no preferred direction, we expect the power spectrum to be independent of m. Finally, we define the angular power spectrum C_l by $\langle \Theta_{lm}^* \Theta_{l'm'} \rangle = \delta_{ll'} \delta_{mm'} C_l$. Here the brackets denote an ensemble average over skies with the same cosmology. The best estimate of C_l is then from the average over m.

Because there are only the $(2l + 1)$ modes with which to detect the power at multipole l, there is a fundamental limit in determining the power. This is known as the cosmic variance (just the variance on the variance from a finite number of samples):

$$\frac{\Delta C_l}{C_l} = \sqrt{\frac{2}{2l + 1}}. \qquad 2.$$

The full uncertainty in the power in a given multipole degrades from instrumental noise, finite beam resolution, and observing over a finite fraction of the full sky, as shown below in Equation 9.

For historical reasons, the quantity that is usually plotted, sometimes termed the TT (temperature-temperature correlation) spectrum, is

$$\Delta T^2 \equiv \frac{l(l+1)}{2\pi} C_l T^2_{CMB},$$ 3.

where T_{CMB} is the blackbody temperature of the CMB. This is the variance (or power) per logarithmic interval in l and is expected to be (nearly) uniform in inflationary models (scale invariant) over much of the spectrum. This normalization is useful in calculating the contributions to the fluctuations in the temperature in a given pixel from a range of l values:

$$\Delta T^2 = \int_{l_{min}}^{l_{max}} \frac{(2l+1)}{4\pi} C_l T^2_{CMB} \, dl.$$ 4.

1.5. Current Understanding of the Temperature Field

Figure 2 shows the current understanding of the temperature power spectrum (from herewith we redefine C_l to have K^2 units by replacing C_l with $C_l T^2_{CMB}$). The region below $l \approx 20$ indicates the initial conditions. These modes correspond to Fourier modes at the time of decoupling, with wavelengths longer than the horizon scale. Note that were the sky describable by random white noise, the C_l spectrum would be flat and the TT power spectrum, defined by Equation 3, would have risen in this region like l^2. The (pleasant) surprise was the observation of finite power at these superhorizon scales. At high l values, there are acoustic oscillations, which are damped at even higher l values. The positions and heights of the acoustic-oscillation peaks reveal fundamental properties about the geometry and composition of the Universe, as we discuss below.

1.6. Acoustic Oscillations

The CMB data reveal that the initial inhomogeneities in the Universe were small, with overdensities and underdensities in the dark matter, protons, electrons, neutrinos, and photons, each having the distribution that would arise from a small adiabatic compression or expansion of their admixture. An overdense region grows by attracting more mass, but only after the entire region is in causal contact.

We noted that the horizon at decoupling corresponds today to approximately $1°$ on the sky. Only regions smaller than this had time to compress before decoupling. For sufficiently small regions, enough time elapses that compression continues until the photon pressure is sufficient to halt the collapse. Then the region expands. In fact, an oscillation is set up: The relativistic fluid of photons is coupled to the electrons via Thomson scattering, and the protons follow the electrons to keep a charge balance. Inflation provides the initial conditions—zero velocity.

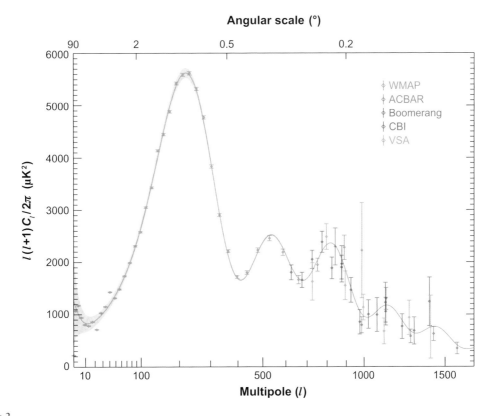

Angular scale (°)

Figure 2

The TT power spectrum. Data from the Wilkinson Microwave Anisotropy Probe (WMAP)
(8) and high-l data from other experiments are shown, in addition to the best-fit cosmological
model to the WMAP data alone. Note the multipole scale on the bottom and the angular scale
on the top. Figure courtesy of the WMAP science team.

Decoupling preserves a snapshot of the state of the photon fluid at that time.
Excellent pedagogical descriptions of the oscillations can be found at **http://
background.uchicago.edu/whu/**. Other useful pages are **http://wmap.gsfc.nasa.
gov/space.mit.edu/home/tegmark/index.html** and **http://www.astro.ucla.edu/
~wright/intro.html**. Perturbations of particular sizes may have undergone (*a*) one
compression, (*b*) one compression and one rarefaction, (*c*) one compression, one rarefaction, and one compression, and so on. Extrema in the density field result in
maxima in the power spectrum.

Consider a standing wave permeating space with frequency ω and wave number k,
where these are related by the velocity of displacements (the sound speed, $v_s \approx \frac{c}{\sqrt{3}}$)
in the plasma: $\omega = kv_s$. The wave displacement A_k for this single mode can then
be written as $A_k(x, t) \propto \sin(kx)\cos(\omega t)$. The displacement is maximal at time t_{dec} of
decoupling for $k_{TT}v_s t_{dec} = \pi, 2\pi, 3\pi \ldots$. We add the TT subscript to label these wave
numbers associated with maximal autocorrelation in the temperature. Note that even
in this tightly coupled regime, the Universe at decoupling was quite dilute, with a

physical density of less than 10^{-20} g cm^{-3}. Because the photons diffuse—their mean free path is not infinitely short—this pattern does not go on without bound. The overtones are damped, and in practice only five or six such peaks will be observed, as seen in **Figure 2**.

1.7. How Spatial Modes Look Like Angular Anisotropies

To help explain these ideas, we reproduce a few frames from an animation by W. Hu. **Figure 3** shows a density fluctuation on the sky from a single k mode and how it appears to an observer at different times. The figure shows the particle horizon just after decoupling. This represents the farthest distance one could in principle see—approximately the speed of light times the age of the Universe. An observer at the center of the figure could not by any means have knowledge of anything outside this region. Of course, just after decoupling, the observer could see a far shorter distance. Only then could light propagate freely.

The subsequent frames show how the particle horizon grows to encompass more corrugations of the original density fluctuation. At first the observer sees a dipole, later a quadrupole, then an octopole, and so on, until the present time when that single mode in density inhomogeneities creates very high multipoles in the temperature anisotropy.

It is instructive to think of how the temperature observed today at a spot on the sky arises from the local moments in the temperature field at the time of last scattering. It is only the lowest three moments that contribute to determining the anisotropies. The monopole terms are the ones transformed into the rich angular spectra. The dipole terms also have their contribution: The motion in the fluid oscillations results in Doppler shifts in the observed temperatures. Polarization, we see below, comes from local quadrupoles.

1.7.1. Inflation revisited. Inflation is a mechanism whereby fluctuations are created without violating causality. There does not seem to be a better explanation for the

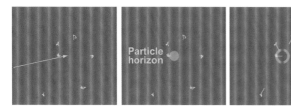

Figure 3

The signature of one frozen mode after decoupling. These frames show one superhorizon temperature mode just after decoupling, with representative photons last scattering and heading toward the observer at the center. (*Left to right*) Just after decoupling; the observer's particle horizon when only the temperature monopole can be detected; some time later when the quadrupole is detected; later still when the 12-pole is detected; and today, a very high, well-aligned multipole from just this single mode in k space is detected. Figure courtesy of W. Hu.

observed regularities. Nevertheless, Wolfgang Pauli's famous statement about the neutrino comes to mind: "I have done a terrible thing: I have postulated a particle that cannot be detected!"

Sometimes it seems that inflation is an idea that cannot be tested, or tested incisively. Of course Pauli's neutrino hypothesis did test positive, and similarly there is hope that the idea of inflation can reach the same footing. Still, we have not (yet) seen any scalar field in nature. We discuss what has been claimed as the smoking gun test of inflation—the eventual detection of gravity waves in the CMB. However, will we ever know with certainty that the Universe grew in volume by a factor of 10^{63} in something like 10^{-35} s?

1.8. CMB Polarization

Experiments have now shown that the CMB is polarized, as expected. Researchers now think that the most fruitful avenue to fundamental physics from the CMB will be in precise studies of the patterns of the polarization. This section treats the mechanisms responsible for the generation of the polarization and how this polarization is described.

1.8.1. How polarization gets generated. If there is a quadrupole anisotropy in the temperature field around a scattering center, even if that radiation is unpolarized, the scattered radiation will be as shown in **Figure 4**: A linear polarization will be

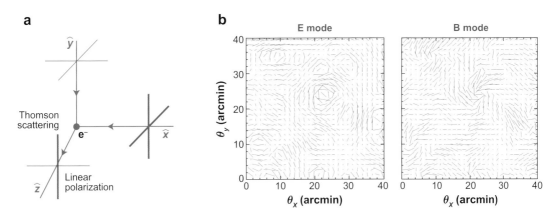

Figure 4

Generation of polarization. (*a*) Unpolarized but anisotropic radiation incident on an electron produces polarized radiation. Intensity is represented by line thickness. To an observer looking along the direction of the scattered photons ($\hat{z}$), the incoming quadrupole pattern produces linear polarization along the $\hat{y}$-direction. In terms of the Stokes parameters, this is $Q = (E_x{}^2 - E_y{}^2)/2$, the power difference detected along the $\hat{x}$- and $\hat{y}$-directions. Linear polarization needs one other parameter, corresponding to the power difference between 45° and 135° from the x-axis. This parameter is easily shown to be Stokes $U = E_x E_y$. (*b*) E and B polarization patterns. The length of the lines represents the degree of polarization, while their orientation gives the direction of maximum electric field. Frames courtesy of W. Hu.

generated. The quadrupole is generated during decoupling, as shown in **Figure 3**. Because the polarization arises from scattering but said scattering dilutes the quadrupole, the polarization anisotropy is much weaker than that in the temperature field. Indeed with each scatter on the way to equilibrium, the polarization is reduced. Any remaining polarization is a direct result of the cessation of scattering. For this reason, the polarization peaks at higher l values than does the temperature anisotropy. The local quadrupole on scales that are large in comparison to the mean free path is diluted from multiple scattering.

1.8.2. The E and B polarization fields. The polarization field is both more complicated and richer than the temperature field. At each point in the sky, one must specify both the degree of polarization and the preferred direction of the electric field. This is a tensor field that can be decomposed into two types, termed E and B, which are, respectively, scalar and pseudoscalar fields, with associated power spectra. Examples of these polarization fields are depicted schematically in **Figure 4**. The E and B fields are more fundamental than the polarization field on the sky, whose description is coordinate-system dependent. In addition, E modes arise from the density perturbations (which do not produce B modes) that we describe, whereas the B modes come from the tensor distortions to the space-time metric (which do have a handedness). We mention here that the E and B fields are nonlocal. Their extraction from measurements of polarization over a set of pixels, often in a finite patch of sky, is a well-developed but subtle procedure (see Section 3.3).

The peaks in the EE (E-polarization correlated with itself) spectrum should be 180° out of phase with those for temperature: Polarization results from scattering and thus is maximal when the fluid velocity is maximal. Calculating the fluid velocity for the mode in Section 1.6, we find $k_{EE} v_s t_{dec} = \pi/2, 3\pi/2, 5\pi/2 \ldots$, defining modes with maximal EE power. The TE (E-polarization correlated with the temperature field) spectrum—how modes in temperature correlate with those with E polarization—is also of cosmological interest, with its own peak structure. Here we are looking at modes that have a maximum at decoupling in the product of their temperature and E-mode polarization (or velocity). Similarly, the appropriate maxima (which in this case can be positive or negative) are obtained when $k_{TE} v_s t_{dec} = \pi/4, 3\pi/4, 5\pi/4 \ldots$. Thus, between every peak in the TT power spectrum there should be one in the EE, and between every TT and EE pair of peaks there should be one in the TE.

1.8.3. Current understanding of polarization data. **Figure 5** shows the EE results in addition to the expected power spectra in the standard cosmological model. Measurements of the TE cross correlation are also shown. The pattern of peaks in both power spectra is consistent with what was expected. What was unexpected was the enhancement at the lowest l values in the EE power spectrum. This is discussed in the next section.

The experiments reported in **Figure 5**, with 20 or fewer detectors, use a variety of techniques and operate in different frequency ranges. This is important in dealing with astrophysical foregrounds (see Section 2) that have a different frequency dependence

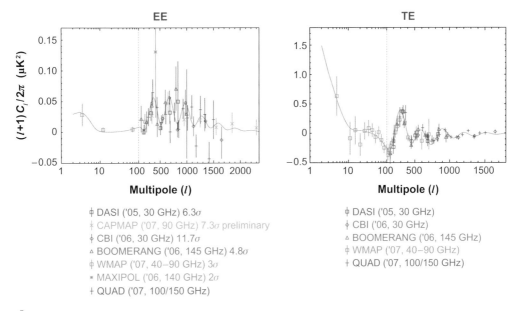

EE

TE

Multipole (*l*)

⏀ DASI ('05, 30 GHz) 6.3σ
✳ CAPMAP ('07, 90 GHz) 7.3σ preliminary
⏀ CBI ('06, 30 GHz) 11.7σ
△ BOOMERANG ('06, 145 GHz) 4.8σ
⏀ WMAP ('07, 40–90 GHz) 3σ
✕ MAXIPOL ('06, 140 GHz) 2σ
+ QUAD ('07, 100/150 GHz)

⏀ DASI ('05, 30 GHz)
⏀ CBI ('06, 30 GHz)
△ BOOMERANG ('06, 145 GHz)
⏀ WMAP ('07, 40–90 GHz)
+ QUAD ('07, 100/150 GHz)

Figure 5

Measurements of EE and TE power spectra together with the WMAP best-fit cosmological model. The names of the experiments, their years of publication, and the frequency ranges covered are indicated, as well as the number of standard deviations with which each experiment claims a detection. Note the change from logarithmic to linear multipole scale at $l = 100$ and that to display features in the very low l range, we plot $(l + 1)C_l/2\pi$.

from that of the CMB. Limits from current experiments on the B-mode power are now at the level of 1–10 μK^2, far from the expected signal levels shown in **Figure 6**. The peak in the power spectrum (for the gravity waves) is at $l \approx 100$, the horizon scale at decoupling. The reader may wonder why the B modes fall off steeply above this scale and show no acoustic oscillations. The reason is simple: A tensor mode will give, for example, a compression in the x-direction followed by a rarefaction in the y-direction, but will not produce a net overdensity that would subsequently contract. In the final section we discuss experiments with far greater numbers of detectors aimed specifically at B-mode science. Note that such gravity waves have frequencies today of order 10^{-16} Hz. However, if their spectrum approximates one of scale invariance, they would in principle be detectable at frequencies nearer 1 Hz, such as in the LISA experiment. This is discussed more fully in Reference 10.

1.9. Processes after Decoupling: Secondary Anisotropies

In this section we briefly discuss three important processes after decoupling: rescattering of the CMB in the reionized plasma of the Universe, lensing of the CMB through gravitational interactions with matter, and scattering of the CMB from hot gas in Galaxy clusters. Although these can be considered foregrounds perturbing the primordial information, each can potentially provide fundamental information.

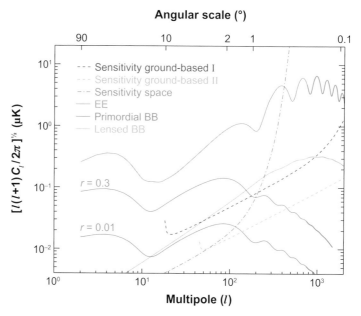

Angular scale (°)

Figure 6

CMB polarization power spectra and estimated sensitivity of future experiments. The solid curves show the predictions for the E- and B-mode power spectra. The primordial B-mode power spectrum is shown for $r = 0.3$ and $r = 0.01$. The predicted B-mode signal power spectrum due to the distortion of E modes by weak gravitational lensing is also shown. Estimated statistical sensitivities for a new space mission (*pink line*) and two sample ground-based experiments, as considered in Reference 9, each with 1000 detectors operating for one year with 100% duty cycle (*dark and light blue lines*), are shown. Experiment I observes 4% of the sky, with a 6-arcmin resolution; experiment II observes 0.4% of the sky, with a 1-arcmin resolution.

1.9.1. Reionization. The enhancement in the EE power spectrum at the very lowest l values in **Figure 5** is the signature that the Universe was reionized after decoupling. This is a subject rich in astrophysics, but for our purposes it is important in that it provides another source for scattering and hence detection of polarization. From the Wilkinson Microwave Anisotropy Probe (WMAP) polarization data (11), one can infer an optical depth of order 10%, the fraction of photons scattering in the reionized plasma somewhere in the region of $z = 10$. This new scattering source can be used to detect the primordial gravity waves. The signature will show up at very low l values, corresponding to the horizon scale at reionization. **Figure 6** shows that the region $l = 4$–8 should have substantial effects from gravity waves. Most likely, the only means of detecting such a signal is from space, and even from there it will be very difficult.

The polarization anisotropies for this very low l region are comparable to what is expected from the surface of last scattering ($l \approx 100$). There are disadvantages to each signature. At the lowest l values, galactic foregrounds are more severe, there are fewer modes in which to make a detection, and systematic errors are likely greater.

At the higher values, there is a foreground that arises from E modes turning into B modes through gravitational lensing (the topic of the next section). Clearly, it will be important to detect the two signatures with the right relative strengths at these two very different scales.

1.9.2. Lensing of the CMB. Both the temperature and polarization fields will be slightly distorted (lensed) when passing collapsing structures in the late Universe. The bending of light means that one is not looking (on the last scattering surface) where one thinks. Although lensing will affect both the polarization and T fields, its largest effect is on the B field, where it shifts power from E to B. Gravitational distortions, although preserving brightness, do not preserve the E and B nature of the polarization patterns.

Figure 6 also shows the expected power spectrum of these lensed B modes. Because this power is sourced by the E modes, it roughly follows their shape, but with ΔT suppressed by a factor of 20. The peak structure in the E modes is smoothed, as the structures doing the lensing are degree scale themselves. Owing to the coherence of the lensing potential for these modes, there is more information than just the power spectrum, and work is ongoing to characterize the expected cross correlation between different multipole bands. This signal should be detectable in next-generation polarization experiments.

For our purposes, the most interesting aspect of this lensing is the handle it can potentially give on the masses of the neutrinos, as more massive neutrinos limit the collapse of matter along the CMB trajectories. All other parameters held fixed, there is roughly a factor-of-two change in the magnitude of the B signal for a 1-eV change in the mean neutrino mass.

1.9.3. CMB scattering since reionization. At very small angular scales—l values of a few thousand, way beyond where the acoustic oscillations are damped—there are additional effects on the power spectra that result from the scattering of CMB photons from electrons after the epoch of reionization, including scattering from gas heated from falling deep in the potential wells of Galaxy clusters (the Sunyaev-Zel'dovich, or SZ, effect). These nonlinear effects are important as they can help in untangling (*a*) when the first structures formed and (*b*) the role of dark energy.

1.10. What We Learn from the CMB Power Spectrum

In this section, we show how the power spectrum information is used to determine important aspects of the Universe. This is normally known as parameter estimation, where the parameters are those that define our cosmology. The observable power spectrum is a function of at least 11 such basic parameters. As we discuss below, some are better constrained than others.

First, there are four parameters that characterize the primordial scalar and tensor fluctuation spectra before the acoustic oscillations, each of which is assumed to follow a power law in wave number. These four are the normalization of the scalar fluctuations (A_s), the ratio of tensor to scalar fluctuations r, and the spectral indices for both (historically denoted with $n_s - 1$ and n_t). Second, there is one equation-of-state parameter (w) that is the ratio of the pressure of the dark energy to its energy density, and one parameter that gives the optical depth (τ) from the epoch of reionization. Finally, there are five parameters that characterize the present Universe: its rate of expansion (Hubble constant, with $H_0 = h \cdot 100$ km s^{-1} Mpc^{-1}), its curvature (Ω_k), and its composition (baryon density, matter density, and dark energy density). The latter three are described in terms of energy densities with respect to the critical density normalized to the present epoch: $\omega_b = \Omega_b h^2$, $\omega_m = \Omega_m h^2$, and $\omega_\Lambda = \Omega_\Lambda h^2$. Just 10 of these are independent as $\Omega_m + \Omega_\Lambda + \Omega_k = 1$.

Even though the CMB data set itself consists of hundreds of measurements, they are not sufficiently orthogonal with respect to the 10 independent parameters for each to be determined independently; there are significant degeneracies. Hence, it is necessary to make assumptions that constrain the values of those parameters upon which the data have little leverage. In some cases, such prior assumptions (priors) can have large effects on the other parameters, and there is as yet no standard means of reporting results.

Several teams have done analyses [WMAP (11, 12), CBI (13), Boomerang (14), see also Reference 15]. Here we first discuss the leverage that the CMB power spectra have on the cosmological parameters. Then we give a flavor for the analyses, together with representative results. We consider analyses, done by the several teams, with just the six most important parameters: ω_b, ω_m, A_s, n_s, τ, and h, where the other five are held fixed. For this discussion we are guided by Reference 12.

Completely within CMB data, there is a geometrical degeneracy between Ω_k, a contribution to the energy density from the curvature of space, and Ω_m. However, taking a very weak prior of $h > 0.5$, the WMAP team, using just their first-year data, determined that $\Omega_k = 0.03 \pm 0.03$, that is, no evidence for curvature. We assume $\Omega_k = 0$ unless otherwise noted. This conclusion has gotten stronger with the three-year WMAP data together with other CMB results, and it is a prediction of the inflationary scenario. Nevertheless, we emphasize that it is an open experimental issue.

1.10.1. The geometry of the Universe. The position of the first acoustic peak reveals that the Universe is flat or nearly so. As we describe above, the generation of acoustic peaks is governed by the (comoving) sound horizon at decoupling, r_S (i.e., the greatest distance a density wave in the plasma could traverse, scaled to today's Universe). The sound horizon depends on ω_m, ω_b, and the radiation density, but not on H_0, Ω_k, ω_Λ, or the spectral tilt n_s. The peak positions versus angular multipole are then determined by $\Theta_A = r_S d_A^{-1}$, where the quantity d_A, the angular diameter distance, is the distance that properly takes into account the expansion history of the Universe between decoupling and today so that when d_A is multiplied by an observed angle, the result is the feature size at the time of decoupling. In a nonexpanding Universe, this would simply be the physical distance. The expression depends on the (evolution of the) content of the Universe. For a flat Universe, we have

$$d_A = \int_0^{z_{dec}} \frac{H_0^{-1} dz}{\sqrt{\Omega_r (1+z)^4 + \Omega_m (1+z)^3 + \Omega_\Lambda}}. \qquad 5.$$

In this expression, Ω_r indicates the (well-known) radiation density, and the dilutions of the different components with redshift z, between decoupling and the present, enter explicitly.

1.10.2. Fitting for spectral tilt, matter, and baryon content. It is easy to see how one in principle determines spectral tilt. If one knew all the other parameters, then the tilt would be found from the slope of the power spectrum after the removal of the other contributions. However, there is clearly a coupling to other parameters. Experiments with a very fine angular resolution will determine the power spectrum at very high l values, thereby improving the measurement of the tilt.

Here we discuss the primary dependences of the acoustic peak heights on ω_m and ω_b. Increasing ω_m decreases the peak heights. With greater matter density, the era of equality is pushed to earlier redshifts, allowing the dark matter more time to form deeper potential wells. When the baryons fall into these wells, their mass has less effect on the development of the potential so that the escaping photons are less redshifted than they would be, yielding a smaller temperature contrast. As to ω_b, increasing it decreases the second peak but enhances that of the third because the inertia in the photon-baryon fluid is increased, leading to hotter compressions and cooler rarefactions (16).

The peak-height ratios give the three parameters n_s, ω_m, and ω_b, with a precision just short of that from a full analysis of the power spectrum (discussed in Section 3.4.4). Following WMAP, we define the ratio of the second to the first peak by H_2^{TT}, the ratio of the third to the second peak by H_3^{TT}, and the ratio of the first to the second peak in the polarization-temperature cross-correlation power spectrum by H_2^{TE}. **Table 1** shows how the errors in these ratios propagate into parameter errors. We see that all the ratios depend strongly on n_s, and that the ratio of the first two peaks depends strongly on ω_b but is also influenced by ω_m. For H_3^{TT}, the relative dependences on ω_b and ω_m are reversed. Finally, the baryon density has little influence on the ratio of the TE peaks. However, increasing ω_m deepens potential wells, increasing fluid velocities and the heights of all polarization peaks.

Table 2 lists the results from six-parameter fits to the power spectrum from several combinations of CMB data with and without complementary data from other sectors. The table includes results from Reference 14, which included most CMB data available at time of publication, and from even more recent analyses by WMAP (8).

Table 1   Matrix of how errors in the peak ratios (defined in text) relate to the parameter errors

	Δn_s	$\dfrac{\Delta \omega_b}{\omega_b}$	$\dfrac{\Delta \omega_m}{\omega_m}$
$\Delta H_2^{TT} / H_2^{TT}$	0.88	−0.67	0.039
$\Delta H_3^{TT} / H_3^{TT}$	1.28	−0.39	0.46
$\Delta H_2^{TE} / H_2^{TE}$	−0.66	0.095	0.45

Table 2   Cosmological parameters from the CMB

Symbol	WMAP1	WMAP3	WMAP3 + other CMB	CMB + LSS	WMAP3 + SDSS
$\Omega_b h^2$	0.024 ± 0.001	0.02229 ± 0.00073	0.02232 ± 0.00074	$0.0226^{+0.0009}_{-0.0008}$	$0.02230^{+0.00071}_{-0.00070}$
$\Omega_m h^2$	0.14 ± 0.02	$0.1277^{+0.0080}_{-0.0079}$	0.1260 ± 0.0081	0.143 ± 0.005	$0.1327^{+0.0063}_{-0.0064}$
h	0.72 ± 0.05	$0.732^{+0.031}_{-0.032}$	$0.739^{+0.033}_{-0.032}$	$0.695^{+0.025}_{-0.023}$	0.710 ± 0.026
τ	$0.166^{+0.076}_{-0.071}$	0.089 ± 0.030	$0.088^{+0.031}_{-0.032}$	$0.101^{+0.051}_{-0.044}$	$0.080^{+0.029}_{-0.030}$
n_S	0.99 ± 0.04	0.958 ± 0.016	0.951 ± 0.016	0.95 ± 0.02	$0.948^{+0.016}_{-0.015}$

Results from six-parameter fits to CMB data, assuming a flat Universe and not showing the scalar amplitude A_s. Shown are results from first-year WMAP data, three-year WMAP data, and WMAP data combined with the bolometric experiments ACBAR and Boomerang. Fits using data from CBI and VSA (using coherent amplifiers) were also made, with consistent results. Also shown are results using LSS data with CMB data available in 2003, and from adding LSS data [from the Sloan Digital Sky Survey (SDSS)] to the WMAP3 data set (11). See Section 1.11 for appropriate references. $\Omega_b h^2$, baryon density; $\Omega_m h^2$, matter density; h, Hubble parameter; τ, optical depth; n_S, spectral index.

1.11. Discussion of Cosmological Parameters

The overall conclusions from the analysis of the peak structure are not dramatically different from those drawn from a collection of earlier ground- and balloon-based experiments. Still, WMAP's first data release put the reigning cosmological model on much stronger footing. Few experiments claimed systematic errors on the overall amplitude of their TT measurements less than 10%; WMAP's errors are less than 0.5%. The overall amplitude is strongly affected by the reionization. With full-sky coverage, WMAP determined the power spectrum in individual l bins with negligible correlations. Now with the WMAP three-year data, results from higher-resolution experiments, and results on EE polarization, we are learning even more.

Remarkably, CMB data confirm the baryon density deduced from Big Bang nucleosynthesis, from processes occurring at approximately 1 s after the Big Bang: $\Omega_b h^2 = 0.0205 \pm 0.0035$. The determination of the nonzero density of dark matter at approximately 300,000 years reinforces the substantial evidence for dark matter in the nearby Universe. Finally, the flat geometry confirms the earlier (supernova) evidence of a dark energy component.

With temperature data alone, there is a significant degeneracy in parameter space, which becomes apparent when one realizes that there are just five key features in the power spectrum (at least with today's precision) to which one is fitting six parameters: the heights of three peaks, the location of the first peak, and the anisotropy on very large scales. The degeneracy can be understood as follows. The peak heights are normalized by the combination $A_s e^{-2\tau}$. Thus, both these parameters can increase in a way that leaves the peak heights unchanged, increasing the power on scales larger than the horizon at reionization. Increasing n_s can restore the balance but can also decrease the second peak. That peak can be brought back up by decreasing ω_b. WMAP broke this degeneracy in its first-year release with a prior requiring $\tau < 0.3$. The new EE data from WMAP, in their sensitivity to reionization, break this degeneracy without the need for a prior.

Table 2 shows that the Hubble constant has been robust and in good agreement with determinations from Galaxy surveys. With better measurements of the peaks, the baryon and matter densities have moved systematically, but within error. The optical depth has decreased significantly and is now based upon the EE, rather than TE, power in the lowest l range. This change is coupled to a large change in the scalar amplitude. Finally, evidence for a spectral tilt ($n_s \neq 1$) is becoming more significant. As this is predicted by the simplest of inflationary scenarios, it is important and definitely worth watching.

The first-year WMAP data confirmed the COBE observation of unexpectedly low power in the lowest multipoles. The WMAP team reported this effect to be more significant than a statistical fluctuation, and lively literature on the subject followed. It is it clear that the quadrupole has little power and appears to be aligned with the octopole. However, the situation is unclear in that the quadrupole lines up reasonably well with the Galaxy itself, and there is concern that the cut on the WMAP data to remove the Galaxy then reduced the inherent quadrupole power. The anomaly has been reduced with the three-year data release, with improvements to the analysis, particularly in the lowest multipoles.

Table 2 also gives results from fitting CMB data with data from other cosmological probes, in particular large-scale structure (LSS) data in the form of three-dimensional Galaxy power spectra (the third dimension is redshift). Such spectra extend the lever arm in k space, allowing a more incisive determination of any possible spectral tilt, n_s. However, there are potential biases with the Galaxy data. In particular, the galaxies may not be faithful tracers of the dark matter density. Already before the three-year WMAP data release, including LSS data with CMB data favored an optical depth closer to its current value and provided evidence of spectral tilt. With three-year WMAP data and the Sloan Digital Sky Survey Galaxy survey data, the significance of a nonzero tilt is near the 3σ level. This is a vigorously debated topic. There are other LSS surveys that give similar and nearly consistent results, yet the systematic understanding is not at the level where combining all such surveys makes sense.

1.11.1. Beyond the six basic parameters. With the LSS data, one can obtain information on other parameters that were held fixed. In particular, relaxing the constraint on Ω_k, one finds consistency with a flat Universe to the level of approximately 0.04 (with CMB data alone) and 0.02 (using LSS data) (see Reference 14). Using WMAP and other surveys, constraints as low as 0.015 are obtained with some sets, giving slight indications for a closed Universe ($\Omega_k < 0$).

There is sensitivity to the fraction of the dark matter that resides in neutrinos: f_ν. The neutrino number density (in the standard cosmological model) is well known; a mean neutrino mass of 0.05 eV corresponds to Ω_ν of approximately 0.001. The current limits are $\bar{M}_\nu < 1$ eV from the CMB alone and $\bar{M}_\nu < 0.4$ eV when including Galaxy power spectra (14).

One can also extract information about the dark energy equation-of-state parameter w. If dark energy is Einstein's cosmological constant, then $w \equiv -1$. Because w affects the expansion history of the Universe at late times, the associated effects on power spectra then give a measure of w. Using all available CMB data, Reference 14 finds $w = -0.86^{+0.35}_{-0.36}$. However, by including both the Galaxy power spectra and SN1A data, the stronger constraint $w = -0.94^{+0.093}_{-0.097}$ is derived. WMAP, using its own data and another collection of LSS data together with supernova data, finds $w = -1.08 \pm 0.12$, where in this fit they also let Ω_k float.

Finally, we want to mention a new effect, even if outside the domain of the CMB— baryon oscillations. In principle, one should be able to see the same kind of acoustic oscillations in baryons (galaxies) seen so prominently in the radiation field. If so, this will provide another powerful measure of the effects of dark energy at late times, specifically the time when its fraction is growing and its effects in curtailing structure formation are the largest. This effect has recently been seen (17) at the level of 3.4σ, and new experiments to study this far more precisely are being proposed. This is an excellent example of how rapidly the field of observational cosmology is developing. In the wonderful textbook *Modern Cosmology* by Scott Dodelson (2, p. 209), Dodelson states that this phenomenon would only be "barely (if at all) detectable."

Before turning to a discussion of the problem of astrophysical foregrounds, we mention that currently the utility of ever more precise cosmological-parameter determination is, like in particle physics, not that we can compare such values with theory but rather that we can either uncover inconsistencies in our modeling of the physics of the Universe or gain ever more confidence in such modeling.

2. FOREGROUNDS

Until now, we have introduced the features of the CMB, enticing the reader with its promises of fascinating insights to the very early Universe. Now we turn our attention toward the challenge of actually studying the CMB, as its retrieval is not at all an easy endeavor. Instrumental noise and imperfections could compromise measurements of the tiny signals (see Section 3). Even with an ideal receiver, various astrophysical or atmospheric foregrounds could contaminate or even suppress the CMB signal. In this section we first give an overview of the relevant foregrounds, then describe the options for foreground removal and estimate their impact.

2.1. Overview

One may be tempted to observe the CMB at its maximum, approximately 150–200 GHz. However, atmospheric, galactic, or extragalactic foregrounds, which have their own dependences on frequency and angular scale, may dominate the total signal, so the maximum may not be the best choice.

The main astrophysical foregrounds come from our own Galaxy, from three distinct mechanisms: synchrotron radiation; radiation from electron-ion scattering, usually referred to as free-free emission; and dust emission. **Figure 7** displays full-sky intensity maps for the main foreground components as derived from WMAP data at microwave frequencies where the bright Galaxy is clearly dominating the pictures. Each component is shown for the WMAP frequency channel where it is dominant.

Figure 8 compares the expected CMB signal as a function of frequency to the rms of WMAP foreground maps on an angular scale of 1°. The ordinate axis records antenna temperature (see Section 3.2.1). An optimal observing frequency range with

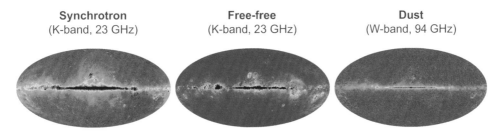

Synchrotron
(K-band, 23 GHz)

Free-free
(K-band, 23 GHz)

Dust
(W-band, 94 GHz)

Figure 7

Unpolarized foreground maps in Galactic coordinates, derived from WMAP. Each map is shown at the WMAP frequency band in which that foreground is dominant. The color scale for the temperature is linear, with maxima set at approximately 5 mK for K-band and 2.5 mK for W-band. Images courtesy of the WMAP science team.

the highest ratio of CMB to foreground signal is in the region around 70 GHz (often termed the cosmological window).

Much less is known about the polarization of foregrounds. Information is extrapolated mostly from very low or very high frequencies or from surveys of small patches. **Figure 8b** shows an analog figure for the polarization fluctuations as estimated from WMAP three-year data on an angular scale of approximately $2°$ ($l = 90$), where the signal from gravitational waves is maximal. The dust estimate has some limitations because the WMAP frequency channels do not extend to the high frequencies where the dust is expected to dominate the foregrounds.

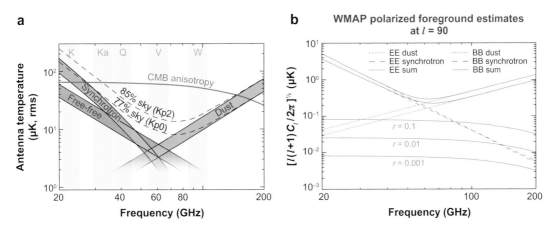

Figure 8

Frequency dependence of foregrounds recorded in antenna temperature. (*a*) The rms on angular scales of $1°$ for the unpolarized CMB compared with that from foregrounds extracted from the WMAP data (18). The WMAP frequency bands (K, Ka, Q, V, W) are overlaid as light bands. These plots are for nearly full sky; the total foregrounds are shown as dashed lines for two different sky cuts. Figure courtesy of the WMAP science team. (*b*) A similar plot of the expected polarization level of foregrounds at $l = 90$ in comparison with that from primordial B modes (which peak around $l = 90$) for different values of r following formula 25 in Reference 19. Again, these estimates are for observations covering most of the sky.

The expected B-mode signal is smaller than the estimated foreground signal even for $r = 0.1$. However, almost the full sky was used for the estimate, whereas recent studies (20, 21) using lower-frequency data and WMAP data indicate that the polarization of synchrotron radiation on selected clean patches can be significantly smaller. Thus, the optimal frequency window will shift depending on which region is observed. After discussing possible foreground effects from Earth's atmosphere, we briefly review what is known about the dominant sources of galactic and extragalactic foregrounds.

2.1.1. Atmospheric effects. The atmosphere absorbs short-wavelength radiation, but fortunately has transmission windows in the range of visible light and microwave radiation. Absorption lines from oxygen (around 60 and 120 GHz) and water vapor (20 and 180 GHz) limit the access to the microwave sky, and, in particular, clouds and high water vapor can compromise ground-based observations. Thermal emission from the atmosphere can add significantly to the observed signal for ground-based experiments (depending on the observing site and the frequency, from 1–40 K) and, together with the instrumental noise and/or thermal emission from warm optical components, can make for the major part of the detected power (see also Section 3.2.1). The observing strategy needs to be designed in a way that allows a proper removal of the varying atmospheric contribution without a big impact on the signal extraction (see also Sections 3.3 and 3.4.1).

Although thermal emission from the atmosphere is unpolarized, the Zeeman splitting of oxygen lines in Earth's magnetic field leads to polarized emission, which is dominantly circularly polarized. Although the CMB is not expected to be circularly polarized, Hanany & Rosenkranz (22) showed that for large angular scales, $l \approx 1$, a 0.01% circular-to-linear polarization conversion in the instrument could produce a signal more than a factor of two higher than the expected gravitational wave B-mode signal if r were small, that is, if $r = 0.01$.

In addition, backscattering of thermal radiation from Earth's surface from ice crystal clouds in the upper troposphere may give signals on the order of micro-Kelvin size (23), again larger than the expected B-mode signal. Although the polarized signal from oxygen splitting would be fixed in Earth's reference frame, and thus could be separated from the CMB, the signal from such ice clouds would reflect the varying inhomogeneous cloud distribution and thus be hard to remove.

2.1.2. Galactic synchrotron radiation. Synchrotron radiation is something familiar to particle physicists, mostly from storage rings where some of the energy meant to boost the particle's energy will be radiated away. The same effect takes place in galactic accelerators, with cosmic-ray electrons passing through the galactic magnetic field. In contrast to the particle physics case, where electrons of energies of a few GeV pass magnetic fields of up to a few 1000 G, we are dealing here with electrons in a galactic field of only a few micro-Gauss.

This component of the foreground radiation is dominant at frequencies below 70 GHz, and its intensity characteristics have been studied at frequencies up to 20 GHz, The frequency and angular dependence both follow power laws $T \propto \nu^{-\beta}$, with a position- and frequency-dependent exponent that varies between 2 and 3.

Theoretically, a high degree of sychrotron polarization (>75%) is expected, but low-frequency data imply much lower values. However, at low frequencies, Faraday rotation—where light traversing a magnetized medium has its left and right circular polarized components travel at different speeds—reduces the polarization.

2.1.3. Galactic dust.

Interstellar dust emits mainly in the far infrared and thus becomes relevant for high frequencies ($\nu > 100$ GHz). The grain size and dust temperature determine the properties of the radiation, where the intensity follows a power law $T \propto T_0 \nu^{\beta}$, with the spectral index $\beta \approx 2$ and with both T_0 and β varying over the sky. Using far-infrared data from COBE, Finkbeiner et al. (24) (FDS) provided a model for the dust emission consisting of two components of different temperature and emissivity ($T = 9.4/16$ K, $\beta = 1.67/2.7$).

There are also indications for another component in the dust emission, as seen through cross correlation of the CMB and far-infrared data. Its spectral index is consistent with free-free emission, but it is spatially correlated with dust. This anomalous dust contribution could derive from spinning dust grains. However, current data do not provide a conclusive picture, and additional data in the 5–15-GHz range are needed to better understand this component (25).

In 2003, the balloon-borne experiment ARCHEOPS reported 5% to 20% polarization of the submillimeter diffuse galactic dust emission, providing the first large coverage maps of polarized galactic submillimeter emission at 13′ resolution (26). More recently, they also published submillimeter polarization limits at large angular scales, which when extrapolated to 100 GHz are still much larger than the expected gravitational wave signal for $r = 0.3$ (27).

2.1.4. Free-free emission.

Electron-ion scattering leads to radiation that is, in this context, termed free-free emission, whereas in the high-energy lab, it is better known as bremsstrahlung. This component does not dominate the foregrounds at any radio frequency. Sky maps of free-free emission can be approximated using measurements of the Hα emission (from the hydrogen transition from $n = 3$ to $n = 2$), which traces the ionized medium. The thermal free-free emission follows a power law $T \propto \nu^{-\beta}$, where $\beta \approx 2$. This foreground is not polarized.

2.1.5. Point sources.

Known extragalactic point sources are a well-localized contaminant and easily removable. However, the contribution from unresolved point sources can severely affect measurements: for example, the recent discussion of their impact on the determination of n_s from WMAP data (28). Point sources impact CMB measurements mostly at high angular scales and low frequencies. For low frequencies, their contribution may still be larger than the signal expected from gravitational waves.

2.2. Foreground Removal

Understanding and removing foregrounds are most critical for the tiny polarization signals. The different frequency dependences of the CMB and galactic foregrounds provide a good handle for foreground removal using multifrequency measurements.

For the polarization analysis, methods where little or no prior information is required are the most useful for now. A promising strategy is the Independent Component Analysis, which has already been applied to several CMB temperature data sets (including COBE, BEAST, and WMAP) and for which formalism has also been developed to cope with polarization data. The foreground and CMB signals are assumed to be statistically independent, with at least one foreground component being non-Gaussian. Then the maximization of a specific measure of entropy is used to disentangle the independent components. Stivoli et al. (29) demonstrated a successful cleaning of foregrounds using simulated data. Verde et al. (30) estimated the impact of foregrounds independent of removal strategy, considering different degrees of effectiveness in cleaning. A 1% level of residual foregrounds, in their power spectrum, was found to be necessary to obtain a 3σ detection of $r = 0.01$ from the ground.

Because all current studies rely on untested assumptions about foregrounds, they need to be justified with more data. Moreover, none of the studies to date takes into account the impact of foregrounds in the presence of lensing and instrumental systematics. Work is needed on both the experimental and theoretical side to obtain a more realistic picture of the foregrounds and their impact.

3. METHODS OF DETECTION

We have argued that in the patterns of the CMB lies greatness; here we outline the essential ingredients for measuring CMB anisotropies. The fundamental elements for detecting microwave emission from the celestial sphere are optics and receivers. The optics comprises telescopes and additional optical elements that couple light into the receivers. The receivers transduce the intensity of the incoming microwave radiation into voltages that can be digitized and stored. Two other CMB experiment requirements are fidelity control (calibration and rejection of spurious signals) and optimized strategies for scanning the telescope beams across the rotating sky. Below, after a general introduction to the problem, we elaborate on these topics and culminate with an overview of data analysis techniques.

3.1. The CMB Experiment Basics

All CMB experiments share certain characteristics. Some main optical element determines the resolution of the experiment. This main optical element may be a reflecting telescope with a single parabolic mirror, or one with two or more mirrors; it may be a refracting telescope using dielectric lenses; it may be an array of mirrors configured as an interferometer; it may be just a horn antenna.[2] In most cases, additional

---

[2]A horn antenna is waveguide flared to the appropriate aperture for the desired resolution; these were used in the COBE satellite instrument that made the first detection of CMB anisotropy.

optical elements are required to bring the light to the receiver. Examples include Dewar windows, lenses, filters, polarization modulators, and feedhorns (which are horn antennas used to collect light from telescopes). Typically these coupling optics are small enough that they can be maintained at cryogenic temperatures to reduce their thermal emission and lossiness.

The low-noise receivers are nearly always cryogenic and divide into two types, described below. Spatial modulation of the CMB signal on timescales of less than one minute is critical to avoid slow drifts in the responsivity of the receivers, and may proceed by movement of the entire optical system, or by moving some of its components while others remain fixed. Large ground screens surround most experiments to shield the receivers from the 300-K radiation from Earth. Typically, the thermal environment of the experiment must be well regulated for stability of the receiver responsivity and to avoid confusion of diurnal effects from the environment with the daily rotation of the celestial signal. Earth-bound experiments suffer the excess noise from the atmosphere, as well as its attenuation of the signal, and must contend with 2π of the 300-K radiation. Balloon-borne experiments suffer less atmosphere, but must be shielded from the balloon's thermal radiation and typically have limited lifetimes (1–20 days). Long flights usually require constant shielding from the Sun during the long austral summer day. Space missions have multiple advantages: no atmosphere, Earth filling a much less solid angle, a very stable thermal environment, and a longer lifetime than current balloon missions.

3.2. The Detection Techniques

Although to fully describe the CMB anisotropies requires their spatial power spectra (which happily are not white), a useful order-of-magnitude number is that the rms of the CMB sky when convolved to $10°$ scales is approximately 30 μK, and approximately 70 μK for $0.7°$ scales (the first acoustic peak). This rms of the CMB temperature is some 20 ppm of the 2.7-K background, the polarization E modes are 20 times lower, and the primordial B-mode rms is predicted to be 50 ppb or less.

A microwave receiver measures one or more of the Stokes parameters of the radiation incident on it. Two classes of low-noise receivers may be identified: coherent receivers, in which phase-preserving amplification of the incident field precedes detection of its intensity, and incoherent receivers, in which direct measurement of the intensity of the incident field is performed.

In coherent receivers, the incident field is piped around transmission lines as a time-varying voltage. That voltage is amplified in transistor amplifiers, and then the signal is eventually detected when it passes through a nonlinear element (such as a diode) with an output proportional to the square of the incident field strength. The critical element in the coherent receiver is not the detector but typically the transistor amplifier, which must be a low-noise amplifier. In cases where transistor amplifiers are not available at high enough frequencies, the first and most critical element in the receiver is a low-noise mixer, which converts the frequency of the radiation to lower frequencies, where low-noise amplifiers are available.

For the CMB, the most widely used incoherent detector to date is the bolometer. A bolometer records the intensity of incident radiation by measuring the temperature rise of an isolated absorber of the radiation. A promising effort is underway to develop receivers in which the bolometers are coupled to transmission lines, where they can serve as the very low-noise detectors in what otherwise looks like a coherent receiver.

3.2.1. Calibration, Kelvins, and system temperature. A microwave receiver outputs a voltage proportional to the intensity I of the incident radiation over some effective bandwidth $\Delta \nu$ centered on frequency ν_0. The output is calibrated in temperature units through observation of blackbody sources. The polarization anisotropies of the CMB are also described in temperature units. This follows because the Stokes parameters Q and U have the same units as the intensity I. There is a factor of two to keep track of: The usual definition of I sums the intensities from two orthogonal polarization modes. Note that the antenna temperature T_A is defined by the approximation $I \propto T_A$. Only in the Rayleigh-Jeans regime of a blackbody does $T_A \approx T$, but it is a convenient measure for comparing the effects of various foregrounds and other contaminants.

Microwave receivers are sensitive to the total intensity of the incident radiation over the bandpass. The incident electric field can be considered as a sum of incoherent (i.e., uncorrelated) sources, each with intensity that can be associated to a temperature in the Rayleigh-Jeans limit. Thus, we can define the system temperature to describe the input power to the receiver:

$$T_{sys} = T_{CMB} + T_{fg} + T_{atm} + T_{gnd} + T_{opt} + T_n, \qquad 6.$$

where we have included terms for the CMB, foregrounds, atmosphere, 300-K emission from the ground, emission from the warm optics, and receiver, respectively. We do not note explicitly that the extra-atmospheric signals are attenuated slightly as they pass through the atmosphere. At good sites such as the Atacama Desert in Chile or the South Pole, this effect is small for $\nu = 110$ GHz. We also neglect absorption in the optics, although in bolometer systems, this effect can be large. Note that when describing bolometer receivers, it is more common to leave the sum in units of power, as we see below.

3.2.2. Sensitivity and noise. Imagine a CMB experiment that scans across a small enough region of sky that the sky curvature may be neglected, recording the temperature of each of N beam-sized patches of sky a single time into a vector d. The error on each measurement is σ_e. Let us first take the (unattainable) case where $\sigma_e \ll 1$ nK. In that case, σ_d^2 measures the variance of the CMB itself. If the CMB power spectrum were white with average level ΔT^2, for example, the variance could be crudely estimated as

$$\sigma_d^2 \approx \Delta T^2 \frac{\delta \ell}{\ell_c}, \qquad 7.$$

where ℓ_c is some average ℓ in the region $\delta \ell$ between $(x_b)^{-1}$ and $(N x_b)^{-1}$, with x_b the diameter of the beam-sized patch in radians. Typical numbers might give $\delta \ell / \ell_c \approx 1$, and $\Delta T = 60$ μK $\approx \sigma_d$. Thus, for σ_e so small, one could estimate the Fourier modes (or the C_ℓ themselves) directly and trace out details of the spectrum over the range $\delta\ell$.

To consider a more realistic case, we note that CMB receivers are characterized according to their sensitivity S in units of K$\sqrt{s}$. The variance σ_T^2 of a series of measurements, each resulting from an integration time τ, is then found by $\sigma_T^2 = S/\tau$. A typical value for the sensitivity of a single receiver is 500 μK/$\sqrt{s}$, so that after 10 min on each patch, one might attain $\sigma_e \approx 20$ μK and then detect the CMB signal excess variance in the data. Typical variances in the CMB E polarization are $(4\,\mu K)^2$, so these experiments must integrate for hundreds of hours and/or use hundreds of receivers. The tougher sensitivity requirements for B modes are described in Section 4.

In cases where the mean photon mode occupancy n_0 for the input radiation is large, $n_0 = (e^x - 1)^{-1} \gg 1$ with $x = h\nu/(kT_{sys})$, we can describe classically the sensitivity of an ideal direct receiver of bandwidth $\Delta\nu$ with the Dicke equation: $S = T_{sys}/\sqrt{\Delta\nu}$. We account for more complex receivers below. Most coherent receivers operate in the limit $n_0 \gg 1$. In cases where $n_0 \ll 1$ (low T_{sys} and high frequency ν), the limitation to sensitivity in ideal bolometric receivers comes from photon shot noise (counting statistics), and the sensitivity depends on $\sqrt{T_{sys}}$ in cases where the bolometer thermal noise is negligible. We return to the noise in bolometers below.

The sensitivity of the receiver captures its best features succinctly. It is also instructive to look at the spectra of postdetection signal noise from the receivers. The scanning of the telescope translates the CMB anisotropy signal to variations in time that lie atop the intrinsic postdetection noise. A typical noise power spectrum is shown in **Figure 9**. This spectrum shows a white-noise level at frequencies $f > 0.01$ Hz; the sensitivity is measured from the white noise. The spectrum also shows characteristic low-frequency noise approximately proportional to $1/f$, which can be parametrized by the frequency f_c where the $1/f$ and white-noise powers are equal. Such $1/f$ noise is ubiquitous; the atmosphere itself has a $1/f$ spectrum. In principal, this noise is quite serious, as it contributes more and more to the variance with longer integration times. In practice, experiments are designed to modulate the signal at frequencies $f > f_c$. Usually the receiver includes a low-pass filter that limits the bandwidth of the postdetection data stream to f_{Ny}. Then these data are Nyquist sampled and digitized at a rate of $2f_{Ny}$ to avoid aliasing high-frequency noise. Roughly speaking, the postdetection bandwidth of interest is usually between 0.01–0.1 Hz and 50–200 Hz.

3.2.3. Coherent receivers.
A key advantage of coherent receivers is that they can be configured so that their output is the correlation of two input signals. We devote the bulk of our discussion here to this topic.

Coherent-receiver noise spectra. The noise properties of amplifiers include both intrinsic output voltage fluctuations, present in principle even in the absence of input signal[3] and characterized by T_n, and fluctuations in the gain coefficient (the factor by

---

[3]We remind the reader that for the frequencies of interest, "absence of input signal" is a difficult requirement because a resistive termination emits thermal radiation, whereas a shorted input reflects back any thermal radiation emitted from the amplifier input.

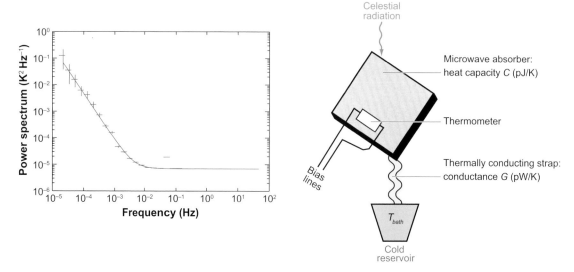

Bolometer

Celestial radiation

Microwave absorber: heat capacity C (pJ/K)

Thermometer

Thermally conducting strap: conductance G (pW/K)

Bias lines

T_{bath}

Cold reservoir

Figure 9

(*Left*) Raw postdetection noise power spectrum from a correlation polarimeter during scans of the sky, illustrating three interesting features: its white-noise level at frequency $f > 0.01$ Hz, a peak corresponding to the scan period of 21 s, and the characteristic low-frequency noise with slope $1/f$. Figure courtesy of C. Bischoff and the CAPMAP team (31). (*Right*) Sketch of a bolometer, indicating that celestial radiation is absorbed in a material with heat capacity C coupled to a cold reservoir at temperature T_{bath} by a strap with low thermal conductance G. A thermometer measures the power delivered to the absorber by recording its temperature increase.

which the input voltage is amplified). An extension of the Dicke equation reads

$$\delta T = a\,T_{sys}\sqrt{\frac{1}{\Delta\nu\tau} + \left(\frac{\delta g}{g}\right)^2},$$

8.

where the coefficient a depends on the exact configuration of the receiver; $a = 1$ for a direct receiver in which the input is amplified and then detected. For the correlation polarimeters described below, $a = \sqrt{2}$.

Signal correlation. Two signals are correlated if the time average of their product is nonzero. In 1952, Ryle (32) introduced the concept of phase switching: periodically introducing a half-wavelength phase lag into one of the two signal lines, which changes the sign of the correlation product. Then, one can sum the two inputs and square them in a detector diode. The correlation product is now modulated at the phase-switching rate and may be recovered. The correlation can also be measured in analog correlators (multipliers) or by digitization of the inputs, followed by multiplication.

Correlation receivers. Correlation receivers have been used in several ways to measure CMB signals. We describe two methods below. The advantages of correlation include reduced sensitivity to $1/f$ variations in amplifiers (because phase switching with diode switches can be effected at kilo-Hertz rates), reduced sensitivity to gain fluctuations (because those fluctuations multiply only the small correlated signal rather than the large, common-mode intensity signal), and the ability to access all four of the Stokes parameters that describe the full polarization state of the radiation using only two inputs.

Interferometers. Perhaps the best-known example of signal correlation in astronomy lies in the interferometer, which uses the correlation of signals from two spatially separated telescopes to measure Fourier modes of the celestial radiation in a limited field of view. The interferometer's receiver detects a slice of the interference pattern that arises when the two input signals originate with phase coherence, much as the screen in a Young's diffraction experiment does. The angular resolution is determined by the spacing between the telescopes, whereas the size of the individual telescopes limits the field of view. The resolution of an interferometer may be described in terms of its synthesized beamwidth, σ_b. Because the CMB is an extended source, nearly all the Fourier modes in the field of view are needed to fully characterize it. All the Fourier modes can be measured only if the individual telescopes are close packed so that they fill all the space required for an equivalent single dish of resolution σ_b.

To recover all four Stokes parameters, the telescope optics usually includes circular polarizers so that a given receiver amplifies either the left (L) or right (R) circular polarization state. Both DASI and CBI used quarter-wave plates in the optics to periodically reverse the polarization state, $L \leftrightarrow R$, so that all possible combinations of correlations (LR, LL, RR, and RL) could be measured from each pair of telescopes.

Correlation polarimeters. In a correlation polarimeter, the incident signal enters through an azimuthally symmetric feedhorn and is separated into two polarization states (either two orthogonal linear polarization states or the L and R circular polarization states) before amplification in high-electron-mobility-transistor amplifiers. After amplification, the two polarization states are correlated so that the output is proportional to a linear Stokes parameter: $V_{out} \propto \pm E_x E_y \propto \pm U$, with the modulation provided by the phase switch. **Figure 9** shows the stability of a phase-switched correlation polarimeter, switched at 4 kHz. The high-electron-mobility-transistor amplifiers in this polarimeter have $1/f$ knees at 1 kHz and cannot be used without this rapid modulation. The sensitivity of this polarimeter is found from the Dicke equation with $a = \sqrt{2}$. If L and R states are used, then the second linear Stokes parameter can also be obtained by phase shifting R by an extra $\pi/2$ and then correlating it with L. The correlation can come about either through direct multiplication of the two polarization states or through the Ryle technique. The QUIET project uses the latter method, in which L and R are sent into a compact module that mounts onto a circuit board. The correlation LR is constructed by differencing the squares of the sum and difference terms $L + R$ and $L - R$; U is found by similar means. Phase switches at 4 kHz modulate the outputs, which are read out on module pins; other module pins are used to bias the amplifiers.

3.2.4. Incoherent detectors. A number of clever ideas beyond the scope of this review are being pursued for tailoring bolometers to search for CMB polarization. One recent great success was the advent of the polarization-sensitive bolometer. Another avenue is using wafer-level silicon fabrication techniques to produce arrays of hundreds of detectors at once, sometimes coupling the detectors directly to planar antennas or antenna arrays (antenna-coupled bolometers). Below we focus on the rudiments of the transition edge sensor (TES) bolometer.

Figure 9 depicts the critical elements of a bolometer. Celestial radiation impinges on a microwave absorber with heat capacity C. The absorber is connected to a cold reservoir at temperature T_{bath} by a thermal strap of heat conductance G. A fluctuation in the intensity of the celestial radiation warms the absorber slightly. This temperature change is recorded by a thermometer, typically a material with a large-magnitude logarithmic derivative $d \log R / d \log T \equiv \alpha$ of resistance with respect to temperature.

Transition edge sensor bolometers. The TES is a superconductor maintained at its critical temperature T_c. We describe the TES as being on (or in) the transition when its resistance is between zero and the normal resistance R_n. For CMB devices optimized to work with $T_{bath} \approx 300$ mK, a typical critical temperature is $T_c \approx 400$–500 mK. The width of the transition is typically a percent of T_c. Many TESs comprise a bilayer: a thin layer of natural superconductor topped with normal metal. The T_c and R_n of bilayers can be controlled by varying the two thicknesses.

The TES is operated in series with an ideal inductor L. To voltage bias the TES at operating resistance R_0, a small shunt resistance $R_{sh} \ll R_0$ is placed in parallel with the TES/inductor combination and fed with a bias current. Fluctuations in R_0 cause current fluctuations through the inductor. Flux through the inductor is linked to a superconducting quantum interference device (SQUID), which serves as a current amplifier. To prevent the SQUID current noise from contributing significantly, the operating resistance R_0 is kept low: $R_0 < 1\ \Omega$.

The voltage bias causes the TES resistor to dissipate heat. Rough order-of-magnitude values for TES bolometers used for the CMB may be $G \approx 30$ pW K^{-1} and $C \approx 0.3$ pJ K^{-1}. The thermal timescale for the bolometer to change temperature in response to a fluctuation in the photon power is $\tau_{tb} = C/G \approx 10$ ms, corresponding to a one-pole low-pass filter at $f = (2\pi\tau)^{-1} \approx 15$ Hz. Conceptually, we note that when a fluctuation in the photon noise warms the bolometer, the TES resistance increases, as $\alpha > 0$. Because the TES is voltage biased, the increase in resistance lowers the Joule heating power V^2/R flowing into the bolometer. These electrical changes can occur more quickly than the thermal timescale τ_{tb}. This is the phenomenon of electrothermal feedback, which speeds up the TES bolometer and also stabilizes it so that it stays in its transition.

TES bolometer noise spectra. In Section 3.2.2 we considered the sensitivity of an instrument in the case of photon occupancy number $n \ll 1$ and mentioned that the sensitivity is then proportional to $\sqrt{T_{sys}}$, or, more appropriately because $n \ll 1$ contraindicates the Rayleigh-Jeans approximation, to $\sqrt{P}$, where P is the power in the incident radiation field, also known as the photon power. In the general case for CMB experiments, $n \approx 1$, and so a more general formula is required (see Reference 33 for a discussion).

For bolometers, the receiver sensitivity is usually described in terms of the noise-equivalent power (NEP) that can be measured in a postdetection bandwidth of 1 Hz. NEP properly has units not of power, but of $W/\sqrt{Hz}$. A bandwidth of 1 Hz is equivalent to a half-second of integration time. Besides this factor, converting to units of $\mu K \sqrt{s}$ appropriate for detecting CMB fluctuations requires the appropriate derivative to convert power to thermodynamic temperature and requires referencing the NEP to the entrance of the optics. (In practice, the second point means correcting for the efficiency $\eta < 1$ in a high-frequency system.)

The total NEP^2 for a receiver can be found by summing up the squares of NEPs from different terms, of which the contribution from photons, NEP_γ, is only one. When NEP_γ dominates the sum, the detector is said to be background limited. Intrinsic sources of noise in bolometer systems include thermal noise from the photons transporting heat to the cold bath ($NEP_{th} \approx \sqrt{2kT_0^2 G}$), Johnson noise power in the bolometer resistance (which is reduced significantly by the electrothermal feedback, as described in Reference 34), back-end noise (from the SQUIDs), and occasional other unexplained sources of noise, including $1/f$ noise.

The inductor in series with the TES resistor serves not only to couple the TES current to the SQUID output, but also to provide a Nyquist filter with time constant L/R_0. The L/R_0 time constant low pass filters the signals, and the noise power, before they emerge from the cryostat. Now that we have reviewed the experimental basics and detectors used for CMB measurements, we turn to the discussion of how to choose an observing strategy for an experiment.

3.3. Observing Strategies

We are surrounded by a bath of CMB photons, and with all directions being equal it seems that only astrophysical foregrounds determine the choice for which parts of the sky should be observed. The size of the observed patch, together with the angular resolution, determines the accessible angular scales. However, the size, shape, and uneven coverage of the observed region, and even the way in which it is scanned, all impact the determination of the CMB power spectrum so that an optimization of the observing strategy requires more sophisticated considerations. A pedagogical illustration of the choices and their impacts is given in Reference 35. We mention here the most important issues that must be taken into account in developing an observing strategy.

Including the cosmic variance (see Equation 2), the achievable precision in the power spectrum (C_l) can be expressed in the following form:

$$\frac{\Delta C_l}{C_l} = \sqrt{\frac{2}{2l+1}} \left(\frac{1}{\sqrt{f_{sky}}} + \frac{4\pi (\Delta T_{exp})^2}{C_l} \sqrt{f_{sky}} e^{l^2 \sigma_b^2} \right). \qquad\qquad 9.$$

Here, f_{sky} represents the observed fraction of the sky, ΔT_{exp} the total experimental sensitivity (ΔT combining all detectors for the duration of the run), and σ_b the width of the beam. Finite beam resolution degrades the sensitivity progressively more at higher l values. For example, the WMAP beam of approximately $0.2°$ limits sensitivity to below $l \approx 600$; smaller beams are needed to study much finer scales. In general, the impact of the scanning strategy on the measurement of the different C_l is summarized in a window function that describes the weight with which each C_l contributes to the measured temperatures. A crude estimate of the lowest accessible l is $l_{min} \approx 1/\Delta\Theta$, with Θ being the angular extent of the survey. Limited sky coverage leads to correlations in the power at different multipoles l. Therefore, the power spectrum is usually reported in largely uncorrelated l bands of width Δl, which should be larger than l_{min}. The error on the average C_l in an l band then follows from Equation 9, with a prefactor of $1/\sqrt{\Delta l}$.

For an experiment to optimally take advantage of its sensitivity ΔT, the size of the observed sky patch should be chosen such that the contributions to the uncertainty in C_l from sample variance (first summand) and from noise (second summand) are roughly equal. This leads to $4\pi f_{sky} = C_l e^{-l^2 \sigma^2} (\Delta T)^{-2}$. Another way to express this is that the ideal patch size is such that the signal-to-noise ratio per pixel $\sigma_{CMB}/(\Delta T_{pix})$ is unity, where σ_{CMB} is the expected CMB fluctuation in a beam-sized pixel and ΔT_{pix} the experimental error on the temperature in that pixel.

A polarization analysis always suffers from small survey sizes because E and B modes are not local fields, and thus a small patch contains ambiguous modes. The power spectrum estimation can then lead to E-to-B leakage and distort the B-mode measurements. Because the CMB B modes are an order of magnitude smaller than the E modes, even a small leakage can significantly affect the measurements. However, Smith (36) has already demonstrated a method for avoiding significant leakage.

In an optimal experiment, the observed patch is covered homogeneously, which maximizes the sensitivity to the power spectrum for a given integration time. It is beneficial to cover any pixel in the observed patch with different receivers, and also with different receiver orientations, especially—but not only—for the polarization analysis. This cross linking provides robustness against effects from time-stream filters in the data processing (see Section 3.4.1) and enables various systematic studies where instrumental effects or pickup of signals from the ground can be distinguished from real signals on the sky (see Reference 37 for another view on this).

Using a purely azimuthal scan on a patch, the atmospheric contribution is constant, but polarization-sensitive receivers detect different combinations of Q and U at different times, leading to pixels having nonuniform weights for Q and U. This also compromises the disentanglement of E and B modes, reducing sensitivity. A more symmetric distribution is achieved by scanning, for example, in a ring pattern, resulting in much more uniform Q/U sensitivity and enabling many systematic cross checks with a high degree of symmetry. However, a changing atmospheric contribution along the ring must then be taken into account.

For any experiment, the scan speed should be as fast as possible to decrease the effects of $1/f$ drifts (see Section 3.2.2), but upper limits are posed by the beam resolution, maximum sampling speed, mechanical constraints of the telescope, and detector time constants. Typical scanning speeds are tens of arcminutes per second. Typical sampling at 10–100 Hz then provides several samples per beam. Given all the above considerations, it is clear that an optimization of the observing strategy is not trivial and any scan strategy will include some compromises. So far we have discussed the challenges in acquiring useful CMB data. In the next section we introduce the obstacles still lying ahead in the processing of data to enable access to its cosmological treasures.

3.4. Techniques of Data Analysis

In CMB experiments, data are accumulated continuously over a duration ranging from a few hours (balloon experiments) up to a few years (space missions). The data volume for past and current experiments was at most a few hundred gigabytes. Even with planned expansions to arrays of ~ 1000 receivers, the data streams will amount to a fraction of typical current high-energy physics experiments: 10–50 Tbytes per year. Although for high-energy physics experiments the data records can be split into different categories of interest and the analyses are usually performed on specific selections of the data with an event-wise treatment, here all of the acquired data contain the same signal and the signal's extraction becomes possible only with long integration times, requiring the processing of all data in the same analysis. Typically the CMB analysis can be divided into four main steps, which are described in the following sections.

3.4.1. Filtering and cutting.

Imaging the CMB requires careful cleaning of the data. Selection of data without instrumental failures or bad weather (for ground-based experiments) comprises the first filter on the data. The detector time streams contain long-term drifts of the detector responsivities and often of the atmosphere, which must be eliminated. Removal of the mean of the data over short time stretches (10–100 s) or high-pass filtering is typically used to reduce the effects of such drifts. **Figure 10** demonstrates the impact of drifts on simulations for the Planck experiment, where the left map was produced without any filtering and the right map represents the same data but with a destriping algorithm applied. The scales of the maps are noticeably different and so is the visible pattern, which emphasizes the need for such filtering.

Other components in the data that must be removed are the effects of ground pickup from the 300-K Earth and changing atmospheric contributions during the scan. These signals are fixed to the reference frame of Earth and so are distinguishable from the celestial CMB signals. Even though any such data filtering removes some

-3000 ▮▮▮▮▮ 3000   -300 ▮▮▮▮▮ 300

μK CMB   **μK CMB**

Figure 10

Effect of destriping on simulated sky maps. (*Left*) Map from a raw time stream. (*Right*) Map after applying a destriping algorithm (note the different scales). This simulation was done for the Planck High Frequency Instrument (38).

cosmological information, observing strategies are devised so that the filtering does not significantly compromise the sensitivity.

3.4.2. Reduction to maps. The most intuitive representation of the CMB signal is a map on the sky. If done properly, this provides orders-of-magnitude compression of the data where no cosmological information is lost (for current typical experiments, the time streams have $N_t \sim 10^7$ samples—this for 10 or more detectors—whereas the number of sky pixels is $N_p \sim 10^5$–10^6).

For an ideal time stream containing only the CMB signal on top of white noise, a map can easily be produced by averaging all observations that fall into a given pixel on the sky. This makes for a robust and fast algorithm, which scales linearly with the number of data samples. However, in a real experiment the filtering of the time streams introduces correlations that make this simple approach fail.

The optimal map estimate from the time stream is produced by maximizing the likelihood for the data given a certain noise model. The likelihood problem can be solved analytically, requiring the inverse of the covariance matrix in the time domain, as well as the calculation and inversion of the covariance matrix in pixel space. Whereas the time-stream covariance matrix is sparse, the covariances in the pixel domain have no special structure and the matrix inversion dominates the processing time, scaling with N_{pix}^3. Parallelization of this procedure is possible. Iterative procedures are another approach and can reduce the required amount of computing power to scale as $N_{it}N_t \log N_t$, where N_{it} is the number of iterations. For those procedures, Monte Carlo methods are required to estimate the covariance, with a possible loss in sensitivity.

3.4.3. Power spectrum estimation. Although the power spectrum could in principle be estimated directly from the time stream, it is more efficient to first produce a map and from it determine the power spectrum. The likelihood can then be expressed in the following form:

$$\mathcal{L} = \frac{1}{\sqrt{\det(C)}} \exp\{-\mathbf{x}^{T} C^{-1} \mathbf{x}/2\}, \qquad 10.$$

where $\mathbf{x}$ represents the map vector and C the pixel-pixel covariance matrix in which the cosmological information is imbedded. The map contains contributions from both signal and receiver noise, and, similarly, the covariance matrix comprises signal added to noise. The likelihood is a measure of how well the scatter as seen in the data agrees with that expected from the combination of noise and the CMB signal. Maximizing the likelihood by setting $\frac{\partial L}{\partial C_l} = 0$ leads to an equation that requires an iterative solution. A common and powerful method for likelihood estimations exploits a quadratic estimator, which uses for the error matrix an ensemble average (Fisher matrix) instead of the full curvature (see Reference 39 for details) and reaches convergence within only a few iterations. The maximum likelihood method offers an optimal evaluation of the power spectrum. However, its large matrix operations become impractical for upcoming experiments with fine-resolution maps of large fractions of the sky ($>10^6$ pixels). Methods that exploit Monte Carlo simulations to approximate the analytical solution are thus an attractive option for the evaluation of the power spectrum and have become popular [e.g., the pseudo-C_l method (40)]. The use of Monte Carlo simulations for the evaluation of real data is a familiar concept to the high-energy physics physicist and is advantageous in that it enables a relatively easy treatment of various instrumental and processing artifacts or distortions.

3.4.4. Cosmological parameter estimation. As described in Section 1, the development and contents of the Universe determine the characteristics of the CMB anisotropies and with that the shape of the CMB power spectrum. Current software tools such as *cmbfast* or *camb* calculate the expected power spectrum from a given set of cosmological parameters within a few seconds so that for a given measured power spectrum, the likelihood for different parameter values can be evaluated reasonably quickly. Still, the likelihood evaluation on a fine grid in the multidimensional parameter space requires huge computing resources so that the problem is typically approached via Markov chain Monte Carlo methods. A Markov chain begins with the evaluation of the likelihood at a specific point in the parameter space where its value influences the next point for the likelihood evaluation. Repeating this leads to a sample density in parameter space proportional to the likelihood where projections onto one- or two-dimensional subspaces result in the marginalized likelihoods. By now, software packages such as *cosmomc* are available and in use by the CMB community.

4. FUTURE PROSPECTS

Here we discuss experiments needed to detect the B-mode signals of lensing at large l values and of gravity waves at intermediate and small l values. **Table 3** shows the sensitivity required to make 3σ detections of several target signals. Estimates come from Equation 9, either for the full sky or for a smaller patch where the balance between sample variance and detector noise is optimized. We give expressions for the

Table 3 Required sensitivities to detect B-mode polarization signals

Signal	ΔT_{cos} (nK)	ΔT_{Gal} (nK)	ΔT_{Lens} (nK)	$\Delta T_{exp}(3\sigma)$(nK)	N_{WMAP}^{opt}
lensing, $l = 1000$	300	150	–	35	1.5
$r = 0.1$ (SLS)	83	295	50	9.8	20
$r = 0.01$ (SLS)	26	295	50	3.1	200
$r = 0.1$ (reion.)	54	780	–	6	50
$r = 0.01$ (reion.)	17	780	–	1.9	500

For detecting the lensing signal at $l = 1000$ and the gravity wave at $r = 0.1$ and 0.01, using either the surface of last scattering or the reionized plasma, we give the magnitude of the cosmological signal at its peak, the size of the galactic contamination, the magnitude of the lensing contaminant, the derived total experimental sensitivity to detect the signal (neglecting foregrounds) at 3σ, and the corresponding increase over that achieved with WMAP in 1 year. The foreground estimates are taken from the WMAP empirical full-sky relation, evaluated at 90 GHz at the appropriate l value. For small, selected patches of sky, foreground contamination will be significantly smaller, perhaps by an order of magnitude. Thus, the lensing signal should be observable from the ground with just eight WMAPs, for example, an array with eight times the number of detectors in WMAP, each with the same sensitivity, observing for one year and so on. The signatures of gravity waves from the surface of last scattering (SLS) should then also be detectable, even in the presence of foregrounds, although contamination from lensing and residual foreground levels will reduce sensitivity from what is shown in the table. Gravity wave signals from the reionized plasma have negligible contamination from lensing but, because of the needed full-sky coverage, larger foregrounds to deal with.

sensitivity to a feature in the power spectrum centered at l_0 and with width $\Delta l = l_0$ and for the fraction of the sky that accomplishes the balance: $(\frac{\Delta C_l}{C_l})_{opt} \approx 3\frac{\Delta T_{exp}}{\Delta T_{cos}}$ and $f_{opt} \approx 1/2(\frac{\Delta T_{cos}}{l_0 \Delta T_{exp}})^2$. For our purposes, both the signal from lensing and from primordial gravity waves have approximately this shape. Here ΔT_{cos} is the (peak) cosmological signal of interest and ΔT_{exp} is the total sensitivity of the experiment, summing over all the observing time and all the detectors.

The lensing detection can be accomplished by observing for a year (from the ground at a good site such as the Atacama Desert in Chile or the South Pole) a patch of approximately 1.6 square degrees, with detectors having 1.5 times the WMAP sensitivity. The table also gives the sensitivities required to detect primordial B modes at various levels of $r = T/S$. Foregrounds will be a problem for $r \leq 0.01$, perhaps more so for the detection of the signal from the reionized plasma than from the surface of last scattering. A satellite experiment can detect the signal from the reionized plasma, where the lensing contamination is nearly negligible.

The Planck experiment (2007) has an order of magnitude in temperature sensitivity over WMAP. Polarization sensitivity was not a primary goal. Still, much has gone into making sure the residual systematic uncertainties (and foregrounds) can be understood sufficiently well to allow the extraction of polarization signals around 50 nK, corresponding to $r = 0.05$.

There is a program of experiments over the coming five to eight years. These will involve, progressively, tens, then hundreds, and finally a thousand or more detectors per experiment and will test polarization modulation schemes, effective scan strategies, foreground-removal methods, and algorithms for separating E and B modes.

Experiments with tens of detectors are already underway. The sister experiments QUaD and BICEP observe from the South Pole, using polarization-sensitive bolometers at 100 and 150 GHz. QUaD, with a 4-arcmin beam, is optimized for gravitational lensing, whereas BICEP, at approximately 40 arcmin, is searching for gravitational waves. MBI and its European analog BRAIN are testing the idea of using bolometers configured as an interferometer, and PAPPA is a balloon effort using waveguide-coupled bolometers from the Goddard Space Flight Center. These latter experiments have beams in the range of 0.5° to 1°.

Five initiatives at the level of hundreds of detectors have so far been put forth. Four use TES bolometers at 90, 150, and 220 GHz: CLOVER, the lone European effort, with an 8-arcmin beam; Polarbear, with a 4-arcmin beam; and EBEX and SPIDER, balloon-borne experiments with 8- and 20–70-arcmin beams, respectively (SPIDER and Polarbear will use antenna-coupled devices). The fifth uses coherent detectors at 44 and 90 GHz: QUIET, initially with a 12-arcmin beam, observing from the Atacama Desert. All are dedicated ground-up polarization experiments that build their own optical systems. The ACT and SPT groups, supported by the National Science Foundation, deploy very large telescopes to study both the cosmology of clusters via the SZ effect and fine-scale temperature anisotropies, and will likely propose follow-up polarimeters.

4.1. The Next Satellite Experiment

The reach of the next satellite experiment, termed CMBPOL as defined by the three-agency task force in the United States (36) and termed BPOL in Europe, is to detect the signal from gravity waves limited only by astrophysical foregrounds. Examining **Figure 6**, we see that $r = 0.01$, and possibly lower values, can be reached. We should know a great deal from the suborbital experiments well before the 2018 target launch date. For studying polarization at large scales, where foregrounds pose their greatest challenge, information from WMAP and Planck will be the most valuable.

5. CONCLUDING REMARKS

There is considerable promise for new, important discoveries from the CMB, ones that can take us back to when the Universe's temperature was between the Grand Unified Theory and Planck energy scales. This is particle physics, and while we hope accelerators will provide crucial evidence for, for example, the particle nature of dark matter, exploring these scales seems out of their reach.

In some ways, cosmology has followed the path of particle physics: It has its Standard Model, accounting for all confirmed phenomena. With no compelling theory, parameter values are not of crucial interest. We cannot predict the mass of the top quark, nor can we predict the primordial energy densities. Each discipline is checking consistency, as any discrepancies would be a hint of new physics.

The CMB field is not as mature as particle physics. It needs considerable detector development, even for current experiments. There is rapid progress, and overall sensitivity continues to increase. Foregrounds are certainly not sufficiently known or characterized. There is a great deal of competition in the CMB, like the early days of particle physics before the experiments grew so large that more than one or two teams exploring the same topic worldwide was too costly. For the moment, this is good, as each team brings something unique in terms of control of systematics, frequencies, regions of the sky scanned, and detection technology. However, there is a difference in the way results are reported in the two fields. In the CMB field, typically almost nothing is said about an experiment between when it is funded and when it publishes. Here publishing means that results are announced, multiple papers are submitted and circulated, and often there is a full data release, including not only of raw data and intermediate data products but sometimes support for others to repeat or extend the analyses. The positives of this tradition are obvious. However, one negative is that one does not learn the problems an experiment is facing in a timely manner. There is a degree of secrecy among CMB scientists.

There are other differences. CMB teams frequently engage theorists to perform the final analysis that yields the cosmological significance of the data. Sophisticated analysis techniques are being developed by a set of scientists and their students who do not work with detectors but do generate a growing literature. There are as yet no standardized analysis techniques; effectively each new experiment invents its own. The days appear to be over where the group of scientists that design, build, and operate an experiment can, by themselves, do the full scientific analysis. Another distinction is that there is no one body looking over the field or advising the funding agencies, and private funds sometimes have a major impact. Nearly all CMB scientists are working on multiple projects, sometimes as many as four or five, holding that many grants. More time is spent writing proposals and reports and arranging support for junior scientists, for whom there is little funding outside of project funds. This is certainly not the optimum way to fund such an exciting and promising field.

DISCLOSURE STATEMENT

The authors have a CMB polarization experiment of their own.

ACKNOWLEDGMENTS

The authors have enjoyed many productive conversations with colleagues at their institutions in the course of preparing this review. The authors are particularly grateful for helpful comments from Norman Jarosik, Bernd Klein, Laura La Porta, and Kendrick Smith. We also wish to acknowledge our collaborators in QUIET and CAPMAP for frequent discussions of relevant issues. This work was partially supported by grants from the National Science Foundation: PHY-0355328, PHY-0551142, and ASTR/AST-0506648.

LITERATURE CITED

1. Kamionkowski M, Kosowsky A. *Annu. Rev. Nucl. Part. Sci.* 49:77 (1999)
2. Dodelson S. *Modern Cosmology*. Acad. Press (2003)
3. Penzias AA, Wilson RW. *Astrophys. J.* 142:419 (1965)
4. Partridge RB. *3K: The Cosmic Microwave Background Radiation*. Cambridge, NY: Cambridge Univ. Press (1995)
5. Fixsen DJ, et al. *Astrophys. J.* 483:586 (1996)

6. Feng JL, Rajaraman A, Takayama F. *Phys. Rev. D* 68:063504 (2003)
7. Smoot GF, et al. *Astrophys. J.* 396:1 (1992)
8. Hinshaw G, et al. *Astrophys. J.* In press (2007)
9. Bock J, et al. astro-ph/0604101 (2006)
10. Smith TL, Kamionkowski M, Cooray A. *Phys. Rev. D* 73(2):023504 (2006)
11. Spergel DN, et al. *Astrophys. J.* In press (2007)
12. Page L, et al. *Astrophys. J. Suppl.* 148:233 (2003)
13. Readhead ACS, et al. *Science* 306:836 (2004)
14. MacTavish CJ, et al. *Astrophys. J.* 647:799 (2006)
15. Tegmark M, et al. *Phys. Rev. D* 69:103501 (2004)
16. Hu W, Sugiyama N. *Phys. Rev. D* 51:2599 (1995)
17. Eisenstein DJ, et al. *Astrophys. J.* 633:560 (2005)
18. Bennett CL, et al. *Astrophys. J. Suppl.* 148:97 (2003)
19. Page L, et al. *Astrophys. J.* In press (2007)
20. La Porta L, Burigana C, Reich W, Reich P. *Astron. Astrophys.* 455(2):L9 (2006)
21. Carretti E, Bernardi G, Cortiglioni S. *MNRAS* 373:93 (2006)
22. Hanany S, Rosenkranz P. *New Astron. Rev.* 47(11–12):1159 (2003)
23. Pietranera L, et al. *MNRAS.* In press (2007)
24. Finkbeiner DP, Davis M, Schlegel DJ. *Astrophys. J.* 524:867 (1999)
25. Vaillancourt JE. *Eur. Astron. Soc. Publ. Ser.* 23:147 (2007)
26. Benoit A, et al. *Astron. Astrophys.* 424:571 (2004)
27. Ponthieu N, et al. *Astron. Astrophys.* 444:327 (2005)
28. Huffenberger KM, Eriksen HK, Hansen FK. *Astrophys. J.* 651:81 (2006)
29. Stivoli F, Baccigalupi C, Maino D, Stompor R. *MNRAS* 372:615 (2006)
30. Verde L, Peiris HV, Jimenez R. *J. Cosmol. Astropart. Phys.* 1:19 (2006)
31. Barkats D, et al. *Astrophys. J. Suppl.* 159(1):1 (2005)
32. Ryle M. *Proc. R. Soc. A* 211:351 (1952)
33. Zmuidzinas J. *Appl. Optics* 42:4989 (2003)
34. Mather JC. *Appl. Optics* 21:1125 (1982)
35. Tegmark M. *Phys. Rev. D* 56:4514 (1997)
36. Smith K. *New Astron. Rev.* 50(11–12):1025 (2006)
37. Crawford T. astro-ph/0702608 (2007)
38. Couchout F, et al. CMB and physics of the early universe. **http://pos.sissa.it/cgi-bin/reader/conf.cgi?confid=27** (2006)
39. Bond JR, Jaffe AH, Knox L. *Astrophys. J.* 533:19 (2000)
40. Hivon E, et al. *Astrophys. J.* 567:2 (2002)

Cosmic-Ray Propagation and Interactions in the Galaxy

Andrew W. Strong,[1] Igor V. Moskalenko,[2] and Vladimir S. Ptuskin[3]

[1]Max-Planck-Institut für extraterrestrische Physik, 85741 Garching, Germany; email: aws@mpe.mpg.de

[2]Hansen Experimental Physics Laboratory and Kavli Institute for Particle Astrophysics and Cosmology, Stanford University, Stanford, California 94305; email: imos@stanford.edu

[3]Institute for Terrestrial Magnetism, Ionosphere and Radiowave Propagation of the Russian Academy of Sciences (IZMIRAN), Troitsk, Moscow region 142190, Russia; email: vptuskin@izmiran.ru

Annu. Rev. Nucl. Part. Sci. 2007. 57:285–327

First published online as a Review in Advance on June 18, 2007

The *Annual Review of Nuclear and Particle Science* is online at http://nucl.annualreviews.org

This article's doi: 10.1146/annurev.nucl.57.090506.123011

Copyright © 2007 by Annual Reviews. All rights reserved

0163-8998/07/1123-0285$20.00

Key Words

energetic particles, gamma rays, interstellar medium, magnetic fields, plasmas

Abstract

We survey the theory and experimental tests for the propagation of cosmic rays in the Galaxy up to energies of 10^{15} eV. A guide to the previous reviews and essential literature is given, followed by an exposition of basic principles. The basic ideas of cosmic-ray propagation are described, and the physical origin of its processes is explained. The various techniques for computing the observational consequences of the theory are described and contrasted. These include analytical and numerical techniques. We present the comparison of models with data, including direct and indirect—especially γ-ray—observations, and indicate what we can learn about cosmic-ray propagation. Some important topics, including electron and antiparticle propagation, are chosen for discussion.

Contents

1. INTRODUCTION 286
2. COSMIC-RAY PROPAGATION: THEORY 289
    2.1. Basics and Approaches 289
    2.2. Propagation Equation 291
    2.3. Diffusion 294
    2.4. Convection 296
    2.5. Reacceleration 297
    2.6. Galactic Structure 298
    2.7. Interactions 299
    2.8. Weighted Slabs and Leaky Boxes 299
    2.9. Explicit Models 300
    2.10. GALPROP 301
    2.11. Numerical versus Analytical 303
    2.12. Self-Consistent Models 303
3. CONFRONTATION OF THEORY WITH DATA 304
    3.1. Stable Secondary-to-Primary Ratios 304
    3.2. Unstable Secondary-to-Primary Ratios: Radioactive Clocks 309
    3.3. K-Capture Isotopes and Acceleration Delay 310
    3.4. K-Capture Isotopes and Reacceleration 310
    3.5. Anisotropy 311
    3.6. Diffuse Galactic Gamma Rays 312
    3.7. Antiprotons and Positrons 317
    3.8. Electrons and Synchrotron Radiation 319
    3.9. Time-Dependent and Space-Dependent Effects 320

1. INTRODUCTION

CRs: cosmic rays

Cosmic rays (CRs) are almost unique in astrophysics in that they can be sampled directly, not just observed via electromagnetic radiation. Other examples are meteorites and stardust. CRs provide us with a detailed elemental and isotopic sample of the current (few million years old) interstellar medium (ISM) that is not available in any other way. This makes the subject especially rich and complementary to other disciplines.

CRs have been featured in approximately 15 articles from 1952 to 1989 in the Annual Reviews series. These reviews address heavy nuclei (1), collective transport effects (2), composition (3), and propagation (4). Cox's recent review, *The Three-Phase Interstellar Medium* (5), contains much discussion of CRs as one essential component of the ISM, but inevitably no mention of their propagation. Two recent Annual Review articles (6, 7) extensively discuss the relation of CRs to turbulence; thus, we do not try to cover this.

A basic reference is the book *Astrophysics of Cosmic Rays* (8), which expounds all the essential concepts and is an update of the classic, *The Origin of Cosmic Rays* (9), which laid the modern foundations of the subject, with an updated presentation in Reference 10. Good books for basic expositions include References 11 and 12, and for high energies, Reference 13. A basic text emphasizing theory is Reference 14, while the book *Astrophysics of Galactic Cosmic Rays* (15) gives a valuable overview of the experimental data and theoretical ideas as of 2001. The biannual International Cosmic Ray Conference proceedings (many of which can be found at **http://adswww.harvard.edu/**) are also an essential source of information, especially for the latest news on the subject.

SNR: supernova remnant

Recently, a plethora of reviews have appeared on the subject of CRs above 10^{15} eV (e.g., 16–18); on interactions (19); on experiments and astrophysics (20); and more on astrophysics, propagation, and composition (21–23); in addition to a review of models (24). We recommend Reference 25 and the very up to date References 26 and 27. Therefore, this topic has been excluded here. At the lowest-energy end, we note that MeV particles are nonthermal (even if not relativistic) and must be mentioned in a review of CRs, especially because they are important sources of heating and ionization of the ISM (28). As one example of their far-ranging influence, star formation in molecular clouds may be suppressed by CRs produced in supernova remnants (SNRs) nearby (29).

It is worth distinguishing between two ways of approaching CR propagation: either from the particle point of view, including the spectrum and interactions, or treating the CRs as a weightless, collisionless relativistic gas with pressure and energy and considering it alongside other components of the ISM (30, 31). Both ways of looking at the problem are valid up to a point, but for consistency, a unified approach is desirable and, to our knowledge, has never been attempted. The nearest approaches to this are in References 32 and 33. Most papers exclusively address one or the other aspect. The first approach is required for comparison with observations of CRs (direct and indirect), whereas the second is required for the ISM: stability, heating, and so on (5).

The major recent advances in the field are the high-quality measurements of isotopic composition and element spectra, and observations by γ-ray telescopes, both satellite and ground based. Space does not allow discussion of the observational data here, but the figures give an illustrative overview of what is now available from both direct (**Figures 1–13**) and indirect (γ-ray) measurements (**Figures 14–16**). Concerning the origin of CRs, we follow Cesarsky's review: "We will, for the most part, sidestep this problem" (4). Hence, we omit CR sources, including composition and acceleration; for SNRs as CR sources, the literature can be traced back from the most recent High Energy Stereoscopic System (H.E.S.S.) TeV γ-ray results (34). We also omit solar modulation, Galaxy clusters, and extragalactic CRs. We restrict our attention mostly to our own Galaxy, but mention important information coming from external galaxies (via synchrotron radiation).

We first introduce the theoretical background, and then consider the confrontation of theory with observation. A number of particular topics are selected for further discussion.

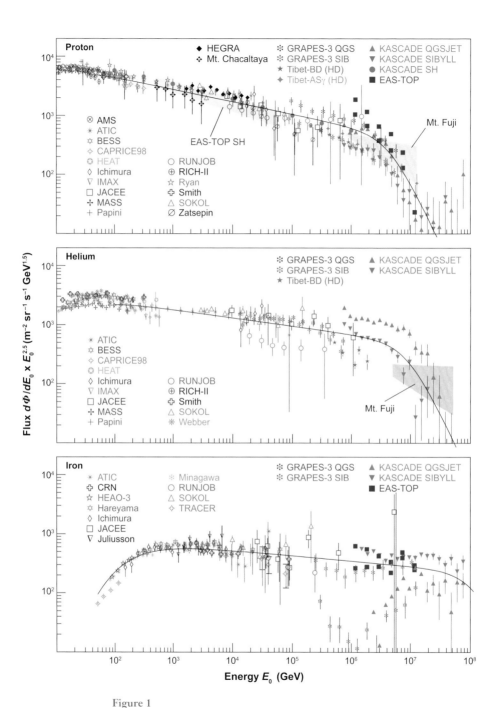

Figure 1

Compilation of spectral data 10^{10}–10^{17} eV for proton, helium, and iron, combining balloon, satellite, and ground-based measurements. Figure adapted with permission from Reference 18 and G. Hörandel (private communication).

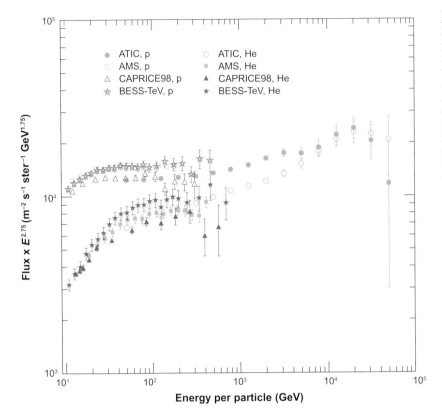

Figure 2

Preliminary spectra of
proton (p) and helium (He)
from ATIC-2, compared
with AMS01, BESS-TeV,
and CAPRICE98. The
ATIC-2 data indicate a
slightly harder spectrum for
He above 1 TeV. Plot taken
from the ATIC
Collaboration (35). Figure
adapted with permission
from Reference 35.

2. COSMIC-RAY PROPAGATION: THEORY

2.1. Basics and Approaches

We present the basic concepts of CR propagation and techniques for relating these
to observational data. Practically all our knowledge of CR propagation comes via
secondary CRs, with additional information from γ-rays and synchrotron radiation.
Note why secondary nuclei in particular are a good probe of CR propagation: The
fact that the primary nuclei are measured (at least locally) means that the secondary
production functions can be computed from primary spectra, cross sections, and interstellar gas densities with reasonable precision. The secondaries can then be propagated and compared with observations.

Since the realization that CRs fill the Galaxy, it has been clear that nuclear interactions imply that their composition contains information on their propagation (44).
A historical event was the arrival of satellite measurements of isotopic Li, Be, and
B in the 1970s (45). Since then the subject has expanded enormously, with models
of increasing degrees of sophistication. The simple observation that the observed
composition of CRs is different from that of solar elements in that rare solar-system
nuclei such as B are abundant in CRs proves the importance of propagation in the

Secondary cosmic rays:
cosmic rays generated by
spallation of primary cosmic
rays on interstellar gas

a

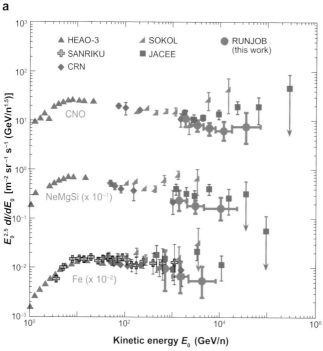

b

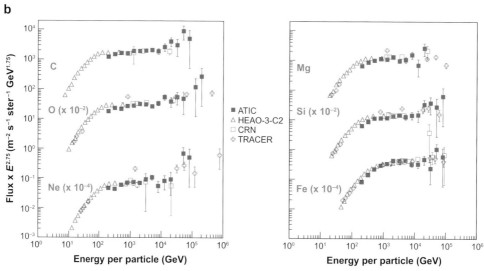

Figure 3

(*a*) Compilation of spectral data for element groups CNO, NeMgSi, and Fe (36) from HEAO-3, SANRIKU, CRN, SOKOL, JACEE, and RUNJOB. Figure adapted with permission from the American Astronomical Society. (*b*) Compilation of separate even-Z elements from preliminary ATIC-2, HEAO-3, CRN, and TRACER. The ATIC-2 data suggest a hardening above 10 TeV. Plot taken from the ATIC Collaboration (35). Figure adapted with permission from Reference 35.

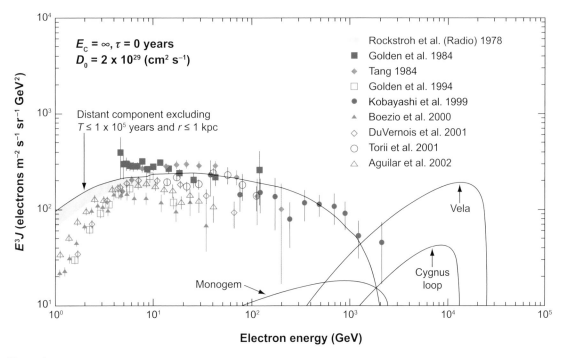

$E_c = \infty$, $\tau = 0$ years
$D_0 = 2 \times 10^{29}$ (cm² s⁻¹)

Rockstroh et al. (Radio) 1978
■ Golden et al. 1984
◆ Tang 1984
□ Golden et al. 1994
● Kobayashi et al. 1999
▲ Boezio et al. 2000
◇ DuVernois et al. 2001
○ Torii et al. 2001
△ Aguilar et al. 2002

Distant component excluding
$T \le 1 \times 10^5$ years and $r \le 1$ kpc

Vela

Monogem

Cygnus
loop

Figure 4

Measurements of the electron spectrum, including AMS01, CAPRICE94, HEAT, and SANRIKU, compared with possible contributions of distant sources and local supernova remnants. Figure adapted from Reference 37 with permission from the American Astronomical Society.

ISM. The canonical "few grams per centimeters squared" of traversed material is one of the widest-known facts of CR physics. At present we believe that the diffusion model with possible inclusion of convection provides the most adequate description of CR transport in the Galaxy at energies below approximately 10^{17} eV. Thus, we begin by presenting this model.

2.2. Propagation Equation

The CR propagation equation for a particular particle species can be written in the general form

$$\frac{\partial \psi(\vec{r}, p, t)}{\partial t} = q(\vec{r}, p, t) + \vec{\nabla} \cdot (D_{xx} \vec{\nabla} \psi - \vec{V} \psi)$$

$$+ \frac{\partial}{\partial p} p^2 D_{pp} \frac{\partial}{\partial p} \frac{1}{p^2} \psi - \frac{\partial}{\partial p} \left[\dot{p} \psi - \frac{p}{3} (\vec{\nabla} \cdot \vec{V}) \psi \right] - \frac{1}{\tau_f} \psi - \frac{1}{\tau_r} \psi, \quad 1.$$

where $\psi(\vec{r}, p, t)$ is the CR density per unit of total particle momentum p at position $\vec{r}$, $\psi(p)dp = 4\pi p^2 f(\vec{p})dp$ in terms of phase-space density $f(\vec{p})$; $q(\vec{r}, p, t)$ is the source term including primary, spallation, and decay contributions; D_{xx} is the spatial

p: particle momentum

Figure 5

Measurements of the
positron spectrum,
including data from
MASS91, AMS01,
CAPRICE94, and HEAT.
Propagation calculations
for interstellar (*dotted
curves*) and modulated
(*solid curves*) spectra are
shown. Red lines,
GALPROP (39); orange
lines, Reference 40. Figure
adapted from Reference
38 with permission from
Astron. Astrophys.

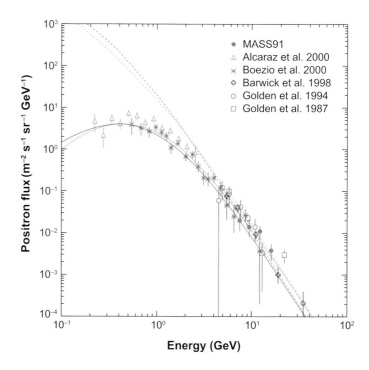

GALPROP: Galactic
Propagation (code)

β: velocity divided by speed
of light

diffusion coefficient; $\vec{V}$ is the convection velocity; diffusive reacceleration is described as diffusion in momentum space and is determined by the coefficient D_{pp}; $\dot{p} \equiv dp/dt$ is the momentum gain or loss rate; τ_f is the timescale for loss by fragmentation; and τ_r is the timescale for radioactive decay.

CR sources are usually assumed to be concentrated near the galactic disk and to have a radial distribution similar to, for example, SNRs. A source injection spectrum and its isotopic composition are required. Usually composition is initially based on solar photospheric elements, but can be determined iteratively from the CR data for later comparison with solar composition. The spallation part of $q(\vec{r}, p, t)$ depends on all progenitor species and their energy-dependent cross sections, and the gas density $n(\vec{r})$. In general, it is assumed that the spallation products have the same kinetic energy per nucleon as the progenitor. K-electron capture and electron stripping can be included via τ_f and q. D_{xx} is, in general, a function of $(\vec{r}, \beta, p/Z)$, where $\beta = v/c$, Z is the charge, and p/Z determines the gyroradius in a given magnetic field. D_{xx} may be isotropic, or more realistically anisotropic, and may be influenced by the CRs (e.g., in wave-damping models). D_{pp} is related to D_{xx} by $D_{pp}D_{xx} \propto p^2$, with the proportionality constant dependent on the theory of stochastic reacceleration (8, 46), as described in Section 2.5. $\vec{V}$ is a function of $\vec{r}$ and depends on the nature of the galactic wind. The term in $\vec{\nabla} \cdot \vec{V}$ represents adiabatic momentum gain or loss in the nonuniform flow of gas, with a frozen-in magnetic field whose inhomogeneities scatter the CRs. τ_f depends on the total spallation cross section and $n(\vec{r})$. $n(\vec{r})$ can be based on surveys of atomic and molecular gas, but can also incorporate small-scale

a

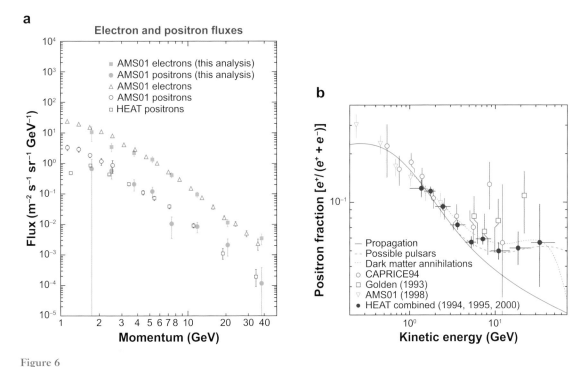

Figure 6

(*a*) Measurements of the electron and positron spectrum by AMS01 and HEAT. Figure adapted with permission from Reference 41. (*b*) $e^+/(e^+ + e^-)$ ratio, including HEAT and AMS01 (42). For the ratio, a propagation calculation (*solid line*) (39), possible contributions from pulsars (*dashed line*), and dark matter annihilations (*dotted line*) are shown. Figure adapted with permission from Phys. Rev. Lett. 93:241102 by Beatty et al. Copyright (2004) by the American Physical Society. **http://scitation.aip.org/getabs/servlet/GetabsServlet?prog= normal&id=PRLTAO000093000024241102000001&idtype=cvips&gifs=yes.**

variations such as the region of low gas density surrounding the Sun. The presence of interstellar helium at approximately 10% of hydrogen by number must be included. Heavier components of the ISM are not important for producing CRs by spallation. This equation treats continuous momentum loss only. Catastrophic losses can be included via τ_f and q. CR electron, positron, and antiproton propagation constitute special cases of this equation, differing only in their energy losses and production rates. The boundary conditions depend on the model. Often, $\psi = 0$ is assumed at the halo boundary, where particles escape into intergalactic space, but this is obviously just an approximation (because the intergalactic flux is not zero) that can be relaxed for models with a physical treatment of the boundary.

Equation 1 is a time-dependent equation. Usually the steady-state solution is required, which can be obtained by either setting $\partial\psi/\partial t = 0$ or following the time dependence until a steady state is reached. The latter procedure is much easier to implement numerically. The time dependence of q is neglected unless effects of

Figure 7

Measurements of the
antiproton spectrum from
BESS, MASS91, and
CAPRICE98, compared
to a propagation
calculation. The dashed
and dotted lines illustrate
the sensitivity of the
calculated (modulated)
antiproton flux to the
normalization of the
diffusion coefficient. LIS,
local interstellar spectrum.
Figure adapted with
permission from
Reference 43.

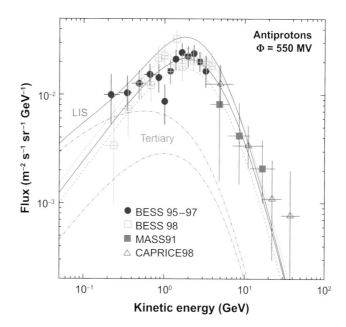

nearby recent sources or the stochastic nature of sources is being studied. By starting with the solution for the heaviest primaries and using this to compute the spallation source for their products, the complete system can be solved, including secondaries, tertiaries, and so on. Then the CR spectra at the solar position can be compared with direct observations, including solar modulation if required.

Source abundances are determined iteratively, comparing propagation calculations with data. For nuclei with very small source abundances, the source values are masked by secondaries and cross-section uncertainties and are therefore difficult to determine. Webber (47) gives a ranking of easy to impossible for the possibility of getting the source abundances using the Advanced Composition Explorer (ACE) data. A recent review of the high-precision abundances from the ACE is in Reference 48, and for Ulysses, in Reference 49. For a useful summary of the various astrophysical abundances relevant to interpreting CR abundances, see Reference 50.

2.3. Diffusion

The concept of CR diffusion explains why energetic charged particles have highly isotropic distributions and why they are well retained in the Galaxy. The galactic magnetic field that tangles the trajectories of particles plays a crucial role in this process. Typical values of the diffusion coefficient found from fitting to CR data are $D_{xx} \sim (3 - 5) \times 10^{28}$ cm^2 s^{-1} at energy $\sim$1 GeV/n, and increase with magnetic rigidity as $R^{0.3} - R^{0.6}$ in different versions of the empirical diffusion model of CR propagation.

On the microscopic level, the diffusion of CRs results from particle scattering on random magnetohydrodynamic (MHD) waves and discontinuities. MHD waves arise

ACE: Advanced
Composition Explorer

B/C: cosmic-ray
boron-to-carbon ratio

in magnetized plasmas in response to perturbations. The effective collision integral for charged energetic particles moving in a magnetic field with small random fluctuations $\delta B \ll B$ can be taken from the standard quasi-linear theory of plasma turbulence (51). The wave-particle interaction is of resonant character, so that an energetic particle is scattered predominantly by those irregularities of magnetic field that have their projection of the wave vector on the average magnetic field direction equal to $k_{11} = \pm s/(r_g \mu)$, where $r_g = pc/ZeB$ is the particle gyroradius and μ is the pitch angle. The integers $s = 0, 1, 2 \ldots$ correspond to cyclotron resonances of different orders. The efficiency of scattering depends on the polarization of the waves and on their distribution in the **k**-space of the wave vectors, where **k** is the wave number. The first-order resonance $s = 1$ is the most important for the isotropic and also for the one-dimensional distribution of random MHD waves along the average magnetic field. In some cases—for calculation of scattering at small μ and of perpendicular diffusion—the broadening of resonances and magnetic mirroring effects should be taken into account. The resulting spatial diffusion is strongly anisotropic locally and goes predominantly along the magnetic field lines. However, strong fluctuations of the magnetic field on large scales $L \sim 100$ pc, where the strength of the random field is several times higher than the average field strength, lead to the isotropization of global CR diffusion in the Galaxy. The rigorous treatment of this effect is not trivial, as the field is almost static and the strictly one-dimensional diffusion along the magnetic field lines does not lead to nonzero diffusion perpendicular to **B** (see Reference 52 and references cited therein).

R: particle magnetic rigidity $= pc/Ze$

Following several detailed reviews of the theory of CR diffusion (8, 14, 53, 54), the diffusion coefficient at $r_g < L$ can be roughly estimated as $D_{xx} \approx (\delta B_{res}/B)^{-2} \, v r_g/3$, where δB_{res} is the amplitude of the random field at the resonant wave number $k_{res} = 1/r_g$. The spectral energy density of interstellar turbulence has a power-law form $w(k)dk \sim k^{-2+a}dk$, $a = 1/3$ over a wide range of wave numbers $1/(10^{20} \text{ cm}) < k < 1/(10^8 \text{ cm})$(see Reference 6), and the strength of the random field at the main scale is $\delta B \approx 5 \ \mu\text{G}$. This gives an estimate of the diffusion coefficient $D_{xx} \approx 2 \times 10^{27} \beta R_{GV}^{1/3}$ cm^2 s^{-1} for all CR particles with magnetic rigidities $R < 10^8$ GV, in fair agreement with the empirical diffusion model (the version with distributed reacceleration). The scaling law $D_{xx} \sim R^{1/3}$ is determined by the value of the exponent $a = 1/3$, typical for a Kolmogorov spectrum. Theoretically, the Kolmogorov-type spectrum may refer only to some part of the MHD turbulence that includes the (Alfvénic) structures strongly elongated along the magnetic field direction and that cannot provide the significant scattering and required diffusion of CRs (55). In parallel, the more isotropic (fast magnetosonic) part of the turbulence, with a smaller value of the random field at the main scale and with the exponent $a = 1/2$ typical for the Kraichnan-type turbulence spectrum, may exist in the ISM (56). The Kraichnan spectrum gives a scaling $D_{xx} \sim R^{1/2}$, which is close to the high-energy asymptotic form of the diffusion coefficient obtained in the plain diffusion version of the empirical propagation model. Thus, the approach based on kinetic theory gives a proper estimate of the diffusion coefficient and predicts a power-law dependence of diffusion on magnetic rigidity. However, the actual diffusion coefficient must be determined with the help of empirical models of CR propagation in the Galaxy.

2.4. Convection

Although the most frequently considered mode of CR transport is diffusion, the existence of galactic winds in many galaxies suggests that convective (or advective) transport may be important. Winds are common in galaxies and can be CR driven (57). CRs play a dynamical role in galactic halos (58, 59). Convection not only transports CRs, but can also produce adiabatic energy losses as the wind speed increases away from the disk. Convection was first considered in Reference 60 and followed up in References 61–65. Both one-zone and two-zone models have been studied: a one-zone model has convection and diffusion everywhere; a two-zone model has diffusion alone up to some distance from the plane, and diffusion plus convection beyond.

Surprisingly, a recent review on galactic winds (66) hardly mentions CRs. In the same volume, Cox's review (5) includes CRs as a basic component. Direct evidence for winds in our own Galaxy seems to be confined to the galactic center region from X-ray images. However, Cox is not sure there is a wind: "A Galactic wind may be occurring, but I do not believe that it carries off a significant fraction of the supernova power from the Solar Neighborhood because it would carry off a similar power in the pervading cosmic rays . . ." (p. 370).

For one-zone diffusion-convection models, a good diagnostic is the energy dependence of the secondary-to-primary ratio. A purely convective transport would have no energy dependence (apart from the velocity dependence of the reaction rate), contrary to what is observed. If the diffusion rate decreases with decreasing energy, any convection will eventually take over and cause the secondary-to-primary ratio to flatten at low energy. This is observed, but convection (proposed in Reference 64 to explain just this effect) does not reproduce, for example, the CR boron-to-carbon ratio (B/C) very well (67). Another test is provided by radioactive isotopes, which effectively constrain the wind speed to <10 km s^{-1} kpc^{-1} for a speed that increases linearly with distance from the disk (67). A value of ≈ 15 km s^{-1} (constant speed wind) is required to fit B/C even in the presence of reacceleration, according to Reference 68, which can be compared to 30 km s^{-1} in the wind model of Reference 69. The latter value implies an energy dependence of the diffusion coefficient that may conflict with CR anisotropy.

Ptuskin et al. (33) studied a self-consistent two-zone model with a wind driven by CRs and thermal gas in a rotating Galaxy. The CR propagation is entirely diffusive in a zone $|z| < 1$ kpc, and diffusive-convective outside. CRs reaching the convective zone do not return, so they act as a halo boundary—with height varying with energy and Galactocentric radius (the distance of a point from the galactic center). It is possible to explain the energy dependence of the secondary-to-primary ratio with this model, and it is also claimed that it is consistent with radioactive isotopes. The effect of a galactic wind on the radial CR gradient has been investigated in Reference 70, in which the authors constructed a self-consistent model with the wind driven by CRs and with anisotropic diffusion. The convective velocities involved in the outer zone are large (100 km s^{-1}), but this model is still consistent with radioactive CR nuclei, which set a much lower limit (67) because this limit is applicable only in the inner zone. Observational support of such models would require direct evidence for a galactic wind in the halo.

2.5. Reacceleration

In addition to spatial diffusion, the scattering of CR particles on randomly moving MHD waves leads to stochastic acceleration, which is described in the transport equation as diffusion in momentum space with some diffusion coefficient D_{pp}. One can estimate it as $D_{pp} = p^2 V_a^2 / (9 D_{xx})$, where the Alfvén velocity V_a is introduced as a characteristic velocity of weak disturbances propagating in a magnetic field (see References 8 and 14 for rigorous formulas).

Distributed acceleration in the entire galactic volume cannot serve as the main mechanism of acceleration of CRs, at least in the energy range $1 - 100$ GeV/n. In this case, the particles of higher energy would spend a longer time in the system, which would result in an increase of the relative abundance of secondary nuclei as energy increases, contrary to observation. This argument does not hold at low energies, where distributed acceleration may be strong, and it may explain the existence of peaks in the ratios of secondary-to-primary nuclei at approximately 1 GeV/n if the distributed acceleration becomes significant at this energy. The process of distributed acceleration in the ISM is also referred to as reacceleration, to distinguish it from the primary acceleration process that occurs in the CR sources. It has been shown (46, 71) that the observed dependence of abundance of secondary nuclei on energy can be explained in the model with reacceleration (*a*) if the CR diffusion coefficient varies as a single power law of rigidity $D_{xx} \sim R^a$, with an exponent $a \sim 0.3$ over the whole energy range (corresponding to particle scattering on MHD turbulence with a Kolmogorov spectrum) and (*b*) if the Alfvén velocity is $V_a \sim 30$ km s^{-1}, which is close to its actual value in the ISM.

In addition to stable secondary nuclei, the secondary K-capture isotopes are useful for the study of possible reacceleration in the ISM (72). The isotopes ^{37}Ar, ^{44}Ti, ^{49}V, and ^{51}Cr and some others decay rapidly by electron capture at low energies, where energetic ions can have an orbital electron. The probability of having an orbital electron depends strongly on energy, and because of this the abundance of these isotopes and of their decay products are strong functions of energy and sensitive to changes of particle energy in the ISM. The first measurements of an energy-dependent decay of ^{49}V and ^{51}Cr in CRs (73) were used to test the rate of distributed interstellar reacceleration (74), but refinement of nuclear production cross sections is required to draw definite conclusions.

The gain of particle energy in the process of reacceleration is accompanied by a corresponding energy loss of the interstellar MHD turbulence. According to calculations (75), the dissipation on CRs may significantly influence the Kraichnan nonlinear cascade of waves at less than 10^{13} cm, and may even terminate the cascade at small scales. This results in a self-consistent change of the rigidity dependence of the diffusion coefficient, with a steep rise of D_{xx} to small rigidities. The scheme explains the high-energy scaling of diffusion $D_{xx} \sim R^{0.5}$ and offers an explanation for the observed energy dependence of secondary-to-primary ratios.

Primary cosmic rays:
cosmic rays accelerated at
sources (for example, in
supernova remnants)

ISRF: interstellar radiation
field

As mentioned above, the data on secondary nuclei provide evidence against strong reacceleration in the entire Galaxy at 1–100 GeV/n. However, the spectra of secondaries can be modified considerably owing to processes in the source regions, with a small total galactic volume filling factor, for the regions where the high-velocity SNR shocks accelerate primary CRs. Two effects could be operating there, and both lead to the production of a component of secondaries with flat energy spectra (76, 77). One effect is the production of secondaries in SNRs by the spallation of primary nuclei that have a flat source energy spectrum close to E^{-2}. Another effect is the direct acceleration by strong SNR shocks of background secondary nuclei residing in the ISM; again the secondaries acquire the flat source energy spectrum. Calculations (77) showed that these effects may produce a flat component of secondary nuclei rising above the standard steep spectrum of secondaries at energies above approximately 100 GeV/n.

2.6. Galactic Structure

Almost all aspects of galactic structure affect CR propagation, but the most important are the gas content for secondary production and the interstellar radiation field (ISRF) and magnetic field for electron energy losses. The magnetic field is clearly also important for diffusion, but the precise absolute magnitude and large-scale structure are less important (at least for CRs below 10^{15} eV) than the turbulence properties.

The distribution of atomic hydrogen is reasonably well known from 21-cm surveys, but the distribution of molecular hydrogen is less well known because it can only be estimated using the CO molecular tracer; the associated scaling factor is hard to determine and may depend on its position in the Galaxy. In fact CR-gas interactions provide one of the best methods for determining the molecular hydrogen content of the Galaxy because of its basic simplicity, as we describe in the section on γ-rays. For more details, refer to Reference 78.

The galactic magnetic field can be determined from pulsar rotation and dispersion measures combined with a model for the distribution of ionized gas. A large-scale field of a few microgauss aligned with spiral arms exists, but there is no general agreement on the details (79). One recent analysis gives a bisymmetric model for the large-scale galactic magnetic field with reversals on arm-interarm boundaries (80). Independent estimates of the strength and distribution of the field can be made by simultaneous analysis of radio synchrotron, CRs, and γ-ray data, and these confirm a value of a few microgauss, increasing toward the inner Galaxy (81). Much effort has gone into constructing magnetic field models to study propagation at energies $>10^{15}$ eV, where the Larmor radius is large enough for the global topology to be important. This is relevant to CR anisotropy and the search for point sources. Because this is excluded from our review, we refer the reader to References 82–84.

The ISRF comes from stars of all types and is processed by absorption and re-emission by interstellar dust. It extends from the far-infrared, through the optical, to the ultraviolet. Computing the ISRF is difficult, but a great deal of new information on the stellar content of the Galaxy and dust is now available to make better models for use in propagation codes (85). The other important radiation field is the cosmic microwave background, with its well-known black-body spectrum.

The local environment around the Sun (86) is also important. For example, the Local Bubble can affect the radioactive nuclei, as described in Section 3.2. Extensive coverage of the local environment including CRs is in Reference 87.

z_h: height of cosmic ray halo in direction perpendicular to galactic plane

2.7. Interactions

This large subject is well covered in the literature. Details of the essential processes with references have been conveniently collected in our series of papers: energy losses of nuclei and electrons (67); bremsstrahlung and synchrotron emission (81); inverse-Compton emission including anisotropic scattering (88); and pion production of γ-rays, electrons, and positrons (39). Pion production was recently studied in great detail using modern particle-physics codes (89, 90), giving spectra harder by 0.05 in the index and γ-ray yields somewhat higher at a few GeV than older treatments. New, more accurate parameterizations will be important for the new generation of CRs and γ-ray experiments. Reference 91 and references therein are useful guides to spallation cross-section measurements and models. A summary of recent advances appear in References 92–94. Accounts of radioactive and K-capture processes can be found in References 73 and 95–98.

2.8. Weighted Slabs and Leaky Boxes

As mentioned at the start of this section, at present we believe that the diffusion model with possible inclusion of convection provides the most adequate description of CR transport in the Galaxy at energies below approximately 10^{17} eV. The closely related leaky-box and weighted-slab formalisms have provided the basis for most literature interpreting CR data.

The leaky-box model uses the simple picture of particles injected by sources distributed uniformly over some volume (box) and escaping from this volume with an escape time independent of position, and a uniform distribution of gas and radiation fields. The leaky-box equations express this mathematically with source and leakage terms. In the leaky-box model, the diffusion and convection terms are approximated by the leakage term, with some characteristic escape time of CRs from the Galaxy. The escape time τ_{esc} may be a function of particle energy (momentum), charge, and mass number if needed, but it does not depend on the spatial coordinates. There are two models where the leaky-box equations can be obtained as a correct approximation to the diffusion model: (a) the model with fast CR diffusion in the Galaxy and particle reflection at the CR halo boundaries with some probability of escape (9), and (b) the flat-halo model ($z_h \ll R$) with thin source and gas disks ($z_{gas} \ll z_h$), which is equivalent formally to the leaky-box model in the case where stable nuclei are considered (10). The nuclear fragmentation is actually determined not by the escape time τ_{esc}, but rather by the escape length in g cm^{-2}: $x = v\rho\tau_{esc}$, where ρ is the average gas density of interstellar gas in a Galaxy with CR distributed over the whole galactic volume, including the halo.

The solution of a system of coupled transport equations for all isotopes involved in the process of nuclear fragmentation is required for studying CR propagation. A powerful method, the weighted-slab technique, which consists of splitting the problem into astrophysical and nuclear parts, was suggested for this problem (9, 99) before the modern computer epoch. The nuclear-fragmentation problem is solved in terms of the slab model, wherein the CR beam is allowed to traverse a thickness x of the interstellar gas and these solutions are integrated over all values of x, weighted with a distribution function $G(x)$ derived from an astrophysical propagation model. In its standard realization (100, 101), the weighted-slab method breaks down for low-energy CRs, where one has strong energy dependence of nuclear cross sections, strong energy losses, and energy-dependent diffusion. Furthermore, if the diffusion coefficient depends on the nuclear species, the method has rather significant errors. After some modification (102), the weighted-slab method becomes rigorous for the important special case of separable dependence of the diffusion coefficient on particle energy (or rigidity) and position, with no convective transport. The modified weighted-slab method was applied to a few simple diffusion models in References 69 and 74. The weighted-slab method can also be applied to the solution of the leaky-box equations. It can easily be shown that the leaky-box model has an exponential distribution of path lengths x (grammage) $G(x) \propto \exp(-x/X)$, with the mean grammage equal to the escape length X.

In a purely empirical approach, one can try to determine the shape of the distribution function $G(x)$ that best fits the data on abundances of stable primary and secondary nuclei (1). The shape of $G(x)$ is close to exponential: $G(x) \propto \exp(-x/X(R, \beta))$, where $\beta =$ velocity of particle/velocity of light, and this justifies the use of the leaky-box model in this case. There are several recent calculations of $G(x)$ (69, 74, 103, 104).

The possible existence of truncation, a deficit at small path lengths (below a few grams per centimeters squared at energies near 1 GeV/n), relative to an exponential path-length distribution, has been discussed for decades (1, 101, 105, 106). The problem was not solved mainly owing to cross-section uncertainties. In a consistent theory of CR diffusion and nuclear fragmentation in the cloudy ISM, the truncation occurs naturally if some fraction of CR sources resides inside dense giant molecular clouds (107).

For radioactive nuclei, the classical approach is to compute the surviving fraction, which is the ratio of the observed abundance to that expected in the case of no decay. Often the result is given in the form of an effective mean gas density, to be compared with the average density in the Galaxy, but this density should not be taken at face value. The surviving fraction can be better related to physical parameters (108). None of these methods can face the complexities of propagation of CR electrons and positrons with their large energy and spatially dependent energy losses.

2.9. Explicit Models

Finally, the mathematical effort required to put the three-dimensional Galaxy into a one-dimensional formalism becomes overwhelming, and it seems better to work in physical space from the beginning. This approach is intuitively simple and easy to interpret. We can call these explicit solutions. The explicit solution approach including secondaries was pioneered in Reference 10, and its application to newer data is described in References 65 and 109, with analytical solutions for two-dimensional diffusion-convection models with a CR source distribution—which, however, had many restrictive approximations to make them tractable (no energy losses, simple gas model). More recently, a two dimensional semi-empirical model that includes energy losses and reacceleration was developed (68, 110). This is a closed-form solution expressed as a Green's function to be integrated over the sources. It incorporates a radial CR source distribution, but the gas model is a simple constant density within the disk. Reference 111 gives an analytical solution for the time-dependent case with a generalized gas distribution but without energy losses. (This shows again the problem of handling both gas and energy losses simultaneously in analytical schemes.)

A myriad sources model (112), which is actually a Green's function method without energy losses, yields similar results to Reference 113 for the diffusion coefficient and halo size. However, applying the no-energy loss case to ACE data is not justified, and authors have pointed out some defects in their formulation (111). A three-dimensional analytical propagation method has been developed (114, 115) with energy loss and reacceleration, going via a path-length distribution, but it cannot handle ionization losses correctly (see section 2.2 in Reference 115). They have no spatial boundaries and rather simplified (exponential) forms for the gas and other distributions. An approach adapted to fine-scale spatial and temporal variations has been described (116). This uses a Green's function without energy losses or a detailed gas model and hence is limited in its application, but is useful for studying the effect of discrete sources.

The most advanced, explicit solution to date is the fully numerical model described in the next section. Even this has limitations in treating some aspects (e.g., when particle trajectories become important at high energies), so one may ask whether a fully Monte Carlo approach (as is commonly used for energies $>10^{15}$ eV) would be better in the future, given increasing computing power. This would allow effects such as field-line diffusion (important for propagation perpendicular to the galactic plane) to be explicitly included. However, it is still challenging. A particle at GeV energies diffusing with a mean free path of 1 pc in a Galaxy with kpc-size halo height takes $\approx 10^6$ scatterings to leave the Galaxy, which even now would need supercomputers to obtain adequate statistics. Hence, we expect the numerical solution of the propagation equations to remain an important approach for the foreseeable future.

2.10. GALPROP

The Galactic Propagation (GALPROP) code (67) was created with the following aims: (*a*) to enable simultaneous predictions of all relevant observations, including CR nuclei, electrons, and positrons; γ-rays; and synchrotron radiation; (*b*) to overcome the limitations of analytical and semi-analytical methods, taking advantage of advances in computing power as CR, γ-ray, and other data become more accurate; (*c*) to incorporate current information on galactic structure and source distributions; and (*d*) to provide a publicly available code as a basis for further expansion. The first

GLAST: Gamma-Ray
Large Area Space Telescope

aim is the most important, the idea being that all data relate to the same system, the Galaxy, and one cannot, for example, allow a model that fits secondary-to-primary ratios while not fitting γ-rays or while being incompatible with the known interstellar gas distribution. There are many simultaneous constraints, and to find one model that satisfies all of them is a challenge that has yet to be met. Upcoming missions should help achieve this goal. GALPROP has been adopted as the standard for diffuse galactic γ-ray emission for NASA's Gamma-Ray Large Area Space Telescope (GLAST) and is also made use of by the AMS, ACE, HEAT, and Pamela collaborations.

We give a very brief summary of GALPROP; for details we refer the reader to the relevant papers (39, 67, 75, 81, 113, 117) and a dedicated Web site (**http://galprop.stanford.edu**). The propagation (Equation 1) is solved numerically on a spatial grid, either in two dimensions with cylindrical symmetry in the Galaxy or in full three dimensions. The boundaries of the model in radius and height, and the grid spacing, are user definable. In addition, there is a grid in momentum; momentum (not, for example, kinetic energy) is used because it is the natural quantity for propagation in Equation 1. Parameters for all processes in Equation 1 can be controlled on input. The distribution of CR sources can be specified, typically to represent SNRs. Source spectral shape and isotopic composition (relative to protons) are input parameters. Interstellar gas distributions are based on current HI (21-cm atomic hydrogen emission) and CO (molecular emission used to trace molecular hydrogen) surveys, and the ISRF is based on a detailed calculation. Cross sections are based on extensive compilations and parameterizations (92). The numerical solution proceeds in time until a steady state is reached; a time-dependent solution is also an option. Starting with the heaviest primary nucleus considered (for example, ^{64}Ni), the propagation solution is used to compute the source term for its spallation products, which are then propagated in turn and so on down to protons, secondary electrons and positrons, and antiprotons. In this way, secondaries, tertiaries, and so on are included. (Production of ^{10}B via the ^{10}Be-decay channel is important and requires a second iteration of this procedure.) GALPROP includes K-capture and electron stripping processes, where a nucleus with an electron (H-like) is considered a separate species because of the difference in lifetime. Because H-like atoms have only one K-shell electron, the K-capture decay half life must be increased by a factor of two compared with the measured half-life value. Primary electrons are treated separately. Normalization of protons, helium, and electrons to experimental data is provided (all other isotopes are determined by the source composition and propagation). Gamma rays and synchrotron emission are computed using interstellar gas data (for pion decay and bremsstrahlung), the ISRF model (for inverse Compton), and the magnetic field model. Spectra of all species on the chosen grid and the γ-ray and synchrotron sky maps are output in a standard astronomical format for comparison with data. Recent extensions to GALPROP include nonlinear wave damping (75) and a dark matter package.

The computing resources required by GALPROP are moderate by today's standards. Although GALPROP has the ambitious goal of being realistic, it is obvious that any such model can be only a crude approximation to reality. Some known limitations are that it applies only to energies below 10^{15} eV (there are no trajectory calculations), it is limited to scales >10 pc (no clumpy ISM; limited by computer power), and the magnetic field is treated as random for synchrotron emission whereas the regular component affects the structure of radio emission. For the cases where these limitations apply, other techniques may be more appropriate and they provide a goal for future developments of GALPROP.

2.11. Numerical versus Analytical

The following expresses the authors' opinion on this matter. The analytical approaches claim to have various advantages as follows:

1. Physical insight: Of course analytical solutions for simple cases give insight into the relations between the quantities involved, and are useful for rough estimates. In fact, the analytical formulae may become so complicated that no insight is gained. In contrast, the numerical models are very intuitive because they explicitly generate the CR distribution over the Galaxy for all species.
2. Equivalence to full solution of propagation equation: This is true only under restrictive conditions, especially involving energy losses and spatially varying densities. Electrons and positrons are beyond analytical methods because their energy losses are spatially dependent and different processes are important in different energy ranges, whereas these particles are an essential component of the CRs.
3. Faster, easier to compute: With today's computers, the issue of speed has become irrelevant, and the implementation of a numerical model is no more difficult than the complicated integrals over Bessel functions and so on.

In summary, we can do no better than a quotation from a paper of 26 years ago:

> It is unclear whether one would wish to go much beyond the generalizations discussed above for an analytically soluble diffusion model. The added insight from any analytic solution over a purely numerical approach is quickly cancelled by the growing complexity of the formulae. With rapidly developing computational capabilities, one could profitably employ numerical solutions....
>
> J.M. Wallace: *Galactic Cosmic-Ray Diffusion with Arbitrary Radial Distributions* (118)

For CR air-shower calculations, analytical methods gave way to numerical ones at least 40 years ago.

2.12. Self-Consistent Models

A few attempts at a self-consistent description of CRs in the Galaxy have been made by including them as a relativistic gas as one component of ISM dynamics. This is obviously much more difficult than the phenomenological models described above, which treat propagation in a prescribed environment. In References 32 and 33, three-dimensional models of the magnetized ISM with a CR-driven wind were made. It is claimed that these models are also consistent with CR secondary-to-primary ratios.

Such a wind has been put forward as a possible explanation of the CR-gradient problem (70). The Parker instability has been reanalyzed recently (119) using anisotropic diffusion (120) and was followed by a CR-driven galactic dynamo model (121), which used an extension of the ZEUS–three-dimensional MHD code (30) including CR propagation and sources. CR propagation in a magnetic field produced by dynamo action of a turbulent flow (31) presents the whole subject from a novel viewpoint. The extension of such approaches to include CR spectra, secondaries, γ-rays, and so on—which would provide a complete set of comparisons with observations—would be very desirable but has not yet been attempted. Another kind of self-consistency is to include the effect of CRs on the diffusion coefficient (75), as described in more detail in Section 2.5.

3. CONFRONTATION OF THEORY WITH DATA

3.1. Stable Secondary-to-Primary Ratios

The reference ratio is almost always B/C because B is entirely secondary. The measurements are better than for other ratios and are available up to 100 GeV. Because C, N, and O are the major progenitors of B, the production cross sections are better known than, for example, in the cases of Be and Li (94, 122).

The usual procedure is to use a leaky-box or weighted-slab formalism with the empirical rigidity dependences $X(R) = (\beta/\beta_0)X_0$ and $(\beta/\beta_0)(R/R_0)^{-\alpha}X_0$ for $R < R_0$ and $R > R_0$, respectively. The break at R_0 is required because B/C is observed to decrease to low energies faster than the β dependence (which just describes the velocity effect on the reaction rate). The source composition depends on the form and parameters of $X(R)$, and vice versa, (because, for example, B is produced by C, N, and O, etc.), so the procedure is iterative, starting from a solar-like composition.

A typical parameter set fitting the data (69) includes $\alpha = 0.54$, $X_0 = 11.8\,\mathrm{g\,cm^{-2}}$, and $R_0 = 4.9$ GV/c, with a source spectrum rigidity index of -2.35. In principle, all other secondary-to-primary ratios should be consistent with the same parameter set. This is generally found to be the case. As a state-of-the-art application of the weighted-slab technique from Reference 102, we again refer to Reference 69. This is applicable to stable nuclei only, but includes energy losses and gains subject to the limitations described in Section 2.8. They apply the method to one-dimensional disk-halo diffusion, convection, turbulent diffusion, and reacceleration models cast in weighted-slab form. **Figure 8** shows B/C and (Sc+Ti+V)/Fe, referred to as sub-Fe/Fe in Reference 69. Clearly, the models cannot be distinguished on the basis of these types of data alone, and they all provide an adequate fit. This shows the

Figure 8

(*a*) Boron-to-carbon ratio (B/C) and (*b*) sub-Fe/Fe data compilation, compared with four models treated by the modified weighted-slab technique. Φ is the cosmic-ray modulation potential in MV. Figure adapted from Reference 69 with permission from the American Astronomical Society.

a

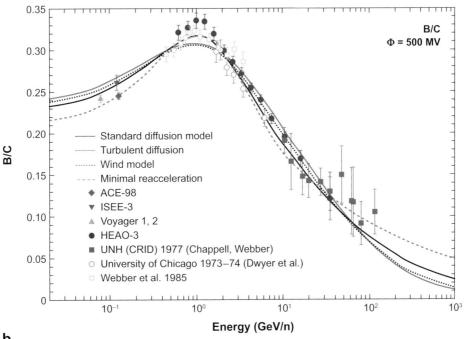

b

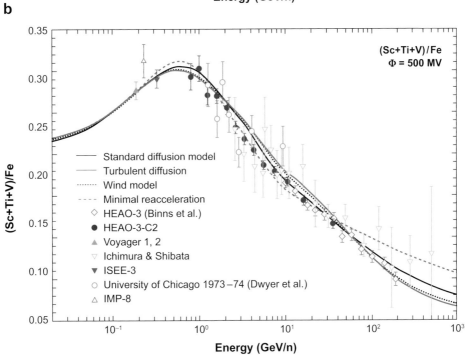

importance of using other species as well as the ones used here. The models can be used to obtain the injection spectrum of primaries, and they find an index of 2.3–2.4 for C and Fe in the energy range 0.5–100 TeV, with the propagated spectrum and data shown in **Figure 9**.

It has been claimed that no break in $X(R)$ is required to fit Voyager 2 outer-heliosphere B/C, N/O, and sub-Fe/Fe data extended to 1.5 GeV, plus High Energy Astrophysical Observatory-3 (HEAO-3) data at higher energies, data, and adopting suitable modulation levels (123). Voyager 2 provides a unique data set because of the lower solar modulation in the outer heliosphere.

We consider now explicit models in the sense of Section 2.9. The same procedure is adopted, with $D_{xx}(R)$ replacing $X(R)$. Again, an ad hoc break in $D_{xx}(R)$ is required in the absence of other mechanisms. Because of the unphysical nature of such a $D_{xx}(R)$, there have been many attempts to find a better explanation, including convection, reacceleration/wave damping, and local sources.

Convection implies an energy-independent escape from the Galaxy so that it dominates at low energies as the diffusion rate decreases, reducing to a simple low-energy asymptotic $X(R) \propto \beta$. However, this does not resemble the observed B/C energy dependence, as it is too monotonic (67). Furthermore, quite severe limits on the convection velocity come from unstable nuclei.

Reacceleration affects the secondary-to-primary ratio, as described in Section 2.5. Many papers have shown how B/C and other ratios can be reproduced with reacceleration at a plausible level, with no ad hoc break in the diffusion coefficient. An example is shown in **Figure 8**. In Reference 94 this model is applied to recent ACE (Li, Be, B, C) data. Because reacceleration must be present at some level if diffusion occurs on moving scatterers (e.g., Alfvén waves), this mechanism is favored but not proven. Direct evidence for reacceleration could come from certain K-capture nuclei (Section 3.4). Note that reacceleration requires a smaller value of α, typically 0.3–0.4, consistent with Kolmogorov turbulence, which helps solve the problems with anisotropy (Section 3.5).

Closely related to reacceleration is wave damping, as described in Section 2.5. This can satisfactorily reproduce B/C, protons, and antiprotons, as shown in **Figure 10**, and also other data (75). The result of this process is a very sharp rise of the diffusion coefficient at rigidities less than approximately 1.5 GV. The Kolmogorov-type dependence is not very successful in this scheme, whereas a Kraichnan-type dependence works better with a high-rigidity asymptotic $D \sim R^{0.5}$ (panel a in **Figure 10**).

Quite a different set of parameters has been proposed (68): $\alpha = 0.7–0.9$ and injection index ≈ 2.0, on the basis of fitting many species simultaneously, and it is suggested to produce the B/C low-energy decrease from convection. Such a large α would give problems for the anisotropy (Section 3.5). A related analysis (110) claims

---

Figure 9

Data compilation for spectra of (*a*) C and (*b*) Fe, compared with four models treated by the modified weighted-slab technique. Φ is the cosmic-ray modulation potential in MV. Figure adapted from Reference 69 with permission from the American Astronomical Society.

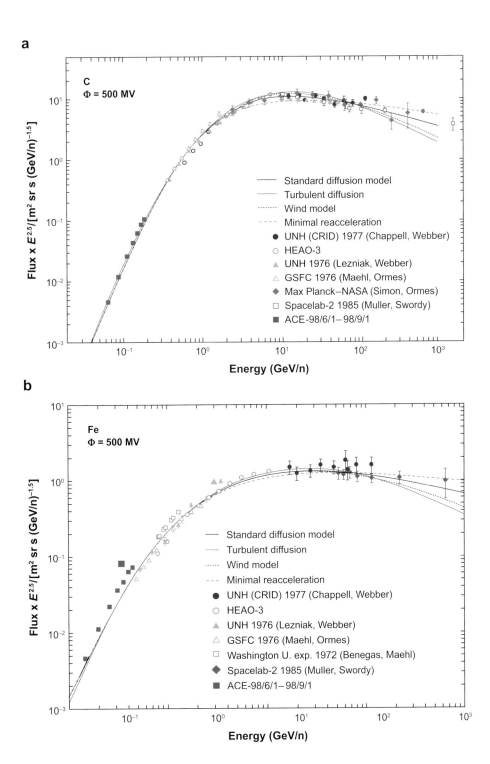

a

C
$\Phi = 500$ MV

Flux × $E^{2.5}$/[m^2 sr s (GeV/n)$^{-1.5}$]

—— Standard diffusion model
—— Turbulent diffusion
······ Wind model
- - - Minimal reacceleration
● UNH (CRID) 1977 (Chappell, Webber)
○ HEAO-3
▲ UNH 1976 (Lezniak, Webber)
△ GSFC 1976 (Maehl, Ormes)
◆ Max Planck–NASA (Simon, Ormes)
□ Spacelab-2 1985 (Muller, Swordy)
■ ACE-98/6/1– 98/9/1

Energy (GeV/n)

b

Fe
$\Phi = 500$ MV

Flux × $E^{2.5}$/[m^2 sr s (GeV/n)$^{-1.5}$]

—— Standard diffusion model
—— Turbulent diffusion
······ Wind model
- - - Minimal reacceleration
● UNH (CRID) 1977 (Chappell, Webber)
○ HEAO-3
▲ UNH 1976 (Lezniak, Webber)
△ GSFC 1976 (Maehl, Ormes)
□ Washington U. exp. 1972 (Benegas, Maehl)
◆ Spacelab-2 1985 (Muller, Swordy)
■ ACE-98/6/1– 98/9/1

Energy (GeV/n)

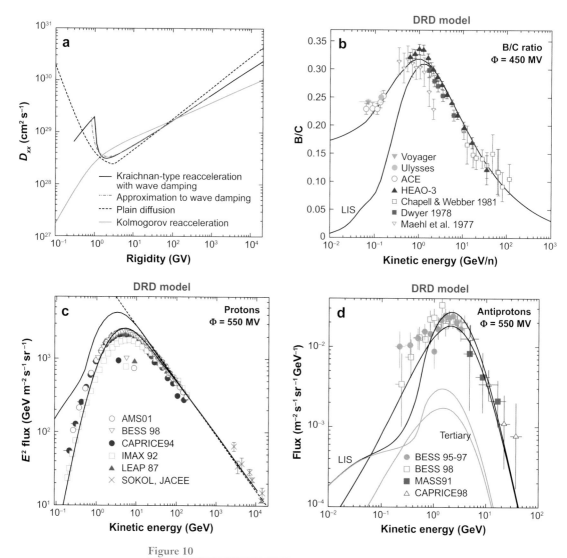

Figure 10

(*a*) Diffusion coefficient in different models (75): plain diffusion (*dashed black line*), Kolmogorov reacceleration (*solid gray line*), and Kraichnan-type reacceleration with wave damping (*solid black line*). (*b*) Boron-to-carbon ratio (B/C). (*c*) Protons. (*d*) Antiprotons in the wave-damping (DRD) model (75). LIS, local interstellar spectrum. Figure adapted with permission from the American Astronomical Society.

to exclude the Kolmogorov-type spectrum. An alternative explanation for the fall off in B/C at low energies invokes a weakly nonlinear (in contrast to quasi-linear) transport theory of CRs in turbulent galactic magnetic fields (124, 125).

Another, simpler, explanation of the B/C energy dependence is the local-source model (104, 126) in which part of the primary CR has an additional local component.

Because the secondary flux must come from the Galaxy at large (the local secondaries being negligible), a steep local primary source will cause B/C to decrease at low energies. The known existence of the Local Bubble containing the Sun, and its probable origin in a few supernovae in the last few million years, makes this plausible, but hard to prove. However, it might be possible if CR composition at low energies were found to have anomalies, indicating a younger age compared to high-energy CR. Davis et al. (104) claim that if B/C is fitted in such a model, then sub-Fe/Fe cannot be fitted by the same model. However, an acceptable fit to this and other data is found in Reference 126 using a diffusion model for the large-scale component.

3.2. Unstable Secondary-to-Primary Ratios: Radioactive Clocks

The five unstable secondary nuclei that live long enough to be useful probes of CR propagation are ^{14}C, ^{10}Be, ^{26}Al, ^{36}Cl, and ^{54}Mn, with properties summarized in References 101, 126, and 127. ^{10}Be is the longest lived and best measured. The theory is presented in Section 2.2. On the basis of these isotopes and updated cross sections (128), the halo height $z_h = 4$–6 kpc, consistent with earlier estimates of 3–7 kpc (98) and 4–12 kpc (67). **Figure 11** compares ^{10}Be/^{9}Be with models, where the ISOMAX ^{10}Be measurements (129) up to 2 GeV (and hence longer decay lifetime) are consistent with the fit to the other data, although the statistics are not very constraining.

The data are often interpreted in terms of the leaky-box model, but this is misleading (108, 127, 131). For the formulae and the detailed procedure for the leaky-box model interpretation, see Reference 132. Luckily, the leaky-box-model surviving fraction can be converted to physically meaningful quantities (131) for a given model. For example, in a simple diffusive halo model, the surviving fraction determines the diffusion coefficient, which can be combined with stable secondary-to-primary ratios to derive the halo size. Typical results are $D_{xx} = (3-5) \times 10^{28}$ cm^2 s^{-1} (at 3 GV) and $z_h = 4$ kpc. We can then compare the leaky-box model's escape time of $\approx 10^7$ yr with the actual time for CRs to reach the halo boundary after leaving their sources, the

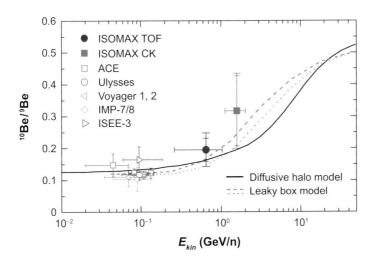

Figure 11
Data on energy-dependence of ^{10}Be/^{9}Be, including ISOMAX, ACE, Ulysses, Voyager, IMP, and ISEE-3 data. The solid black line is a diffusive halo model with 4-kpc scale height using GALPROP (98). The gray lines are leaky-box models (130). Figure adapted from Reference 129 with permission from the American Astronomical Society.

latter being typically an order of magnitude larger. The leaky-box model gas density is typically 0.3 cm^{-3}, compared to the actual average density of 0.03 cm^{-3} for a 4-kpc halo height, again an order-of-magnitude difference.

Because radioactive secondaries travel only a few hundred parsecs before decaying, it is sometimes believed (133) that they cannot give information on the propagation region of CRs. This is somewhat misleading because it is precisely the combination of stable and radioactive data that allows this. The radioactives determine the diffusion coefficient, which then allows the size of the full propagation region to be determined from the stable secondary-to-primary ratio (where the CRs do come from the entire containment region). This of course assumes the diffusion coefficient does not vary greatly from the local region to the full volume.

The effect of the Local Bubble surrounding the Sun on the interpretation of radioactive nuclei was pointed out in Reference 127. The flux of unstable secondaries is reduced because they are not produced in a gas-depleted region around the Sun. Because they will decay before reaching us, this could lead to an overestimate of the halo size if interpreted in a simple diffusive halo model. This effect would be reduced if the diffusion coefficient in the local region were larger than the large-scale value. In fact, the situation is even more complex. According to References 86, 87, and 134, the Sun left the Local Bubble approximately 10^5 years ago after spending several million years inside, and we now live in the collection of local interstellar clouds, with HI density approximately 0.2 cm^{-3} and 35 pc extent. This aspect of the problem for CR propagation has not yet been addressed.

3.3. K-Capture Isotopes and Acceleration Delay

Three isotopes produced in explosive nucleosynthesis, ^{59}Ni (7.6×10^4 y), ^{57}Co (0.74 y), ^{56}Ni (6 d), decay only by K-capture. If acceleration occurs before decay, the decay will be suppressed because the nuclei are stripped. ^{56}Ni is absent as expected, but the other two nuclei are more interesting. Wiedenbeck et al. (135) used ACE data on these nuclei to show that the delay between synthesis and acceleration is long compared with the ^{59}Ni decay time, unless significant ^{59}Co is synthesized in supernovae. Considering theoretical ^{59}Co yields, they conclude on a delay of $\geq 10^5$ years. This is inconsistent with models in which supernovae accelerate their own ejecta, but consistent with acceleration of existing interstellar material. However, the possibility of in-flight electron attachment complicates the analysis (135). For further discussion, see Reference 133. A result from TIGER on Co/Ni at 1–5 GeV/n (136) supports the acceleration delay at higher energies as well.

3.4. K-Capture Isotopes and Reacceleration

There have been analyses using ACE data for ^{49}V, ^{51}V, ^{51}Cr, ^{52}Cr, ^{49}Ti, and other nuclei (73, 74, 137). Niebur et al. (137) used ACE data on ^{37}Ar, ^{44}Ti, ^{49}V, ^{51}Cr, ^{55}Fe, and ^{57}Co, with inconclusive results. They found that although ^{51}V/^{52}Cr was in better agreement with reacceleration models, ^{49}Ti/$^{46+47+48}$Ti gave the opposite result. Jones et al. (74) found that although V/Cr ratios are in slightly better agreement with models that include reacceleration, ratios involving Ti were inconclusive. The main problem is the accuracy of the fragmentation cross sections (126). Discussion of ACE and previous results can be found in section 3 of Reference 133.

3.5. Anisotropy

High isotropy is a distinctive quality of CRs observed on Earth. The global leakage of CRs from the Galaxy and the contribution of individual sources lead to anisotropy, but the trajectories of energetic charged particles are highly tangled by regular and stochastic interstellar magnetic fields that isotropize the CR angular distribution. This makes the direct association of detected CR particles, except for the highest-energy particles, with their sources difficult or even impossible. Observations give the amplitude of the first angular harmonic of anisotropy at the level of $\delta \sim 10^{-3}$ in the energy range 10^{12} to 10^{14} eV, where the most reliable data are available (**Figure 12**) (see References 138 and 139). The angular distribution of particles at lower energies is significantly modulated by the solar wind. The statistics at higher energies are not good enough yet, but the measurements indicate the anisotropy amplitude at a level of a few percent at $10^{16} - 10^{18}$ eV. The data of the Super-Kamiokande-I detector (140) allowed accurate two-dimensional mapping of CR anisotropy at 10^{13} eV [see also the Tibet Air-Shower Array results (138)]. After correction for atmospheric effects, the deviation from the isotropic event rate is 0.1%, with a statistical significance $>5\sigma$ and direction to maximum excess at roughly $\alpha = 75°$ and $\delta = -5°$.

The amplitude of anisotropy in the diffusion approximation is given by the following equation, which includes pure diffusion and convection terms: $\delta = -[3D\nabla f + up(\partial f/\partial p)]/vf$, where f is defined after Equation 1 (see Reference 8). Here, D is the

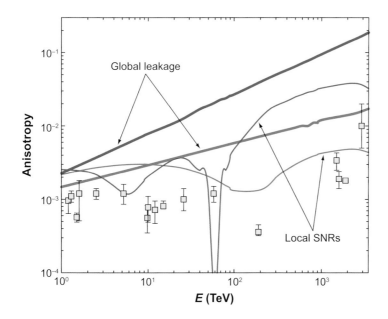

Figure 12

The anisotropy of cosmic rays in the reacceleration (*red curves*) and plain diffusion (*blue curves*) models. Shown separately are the effects of the global leakage from the Galaxy (*thick lines*) and the contribution from local supernova remnants (*thin lines*). The collection of data on cosmic-ray anisotropy (*yellow squares*) is taken from Reference 141, where the references to individual experiments can be found.

diffusion tensor, and it is assumed that the magnetic inhomogeneities that scatter CR particles are frozen in the background medium, moving with velocity $u \ll v$, which gives rise to the convection term (also known as the Compton-Getting term).

The Compton-Getting anisotropy is equal to $(\gamma + 2)u/c$ for ultrarelativistic CRs with a power-law spectrum $I(E) \sim p^2 f(p) \sim E^{-\gamma}$. The motion of the solar system through the local ISM produces the constant term in the energy dependence of the anisotropy $\sim 4 \times 10^{-4}$, with a maximum intensity in the general direction of the galactic center region, which does not agree with the data at $10^{12}-10^{14}$ eV, which point to an excess intensity from the anticenter hemisphere. The convection effect is outweighed by the diffusion anisotropy, which is due to the nonuniform distribution of CR in the Galaxy. The systematic decrease of CR intensity to the periphery of the Galaxy and the CR fluctuations produced by nearby SNRs are almost equally important for the formation of the local CR gradient.

Calculations of CR anisotropy (142) are illustrated in **Figure 12**, where the effect of global leakage from the Galaxy and the overall contribution of known SNRs with distances up to 1 kpc are shown separately. Two basic versions of the flat-halo diffusion model—the plain diffusion model with $D \sim E^{0.54}$ and the reacceleration model with $D_{xx} \sim E^{0.3}$—were used in the calculations. The values of D_{xx} were taken from Reference 69. The diffusion from a few nearby SNRs with different ages and distances results in a nonmonotonic dependence of anisotropy on energy. The flux from the Vela SNR probably dominates over other SNRs at energies below approximately 6×10^{13} eV in the plain diffusion model and below approximately 10^{14} eV for the reacceleration model. The results of these calculations indicate that the diffusion model with reacceleration is compatible with the data on CR anisotropy within a factor of approximately three. However, it seems that the plain diffusion model, with its relatively strong dependence of diffusion on energy, predicts a too-large anisotropy at $E > 10^{14}$ eV. Note that CR diffusion here is assumed to be isotropic, which is a seriously simplifying assumption, but difficult to avoid because of the complicated and unknown detailed structure of the galactic magnetic field. The presence of a large-scale random magnetic field justifies the approximation of isotropic diffusion on scales larger than a few hundred parsecs, but the anisotropy is a characteristic that is very sensitive to the local surroundings of the solar system, including the direction of magnetic field and the value of the diffusion tensor (see discussion in Reference 8).

3.6. Diffuse Galactic Gamma Rays

Gamma rays (above approximately 100 MeV) from the ISM hold great promise for CR studies because we can observe them throughout the Galaxy, not just from the local region of direct CR measurements. The complementarity of γ-rays and direct CR measurements can be exploited to learn about CR origin and propagation. Gamma rays are produced in the ISM by interactions of CR protons, He (π^0-decay), and electrons (bremsstrahlung) with gas, and electrons with the ISRF via inverse-Compton scattering. For details of the processes, the reader is referred to References 39, 81, 88, and 113. Additional astronomical material comes into play, such as the distribution of atomic and molecular gas, and the ISRF. In fact, γ-rays provide an important independent handle on molecular hydrogen and its relation to its CO molecular tracer, which has taken its place beside more traditional determinations.

CGRO: Compton Gamma-Ray Observatory

EGRET: Energetic Gamma-Ray Experiment Telescope

Historically, observations started with the OSO-III satellite in 1968, followed by the SAS-2 (1972), the COS-B (1975–1982), and the Compton Gamma-Ray Observatory (CGRO) (1991–2000). Each of these experiments represented a significant leap forward with respect to its predecessor. SAS-2 established the existence of emission from the ISM and allowed a first attempt to derive the CR distribution and also constrain the CR halo size (143). With COS-B, the CR distribution could be better derived and it was found not to follow the canonical distribution of SNRs, which posed a problem for the SNR origin of CRs. A relatively dependable value for the $CO-H_2$ relation was derived. With the Energetic Gamma-Ray Experiment Telescope (EGRET) and COMPTEL on the CGRO, the improvement in data quality was sufficient to allow such studies to be performed in much greater detail. There is now so much relevant information and theory that a rather realistic approach is justified and indeed necessary. At this point, the best approach seems to be explicit modeling of the high-energy Galaxy, putting in concepts from CR sources and propagation, galactic structure, and so on. The idea of a single model to reproduce both CR and γ-ray (and other) data simultaneously arose naturally and is the goal of the GALPROP project (see Section 2.10). For a recent review, see Reference 144.

To illustrate the current state of the art, we show spectra and profiles from a recent GALPROP model compared with CGRO/EGRET and COMPTEL data. These are based on References 113 and 145, where full details can be found. The first model is based simply on the directly measured CR spectra together with a radial gradient in the CR sources. It is immediately clear that the spectrum is not well predicted, being below the EGRET data for energies above 1 GeV. However, remembering that this is an unfitted prediction, it does show that the basic assumption that γ-rays are produced in CR interactions is correct. The factor of two difference tells us something about the remaining uncertainties. The extent to which the CR spectra have to be modified to get a good fit that includes the excess in the GeV region is shown in **Figure 13**, and the resulting γ-ray-model prediction in **Figure 14**. The difference between the directly observed and modified spectra is within plausible limits, considering solar modulation and spatial fluctuations in the Galaxy on scales >100 pc. The model has been tuned to fit the CR secondary antiprotons also produced in proton-proton interactions, on the assumption that these antiprotons have smaller spatial fluctuations than protons in the Galaxy. A detailed justification is lacking, however, so this so-called optimized model is just an existence proof rather than a conclusive result.

Other more drastic modifications to the CR spectrum have been proposed (78) as follows:

- A very hard electron injection spectrum, which could be possible invoking large fluctuations due to energy losses and the stochastic nature of SNRs in space and time: The solar vicinity is not necessarily a typical place for electron measurements to be representative (81, 146). However, the variations required are even larger than can reasonably be expected (113), so this model seems unlikely.

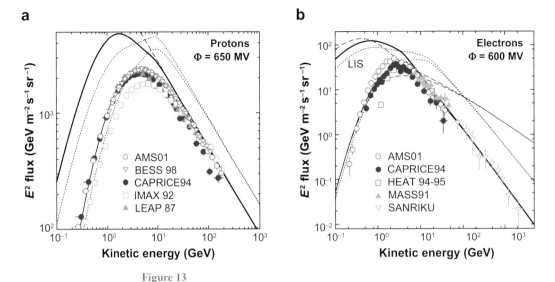

Figure 13

Directly observed (*solid lines*) and modified (*dotted lines*) proton and electron spectra. The modified spectra are deduced from fits to antiproton and γ-ray data (113). Data shown are from AMS01, BESS, CAPRICE94, IMAX, LEAP, HEAT, MASS91, and SANRIKU. LIS, local interstellar spectra. Figure adapted with permission from the American Astronomical Society.

- A hard proton spectrum, again invoking spatial variations in the Galaxy so the solar vicinity is not typical: This is more difficult than for electrons because the proton energy losses are negligible. It turns out this possibility can be ruled out on the basis of antiproton measurements. Too many antiprotons would be produced by the same protons that generate the γ-rays (78, 113, 147). A related suggestion invokes the dispersion in the radio spectral indices of SNRs, which indicates a dispersion in electron indices and, if assumed to apply to CR protons (148), could produce the GeV excess. This should also be tested against antiprotons.

It could be that a completely different source of the excess GeV γ-rays is present, and the possibility of a dark matter origin was pursued (150) but found to produce an excess of CR antiprotons (151). We will not enter this debate here because it is not related directly to the problem of CR propagation. It does, however, show how a good understanding of the CR-induced γ-rays in the Galaxy is essential to the study of potentially more fundamental physics (152).

The angular distribution of γ-rays provides an essential test for any model. The problem with the large expected gradient from SNRs is critical. The distribution of SNRs is hard to measure because of selection effects, so this problem could be safely ignored in the past. However, now the distribution of pulsars can be determined with reasonable accuracy, and this should trace SNRs because supernovae are pulsar progenitors. The pulsar gradient is indeed larger than originally deduced from γ-rays, as shown in **Figure 15**. The distribution of SNRs in external galaxies shows a similar

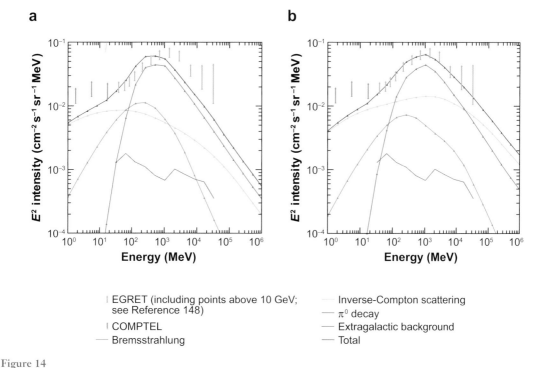

a

b

EGRET (including points above 10 GeV; see Reference 148)

COMPTEL

Bremsstrahlung

Inverse-Compton scattering

π^0 decay

Extragalactic background

Total

Figure 14

Gamma-ray spectrum of the inner Galaxy ($330° < l < 30°$, $|b| < 5°$) for models based on the directly observed cosmic-ray spectra and modified spectra shown in **Figure 13**. l is galactic longitude; b is galactic latitude. This is an update of the spectra shown in Reference 113.

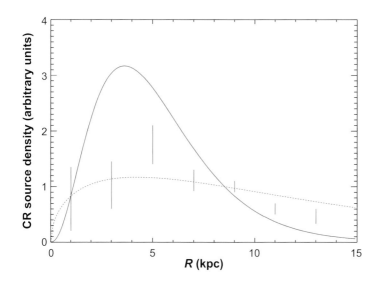

Figure 15

Possible cosmic-ray (CR) source distributions as a function of a Galactocentric radius R. Pulsars (*blue line*) and supernova remants (*vertical bars*) as deduced from γ-rays assuming a constant H_2-to-CO relation (*red line*) (145).

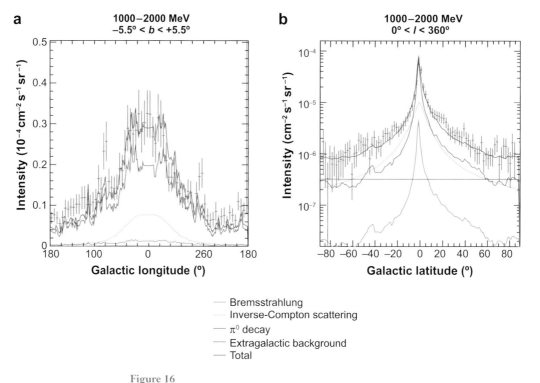

a 1000–2000 MeV
−5.5° < *b* < +5.5°

b 1000–2000 MeV
0° < *l* < 360°

— Bremsstrahlung
— Inverse-Compton scattering
— π^0 decay
— Extragalactic background
— Total

Figure 16

Gamma-ray longitude and latitude profiles, for models with pulsar source distribution and H_2-to-CO relation, which varies with the Galactocentric radius (145).

gradient to pulsars (153). The problem therefore remains, but a possible solution may be found in another uncertainty—the galactic distribution of molecular hydrogen. Because we must rely on the CO tracer molecule, any variation in the relation of CO to H_2 will affect the interpretation of the γ-ray data. Strong et al. (145) noted that there is independent evidence for an increase in the ratio of H_2 to CO with a Galactocentric radius, related to galactic metallicity gradients, and that this can resolve the problem, allowing CR sources to follow the SNR distribution as traced by pulsars. **Figure 16** shows longitude and latitude profiles based on such a model, showing a satisfactory agreement with EGRET data. However, the magnitude of the variation in the ratio of H_2 to CO, or its relation to metallicity and other effects, is uncertain, so the issue will need to be studied more. Future data from GLAST will help by giving much finer angular resolution and the possibility of better separating the molecular and atomic hydrogen components. Breitschwerdt et al. (70) proposed another possible solution to the gradient puzzle in terms of a radially dependent galactic wind (see Sections 2.4 and 2.12).

Note the large contribution from the inverse-Compton emission to the spectra and profiles (**Figure 14** and **Figure 16**). This is the reason why the predicted emission is in good agreement with the EGRET data right up to the highest latitudes (where the gas-related pion-decay emission is small). Therefore, γ-rays constrain both protons and electrons, the different angular distributions being clearly distinguishable. The high-latitude inverse-Compton emission is independent evidence for the existence of a CR halo extending up to several kiloparsecs above the plane, deduced from radioactive CR isotopes as explained in Section 3.2. Interestingly, secondary electrons and positrons make a significant contribution to the lower-energy bremsstrahlung and inverse-Compton γ-rays (113). Thus, we can in principle observe secondaries from all over the Galaxy, complementary to the local direct measurements.

Direct measurements: measurements on cosmic rays with detectors in balloons and satellites

3.7. Antiprotons and Positrons

The spectrum and origin of antiprotons in CRs has been a matter of active debate since the first reported detections in balloon flights (154, 155). There is a consensus that most of the CR antiprotons observed near Earth are secondaries (156). Owing to the kinematics of secondary production, the spectrum of antiprotons has a unique shape that distinguishes it from other CR species. It peaks at approximately 2 GeV, decreasing sharply toward lower energies. In addition to secondary antiprotons, there are possible sources of primary antiprotons. Those most often discussed are dark matter particle annihilation and evaporation of primordial black holes.

In recent years, new data with large statistics on both low- and high-energy antiproton fluxes have become available (157–163), thanks mostly to continuous improvements of the BESS instrument and its systematic launches every one to two years. This allows one to test models of CR propagation and heliospheric modulation. Additionally, accurate calculation of the secondary antiproton flux predicts the background for searches for exotic signals such as weakly interacting massive particle (WIMP) annihilation.

Despite numerous efforts and overall agreement on the secondary nature of the majority of CR antiprotons, published estimates of the expected flux differ significantly (see, e.g., figure 3 in Reference 159). Calculation of the secondary antiproton flux is a complicated task. The major sources of uncertainties are threefold: (a) incomplete knowledge of cross sections for antiproton production, annihilation, and scattering; (b) parameters and models of particle propagation in the Galaxy; and (c) modulation in the heliosphere. Although the interstellar antiproton flux is affected only by uncertainties in the cross sections and propagation models, the final comparison with experiment can be made only after correcting for the solar modulation. Besides, the spectra of CR nucleons have been directly measured only inside the heliosphere, whereas we need to know the spectrum outside, in interstellar space, to correctly compute the antiproton production rate.

Moskalenko et al. (117) recently showed that accurate antiproton measurements during the last solar minimum (1995–1997) (159) are inconsistent with existing propagation models at the ~40% level at approximately 2 GeV, while the stated measurement uncertainties in this energy range are now ~20%. The conventional models based on (a) local CR measurements, (b) simple energy dependence of the diffusion coefficient, and (c) uniform CR source spectra throughout the Galaxy fail to reproduce simultaneously both the secondary-to-primary nuclei ratio and antiproton flux.

A reacceleration model designed to match secondary-to-primary nuclei ratios produces too few antiprotons because the diffusion coefficient is too large. Models without reacceleration can reproduce the antiproton flux, but cannot explain the low-energy decrease in the secondary-to-primary nuclei ratios. To be consistent with both, the introduction of breaks in the diffusion coefficient and the injection spectrum is required, which would suggest new phenomena in particle acceleration and propagation. A solution in terms of propagation models requires a break in the diffusion coefficient at a few GeV, which has been interpreted as a change in the propagation mode (117). This calculation employs a modern steady-state drift model of propagation in the heliosphere to predict the proton and antiproton fluxes near Earth for different modulation levels and magnetic polarity.

If our local environment influences the spectrum of CRs, then it is possible to solve the problem by invoking a fresh, unprocessed nuclei component at low energies (126), which may be produced in the Local Bubble. The idea is that primary CRs such as C and O have a local low-energy component, whereas secondary CRs such as B are produced Galaxy wide over the confinement time of 10–100 Myr. In this way, an excess of B, which appears when propagation parameters are tuned to match the $\bar{p}$ data, can be eliminated by an additional local source of C (see Section 3.1). The model appears to describe a variety of CR data, but at the cost of additional parameters.

A consistent $\bar{p}$ flux can be obtained in reacceleration models if there are additional sources of protons $\leq$20 GeV (164). This energy is above the $\bar{p}$ production threshold and effectively produces $\bar{p}$s at $\leq$2 GeV. The intensity and spectral shape of this component can be derived by combining restrictions from $\bar{p}$s and diffuse γ-rays.

Clearly, accurate measurements of $\bar{p}$ flux are the key to testing current propagation models. The GeV excess in diffuse γ-rays discussed in the sections above may be interpreted in terms of fluctuations in the CR density in the Galaxy, and if so, the antiproton and γ-ray excesses have the same origin (113). If new measurements confirm the $\bar{p}$ excess, current propagation and/or modulation models will be challenged. If not, it will be evidence that the reacceleration model is currently the best one to describe the data. Here we must wait for the next BESS-Polar flight, which should bring much larger statistics, and for Pamela (165), currently in orbit.

Positrons in CRs were discovered in 1964 (166). The ratio of positrons to electrons is approximately 0.1 above $\sim$1 GeV, as measured near Earth. Secondary positrons in CRs are mostly the decay products of charged pions and kaons (π^+, K^+) produced in CR interactions with interstellar gas. In general, the calculation of the positron production by CRs agrees with the data and indicates that the majority of CR positrons are secondary (39, 167). Some small fraction of positrons may also come from sources (168) such as pulsar winds (169) or WIMP annihilations (170).

In interstellar space, the secondary positron flux below $\sim$1 GeV is comparable to the electron flux, which makes positrons non-negligible contributors to the diffuse γ-ray emission in the MeV range (113). The contribution of secondary positrons and electrons in diffuse galactic γ-ray emission models essentially improves the agreement with the data, making secondary positrons effectively detectable in γ-rays.

The accuracy of spacecraft and balloon-borne experiments has increased substantially during the past decade. The data taken by various instruments, including AMS,

CAPRICE, and HEAT (41, 171–173), are in agreement with each other within the error bars, which are, however, still quite large. A recent measurement of CR positron fraction by the HEAT instrument (42) indicates an excess (feature) between 5 and 7 GeV at the level of ~3σ over the smooth prediction of a propagation model (39). When all the HEAT flights are combined, the data show an excess above ~8 GeV, with the most significant point at ~8 GeV (42). The Pamela satellite (165) currently in orbit will have very good positron statistics.

3.8. Electrons and Synchrotron Radiation

CR electrons require a separate treatment from nuclei because of their rapid energy losses and their link with synchrotron radiation. Their low energy density compared with nuclei ($\approx 1\%$) is not yet understood. Standard SNR shock acceleration does not predict such a ratio, and it is normally a free parameter in models (e.g., Reference 174). Direct measurements extend from MeV to TeV. At low energies, solar modulation is so large that the interstellar fluxes are completely unknown. At TeV energies, the statistics are very poor, but new experiments are in progress. Young SNRs contain TeV electrons, as shown by their nonthermal X-ray emission, and energies up to 100 TeV are possible (175). The rapid energy losses mean that contributions from old nearby SNRs such as Loop 1, Vela, the Cygnus Loop, and MonoGem could make an important contribution to the local electron spectrum above 100 GeV and dominate above 1 TeV (see **Figure 4**) (37). The rapid energy losses of electrons also produce time- and space-dependent effects, as described in the next section.

The essential propagation effects are seen in **Figure 13**. At energies below 1 GeV, the interstellar spectrum is flatter than the injection spectrum owing to Coulomb losses. There is an intermediate energy range where the spectrum is similar to that at injection, and at high energies it steepens owing to inverse Compton and synchrotron losses and energy-dependent diffusion. For propagation models, we need an injection spectrum that reproduces the observations. A typical spectral index is 2.4, similar to nuclei, which produces the observed high-energy slope of 3.3 (113). The synchrotron spectral index is $\beta_\nu = (\gamma - 1)/2$ for an electron index γ. The observed $\beta_\nu = 0.6 - 1$, increasing with frequency, implies $\gamma = 2.4 - 3$, increasing with energy, in satisfactory agreement. Detailed geometrical models of the galactic synchrotron emission have been constructed (176, 177). A more physical interpretation requires propagation and magnetic field modeling, as, for example, in Reference 81. Combined with γ-ray data, this allows the magnetic field to be determined as a function of a Galactocentric radius independent of other techniques (81). Since the work reported in Reference 81, a great deal of new radio data have become available both from new ground-based surveys and from balloons and satellites for cosmology (e.g., WMAP). Because galactic synchrotron is an essential foreground for cosmic microwave background studies, there is a large overlap of interest between these fields. Exploitation of these data for CR studies has only just begun, but is expected to accelerate with the forthcoming Planck mission.

Radio continuum observations of other galaxies provide a complementary view of electrons via synchrotron radiation (178, 179). The classical case is the edge-on galaxy NGC891 (180, 181), which has a nonthermal halo extending to several kilo-parsecs. This helped give credence to the idea of a large halo in our Galaxy. More recent observations of this and other edge-on galaxies confirm large halos (182). The spectral index variations due to energy losses provide a test of propagation in principle, although this has not been very fruitful up until now. This effect can also be measured in starburst galaxies (183). Face-on galaxies such as M51 show marked spiral structure in the continuum (178), but this is probably mainly a magnetic field effect. It is not possible to separate CR and magnetic field variations reliably, although this is often attempted assuming equipartition of energy between CRs and field, which although having no firmly established physical basis, nevertheless gives field values quite similar to other methods (179, 184).

The tight radio-continuum/far-infrared correlation for galaxies (185–187) is worth mentioning here. It holds both within galaxies (down to 100 pc or less) and from galaxy to galaxy, over several orders of magnitude in luminosity. It presumably contains clues to CR origin and propagation. The CR calorimeter (188) is the simplest interpretation. Simply relating both CR production and ultraviolet (reprocessed to far-infrared) to star formation rates is apparently insufficient to explain the close correlation, and this has given rise to several different interpretations, including hydrostatic regulation, without being very conclusive (185–187).

3.9. Time-Dependent and Space-Dependent Effects

CR propagation is usually modeled as a spatially smooth, temporally steady-state process. Because of the rapid diffusion and long containment timescales in the Galaxy, this is usually a sufficient approximation. However, there are cases where it breaks down. The rapid energy loss of electrons above 100 GeV and the stochastic nature of their sources produce spatial and temporal variations. Supernovae are stochastic events and each SNR source of CRs lasts only perhaps $10^4–10^5$ years, so that they leave their imprint on the distribution of electrons. This leads to large fluctuations in the CR electron density at high energies, so that the electron spectrum measured near the Sun may not be typical (113). A statistical calculation of this effect can be found in References 146 and 189, and for local SNRs in Reference 37. These effects are much smaller for nucleons because in this case there are essentially no energy losses, but they can still be important (190). For the theory of CR fluctuations for a Galaxy with random SNR events, see Reference 142 and Section 3.5. Such effects can influence the B/C ratio (111, 116) mainly through variations in the primary spectra. The Local Bubble can also have an effect on the energy dependence of B/C (126). Dispersion in CR injection spectra from SNRs may cause the locally observed spectrum to deviate from the average (148).

SUMMARY POINTS

1. Presently, progress in CR research, with the new generation of detectors for both direct and indirect measurements, is rapid. However, it is still hard to

pin down particular theories. Even now the origin of the nucleonic component is not proven (34), although SNRs are the leading candidates.

2. Considering all relevant data, both direct (particles) and indirect (γ-ray,
synchrotron) measurements, is important.

3. Increase in computing power has made many of the old approximations for
interpreting CR data unnecessary.

4. New high-quality data will require detailed numerical models.

FUTURE ISSUES

We end by listing some of the open questions regarding CR propagation, which
may be answered with observations in the near future.

1. The interpretation of the energy-dependence of the secondary-to-primary
ratio, which requires accurate measurements at both low and high energies,
is an open question.

2. The size of the propagation region—the existence of an extended
halo—requires measurements of radioactive species over a broad energy
range.

3. What are the relative roles of diffusion, convection, and reacceleration?

4. What is the importance of local sources to the primary CR flux?

5. What is the origin of the GeV excess in γ-rays relative to the prediction
based on locally observed CRs?

6. Is the CR source distribution similar to SNRs?

7. Are positrons and antiprotons fully explained as secondaries from primary
CRs, or is there a perhaps exotic excess?

8. What is the relation of CR-dynamical models of the Galaxy to CR propagation theory?

DISCLOSURE STATEMENT

The authors are not aware of any biases that might be perceived as affecting the
objectivity of this review.

ACKNOWLEDGMENTS

I.V.M. acknowledges partial support from a NASA Astronomy and Physics Research
and Analysis program grant.

LITERATURE CITED

1. Shapiro MM, Silberberg R. *Annu. Rev. Nucl. Part. Sci.* 20:323 (1970)
2. Wentzel DG. *Annu. Rev. Astron. Astrophys.* 12:71 (1974)
3. Simpson JA. *Annu. Rev. Nucl. Part. Sci.* 33:323 (1983)
4. Cesarsky CJ. *Annu. Rev. Astron. Astrophys.* 18:289 (1980)
5. Cox DP. *Annu. Rev. Astron. Astrophys.* 43:337 (2005)
6. Elmegreen BG, Scalo J. *Annu. Rev. Astron. Astrophys.* 42:211 (2004)
7. Scalo J, Elmegreen BG. *Annu. Rev. Astron. Astrophys.* 42:275 (2004)
8. Berezinskii VS, Bulanov SV, Dogiel VA, Ptuskin VS. *Astrophysics of Cosmic Rays*, ed. VL Ginzburg. Amsterdam: North-Holland (1990)
9. Ginzburg VL, Syrovatskii SI. *The Origin of Cosmic Rays*. New York: Macmillan (1964)
10. Ginzburg VL, Ptuskin VS. *Rev. Mod. Phys.* 48:161 (1976)
11. Hillas AM. *Cosmic Rays. The Commonwealth and International Library, Selected Readings in Physics*. Oxford: Pergamon (1972)
12. Gaisser TK. *Cosmic Rays and Particle Physics*. Cambridge/New York: Cambridge Univ. Press (1990)
13. Stanev T. *High Energy Cosmic Rays. Praxis Books in Astrophysics and Astronomy*. Chichester, UK: Springer (2004)
14. Schlickeiser R. *Cosmic Ray Astrophysics. Astronomy and Astrophysics Library*. Berlin: Springer (2002)
15. Diehl R, Parizot E, Kallenbach R, von Steiger R, eds. *The Astrophysics of Galactic Cosmic Rays*. Dordrecht: Kluwer (2002)
16. Nagano M, Watson AA. *Rev. Mod. Phys.* 72:689 (2000)
17. Cronin JW. *Nucl. Phys. B Proc. Suppl.* 138:465 (2005)
18. Hörandel JR. astro-ph/0508014 (2005); Hörandel JR. *J. Phys. Conf. Ser.* 47:132 (2006); Hörandel JR. astroph-0702370 (2007)
19. Engel R. *Nucl. Phys. B Proc. Suppl.* 151:437 (2006)
20. Stanev T. In *AIP Conf. Proc.*, ed. Z Parsa, 698:357. New York: AIP (2004)
21. Berezinsky V, Gazizov A, Grigorieva S. *Nucl. Phys. B Proc. Suppl.* 136:147 (2004)
22. Berezinsky V, Gazizov A, Kachelrieß M. *Phys. Rev. Lett.* 97:231101 (2006)
23. Hooper D, Sarkar S, Taylor AM. astro-ph/0608085 (2006)
24. Hörandel JR. *Astropart. Phys.* 21:241 (2004)
25. Hillas AM. astro-ph/0607109 (2006)
26. Gaisser TK. astro-ph/0608553 (2006)
27. Kampert KH. astro-ph/0611884 (2006)
28. Webber WR. *Astrophys. J.* 506:329 (1998)
29. Fatuzzo M, Adams FC, Melia F. *Astrophys. J.* 653:L49 (2006)
30. Hanasz M, Lesch H. *Astron. Astrophys.* 412:331 (2003)
31. Snodin AP, Brandenburg A, Mee AJ, Shukurov A. *MNRAS* 373:643 (2006)
32. Zirakashvili VN, Breitschwerdt D, Ptuskin VS, Völk HJ. *Astron. Astrophys.* 311:113 (1996)
33. Ptuskin VS, Völk HJ, Zirakashvili VN, Breitschwerdt D. *Astron. Astrophys.* 321:434 (1997)

34. Aharonian F, et al. *Astron. Astrophys.* 449:223 (2006)
35. Panov AD, et al. astro-ph/0612377 (2006), Panov AD, et al. *Proc. Russ. Cosmic Ray Conf.*, 71:494–97. New York: Bull. Russ. Acad. Sci. Phys. (2007)
36. Derbina VA, et al. *Astrophys. J.* 628:L41 (2005)
37. Kobayashi T, Komori Y, Yoshida K, Nishimura J. *Astrophys. J.* 601:340 (2004)
38. Grimani C, et al. *Astron. Astrophys.* 392:287 (2002)
39. Moskalenko IV, Strong AW. *Astrophys. J.* 493:694 (1998)
40. Stephens SA. *Adv. Space Res.* 27:687 (2001)
41. Gast H, Olzem J, Schael S. astro-ph/0605254 (2006)
42. Beatty JJ, et al. *Phys. Rev. Lett.* 93:241102 (2004)
43. Moskalenko IV, Strong AW, Ormes JF, Mashnik SG. *Adv. Space Res.* 35:156 (2005)
44. Bradt HL, Peters B. *Phys. Rev.* 80:943 (1950)
45. Garcia-Munoz M, Mason GM, Simpson JA. *Astrophys. J.* 201:L145 (1975)
46. Seo ES, Ptuskin VS. *Astrophys. J.* 431:705 (1994)
47. Webber WR. *Space Sci. Rev.* 86:239 (1998)
48. Wiedenbeck ME, et al. *Space Sci. Rev.* 99:15 (2001)
49. Connell JJ. *Space Sci. Rev.* 99:41 (2001)
50. Binns WR, et al. *Astrophys. J.* 634:351 (2005)
51. Kennel C, Engelmann F. *Phys. Fluids* 9:2377 (1966)
52. Casse F, Lemoine M, Pelletier G. *Phys. Rev. D* 65:023002 (2002)
53. Jokipii JR. *Rev. Geophys. Space Phys.* 9:27 (1971)
54. Toptygin IN. *Cosmic Rays in Interplanetary Magnetic Fields*. Dordrecht: Reidel (1985)
55. Goldreich P, Sridhar S. *Astrophys. J.* 438:763 (1995)
56. Yan H, Lazarian A. *Astrophys. J.* 614:757 (2004)
57. Breitschwerdt D, Komossa S. *Astrophys. Space Sci.* 272:3 (2000)
58. Breitschwerdt D, Völk HJ, McKenzie JF. *Astron. Astrophys.* 245:79 (1991)
59. Breitschwerdt D, McKenzie JF, Völk HJ. *Astron. Astrophys.* 269:54 (1993)
60. Jokipii JR. *Astrophys. J.* 208:900 (1976)
61. Owens AJ, Jokipii JR. *Astrophys. J.* 215:677 (1977)
62. Owens AJ, Jokipii JR. *Astrophys. J.* 215:685 (1977)
63. Jones FC. *Astrophys. J.* 222:1097 (1978)
64. Jones FC. *Astrophys. J.* 229:747 (1979)
65. Bloemen JBGM, Dogel' VA, Dorman VL, Ptuskin VS. *Astron. Astrophys.* 267:372 (1993)
66. Veilleux S, Cecil G, Bland-Hawthorn J. *Annu. Rev. Astron. Astrophys.* 43:769 (2005)
67. Strong AW, Moskalenko IV. *Astrophys. J.* 509:212 (1998)
68. Maurin D, Taillet R, Donato F. *Astron. Astrophys.* 394:1039 (2002)
69. Jones FC, Lukasiak A, Ptuskin V, Webber W. *Astrophys. J.* 547:264 (2001)
70. Breitschwerdt D, Dogiel VA, Völk HJ. *Astron. Astrophys.* 385:216 (2002)
71. Simon M, Heinrich W, Mathis KD. *Astrophys. J.* 300:32 (1986)
72. Letaw JR, Adams JH Jr., Silberberg R, Tsao CH. *Astrophys. Space Sci.* 114:365 (1985)

73. Niebur SM, et al. In *AIP Conf. Proc.*, ed. RA Mewaldt, et al., 528:406. New York: AIP (2000)

74. Jones FC, Lukasiak A, Ptuskin VS, Webber WR. *Adv. Space Res.* 27:737 (2001)

75. Ptuskin VS, et al. *Astrophys. J.* 642:902 (2006)

76. Wandel A, et al. *Astrophys. J.* 316:676 (1987)

77. Berezhko EG, et al. *Astron. Astrophys.* 410:189 (2003)

78. Moskalenko IV, Strong AW, Reimer O. In *ASSL*, ed. KS Cheng, GE Romero, 304:279. Dordrecht: Kluwer (2004)

79. Beck R. *Space Sci. Rev.* 99:243 (2001)

80. Han JL, et al. *Astrophys. J.* 642:868 (2006)

81. Strong AW, Moskalenko IV, Reimer O. *Astrophys. J.* 537:763 (2000)

82. Alvarez-Muñiz J, Engel R, Stanev T. *Astrophys. J.* 572:185 (2002)

83. Takami H, Yoshiguchi H, Sato K. *Astrophys. J.* 639:803 (2006)

84. Kachelriess M, Serpico PD, Teshima M. astro-ph/0510444 (2005)

85. Moskalenko IV, Porter TA, Strong AW. *Astrophys. J.* 640:L155 (2006)

86. Frisch PC, Slavin JD. astro-ph/0606743 (2006)

87. Frisch PC, ed. *Astrophysics and Space Science Library*. Dordrecht: Springer (2006)

88. Moskalenko IV, Strong AW. *Astrophys. J.* 528:357 (2000)

89. Kamae T, et al. *Astrophys. J.* 647:692 (2006)

90. Kelner SR, Aharonian FA, Bugayov VV. *Phys. Rev. D* 74:034018 (2006)

91. Müller D, et al. *Space Sci. Rev.* 99:353 (2001)

92. Mashnik SG, et al. *Adv. Space Res.* 34:1288 (2004)

93. Moskalenko IV, Strong AW, Mashnik SG. In *AIP Conf. Proc.*, ed. RC Haight, et al., 769:1612. New York: AIP (2005)

94. de Nolfo GA, et al. *Adv. Space Res.* 38:1558 (2006)

95. Endt PM, Van Der Leun C. *Nucl. Phys. A* 310:1 (1978)

96. Martínez-Pinedo G, Vogel P. *Phys. Rev. Lett.* 81:281 (1998)

97. Wuosmaa AH, et al. *Phys. Rev. Lett.* 80:2085 (1998)

98. Strong AW, Moskalenko IV. *Adv. Space Res.* 27:717 (2001)

99. Davis LJ. *Proc. 6th Int. Cosmic Ray Conf. (Moscow)* 3:220 (1960)

100. Protheroe RJ, Ormes JF, Comstock GM. *Astrophys. J.* 247:362 (1981)

101. Garcia-Munoz M, et al. *Astrophys. J. Suppl.* 64:269 (1987)

102. Ptuskin VS, Jones FC, Ormes JF. *Astrophys. J.* 465:972 (1996)

103. Stephens SA, Streitmatter RE. *Astrophys. J.* 505:266 (1998)

104. Davis AJ, et al. In *AIP Conf. Proc.*, ed. RA Mewaldt, et al., 528:421. New York: AIP (2000)

105. Webber WR. *Astrophys. J.* 402:188 (1993)

106. Duvernois MA, Simpson JA, Thayer MR. *Astron. Astrophys.* 316:555 (1996)

107. Ptuskin VS, Soutoul A. *Astron. Astrophys.* 237:445 (1990)

108. Ptuskin VS, Soutoul A. *Astron. Astrophys.* 337:859 (1998)

109. Webber WR, Lee MA, Gupta M. *Astrophys. J.* 390:96 (1992)

110. Maurin D, Donato F, Taillet R, Salati P. *Astrophys. J.* 555:585 (2001)

111. Taillet R, et al. *Astrophys. J.* 609:173 (2004)

112. Higdon JC, Lingenfelter RE. *Astrophys. J.* 582:330 (2003)
113. Strong AW, Moskalenko IV, Reimer O. *Astrophys. J.* 613:962 (2004)
114. Shibata T, Hareyama M, Nakazawa M, Saito C. *Astrophys. J.* 612:238 (2004)
115. Shibata T, Hareyama M, Nakazawa M, Saito C. *Astrophys. J.* 642:882 (2006)
116. Büsching I, et al. *Astrophys. J.* 619:314 (2005)
117. Moskalenko IV, Strong AW, Ormes JF, Potgieter MS. *Astrophys. J.* 565:280 (2002)
118. Wallace JM. *Astrophys. J.* 245:753 (1981)
119. Hanasz M, Lesch H. *Astrophys. J.* 543:235 (2000)
120. Giacalone J, Jokipii JR. *Astrophys. J.* 520:204 (1999)
121. Hanasz M, Kowal G, Otmianowska-Mazur K, Lesch H. *Astrophys. J.* 605:L33 (2004)
122. Moskalenko IV, Mashnik SG. *Proc. 28th Int. Cosmic Ray Conf. (Tsukuba)* 4:1969 (2003)
123. Webber WR, McDonald FB, Lukasiak A. *Astrophys. J.* 599:582 (2003)
124. Shalchi A, Bieber JW, Matthaeus WH, Qin G. *Astrophys. J.* 616:617 (2004)
125. Shalchi A, Schlickeiser R. *Astrophys. J.* 626:L97 (2005)
126. Moskalenko IV, Strong AW, Mashnik SG, Ormes JF. *Astrophys. J.* 586:1050 (2003)
127. Donato F, Maurin D, Taillet R. *Astron. Astrophys.* 381:539 (2002)
128. Moskalenko IV, Mashnik SG, Strong AW. *Proc. 27th Int. Cosmic Ray Conf. (Hamburg)* 5:1836 (2001)
129. Hams T, et al. *Astrophys. J.* 611:892 (2004)
130. Streitmatter RE, Stephens SA. *Adv. Space Res.* 27:743 (2001)
131. Ptuskin VS, Soutoul A. *Space Sci. Rev.* 86:225 (1998)
132. Yanasak NE, et al. *Astrophys. J.* 563:768 (2001)
133. Mewaldt RA, et al. *Space Sci. Rev.* 99:27 (2001)
134. Mueller HR, Frisch PC, Florinski V, Zank GP. astro-ph/0607600 (2006)
135. Wiedenbeck ME, et al. In *AIP Conf. Proc.*, ed. RA Mewaldt, et al., 528:363. New York: AIP (2000)
136. de Nolfo GA, et al. *Proc. 29th Int. Cosmic Ray Conf. (Pune)* 3:61 (2005)
137. Niebur SM, et al. *Proc. 27th Int. Cosmic Ray Conf. (Hamburg)* 5:1675 (2001)
138. Amenomori M, et al. *Astrophys. J.* 626:L29 (2005)
139. Tada J, et al. *Nucl. Phys. B Proc. Suppl.* 151:485 (2006)
140. Guillian G. *Proc. 29th Int. Cosmic Ray Conf. (Pune)* 6:85 (2005)
141. Ambrosio M, et al. *Phys. Rev. D* 67:042002 (2003)
142. Ptuskin VS, Jones FC, Seo ES, Sina R. *Adv. Space Res.* 37:1909 (2006)
143. Stecker FW, Jones FC. *Astrophys. J.* 217:843 (1977)
144. Moskalenko IV, Strong AW. In *AIP Conf. Proc.*, ed. T Bulik, et al., 801:57. New York: AIP (2005)
145. Strong AW, et al. *Astron. Astrophys.* 422:L47 (2004)
146. Pohl M, Esposito JA. *Astrophys. J.* 507:327 (1998)
147. Moskalenko IV, Strong AW, Reimer O. *Astron. Astrophys.* 338:L75 (1998)

148. Büsching I, Pohl M, Schlickeiser R. *Astron. Astrophys.* 377:1056 (2001)

149. Strong AW, et al. *Astron. Astrophys.* 444:495 (2005)

150. de Boer W, et al. *Astron. Astrophys.* 444:51 (2005)

151. Bergström L, Edsjö J, Gustafsson M, Salati P. *J. Cosmol. Astropart. Phys.* 5:6 (2006)

152. Moskalenko IV, et al. astro-ph/0609768 (2006)

153. Sasaki M, Breitschwerdt D, Supper R. *Astrophys. Space Sci.* 289:283 (2004)

154. Golden RL, et al. *Phys. Rev. Lett.* 43:1196 (1979)

155. Bogomolov EA, et al. *Proc. 19th Int. Cosmic Ray Conf. (La Jolla)* 2:362 (1985)

156. Mitchell JW, et al. *Phys. Rev. Lett.* 76:3057 (1996)

157. Hof M, et al. *Astrophys. J.* 467:L33 (1996)

158. Grimani C, et al. *Astron. Astrophys.* 392:287 (2002)

159. Orito S, et al. *Phys. Rev. Lett.* 84:1078 (2000)

160. Bergström D, et al. *Astrophys. J.* 534:L177 (2000)

161. Sanuki T, et al. *Astrophys. J.* 545:1135 (2000)

162. Maeno T, et al. *Astropart. Phys.* 16:121 (2001)

163. Asaoka Y, et al. *Phys. Rev. Lett.* 88:051101 (2002)

164. Moskalenko IV, Strong AW, Mashnik SG, Ormes JF. *Proc. 28th Int. Cosmic Ray Conf. (Tsukuba)* 4:1921 (2003)

165. Picozza P, et al. astro-ph/0608697 (2006)

166. de Shong JA, Hildebrand RH, Meyer P. *Phys. Rev. Lett.* 12:3 (1964)

167. Protheroe RJ. *Astrophys. J.* 254:391 (1982)

168. Coutu S, et al. *Astropart. Phys.* 11:429 (1999)

169. Chi X, Cheng KS, Young ECM. *Astrophys. J.* 459:L83 (1996)

170. Baltz EA, Edsjö J. *Phys. Rev. D* 59:023511 (1999)

171. Aguilar M, et al. (AMS Collab.) *Phys. Rep.* 366:331 (2002)

172. Boezio M, et al. *Astrophys. J.* 532:653 (2000)

173. DuVernois MA, et al. *Astrophys. J.* 559:296 (2001)

174. Ellison DC. *Space Sci. Rev.* 99:305 (2001)

175. Allen G. *Proc. 26th Int. Cosmic Ray Conf. (Salt Lake City)* 3:480 (1999)

176. Beuermann K, Kanbach G, Berkhuijsen EM. *Astron. Astrophys.* 153:17 (1985)

177. Broadbent A, Haslam TCG, Osborne LJ. *Proc. 21st Int. Cosmic Ray Conf. (Adelaide)* 3:229 (1990)

178. Beck R. In *The Magnetized Plasma in Galaxy Evolution*, ed. KT Chyzy, et al., p. 193. Kraków: Jagiellonian Univ. (2005)

179. Beck R. astro-ph/0603531 (2006)

180. Allen RJ, Sancisi R, Baldwin JE. *Astron. Astrophys.* 62:397 (1978)

181. Heald GH, Rand RJ, Benjamin RA, Bershady MA. *Astrophys. J.* 647:1018 (2006)

182. Dumke M, Krause M. In *LNP*, ed. D Breitschwerdt, et al., 506:555. Berlin: Springer-Verlag (1998)

183. Heesen V, Krause M, Beck R, Dettmar RJ. In *The Magnetized Plasma in Galaxy Evolution*, ed. KT Chyzy, et al., p. 156. Kraków: Jagiellonian Univ. (2005)

184. Beck R, Krause M. *Astron. Nachrichten* 326:414 (2005)

185. Niklas S, Beck R. *Astron. Astrophys.* 320:54 (1997)

186. Murgia M, et al. *Astron. Astrophys.* 437:389 (2005)

187. Paladino R, et al. *Astron. Astrophys.* 456:847 (2006)
188. Völk HJ. *Astron. Astrophys.* 218:67 (1989)
189. Strong AW, Moskalenko IV. *Proc. 27th Int. Cosmic Ray Conf. (Hamburg)* 5:1964 (2001)
190. Erlykin AD, Wolfendale AW. *J. Phys. G: Nucl. Part. Phys.* 28:359 (2002)

Related Resources

GALPROP Web site: **http://galprop.stanford.edu**
GLAST Web site: **http://glast.stanford.edu**

An Introduction to Effective Field Theory

C.P. Burgess

Department of Physics & Astronomy, McMaster University, L8S 4M1 Ontario,
Canada, and Perimeter Institute for Theoretical Physics, N2L 2Y5 Ontario, Canada;
email: cburgess@physics.mcmaster.ca

Annu. Rev. Nucl. Part. Sci. 2007. 57:329–62

First published online as a Review in Advance on
June 27, 2007

The *Annual Review of Nuclear and Particle Science* is
online at http://nucl.annualreviews.org

This article's doi:
10.1146/annurev.nucl.56.080805.140508

Copyright © 2007 by Annual Reviews.
All rights reserved

0163-8998/07/1123-0329$20.00

Key Words

low-energy approximation, power counting

Abstract

This review summarizes effective field theory techniques, which are
the modern theoretical tools for exploiting the existence of hierar-
chies of scale in a physical problem. The general theoretical framework is described and evaluated explicitly for a simple model. Power-
counting results are illustrated for a few cases of practical interest,
and several applications to quantum electrodynamics are described.

Contents

1. INTRODUCTION 330
   1.1. A Toy Model 331
2. GENERAL FORMULATION 332
   2.1. The One-Particle-Irreducible and One-Light-Particle-Irreducible
        Actions 333
   2.2. The Wilson Action 342
3. POWER COUNTING 347
   3.1. A Class of Effective Interactions 347
   3.2. Power-Counting Rules 347
   3.3. The Effective Lagrangian Logic 349
4. APPLICATIONS 350
   4.1. Quantum Electrodynamics 351
   4.2. Power-Counting Examples: QCD and Gravity 359

1. INTRODUCTION

It is a basic fact of life that nature comes to us in many scales. Galaxies, planets, aardvarks, molecules, atoms, and nuclei have very different sizes, and are held together with very different binding energies. Happily enough, we do not need to understand what is going on at all scales at once to figure out how nature works at a particular scale. Like good musicians, good physicists know which scales are relevant for which compositions.

The mathematical framework we use to describe nature—quantum field theory—shares this basic feature of nature: It automatically limits the role smaller-distance scales can play in the description of larger objects. This property has many practical applications because a systematic identification of how scales enter into calculations provides an important tool for analyzing systems that have two very different scales, $m \ll M$. In these systems, it is usually profitable to expand quantities in powers of the small parameter, m/M, and the earlier this is done in a calculation, the more it is simplified.

This review provides a practical introduction to the technique of effective field theory, which is the main modern tool for exploiting the simplifications that arise for systems exhibiting a large hierarchy of scales (1–4). The goal is to provide an overview of the theoretical framework, with an emphasis on practical applications and concrete examples. The intended audience is assumed to be knowledgeable in the basic techniques of quantum field theory, including its path-integral formulation.

Although it is not the main focus of this review, it is hoped that one of the more satisfying threads running throughout is the picture that emerges of the physics that underlies the technique of renormalization. Renormalization is a practice that used to be widely regarded as distasteful, and so was largely done in the privacy of one's own home. That has all changed. As used in effective field theories, renormalizing is not only respectable, it is often the smart thing to do when extracting the dependence of physical quantities on large logarithms of scale ratios, $\sim \log(M/m)$.

Another attractive conceptual spin-off of effective field theory techniques is the understanding they provide of the physical interpretation of nonrenormalizable theories, such as Einstein's general theory of relativity. Although much has been made about the incompatibility of gravity and quantum mechanics, the quantization of nonrenormalizable theories can make perfect sense provided they are applied only to low-energy predictions.

1.1. A Toy Model

To make the discussion as concrete as possible, consider a system involving two spinless particles, l and H, with one, l, being very light compared with the other, H. Taking the classical action for the system to be the most general renormalizable one consistent with the discrete symmetry $l \to -l$ gives

$$S_c[l, H] = - \int d^4x \left[\frac{1}{2}(\partial_\mu l \partial^\mu l + \partial_\mu H \partial^\mu H) + V(l, H) \right], \qquad 1.$$

where the interaction potential is

$$V(l, H) = \frac{1}{2}m^2 l^2 + \frac{1}{2}M^2 H^2 + \frac{g_l}{4!}l^4 + \frac{g_b}{4!}H^4 + \frac{g_{lb}}{4}l^2 H^2 + \frac{\tilde{m}}{2}l^2 H + \frac{\tilde{g}_b M}{3!}H^3. \qquad 2.$$

We use units $\hbar = c = k_B = 1$ and adopt the "mostly plus" metric signature. M and m denote the two particle masses, and we suppose the three dimensionful quantities—m, $\tilde{m}$, and M—satisfy $M \gg m, \tilde{m}$, in order to ensure a large hierarchy of scales.

Now imagine computing a low-energy physical process in this model, which for simplicity we take to be two-body $l - l$ scattering at center-of-mass (CM) energies much smaller than the heavy-particle mass: $E_{\rm cm} \ll M$. The Feynman graphs that give rise to this scattering at tree level are given in **Figure 1**.

The S-matrix element that follows from these graphs may be written as

$$S(p_1, p_2; p_3, p_4) = i(2\pi)^4 \delta^4(p_1 + p_2 - p_3 - p_4) \quad \mathcal{A}(p_1, p_2, p_3, p_4), \qquad 3.$$

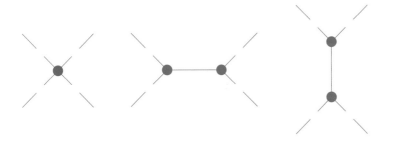

Figure 1

The Feynman graphs that contribute to two-body light-particle scattering at tree level in the toy model. Solid (dashed) lines represent the heavy (light) scalar.

where

$$A(p_1, p_2, p_3, p_4) = -g_l + \tilde{m}^2 \left[\frac{1}{(p_1 - p_3)^2 + M^2} + \frac{1}{(p_1 - p_4)^2 + M^2} \right.$$
$$\left. + \frac{1}{(p_1 + p_2)^2 + M^2} \right]$$
$$\approx -g_l + \frac{3\tilde{m}^2}{M^2} + \frac{4\tilde{m}^2 m^2}{M^4} + \mathcal{O}(M^{-6}). \qquad 4.$$

This last (approximate) equality assumes momenta and energies to be much smaller than M. The final line also uses the identity $(p_1 - p_4)^2 + (p_1 - p_3)^2 + (p_1 + p_2)^2 = -4m^2$, which follows from the mass-shell condition $p_i^2 = -m^2$.

Two points should be emphasized about this last expression. First, this scattering amplitude (and all others) simplifies considerably in the approximation that powers of m/M, $\tilde{m}/M$, and E_{cm}/M may be neglected. Second, the result up to $O(M^{-4})$ could be obtained for a theory involving only l particles interacting through the following effective potential:

$$V_{\text{eff}}^{(4)} = \frac{1}{4!} \left(g_l - \frac{3\tilde{m}^2}{M^2} - \frac{4\tilde{m}^2 m^2}{M^4} \right) l^4. \qquad 5.$$

Even more interesting, the $O(M^{-4})$ contribution to all other observables involving low-energy l scattering is completely captured if the following terms are added to the above potential:

$$V_{\text{eff}}^{(6)} = \frac{\tilde{m}^2}{M^4} \left(\frac{g_{lh}}{16} - \frac{g_l}{6} \right) l^6. \qquad 6.$$

There are several reasons why it is useful to establish at the outset that the low-energy interactions among light particles are described, in the large-M limit, by some sort of effective interactions such as Equations 5 and 6. Most prosaically, it is obviously much easier to calculate more complicated processes starting from Equations 5 and 6 than by computing the full result using Equations 1 and 2, and only then expanding in powers of $1/M$. For more difficult calculations, such as the cross section for the reaction $6l \rightarrow 12l$, such ease of calculation can be the difference that decides whether a computation is feasible. Furthermore, knowledge of the possible types of effective interactions at a given order in $1/M$ may guide us to identify which low-energy observables are most (or least) sensitive to the properties of the heavy particles.

Fortunately, it appears to be a basic property of quantum field theories that so long as a large hierarchy of masses exists, a low-energy description in terms of a collection of effective interactions is indeed possible. The point of this review is to sketch why this is so, what may be said about its properties, and how to compute it (if possible) from a full theory of both light and heavy particles.

2. GENERAL FORMULATION

A variety of observables could be used to illustrate effective field theory techniques, but because these can be computed from knowledge of the various correlation functions of the theory, it is convenient to phrase the discussion in terms of the generating functional for these correlation functions. In particular, we focus on the generator Γ of one-particle-irreducible (1PI) correlation functions. Although this quantity is often termed the effective action, particularly in older references, we reserve this name for another quantity of more direct interest, which we discuss below.

2.1. The One-Particle-Irreducible and One-Light-Particle-Irreducible Actions

We start by reviewing the standard definition for the generating functional. Consider a theory whose fields are denoted generically by ϕ. Our interest in this theory is in the correlation functions of these fields, as other physical quantities can be generically constructed from these. These correlations may be obtained by studying the response of the theory to the application of an external field, $J(x)$, which couples to $\phi(x)$.

For instance, a path-integral definition of the correlation function would be

$$\langle \phi(x_1) \cdots \phi(x_k) \rangle_J \equiv e^{-iW[J]} \int \mathcal{D}\phi [\phi(x_1) \cdots \phi(x_k)] \exp \left\{ i \int d^4x [\mathcal{L} + J\phi] \right\}, \qquad 7.$$

where $\mathcal{L}$ denotes the Lagrangian density that describes the system's dynamics, and the quantity $W[J]$ is defined by

$$\exp\{iW[J]\} = \int \mathcal{D}\phi \exp \left\{ i \int d^4x [\mathcal{L}[\phi] + J\phi] \right\}. \qquad 8.$$

$W[J]$ generates the connected correlations of the operator ϕ, in the sense that

$$\langle \phi(x_1) \cdots \phi(x_k) \rangle_{c,J} = (-i)^{k-1} \frac{\delta^k W}{\delta J(x_1) \cdots \delta J(x_k)}. \qquad 9.$$

This can be taken to define the connected part, but it also agrees with the usual graphical sense of connectedness. When this average is evaluated at $J = 0$, it coincides with the covariant time-ordered—more properly, T^*-ordered—vacuum expectation value of ϕ.

2.1.1. The one-particle-irreducible generator. One-particle-reducible graphs are defined as those connected graphs that can be broken into two disconnected parts by simply cutting a single internal line. 1PI graphs are those connected graphs that are not one-particle reducible.

A nongraphical formulation of one-particle reducibility of this sort can be had by performing a Legendre transformation on the functional $W[J]$ (5). With this choice, if the mean field φ is defined by

$$\varphi(J) \equiv \frac{\delta W}{\delta J} = \langle \phi(x) \rangle_J, \qquad 10.$$

then the Legendre transform of $W[J]$ is defined to be the functional $\Gamma[\varphi]$, where

$$\Gamma[\varphi] \equiv W[J(\varphi)] - \int d^4x \, \varphi \, J. \qquad 11.$$

Here we imagine $J(\varphi)$ to be the external current required to obtain the expectation value $\langle \phi \rangle_J = \varphi$, and that may be found, in principle, by inverting Equation 10. For simplicity, we assume the inversion of Equation 10 is possible. If $\Gamma[\varphi]$ is known, $J(\varphi)$ can be found by directly differentiating the defining equation for $\Gamma[\varphi]$, which gives

$$\frac{\delta\Gamma[\varphi]}{\delta\varphi} + J = 0. \qquad 12.$$

This last equation implies another important property for $\Gamma[\varphi]$: Its stationary point specifies the expectation value of the original operator $\langle\phi(x)\rangle_{J=0}$. This can be seen from Equations 10 and 12 above. For time-independent field configurations, this argument can be sharpened to show that $\Gamma[\varphi]$ is the minimum expectation value of the system's Hamiltonian, given that the expectation of the field $\phi(x)$ is constrained to equal $\varphi(x)$ (6).

A graphical representation for $\Gamma[\varphi]$ can be obtained by setting up a path-integral formulation for it. An expression for $\Gamma[\varphi]$ as a path integral is found by combining the definitions of Equations 8 and 11:

$$\exp\{i\Gamma[\varphi]\} = \int \mathcal{D}\phi \exp\left\{i \int d^4x[\mathcal{L}(\phi) + J(\phi - \varphi)]\right\}$$
$$= \int \mathcal{D}\hat{\phi} \exp\left\{i \int d^4x[\mathcal{L}(\varphi + \hat{\phi}) + J\hat{\phi}]\right\}. \qquad 13.$$

At first sight, this last equation is a doubtful starting point for calculations because it gives only an implicit expression for $\Gamma[\varphi]$. Equation 13 is only implicit because the current J, appearing on the right, must itself be given as a function of φ using Equation 12 $[J = -(\delta\Gamma/\delta\varphi)]$, which again depends on $\Gamma[\varphi]$. The implicit nature of Equation 13 turns out not to be an obstacle for computing $\Gamma[\varphi]$, however. To see this, recall that a saddle-point evaluation of Equation 8 gives $W[J]$ as the sum of all connected graphs constructed using vertices and propagators built from the classical Lagrangian $\mathcal{L}$ and having the currents J as external lines. However, $\Gamma[\varphi]$ just differs from $W[J]$ by subtracting $\int d^4x \, J\varphi$ and evaluating the result at the specific configuration $J[\varphi] = -(\delta\Gamma/\delta\varphi)$. This merely lops off all of the one-particle-reducible graphs, ensuring that $\Gamma[\varphi]$ is given by summing 1PI graphs.

Semiclassically, this gives the following result for $\Gamma[\varphi]$:

$$\Gamma[\varphi] = S[\varphi] + \Gamma_1[\varphi] + (\text{Feynman graphs}). \qquad 14.$$

Here, the leading (tree-level) term is the classical action $S = \int \mathcal{L}(\varphi)d^4x$. The next-to-leading result is the usual one-loop functional determinant

$$i\Gamma_1 = \mp\frac{1}{2} \log \det\left[\delta^2 S/\delta\phi(x)\delta\phi(y)\right]_{\phi=\varphi} \qquad 15.$$

of the quadratic part of the action expanded about the configuration $\phi = \varphi$. The sign $\mp$ is negative for bosonic fields and positive for fermionic ones. Finally, the contribution of the Feynman graphs denotes the sum of all possible graphs that (*a*) involve two or more loops, (*b*) have no external lines [with φ-dependent internal lines constructed by inverting $\delta^2 S/\delta\phi(x)\delta\phi(y)$ evaluated at $\phi = \varphi$], and (*c*) are 1PI.

2.1.2. Light-particle correlations. We now specialize this general framework to the specific model described above that contains the two scalar fields, working in the limit where one is much more massive than the other. If we couple an external current j to l, and another external current J to H, then the generator of 1PI correlations in this model, $\Gamma[\ell, h]$, depends on two scalar variables, $\ell = \langle l \rangle_{jJ}$ and $h = \langle H \rangle_{jJ}$. The correlation functions it generates can be used to construct general scattering amplitudes for both l and H particles by using standard techniques.

Our interest is in determining the dependence on the heavy mass M of low-energy observables, where low energy here means those observables for which all particles involved have CM momenta and energies that satisfy $p, E \ll M$. If only l particles are initially present in any scattering process at such low energies, then no heavy particles can ever appear in the final state because there is not enough energy available for their production. As a result, the heavy-field correlations can be ignored, and it suffices to consider only the generator of 1PI correlations exclusively for the light fields in the problem.

Because H correlators are of no interest to us, we are free to set $J = 0$ when evaluating $\Gamma[\ell, h]$, as we never need to differentiate with respect to J. As the above sections show, the condition $J = 0$ is equivalent to evaluating $\Gamma[\ell, h]$ with its argument h evaluated at the stationary point $h = \bar{h}(\ell)$, where $\delta\Gamma/\delta h$ vanishes.

Furthermore, if only low-energy observables are of interest, we may also restrict the external current j to be slowly varying in space, so that its Fourier transform has support only on low-momentum modes for which $p, E \ll M$. As above for J and h, the vanishing of these j modes corresponds to using the condition $\delta\Gamma/\delta\ell = 0$ to eliminate the high-frequency components of ℓ in terms of the low-frequency components. The restriction of $\Gamma[\ell, h]$ with these two conditions, denoted $\gamma[\ell]$, in principle contains all the information required to compute any low-energy observable.

How is $\gamma[\ell]$ computed? Recall that $\Gamma[\ell, h]$, obtained by performing a Legendre transformation on both j and J, is obtained by summing over all 1PI graphs in the full theory. But $\gamma[\ell]$ differs from $\Gamma[\ell, h]$ only by setting both J and the short-wavelength components of j to zero (rather than to the configurations $J = -\delta\Gamma/\delta h$ and $j = -\delta\Gamma/\delta\ell$). Because the currents are responsible for cancelling out the one-particle reducible graphs in $\Gamma[\ell, h]$, setting currents to vanish simply means that this cancellation does not take place.

We see from this that $\gamma[\ell]$ is given by the sum of one-light-particle-irreducible (1LPI) graphs: That is, the graphs that contribute to $\gamma[\ell]$ are 1PI only with respect to cutting low-momentum, low-frequency, light-particle l lines, but are one-particle reducible with respect to cutting high-momentum l lines or H lines having any momentum.

To make these manipulations concrete, the remainder of this section derives explicit expressions for both $\Gamma[\ell, h]$ and $\gamma[\ell]$ in the toy model, working to the tree-level and one-loop approximations.

2.1.3. Tree-level calculation. Suppose both $\Gamma[\ell, h]$ and $\gamma[\ell]$ are computed approximately within a semiclassical loop expansion: $\Gamma = \Gamma^t + \Gamma^{1-\text{loop}} + \cdots$ and $\gamma = \gamma^t + \gamma^{1-\text{loop}} + \cdots$. Explicit formulae for both are particularly simple at tree level because the tree-level approximation Γ^t to Γ is given simply by evaluating the classical action at $l = \ell$ and $H = h$: $\Gamma^t[\ell, h] = S[\ell, h]$. The tree-level approximation for

$\gamma^t[\ell]$ is therefore obtained by solving the classical equations of motion for h—and the high-frequency part of ℓ—as a function of the low-frequency part of ℓ and then substituting this result, $\overline{h}_t(\ell)$, back into the classical action.

We next compute the resulting expression for $\gamma^t[\ell]$ in the limit that M is much larger than all the other scales of interest. The classical equation of motion for h, which is obtained from Equations 1 and 2, is

$$\Box h - M^2 h - \frac{1}{2} g_{lh} \ell^2 h - \frac{g_h}{3!} h^3 - \frac{\tilde{m}}{2} \ell^2 - \frac{\tilde{g}_h M}{2} h^2 = 0, \qquad 16.$$

so the solution $\overline{h}_t(\ell)$ can be written formally as

$$\overline{h}_t(\ell) = \left(\Box - M^2 - \frac{g_{lh}}{2} \ell^2 \right)^{-1} \left(\frac{\tilde{m}}{2} \ell^2 + \frac{g_h}{3!} \overline{h}^3 + \frac{\tilde{g}_h M}{2} \overline{h}^2 \right). \qquad 17.$$

This may be solved perturbatively in powers of g_h, $\tilde{g}_h$, and $1/M$, with the leading contribution obtained by taking $\overline{h} = 0$ on the right. This leads to the explicit solution

$$\overline{h}_t(\ell) = \left[-\frac{1}{M^2} - \frac{1}{M^4} \left(\Box - \frac{g_{lh}}{2} \ell^2 \right) + \cdots \right] \left(\frac{\tilde{m}}{2} \ell^2 \right) + \mathcal{O}(g_h, \tilde{g}_h)$$

$$= -\frac{\tilde{m}}{2M^2} \ell^2 + \frac{g_{lh} \tilde{m}}{4M^4} \ell^4 - \frac{\tilde{m}}{2M^4} \Box(\ell^2) + \mathcal{O}\left(\frac{1}{M^5} \right), \qquad 18.$$

where the leading contributions involving nonzero g_h and $\tilde{g}_h$ contribute at $\mathcal{O}(M^{-5})$.

Substituting this last result back into the classical action gives the tree-level expression for the generating functional $\gamma[\ell]$ as an expansion in powers of $1/M$. We find the result $\gamma^t[\ell] = \gamma_0^t[\ell] + \gamma_2^t[\ell]/M^2 + \gamma_4^t[\ell]/M^4 + \cdots$, with

$$\gamma_0^t[\ell] = \int d^4x \left(-\frac{1}{2} \partial_\mu \ell \partial^\mu \ell - \frac{m^2}{2} \ell^2 - \frac{g_l}{4!} \ell^4 \right),$$

$$\frac{\gamma_2^t[\ell]}{M^2} = \int d^4x \left(\frac{\tilde{m}^2}{8M^2} \ell^4 \right),$$

$$\frac{\gamma_4^t[\ell]}{M^4} = \int d^4x \left(-\frac{\tilde{m}^2}{2M^4} \ell^2 \partial_\mu \ell \partial^\mu \ell - \frac{g_{lh} \tilde{m}^2}{16M^4} \ell^6 \right). \qquad 19.$$

These expressions for $\gamma^t[\ell]$ also have a simple representation in terms of Feynman graphs. All of the M-dependent terms may be obtained by summing connected tree diagrams that have only ℓ particles as external lines and only h particles for internal lines, and then Taylor expanding all the h propagators in powers of $1/M$. The graphs of this type, which contribute up to and including $O(1/M^6)$, are given explicitly in **Figure 2**.

This calculation brings out several noteworthy points:

■ Decoupling: All of the M dependence in $\gamma^t[\ell]$ vanishes as $M \to \infty$, reflecting (at tree level) the general result that particles decouple from low-energy physics in the limit that their mass becomes large.

■ Truncation: The only part of $\gamma^t[\ell]$ that survives as $M \to \infty$ consists of those terms in the classical action that are independent of the heavy field h. That is, at tree level, $\gamma_0^t[\ell]$ is obtained from the classical action $S[\ell, h]$ simply by truncating it to $h = 0$: $\gamma_0^t[\ell] = S[\ell, 0]$. Note that this relies on two assumptions: the absence of $M^3 H$ terms in the potential and the condition $\tilde{m} \ll M$. [Even though $\overline{h}^t(\ell)$

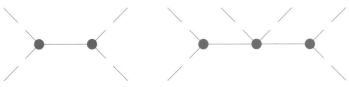

Figure 2

The Feynman graphs that contribute the corrections to the tree-level generating functional $\gamma[\ell]$ to order $1/M^4$. Solid (dashed) lines represent heavy (light) scalars.

would still vanish like $1/M$ if $\tilde{m}$ were $O(M)$, this is not enough to ensure the vanishing of terms such as $M^2(\bar{b}')^2$. This latter condition often breaks down in supersymmetric theories (7).]

- Locality: Although the exact expression for $\bar{b}_t(\ell)$, and hence for $\gamma'[\ell]$, involves nonlocal quantities such as $G(x, x') = \langle x | (-\Box + M^2)^{-1} | x' \rangle$, the entire result becomes local once these are expanded in powers of $1/M$ because of the locality of propagators in this limit:

$$G(x, x') = \int \frac{d^4 p}{(2\pi)^4} \left[\frac{e^{ip(x-x')}}{p^2 + M^2} \right] = \left[\frac{1}{M^2} + \frac{\Box}{M^4} + \cdots \right] \delta^4(x - x'). \qquad 20.$$

This locality is ultimately traceable to the uncertainty principle. The key observation is that the M-dependent interactions in $\gamma[\ell]$ express the effects of virtual heavy particles whose energies, $E_h \geq M$, are much higher than those of any of the light particles whose scatterings are under consideration in the low-energy limit. Indeed, it is precisely the high energy of these particles that precludes their inadvertent production as real final-state particles in any low-energy scattering process, and so guarantees that they only mediate transitions among the light-particle states.

However, the virtual contribution of heavy particles to low-energy processes can nevertheless occur within perturbation theory because the uncertainty principle permits the violation of energy conservation required for their production, provided it takes place only over short enough times, $\Delta t \leq 1/\Delta E_h \leq 1/M$. As a consequence, from the low-energy perspective, the influence of heavy particles appears to be instantaneous (i.e., local in time). The uncertainty principle similarly relates the momentum required to produce heavy virtual particles with the distances over which they can travel, thereby making their influence also local in space. We assume a relativistic dispersion relation ($E^2 = p^2 + m^2$) so states having large momentum also have high energy and must therefore be excluded from the low-energy theory.

The influence at low energies, E, of very massive particles, $M \gg E$, can therefore be generically reproduced to a fixed order in E/M by local interactions involving only the particles present at low energies. It is precisely this locality that makes possible the construction of a low-energy effective theory that accurately describes these virtual effects.

Redundant interactions. At face value, Equation 19 is not precisely the same as the effective potential $V^{(4)} + V^{(6)}$ encountered in Equations 5 and 6 of the introduction. This apparent difference is illusory, however, because the difference can be removed simply by performing a field redefinition.

To see how this works, suppose we have an action $S[\phi^i]$, given as a series in some small parameter ε:

$$S[\phi^i] = S_0[\phi^i] + \varepsilon S_1[\phi^i] + \varepsilon^2 S_2[\phi^i] + \cdots, \qquad 21.$$

where ε could be a small loop-counting parameter or a small energy ratio E/M. Suppose also that somewhere among the interactions appearing in the $O(\varepsilon^n)$ contribution $S_n[\phi^i]$, there is a term, $S_n^R[\phi^i]$, that vanishes when the fields ϕ^i are chosen to satisfy the equations of motion for the lowest-order action $S_0[\phi^i]$. In equations,

$$S_n^R[\phi] = \int d^4x f^i(x) \frac{\delta S_0}{\delta \phi^i(x)}, \qquad 22.$$

where the coefficients $f^i(x)$ are ultralocal functions of the fields and their derivatives at the space-time point x.

The claim is that, to order ε^n, any such interaction, $S_n^R[\phi^i]$, can be removed by performing a field redefinition without altering any of the other terms at lower or equal order, and so can have no physical consequences. The required field redefinition is

$$\phi^i(x) \to \tilde{\phi}^i(x) = \phi^i(x) - \varepsilon^n f^i(x), \qquad 23.$$

because under this redefinition the $O(\varepsilon^n)$ terms in the action vary into

$$S[\phi^i] \to S[\tilde{\phi}^i] = S[\phi^i] - \varepsilon^n \int d^4x \, f^i(x) \frac{\delta S_0}{\delta \phi^i(x)} + O(\varepsilon^{n+1}). \qquad 24.$$

Clearly, to $O(\varepsilon^n)$, the sole effect of this redefinition is simply to cancel $S_n^R[\phi^i]$.

Applying this reasoning to $\gamma[\ell]$ above, note that integrating by parts gives

$$\int d^4x \ell^2 \partial_\mu \ell \partial^\mu \ell = -\frac{1}{3} \int d^4x \ell^3 \, \Box \, \ell, \qquad 25.$$

and so using the lowest-order equations of motion for ℓ derived from $\gamma_0^t[\ell]$—that is, $\Box \ell = m^2 \ell + g_\ell \ell^3/3!$—implies

$$\int d^4x \ell^2 \partial_\mu \ell \partial^\mu \ell = -\frac{1}{3} \int d^4x \left(m^2 \ell^4 + \frac{g_\ell}{3!} \ell^6 \right). \qquad 26.$$

Using this in Equation 19 reproduces the potential of Equations 5 and 6.

2.1.4. One-loop calculation. More can be learned by examining some of the subdominant terms in the loop expansion for $\gamma[\ell]$. At face value, keeping one-loop corrections changes $\gamma[\ell]$ in two different ways: through the one-loop corrections to the functional form of $\Gamma[\ell, b]$, and through the corrections that these imply for the stationary point $\overline{b}(\ell)$. In practice, however, the correction to $\overline{b}(\ell)$ does not contribute to $\gamma[\ell]$ at one loop. To see this, use the expansions $\Gamma = \Gamma^t + \Gamma^{1-\mathrm{loop}} + \cdots$ and $\overline{b} = \overline{b}^t + \overline{b}^{1-\mathrm{loop}} + \cdots$ in the definition for $\gamma[\ell]$. To one-loop order, this gives

$$\begin{aligned}
\gamma[\ell] &= \Gamma^t[\ell, \overline{b}_t + \overline{b}^{1-\mathrm{loop}}] + \Gamma^{1-\mathrm{loop}}[\ell, \overline{b}^t] + \cdots \\
&= \left[\Gamma^t[\ell, \overline{b}^t] + \int d^4x \left(\frac{\delta \Gamma^t}{\delta b} \right)_{\overline{b}^t} \overline{b}^{1-\mathrm{loop}} \right] + \Gamma^{1-\mathrm{loop}}[\ell, \overline{b}^t] + \cdots \\
&= \gamma^t[\ell] + \Gamma^{1-\mathrm{loop}}[\ell, \overline{b}^t] + \cdots,
\end{aligned} \qquad 27.$$

where the second equality follows by expanding Γ^t about $b = \overline{b}^t$, keeping only terms of tree level and one-loop order, and the last equality uses the fact that $\delta \Gamma^t / \delta b = 0$ when evaluated at $\overline{b}^t$, together with the tree-level result $\gamma^t[\ell] = \Gamma^t[\ell, \overline{b}^t(\ell)]$. Equation 27 states that the one-loop approximation to γ is obtained by evaluating $\Gamma^{1-\text{loop}}$ at the tree-level configuration $\overline{b}^t$: That is, $\gamma^{1-\text{loop}}[\ell] = \Gamma^{1-\text{loop}}[\ell, \overline{b}^t(\ell)]$.

To proceed, we require the one-loop contribution $\Gamma^{1-\text{loop}}$, which is given by

$$\Gamma^{1-\text{loop}}[\ell, b] = \frac{i}{2} \log \det \begin{bmatrix} (-\square + V_{\ell\ell})/\mu^2 & V_{\ell b}/\mu^2 \\ V_{\ell b}/\mu^2 & (-\square + V_{bb})/\mu^2 \end{bmatrix}, \qquad 28.$$

where μ is an arbitrary scale required on dimensional grounds, and the matrix of second derivatives of the scalar potential is given by

$$
\begin{aligned}
V &= \begin{pmatrix} V_{\ell\ell} & V_{\ell b} \\ V_{\ell b} & V_{bb} \end{pmatrix} \\
&= \begin{pmatrix} m^2 + \frac{g_l}{2}\ell^2 + \frac{g_{lb}}{2}b^2 + \tilde{m}b & g_{lb}\ell b + \tilde{m}\ell \\ g_{lb}\ell b + \tilde{m}l & M^2 + \tilde{g}_b Mb + \frac{g_b}{2}b^2 + \frac{g_{lb}}{2}\ell^2 \end{pmatrix}. \qquad 29.
\end{aligned}
$$

Evaluating the functional determinant in the usual way gives an expression that diverges in the UV. For the two-scalar theory under consideration, these divergences appear (at one loop) only in the part of $\Gamma[\ell, b]$ that does not depend on derivatives of ℓ or b (i.e., the scalar effective potential). If the divergent terms are written as $\Gamma_{\text{div}} = -\int d^4x V_{\text{div}}$, then V_{div} is

$$V_{\text{div}} = \frac{1}{32\pi^2} \left[C + (V_{\ell\ell} + V_{bb})\Lambda^2 - \frac{1}{2}\left(V_{\ell\ell}^2 + V_{bb}^2 + 2V_{\ell b}^2\right)L \right], \qquad 30.$$

where C is a field-independent divergent constant, and $L = \log\left(\Lambda^2/\mu^2\right)$. Λ is a UV cutoff that has been used to regulate the theory and that is assumed to be sufficiently large compared to all other scales in the problem that all inverse powers of Λ can be neglected.

Alternatively, using dimensional regularization instead of a UV cutoff leads to the same expressions as above, with two changes: (a) all of the terms proportional to C or to Λ^2 are set to zero, and (b) the logarithmic divergence is replaced by $L \to 2/(4-n) = 1/\varepsilon$, where $n = 4 - 2\varepsilon$ is the dimension of space-time.

All the divergences can be absorbed into renormalizations of the parameters of the Lagrangian by defining the following renormalized couplings:

$$
\begin{aligned}
A_R &= A_0 + \frac{1}{32\pi^2}\left[C + (M^2 + m^2)\Lambda^2 - \frac{1}{2}(M^4 + m^4)L \right], \\
B_R &= B_0 + \frac{1}{32\pi^2}\left[(\tilde{m} + \tilde{g}_b M)\Lambda^2 - \left(m^2\tilde{m} + \tilde{g}_b M^3\right)L \right], \\
m_R^2 &= m^2 + \frac{1}{32\pi^2}\left[(g_l + g_{lb})\Lambda^2 - \left(g_l m^2 + g_{lb}M^2 + 2\tilde{m}^2\right)L \right], \\
M_R^2 &= M^2 + \frac{1}{32\pi^2}\left[(g_b + g_{lb})\Lambda^2 - \left(g_{lb}m^2 + \left(g_b + \tilde{g}_b^2\right)M^2 + \tilde{m}^2\right)L \right],
\end{aligned}
$$

$$\tilde{m}_R = \tilde{m} - \frac{1}{32\pi^2}\left(g_l\tilde{m} + g_{lb}\tilde{g}_b M + 4g_{lb}\tilde{m}\right)L,$$

$$(\tilde{g}_b)_R = \tilde{g}_b - \frac{3}{32\pi^2}\left(g_{lb}\frac{\tilde{m}}{M} + g_b\tilde{g}_b\right)L,$$

$$(g_l)_R = g_l - \frac{3}{32\pi^2}\left(g_l^2 + g_{lb}^2\right)L,$$

$$(g_b)_R = g_b - \frac{3}{32\pi^2}\left(g_b^2 + g_{lb}^2\right)L,$$

$$(g_{lb})_R = g_{lb} - \frac{1}{32\pi^2}(g_l + g_b + 4g_{lb})g_{lb}L. \qquad 31.$$

In these expressions, A_0 and B_0 are the coefficients of the terms $A + Bb$, which do not appear in the classical potential but are not forbidden by any symmetries. By not including such terms in the classical potential, we are implicitly choosing A_0 and B_0 to satisfy the renormalization condition $A_R = B_R = 0$, which can be ensured by appropriately shifting the renormalized fields. Similarly, the assumed hierarchy of masses between the light field ℓ and the heavy field b assumes the renormalized quantities satisfy $m_R, \tilde{m}_R \ll M_R$. Note that these conditions are unnatural in that they require that large $O(M)$ renormalization corrections must be cancelled by the bare quantities A_0, B_0, m, and $\tilde{m}$, in addition to the cancellation of any divergent parts. In what follows, we assume that this renormalization has been performed and that all parameters appearing in subsequent expressions are the renormalized quantities— even though the subscript R is not explicitly written.

Once this renormalization has been performed, the remaining expression is finite and its dependence on the heavy mass scale M can be identified. As for the tree-level analysis, it is convenient to organize $\gamma^{1-\text{loop}}[\ell]$ into an expansion in powers of $1/M$. A new feature that arises at one loop is that the dominant term in this expansion now varies as $M^4 \log M$ rather than as M^0 when $M \to \infty$. We therefore write $\gamma^{1-\text{loop}} = \gamma_{-4}^{1-\text{loop}}M^4 + \gamma_{-2}^{1-\text{loop}}M^2 + \gamma_{-1}^{1-\text{loop}}M + \gamma_0^{1-\text{loop}} + \gamma_2^{1-\text{loop}}/M^2 + \cdots$.

Those terms in $\gamma^{1-\text{loop}}[\ell]$ that involve derivatives of ℓ first appear at order $1/M^2$ in this expansion. This result is a special feature of the two-scalar system at one loop. At two loops, or at one loop for other systems, kinetic terms such as $\partial_\mu \ell \, \partial^\mu \ell$ can appear proportional to logarithms of M. To identify all terms larger than $O(1/M^2)$, it therefore suffices to work only with the effective scalar potential. After performing the renormalizations indicated in Equations 31, the one-loop result for the scalar potential becomes

$$\gamma_{\text{pot}}^{1-\text{loop}}[\ell] = -\frac{1}{64\pi^2}\int d^4x \left[V_+^2 \log\left(\frac{V_+}{\mu^2}\right) + V_-^2 \log\left(\frac{V_-}{\mu^2}\right)\right], \qquad 32.$$

where $V_\pm$ denotes the two eigenvalues of the matrix $\mathbf{V}$, which are given explicitly by

$$V_\pm = \frac{1}{2}\left[(V_{\ell\ell} + V_{bb}) \pm \sqrt{(V_{\ell\ell} - V_{bb})^2 + 4V_{\ell b}^2}\right]. \qquad 33.$$

Our interest is in a potential for which $V_{bb} \gg V_{\ell\ell}, V_{\ell b}$, and so these expressions can be simplified to

$$V_+ \approx V_{bb} + \frac{V_{\ell b}^2}{V_{bb}} + \frac{V_{\ell\ell} V_{\ell b}^2}{V_{bb}^2} + O\left(\frac{1}{V_{bb}^3}\right),$$

$$V_- \approx V_{\ell\ell} - \frac{V_{\ell b}^2}{V_{bb}} - \frac{V_{\ell\ell} V_{\ell b}^2}{V_{bb}^2} + O\left(\frac{1}{V_{bb}^3}\right),$$

34.

and

$$\gamma_{\text{pot}}^{1-\text{loop}}[\ell] = -\frac{1}{64\pi^2} \int d^4x \left\{ V_{bb}^2 \log\left(\frac{V_{bb}}{\mu^2}\right) + V_{\ell\ell}^2 \log\left(\frac{V_{\ell\ell}}{\mu^2}\right) + V_{\ell b}^2 \left[1 + 2\log\left(\frac{V_{bb}}{\mu^2}\right)\right] \right.$$

$$\left. + \frac{2 V_{\ell\ell} V_{\ell b}^2}{V_{bb}} \log\left(\frac{V_{bb}}{V_{\ell\ell}}\right) + O\left(\frac{1}{V_{bb}^2}\right) \right\}.$$

35.

Using the explicit expressions given above for the scalar potential, and evaluating the result at $b = \overline{b}'(\ell)$, the first few terms of the $1/M$ expansion at one loop are given by

$$M^4 \gamma_{-4}^{1-\text{loop}} = -\frac{M^4}{64\pi^2} \log\left(\frac{M^2}{\mu^2}\right) \int d^4x,$$

$$M^2 \gamma_{-2}^{1-\text{loop}}[\ell] = -\frac{M^2}{64\pi^2} \left[\log\left(\frac{M^2}{\mu^2}\right) + \frac{1}{2}\right] \int d^4x \, g_{lb} \ell^2,$$

$$M \gamma_{-1}^{1-\text{loop}}[\ell] = +\frac{M}{64\pi^2} \left[\log\left(\frac{M^2}{\mu^2}\right) + \frac{1}{2}\right] \int d^4x \, \tilde{g}_b \tilde{m} \ell^2,$$

$$\gamma_0^{1-\text{loop}} = -\frac{1}{64\pi^2} \int d^4x \left[\left(\frac{g_{lb}^2}{4} \ell^4 + 2\tilde{m}^2 \ell^2\right) \log\left(\frac{M^2}{\mu^2}\right) + \frac{3 g_{lb}^2}{8} \ell^4 + \tilde{m}^2 \ell^2 \right.$$

$$\left. + \left(m^2 + \frac{g_l^2}{2} \ell^2\right)^2 \log\left(\frac{m^2 + \frac{g_l^2}{2} \ell^2}{\mu^2}\right) \right].$$

36.

We see that this shares two crucial properties of the tree-level result:

- Decoupling: Note that, superficially, the effects of the heavy particle no longer appear to vanish as $M \to \infty$. However, all terms that grow as M grows have the same form as the classical Lagrangian, and so they can all be absorbed into finite renormalizations of A_R, m_R^2, and $(g_l)_R$. That is, if we define the new quantities,

$$A'_R = A_R + \frac{M^4}{64\pi^2} \log\left(\frac{M^2}{\mu^2}\right),$$

$$(m_R^2)' = m_R^2 + \frac{1}{64\pi^2} \left\{ M^2 \left(g_{lb} - \tilde{g}_b \frac{\tilde{m}}{M}\right) \left[1 + 2\log\left(\frac{M^2}{\mu^2}\right)\right] + 4\tilde{m}^2 \log\left(\frac{M^2}{\mu^2}\right) \right\},$$

$$(g_l)'_R = (g_l)_R + \frac{3 g_{lb}^2}{32\pi^2} \log\left(\frac{M^2}{\mu^2}\right),$$

37.

then the one-loop contribution $\gamma^{1-\text{loop}}[\ell]$ to the 1LPI generator becomes

$$\gamma^{1-\text{loop}}[\ell] = -\frac{1}{64\pi^2} \int d^4x \left[\frac{3 g_{lb}^2}{8} \ell^4 + \tilde{m}^2 \ell^2 + \left(m^2 + \frac{g_l^2}{2} \ell^2\right)^2 \log\left(\frac{m^2 + \frac{1}{2} g_l^2 \ell^2}{\mu^2}\right) \right]$$

$$+ O\left(\frac{1}{M}\right),$$

38.

where m^2 represents $(m_R^2)'$, and so on.

Clearly, after such renormalizations are performed, all the remaining M dependence vanishes in the limit $M \to \infty$. Provided the values of renormalized couplings are in any case inferred from experiment, all the physical effects of the heavy particle are suppressed for large M, ensuring the heavy particle does decouple from physical observables.

■ Locality: Because the one-loop action $\gamma^{1-\text{loop}}[\ell]$ is the integral over space-time of a quantity evaluated at a single space-time point when expanded in inverse powers of M, it shares the locality of the classical result. The underlying source of this locality is again the uncertainty principle, which precludes violations of energy and momentum conservation over large distances—a result that hinges on our keeping only states defined by their low energy.

2.2. The Wilson Action

We next reorganize the same calculation, with the goal of making the M dependence of physical results manifest from the outset. To this end, suppose we start from the path-integral expression for the 1LPI generating functional $\gamma[\ell]$, derived as Equation 13 above:

$$\exp\{i\gamma[\ell]\} = \int \mathcal{D}l\, \mathcal{D}H \, \exp\left\{i \int d^4x [\mathcal{L}(\ell + l, H) + jl]\right\}, \qquad 39.$$

with the external current j defined by $j[\ell] = -\delta\gamma[\ell]/\delta\ell$. No similar current is coupled to the heavy field H in Equation 39 because our attention is restricted to low-energy processes for which no heavy particles appear in the initial or final states.

2.2.1. The Wilson action.
Now imagine schematically dividing the functional integral into its low-energy and high-energy parts, $\mathcal{D}l\, \mathcal{D}H = [\mathcal{D}l]_{\text{l.e.}} [\mathcal{D}l\, \mathcal{D}H]_{\text{h.e.}}$, relative to some arbitrary intermediate scale λ. (For instance, this might be done by requiring high-energy modes to satisfy $p^2 + m^2 > \lambda^2$ in Euclidean signature.) Using this distinction between low- and high-energy modes, it becomes possible to perform the functional integration over, or to integrate out, the high-energy modes once and for all:

$$\exp\{i\gamma[\ell]\} = \int [\mathcal{D}l]_{\text{l.e.}} \exp\left\{i \int d^4x [\mathcal{L}_W(\ell + l_{\text{l.e.}}) + jl_{\text{l.e.}}]\right\}, \qquad 40.$$

where $S_W = \int d^4x \mathcal{L}_W$ is known as the Wilson action and defined as the result of performing the high-energy part of the functional integral:

$$\exp\{i S_W[\ell + l_{\text{l.e.}}]\} = \int [\mathcal{D}l\, \mathcal{D}H]_{\text{h.e.}} \exp\{i S[\ell + l_{\text{l.e.}}, l_{\text{h.e.}}, H_{\text{h.e.}}]\}. \qquad 41.$$

Equations 40 and 41 are the central definitions from which the calculation of $\gamma[\ell]$ à la Wilson proceeds.

There are two points about these last expressions that bear special emphasis. Note first that $\mathcal{L}_W$ appears in Equation 40 in precisely the same way as would the classical Lagrangian in a theory for which no heavy field existed. Consequently, once $\mathcal{L}_W$ is known, it may be used to compute $\gamma[\ell]$ in the usual way: One must sum over all 1LPI vacuum Feynman graphs using the interactions and propagators for the light fields dictated by the effective Lagrangian density $\mathcal{L}_W$.

The second point, which is what makes Equation 40 so useful, is that because $\mathcal{L}_W$ is computed by integrating only over high-energy modes, the uncertainty principle guarantees that it is local once it is expanded in inverse powers of the heavy scales. Consequently, to the extent that we work only to a fixed order in this expansion, we need not worry that Equation 41 will generate arbitrary nonlocal interactions.

2.2.2. The physics of renormalization.
Equations 40 and 41 share another beautiful feature. Although the Wilson action depends explicitly on the scale λ that is used in its definition, this dependence always drops out of any physical observables. It must drop out because it only arises from our choice to perform the calculation in two steps: first integrating modes heavier than λ and then integrating the lighter modes.

In detail, this cancellation arises because λ enters into the low-energy part of the calculation in two ways. The first way is through the explicit λ dependence of all couplings of the Wilson action S_W. However, λ also enters because all contributions of virtual particles in the low-energy theory have their momenta cutoff at the scale λ. The λ dependence of the couplings in S_W is just what is required to cancel the λs entering through the cutoff.

This entire discussion of λ cancellation induces a strong sense of déja vu because it parallels exactly the traditional renormalization program wherein the regularization dependence of divergent loop integrals is cancelled by introducing regularization-dependent interactions (or counterterms) into the classical action S. This similarity makes it irresistible to regard the original classical action $S[l, H]$ as itself being the Wilson action for a yet more fundamental theory that applies at still higher energies, above the cutoff Λ. Any such Wilson action would be used to compute physical observables in precisely the same way as one traditionally uses the classical action, including the renormalization of all couplings to cancel the cutoff dependence of all observable quantities. The great benefit of adopting this point of view is the insight it gives into the physical nature of this cancellation.

2.2.3. The dimensionally regularized Wilson action.
The Wilson action defined with an explicit cutoff is somewhat cumbersome for practical calculations, for a variety of reasons. Cutoffs make it difficult to keep the gauge symmetries of a problem manifest when there are spin-one gauge bosons (such as photons) in the problem. Cutoffs also complicate our goal of following how the heavy scale M appears in physically interesting quantities such as $\gamma[\ell]$, because they muddy the dimensional arguments used to identify which interactions in S_W contribute to observables order by order in $1/M$.

It is much more convenient to use dimensional regularization, even though dimensional regularization seems to run counter to the entire spirit of a low-energy action by keeping arbitrarily high momenta within the effective theory. This is not a problem in practice, however, because the error we make by keeping such high-momentum modes can always be absorbed into an appropriate renormalization of the effective couplings. This is always possible precisely because our mistake is to keep high-energy modes, whose contributions at low energies can always be represented using local effective interactions. Whatever damage we do by using dimensional regularization to define the low-energy effective action can always be undone by appropriately renormalizing our effective couplings.

We are led to the following prescription for defining a dimensionally regularized effective action in the two-scalar toy model: First, dimensionally regulate the full theory involving both fields l and H, using for convenience the mass-independent $\overline{MS}$ renormalization scheme. At one loop, this amounts to renormalizing as in Equations 31, but with all the quartically and quadratically divergent terms set to zero and substituting $L \to 1/\varepsilon + k$ in the logarithmically divergent terms, where $k = \gamma - \log(4\pi)$ and $\gamma = 0.577215\ldots$ is the Euler-Mascherelli constant. Next, define the effective theory to include only the light field l, also regulated using dimensional regularization. However, rather than using minimal subtraction in the effective theory, we instead renormalize the effective couplings by demanding that they successfully reproduce the low-energy limit of the full theory, using for this purpose any convenient set of observables. Once this matching calculation has been done, the resulting effective theory can be used to compute any other quantities as required. This construction is best understood using the concrete example of the two-scalar model.

2.2.4. Tree-level calculation. Imagine computing both γ and $\mathcal{L}_W$ within the loop expansion: $\gamma = \gamma^t + \gamma^{1-\text{loop}} + \cdots$ and $\mathcal{L}_W = \mathcal{L}_W^t + \mathcal{L}_W^{1-\text{loop}} + \cdots$. At tree level, the distinction between $S_W = \int d^4x \mathcal{L}_W$ and the 1LPI generator γ completely degenerates. This is because the tree-level approximation to $\gamma[\ell]$ is simply obtained by evaluating the integrands of Equations 40 and 41 at the classical saddle point. In the present case, this implies that γ^t is simply given by evaluating $\mathcal{L}_W^t$ at $l = \ell$, leading to

$$\gamma^t[\ell] = \int d^4x \mathcal{L}_W^t(\ell). \tag{42}$$

Similarly, evaluating the path-integral expression of Equation 41 to obtain $\mathcal{L}_W$ in the tree-level approximation entails evaluating the classical action at the saddle point $l = \ell$ and $H = \overline{b}^t(\ell)$. Graphically, this gives the tree-level Wilson action as the sum over all tree graphs that have only heavy particles propagating in their internal lines and only light particles for external lines. Retracing the steps taken in previous sections to compute $\gamma[\ell]$ at tree level gives an explicit expression for $\mathcal{L}_W^t = \mathcal{L}(\ell, \overline{b}^t(\ell))$:

$$
\begin{aligned}
\mathcal{L}_W^t(\ell) &= -\frac{1}{2}\partial_\mu \ell \partial^\mu \ell - \frac{m^2}{2}\ell^2 - \left(\frac{g_l}{4!} - \frac{\tilde{m}^2}{8M^2}\right)\ell^4 \\
&\quad - \left(\frac{\tilde{m}^2}{2M^4}\right)\ell^2 \partial_\mu \ell \partial^\mu \ell - \left(\frac{g_{lb}\tilde{m}^2}{16M^4}\right)\ell^6 + \cdots, \tag{43} \\
&= -\frac{1}{2}\partial_\mu \ell \partial^\mu \ell - \frac{m^2}{2}\ell^2 - \left(\frac{g_l}{4!} - \frac{\tilde{m}^2}{8M^2} - \frac{m^2\tilde{m}^2}{6M^4}\right)\ell^4 \\
&\quad - \left(\frac{g_{lb}\tilde{m}^2}{16M^4} - \frac{g_l\tilde{m}^2}{36M^4}\right)\ell^6 + \cdots, \tag{44}
\end{aligned}
$$

where the freedom to redefine fields has been used, and where ellipses represent terms of higher order in $1/M$ than those displayed. This result also could be obtained by asking which local Lagrangian involving only light fields reproduces the two-body and three-body scattering of the full theory to $O(1/M^4)$.

2.2.5. One-loop calculation.

At one loop, many of the nontrivial features of the Wilson action emerge for the first time. As usual, we assume the renormalized dimensional parameters, which ensure the hierarchy of scales in the two-scalar model satisfies $m \sim \tilde{m} \ll M$.

The most general possible Wilson action, which is local and consistent with the symmetry $l \to -l$, is

$$S_W[\ell] = - \int d^4x \left[a_0 + \frac{a_2}{2} \ell^2 + \frac{a_4}{4!} \ell^4 + \frac{1}{2}(1 + b_2)\partial_\mu \ell \partial^\mu \ell + \cdots \right], \qquad 45.$$

where the ellipses denote higher-dimension interactions, and the constants a_k, b_k, and so on are determined by matching to the full theory, as is now described.

For simplicity, we specialize to matching to $O(M^0)$ because this allows us to specialize to background configurations for which ℓ is space-time independent—that is, $\partial_\mu \ell = 0$. The only effective couplings relevant in this case (more about this in the next section) are a_0, a_2, and a_4, and it is convenient to determine these by requiring that the one-loop result for $\gamma^{1-\text{loop}}[\ell]$, computed using S_W, agrees with the result computed with the full theory, to $O(M^0)$.

The previous section gives the one-loop calculation in the full theory to this order as Equation 35, evaluated at $h = \overline{h}'(\ell) = -(\tilde{m}/2M^2)\ell^2 + O(1/M^4)$. This gives $\gamma^{1-\text{loop}}_{\text{pot}}[\ell] = - \int d^4x V^{1-\text{loop}}$, with $V^{1-\text{loop}}$ as given in Equation 36:

$$V^{1-\text{loop}}(\ell) = \frac{1}{64\pi^2} \left\{ M^4 \log\left(\frac{M^2}{\mu^2}\right) + (g_{lb}M^2 - \tilde{g}_b \tilde{m}M + 2\tilde{m}^2) \left[\log\left(\frac{M^2}{\mu^2}\right) + \frac{1}{2} \right] \ell^2 \right.$$
$$\left. + \frac{g_l^2}{4} \left[\log\left(\frac{M^2}{\mu^2}\right) + \frac{3}{2} \right] \ell^4 + \left(m^2 + \frac{g_l}{2} \ell^2 \right)^2 \log\left(\frac{m^2 + \frac{1}{2} g_l \ell^2}{\mu^2}\right) \right\}, \qquad 46.$$

neglecting $O(1/M)$ terms. With UV divergences regularized in $n = 4 - 2\varepsilon$ dimensions, and renormalized using the $\overline{MS}$ renormalization scheme, the counterterms used to obtain this expression are those of Equation 31:

$$A_R = A_0 - \frac{1}{64\pi^2}(M^4 + m^4)L,$$

$$B_R = B_0 - \frac{1}{32\pi^2}(m^2\tilde{m} + \tilde{g}_b M^3)L,$$

$$m_R^2 = m^2 - \frac{1}{32\pi^2}(g_l m^2 + g_{lb}M^2 + 2\tilde{m}^2)L,$$

$$M_R^2 = M^2 - \frac{1}{32\pi^2}\left(g_{lb}m^2 + (g_b + \tilde{g}_b^2) M^2 + \tilde{m}^2 \right)L,$$

$$\tilde{m}_R = \tilde{m} - \frac{1}{32\pi^2}(g_l\tilde{m} + g_{lb}\tilde{g}_b M + 4g_{lb}\tilde{m})L,$$

$$(\tilde{g}_b)_R = \tilde{g}_b - \frac{3}{32\pi^2}\left(g_{lb}\frac{\tilde{m}}{M} + g_b\tilde{g}_b \right)L,$$

$$(g_l)_R = g_l - \frac{3}{32\pi^2}\left(g_l^2 + g_{lb}^2\right)L,$$

$$(g_b)_R = g_b - \frac{3}{32\pi^2}\left(g_b^2 + g_{lb}^2\right)L,$$

$$(g_{lb})_R = g_{lb} - \frac{1}{32\pi^2}(g_l + g_b + 4g_{lb})g_{lb}L,$$

47.

with $L = 1/\varepsilon + k$.

Repeating the same calculation using the Wilson action (Equation 45) instead leads to

$$V_{\mathrm{eff}}^{1-\mathrm{loop}} = \frac{1}{64\pi^2}\left(\hat{a}_2 + \frac{\hat{a}_4}{2}\ell^2\right)^2 \log\left(\frac{\hat{a}_2 + \frac{\hat{a}_4}{2}\ell^2}{\mu^2}\right),$$

48.

where $\hat{a}_2 = a_2/(1+b_2) \approx a_2(1 - b_2 + \cdots)$ and $\hat{a}_4 = a_4/(1+b_2)^2 \approx a_4(1 - 2b_2 + \cdots)$. Here, the a_k constants have been renormalized in the effective theory, being related to the bare couplings by expressions similar to Equation 47:

$$(a_0)_R = a_0 - \frac{a_2^2}{64\pi^2}L,$$

$$(a_2)_R = a_2 - \frac{a_4 a_2}{32\pi^2}L,$$

47.

$$(a_4)_R = a_4 - \frac{3a_4^2}{32\pi^2}L.$$

49.

The a_k constants are now fixed by performing the UV-finite renormalization required to make the two calculations of $V^t + V^{1-\mathrm{loop}}$ agree. The b_k constants are similarly determined by matching the two-derivative terms in $\gamma^{1-\mathrm{loop}}$, and so on. [For the scalar model under consideration, this gives $b_2 = 0$ to the order of interest, as a single H loop first contributes to the $\partial_\mu \ell \partial^\mu \ell$ term at $O(1/M^2)$.] Requiring the tree-level contribution $V_{\mathrm{eff}}^t = a_0 + \frac{1}{2}a_2\ell^2 + \frac{1}{4!}a_4\ell^4$ to capture the terms in $V^{1-\mathrm{loop}}$ that are missing in $V_{\mathrm{eff}}^{1-\mathrm{loop}}$ gives the required effective couplings in the Wilson action:

$$(a_0)'_R = A + \frac{1}{64\pi^2}M^4 \log\left(\frac{M^2}{\mu^2}\right),$$

$$(a_2)'_R = m^2 + \frac{1}{64\pi^2}(g_{lb}M^2 - \tilde{g}_b\tilde{m}M + 2\tilde{m}^2)\left[2\log\left(\frac{M^2}{\mu^2}\right) + 1\right],$$

50.

$$(a_4)'_R = g_l + \frac{3g_l^2}{32\pi^2}\left[\log\left(\frac{M^2}{\mu^2}\right) + \frac{3}{2}\right].$$

Such finite renormalizations, arising as a heavy particle is integrated out, are termed threshold corrections, and it is through these that the explicit powers of M get into the low-energy theory in dimensional regularization. [Note also that the coefficients of M^2 in these expressions need not agree with those of Λ^2 in a cutoff low-energy theory, revealing the fallacy of using quadratic divergences in an effective theory to track heavy-mass dependence (8).]

3. POWER COUNTING

Clearly from the two-scalar model, effective Lagrangians typically involve a potentially infinite number of interactions corresponding to the ultimately infinite number of terms that can arise once the M dependence of physical observables is expanded in powers of $1/M$. If it were necessary to deal with even a large number of these terms, there would be no real utility in using them in practical calculations.

The most important part of an effective Lagrangian analysis is therefore the identification of which terms in $\mathcal{L}_{\text{eff}}$ are required to compute observables to any given order in $1/M$, and this is accomplished using the power-counting rules of this section.

3.1. A Class of Effective Interactions

To keep the discussion interesting but general, in this section we focus on a broad class of effective Lagrangians that can be written in the following way:

$$\mathcal{L}_{\text{eff}} = f^4 \sum_k \frac{c_k}{M^{d_k}} \mathcal{O}_k \left(\frac{\phi}{v} \right). \tag{51.}$$

In this expression, ϕ is meant to generically represent the fields of the problem, which for simplicity of presentation are taken here to be bosons and so to have the canonical dimension of mass, using fundamental units for which $\hbar = c = 1$. (Treating more general situations that can include fermions is straightforward.) The quantities f, v, and M are all constants also having the dimensions of mass. The index k runs over all the labels of the various effective interactions appearing in $\mathcal{L}_{\text{eff}}$. These are denoted by $\mathcal{O}_k$, and are assumed to have dimension $(\text{mass})^{d_k}$. Because the ratio ϕ/v is dimensionless, this entire dimension is carried by derivatives $\partial \phi$. As a result, d_k simply counts the number of derivatives appearing in the effective interaction $\mathcal{O}_k$.

3.2. Power-Counting Rules

Imagine now computing Feynman graphs using these effective interactions, with the goal of tracking how the result depends on the scales f, v, and M, as well as the mass scale m of the low-energy particles. Consider in particular a graph $\mathcal{A}_E(q)$, involving E external lines whose four momenta are collectively denoted by q. Suppose also that this graph has I internal lines and V_{ik} vertices. The labels i and k indicate two properties of the vertices, with i counting the number of lines that converge at the vertex and k counting the power of momentum that appears in the vertex. Equivalently, i counts the number of powers of the fields ϕ, which appear in the corresponding interaction term in the Lagrangian, and k counts the number of derivatives of these fields that appear there.

3.2.1. Some useful identities. The positive integers I, E, and V_{ik}, which characterize the Feynman graph in question, are not all independent because they are related by the rules for constructing graphs from lines and vertices. One such a relation is obtained by equating the two equivalent ways of counting the number of ends of internal and external lines in a graph. On one hand, because all lines end at a vertex, the number of ends is given by summing over all ends that appear in all the vertices: $\sum_{ik} i V_{ik}$. On the other hand, there are two ends for each internal line and one end for each external line in the graph: $2I + E$. Equating these gives the identity that expresses the conservation of ends:

$$2I + E = \sum_{ik} i V_{ik} \quad \text{(conservation of ends).} \qquad 52.$$

A second useful identity defines the number of loops L for each (connected) graph:

$$L = 1 + I - \sum_{ik} V_{ik} \quad \text{(definition of } L\text{).} \qquad 53.$$

This definition does not come out of thin air; for graphs that can be drawn on a plane, it agrees with the intuitive notion of the number of loops in a graph.

3.2.2. Estimating integrals.

Reading the Feynman rules from the Lagrangian of Equation 51 shows that the vertices in the Feynman graph of interest contribute a factor of

$$(\text{Vertex}) = \prod_{jk} \left[i(2\pi)^4 \delta^4(p) \left(\frac{p}{M} \right)^k \left(\frac{f^4}{v^j} \right) \right]^{V_{jk}}, \qquad 54.$$

where p generically denotes the various momenta running through the vertex. Similarly, each internal line contributes the additional factors:

$$(\text{Internal line}) = \left[-i \int \frac{d^4 p}{(2\pi)^4} \left(\frac{M^2 v^2}{f^4} \right) \frac{1}{p^2 + m^2} \right]^I, \qquad 55.$$

where, again, p denotes the generic momentum flowing through the line. m denotes the mass of the light particles that appear in the effective theory, and it is assumed that the kinetic terms that define their propagation are those terms in $\mathcal{L}_{\text{eff}}$ that involve two derivatives and two powers of the fields ϕ.

As usual for a connected graph, all but one of the momentum-conserving delta functions in Equation 54 can be used to perform one of the momentum integrals in Equation 55. The one remaining delta function left after doing so depends only on the external momenta $\delta^4(q)$ and expresses the overall conservation of four-momentum for the process. Future formulae are less cluttered if this factor is extracted once and for all, by defining the reduced amplitude A by

$$\mathcal{A}_E(q) = i(2\pi)^4 \delta^4(q) A_E(q). \qquad 56.$$

The number of four-momentum integrations left after having used all the momentum-conserving delta functions is then $I - \sum_{ik} V_{ik} + 1 = L$. This last equality uses the definition (Equation 53) of the number of loops L.

To track how the result depends on the scales in $\mathcal{L}_{\text{eff}}$, it is convenient to estimate the results of performing the various multidimensional momentum integrals using dimensional analysis. Because these integrals are typically UV divergent, they must first be regulated, and this is where the use of dimensional regularization pays off. The key observation is that if a dimensionally regulated integral has dimensions, then its size is set by the light masses or external momenta that appear in the integrand. That is, dimensional analysis applied to a dimensionally regulated integral implies

$$\int \cdots \int \left(\frac{d^n p}{(2\pi)^n} \right)^A \frac{p^B}{(p^2 + m^2)^C} \sim \left(\frac{1}{4\pi} \right)^{2A} m^{nA+B-2C}, \qquad 57.$$

with a dimensionless prefactor that depends on the dimension n of space-time, and that may be singular in the limit that $n \to 4$. Here, m represents the dominant scale that appears in the integrand of the momentum integrations. If the light particles appearing as external states in $A_E(q)$ should be massless, or highly relativistic, then the typical external momenta q are much larger than m, and m in the above expression should be replaced by q. Any logarithmic infrared mass singularities that may arise in this limit are ignored here, as our interest is in following powers of ratios of the light and heavy mass scales. q is used as the light scale controlling the size of the momentum integrations in the formulae quoted below.

With this estimate for the size of the momentum integrations, we find the following quantity appears in the amplitude $A_E(q)$:

$$\int \cdots \int \left(\frac{d^4 p}{(2\pi)^4} \right)^L \frac{p^{\sum_{ik} kV_{ik}}}{(p^2 + q^2)^I} \sim \left(\frac{1}{4\pi} \right)^{2L} q^{4L-2I+\sum_{ik} kV_{ik}}, \qquad 58.$$

which, with liberal use of the identities (Equations 52 and 53), gives an estimate for $A_E(q)$:

$$A_E(q) \sim f^4 \left(\frac{1}{v} \right)^E \left(\frac{Mq}{4\pi f^2} \right)^{2L} \left(\frac{q}{M} \right)^{2+\sum_{ik}(k-2)V_{ik}}. \qquad 59.$$

This last formula is the main result, which is used in the various applications considered later. Its utility lies in the fact that it links the contributions of the various effective interactions in the effective Lagrangian (Equation 51) with the dependence of observables on small mass ratios such as q/M. Note in particular that more and more complicated graphs, for which L and V_{ik} become larger and larger, are suppressed in their contributions to observables only if q is much smaller than the scales M and f.

Note also that the basic estimate, Equation 57, would have been much more difficult to do if the effective couplings were defined using a cutoff, λ, because in this case it is the cutoff that would dominate the integral, as it is then the largest external scale. However, knowing this cutoff dependence is less useful because the general arguments of the previous sections show that λ is guaranteed to drop out of any physical quantity.

3.3. The Effective Lagrangian Logic

Power-counting estimates of this sort suggest the following general logic concerning the use of effective Lagrangians:

- Step I: Choose the accuracy (e.g., one part per mille) with which observables, such as $A_E(q)$, are to be computed.

- Step II: Determine the order in the small mass ratios q/M or m/M that must be required to achieve the desired accuracy.
- Step III: Use the power-counting result (Equation 59) to find which terms in the effective Lagrangian are needed to compute to the desired order in q/M and m/M. Equation 59 also determines which order is required in the loop expansion for each effective interaction of interest.
- Step IVa: Compute the couplings of the required effective interactions using the full underlying theory. If this step should prove impossible, owing either to ignorance of the underlying theory or to the intractability of the required calculation, then it may be replaced by Step IVb.
- Step IVb: If the coefficients of the required terms in the effective Lagrangian cannot be computed, then they may instead be regarded as unknown parameters, which must be taken from experiment. Once a sufficient number of observables are used to determine these parameters, all other observables may be unambiguously predicted using the effective theory.

A number of points cry out for comment at this juncture.

- Utility of Step IVb: The possibility of treating the effective Lagrangian phenomenologically, as in Step IVb above, immeasurably broadens the utility of effective Lagrangian techniques, as they need not be restricted to situations for which the underlying theory is both known and simple to calculate. Implicit in such a program is the underlying assumption that there is no loss of generality in working with a local field theory. This assumption has been borne out in all known examples of physical systems. It is based on the conviction that the restrictions implicit in working with local field theories are simply those that follow from general physical principles, such as unitarity and cluster decomposition (9).
- When to expect renormalizability: Because Equation 59 states that only a finite number of terms in $\mathcal{L}_{\text{eff}}$ contributes to any fixed order in q/M, and these terms need appear in only a finite number of loops, it follows that only a finite amount of labor is required to obtain a fixed accuracy in observables. Renormalizable theories represent the special case for which it suffices to work only to zeroth order in the ratio q/M. This can be thought of as the reason why renormalizable theories play such an important role throughout physics.
- How to predict using nonrenormalizable theories: An interesting corollary of the above observations is the fact that only a finite number of renormalizations are required in the low-energy theory to make finite the predictions for observables to any fixed order in q/M. Thus, although an effective Lagrangian is not renormalizable in the traditional sense, it nevertheless is predictive in the same way a renormalizable theory is.

4. APPLICATIONS

We next turn briefly to a few illustrative applications of these techniques.

4.1. Quantum Electrodynamics

The lightest electromagnetically interacting elementary particles are the photon and the electron, and from the general arguments given above we expect that the effective field theory that describes their dominant low-energy interactions should be renormalizable, corresponding to the neglect of all inverse masses heavier than the electron. The most general renormalizable interaction possible, given the electron charge assignment, is

$$\mathcal{L}_{\text{QED}}(A_\mu, \psi) = -\frac{1}{4} F_{\mu\nu} F^{\mu\nu} - \overline{\psi}(\slashed{D} + m_e)\psi, \qquad\qquad 60.$$

where $F_{\mu\nu} = \partial_\mu A_\nu - \partial_\nu A_\mu$ gives the electromagnetic field strength in terms of the electromagnetic potential A_μ and where the electron field ψ is a (four-component) Dirac spinor. The covariant derivative for the electron field is defined by $D_\mu \psi = \partial_\mu \psi + ie A_\mu \psi$, where e is the electromagnetic coupling constant. Note that this is precisely the Lagrangian of quantum electrodynamics (QED). In this we have the roots of an explanation for why QED is such a successful description of electron-photon interactions.

4.1.1. Integrating out the electron.
Many practical applications of electromagnetism involve the interaction of photons with macroscopic electric charge, and current distributions at energies E and momenta p much smaller than the electron mass m_e. As a result, they fall within the purview of low-energy techniques, and so lend themselves to being described by an effective theory that is defined below the electron mass m_e, as is now described.

For present purposes, it suffices to describe the macroscopic charge distributions using an external electromagnetic current J^μ_{em}, which can be considered as an approximate, mean-field, description of a collection of electrons in a real material. This approximation is extremely good for macroscopically large systems, if these systems are only probed by electromagnetic fields whose energies are very small compared to their typical electronic energies. We take the interaction term coupling the electromagnetic field to this current to be the lowest possible dimension interaction:

$$\mathcal{L}_J = -e A_\mu J^\mu_{\text{em}}. \qquad\qquad 61.$$

This coupling is only consistent with electromagnetic gauge invariance if the external current is identically conserved: $\partial_\mu J^\mu_{\text{em}} = 0$, independent of any equations of motion, and the current falls off sufficiently quickly to ensure there is no current flow at spatial infinity. Two practical examples of such conserved configurations are those of (a) a static charge distribution: $J^0_{\text{em}} = \rho(\mathbf{r})$ and $\mathbf{J}_{\text{em}} = 0$, or (b) a static electrical current: $J^0_{\text{em}} = 0$ and $\mathbf{J}_{\text{em}} = \mathbf{j}(\mathbf{r})$, where $\rho(\mathbf{r})$ and $\mathbf{j}(\mathbf{r})$ are localized, time-independent distributions, satisfying $\nabla \cdot \mathbf{j} = 0$.

Our interest is in the properties of electromagnetic fields outside of such distributions, and of tracking in particular the low-energy effects of virtual electrons. The most general effective theory involving only photons that can govern the low-energy limit is

$$\mathcal{L}_4 = \mathcal{L}_{\text{eff}}(A) - e A_\mu J^\mu_{\text{em}}, \qquad\qquad 62.$$

where $\mathcal{L}_{\text{eff}}$ may be expanded in terms of interactions having successively higher dimensions. Writing $\mathcal{L}_{\text{eff}} = \mathcal{L}_4 + \mathcal{L}_6 + \mathcal{L}_8 + \cdots$, we have

$$\mathcal{L}_4 = -\frac{Z}{4} F_{\mu\nu} F^{\mu\nu},$$

$$\mathcal{L}_6 = \frac{a}{m_e^2} F_{\mu\nu} \square F^{\mu\nu} + \frac{a'}{m_e^2} \partial_\mu F^{\mu\nu} \partial^\lambda F_{\lambda\nu},$$

$$\mathcal{L}_8 = \frac{b}{m_e^4} (F_{\mu\nu} F^{\mu\nu})^2 + \frac{c}{m_e^4} \left(F_{\mu\nu} \tilde{F}^{\mu\nu}\right)^2 + (\partial^4 F^2 \text{ terms}), \qquad 63.$$

and so on. In this expression for $\mathcal{L}_8$, $\tilde{F}$ represents the dual field-strength tensor, defined by $\tilde{F}_{\mu\nu} = \frac{1}{2}\varepsilon_{\mu\nu\lambda\rho} F^{\lambda\rho}$, and possible terms involving more derivatives exist but have not been written. A power of $1/m_e$ has been made explicit in the coefficient of each term, as we work perturbatively in the electromagnetic coupling, and m_e is the only mass scale that appears in the underlying QED Lagrangian. With this power extracted, the remaining constants, Z, a, a', b, c, and so on, are dimensionless.

Note that no $\partial^2 F^3$ terms appear because any terms involving an odd number of Fs are forbidden by charge-conjugation invariance, for which $F_{\mu\nu} \to -F_{\mu\nu}$. This is Furry's theorem (10) in its modern low-energy guise.

Note also that we may use the freedom to redefine fields to rewrite the $1/m_e^2$ terms in terms of direct current-current contact interactions. The simplest way to do so is to rewrite these interactions using the lowest-dimension equations of motion, $\partial_\mu F^{\mu\nu} = e J_{\text{em}}^\nu$—and thus also, $\square F^{\mu\nu} = e(\partial^\mu J_{\text{em}}^\nu - \partial^\nu J_{\text{em}}^\mu)$—following the general arguments of above sections. Together with an integration by parts, this allows $\mathcal{L}_6$ to be written as

$$\mathcal{L}_6 = \frac{e^2}{m_e^2}(a' - 2a) J_{\text{em}}^\nu J_{\text{em}\nu}. \qquad 64.$$

This shows in particular that these interactions are irrelevant for photon propagation and scattering. However, things are more interesting if conducting boundaries are present (see Reference 11).

In principle, we may now proceed to determine which of these effective interactions are required when working to a fixed order in $1/m_e^2$ in any given physical observable. Having done so, we may then compute the relevant dimensionless coefficients $Z, a' - 2a, b, c$, and so on and thereby retrieve the low-energy limit of the full QED prediction for the observable in question.

4.1.2. The scattering of light by light.

Inspection of the effective Lagrangian (Equation 63) shows that the simplest interaction to receive contributions to $\mathcal{O}(m_e^{-4})$ is the scattering of light by light, for which the leading contribution to the cross section at CM energies $E_{\text{cm}} \ll m_e$ can be inferred relatively easily.

The first step is to determine precisely which Feynman graphs built from the interactions in the effective Lagrangian of Equation 63 contribute, order by order, in $1/m_e$. For this we may directly use the general power-counting results of the previous section because the effective Lagrangian of Equation 63 is a special case of the form considered in Equation 51, with the appropriate dimensionful constants being $f = M = v = m_e$. Directly using Equation 59 for the E-point scattering amplitude $A_E(q)$ leads in the present case to

$$A_E(q) \sim q^2 m_e^2 \left(\frac{1}{m_e}\right)^E \left(\frac{q}{4\pi m_e}\right)^{2L} \left(\frac{q}{m_e}\right)^{\sum_{ik}(k-2)V_{ik}}. \qquad 65.$$

We may also use some specific information for the QED Lagrangian that follows from the gauge invariance of the problem. In particular, because the gauge potential A_μ only appears in $\mathcal{L}_{\mathrm{eff}}$ through its field strength $F_{\mu\nu}$, all interactions of the effective theory must contain at least as many derivatives as they have powers of A_μ. In equations, $V_{ik} = 0$ unless $k \geq i$. In particular, because $k = 2$ implies $i \leq 2$, we see that the only term in $\mathcal{L}_{\mathrm{eff}}$ having exactly two derivatives is the kinetic term $F_{\mu\nu}F^{\mu\nu}$. Because this is purely quadratic in A_μ, it is not an interaction (rather, it is the unperturbed Lagrangian), and so we may take $V_{ik} = 0$ for $k \leq 2$. A consequence of these considerations is the inequality $\sum_{ik}(k-2)V_{ik} \geq 2$, and this sum equals 2 only if $V_{ik} = 0$ for all $k > 4$, and if $V_{i4} = 1$.

For two-body photon-photon scattering, we may take $E = 4$, and from Equation 65, clearly the minimum power of q/m_e that can appear in $A_4(q)$ corresponds to taking (a) $L = 0$, (b) $V_{ik} = 0$ for $k \neq 4$, and (c) $V_{i4} = 1$ for precisely one vertex for which $k = 4$. This tells us that the only graph relevant for photon-photon scattering at leading order in E_{cm}/m_e is the one shown in **Figure 3**, which uses precisely one of the two vertices in $\mathcal{L}_8$, for which $i = k = 4$, and gives a cross section of order q^6/m_e^8 (where $q \sim E_{\mathrm{cm}}$).

The effective Lagrangian really starts saving work when subleading contributions are computed. For photon-photon scattering, Equation 65 implies that the terms suppressed by two additional powers of E_{cm}/m_e require either $L = 1$ and $\sum(k-2)V_{ik} = 2$ or $L = 0$ and $\sum(k-2)V_{ik} = 4$. However, because all interactions have $k \geq i \geq 4$, no such graphs are possible. This shows that the next-to-leading contribution is down by at least $(E_{\mathrm{cm}}/m_e)^4$ relative to the leading terms.

The leading term comes from using the quartic interactions of $\mathcal{L}_8$ in Equation 63 in the Born approximation. This gives the following differential cross section for

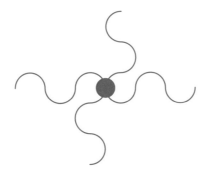

Figure 3

The Feynman graph that contributes the leading contribution to photon-photon scattering in the effective theory for low-energy QED. The vertex represents either of the two dimension-eight interactions discussed in the text.

Figure 4

The leading Feynman graphs in QED that generate the effective four-photon operators in the low-energy theory. Straight (wavy) lines represent electrons (photons).

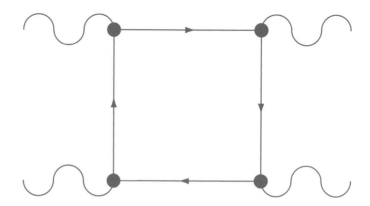

unpolarized photon scattering, in the CM frame:

$$\frac{d\sigma_{\gamma\gamma}}{d\Omega} = \frac{278}{65\pi^2}[(b+c)^2 + (b-c)^2]\left(\frac{E_{cm}^6}{m_e^8}\right)(3+\cos^2\theta)^2\left[1+\mathcal{O}\left(\frac{E_{cm}^4}{m_e^4}\right)\right]. \qquad 66.$$

Here, E_{cm} is the energy of either photon in the CM frame, $d\Omega$ is the differential element of the solid angle for one of the outgoing photons, and θ is the angular position of this solid-angle element relative to the direction of (either of) the incoming photons. Note that, to this point, we have used only the expansion in powers of E_{cm}/m_e and have not also expanded the cross section in powers of $\alpha = e^2/4\pi$.

The final result for QED is reproduced once the constants b and c are computed from the underlying theory, and it is at this point that we first appeal to perturbation theory in α. To lowest order in α, the relevant graph is shown in **Figure 4**. This graph is UV finite so no counterterms are required, and the result is simply

$$b = \frac{4}{7}c = \frac{\alpha^2}{90}, \qquad 67.$$

leading to the standard result (12)

$$\frac{d\sigma_{\gamma\gamma}}{d\Omega} = \frac{139}{4\pi^2}\left(\frac{\alpha^2}{90}\right)^2\left(\frac{E_{cm}^6}{m_e^8}\right)(3+\cos^2\theta)^2\left[1+\mathcal{O}\left(\frac{E_{cm}^4}{m_e^4}\right)\right]. \qquad 68.$$

4.1.3. Renormalization and large logs. We next return to the terms in $\mathcal{L}_{eff}$ that are unsuppressed at low energies by powers of $1/m_e$, for both practical and pedagogical reasons. Pedagogy is served by using this to introduce the decoupling subtraction (DS) renormalization scheme, which is the natural generalization of minimal subtraction to the effective Lagrangian framework. The practical purpose of this example is to demonstrate how classical electromagnetism gives an exact description of photon response at low energies, whose corrections require powers of $1/m_e$ (and not, for example, simply more powers of α).

The only term in the effective Lagrangian of Equation 63 that is not suppressed by powers of $1/m_e$ is the term

$$\mathcal{L}_4 = -\frac{Z}{4}F_{\mu\nu}F^{\mu\nu} - eA_\mu J_{em}^\mu. \qquad 69.$$

The influence of the underlying physics appears here only through the dimensionless parameter Z, whose leading contribution from virtual electrons may be explicitly computed in QED to be

$$Z = 1 - \frac{\alpha}{3\pi} \left[\frac{1}{\varepsilon} + k + \log \left(\frac{m_e^2}{\mu^2} \right) \right], \qquad 70.$$

where we regulate UV divergences using dimensional regularization. k is the constant encountered in Section 2.2.3, which appears universally with the divergence $1/\varepsilon$ in dimensional regularization, and μ is the usual (arbitrary) mass scale introduced in dimensional regularization to keep the coupling constant e dimensionless.

The physical interpretation of Z is found by performing the rescaling $A_\mu = Z^{-\frac{1}{2}} A_\mu^R$, to put the photon kinetic term into canonical form. In this case, we recover the effective theory

$$\mathcal{L}_4 = -\frac{1}{4} F_{\mu\nu} F^{\mu\nu} - e_{\text{phys}} A_\mu J_{\text{em}}^\mu, \qquad 71.$$

where $e \equiv Z^{\frac{1}{2}} e_{\text{phys}}$. We see that to leading order, virtual electrons affect low-energy photon properties only through the value taken by the physical electric charge e_{phys}, and do not otherwise affect electromagnetic properties.

This conclusion has important practical implications concerning the accuracy of the calculations of electromagnetic properties at low energies in QED. It states that the justification for simply using Maxwell's equations to describe photon properties at low energies is the neglect of terms of order $1/m_e$, and not the neglect of powers of α. Thus, for example, even though the classical formulae for the scattering of electromagnetic waves by a given charge distribution are results obtained only in tree-level approximation in QED, corrections to these formulae do not arise at any order in α unsuppressed by powers of E_{cm}/m_e. To zeroth order in $1/m_e$, the sole effect of all higher-loop corrections to electromagnetic scattering is to renormalize the value of α in terms of which all observables are computed. The significance of this renormalization enters once it is possible to measure the coupling at more than one scale because then the logarithmic running of couplings with scale causes real physical effects, in particular encoding the potential dependence of observables on some of the logarithms of large mass ratios.

To see how to extract large logarithms most efficiently, in this section we first contrast two useful renormalization schemes. The first of these is the one defined above, in which all of Z is completely absorbed into the fields and couplings:

$$A_\mu = Z_{\text{phys}}^{-\frac{1}{2}} A_\mu^{\text{phys}}, \quad \text{and} \quad e = Z_{\text{phys}}^{\frac{1}{2}} e_{\text{phys}}, \quad \text{with}$$
$$Z_{\text{phys}} = Z = 1 - \frac{\alpha}{3\pi} \left[\frac{1}{\varepsilon} + k + \log \left(\frac{m_e^2}{\mu^2} \right) \right]. \qquad 72.$$

The subscript phys emphasizes that the charge e_{phys} is a physical observable whose value can be experimentally determined, in principle. For instance, it could be measured by taking a known static charge distribution, containing a predetermined number of electrons, and then using Maxwell's equations to predict the resulting flux of electric field at large distances from these charges. Comparing this calculated flux with the measured flux gives a measurement of e_{phys}.

The alternative scheme of choice for most practical calculations is the $\overline{MS}$ scheme, for which the renormalization is defined as subtracting only the term $1/\varepsilon + k$ in Z. That is,

$$A_\mu = Z_{\overline{MS}}^{-\frac{1}{2}} A_\mu^{\overline{MS}}, \quad \text{and} \quad e = Z_{\overline{MS}}^{\frac{1}{2}} e_{\overline{MS}}, \quad \text{with}$$

$$Z_{\overline{MS}} = 1 - \frac{\alpha}{3\pi}\left[\frac{1}{\varepsilon} + k\right].$$

73.

In terms of this scheme, the effective Lagrangian becomes, to this order in α,

$$\mathcal{L}_{\text{eff}} = -\frac{1}{4}\left[1 - \frac{\alpha}{3\pi}\log\left(\frac{m_e^2}{\mu^2}\right)\right] F_{\mu\nu}^{\overline{MS}} F_{\overline{MS}}^{\mu\nu} - e_{\overline{MS}} A_\mu^{\overline{MS}} J_{\text{em}}^\mu.$$

74.

The $\overline{MS}$ coupling defined in this way is not itself a physical quantity, but is simply a parameter in terms of which the effective Lagrangian is expressed. The physical coupling $\alpha_{\text{phys}} = e_{\text{phys}}^2/4\pi$ and the $\overline{MS}$ coupling $\alpha_{\overline{MS}} = e_{\overline{MS}}^2/4\pi$ are related in the following way:

$$\alpha_{\overline{MS}} = \left(\frac{Z_{\text{phys}}}{Z_{\overline{MS}}}\right)\alpha_{\text{phys}} = \left[1 - \frac{\alpha}{3\pi}\log\left(\frac{m_e^2}{\mu^2}\right)\right]\alpha_{\text{phys}}.$$

75.

A key observation is that, because α_{phys} is a physical quantity, it cannot depend on the arbitrary scale μ. As a result, this last equation implies a μ dependence for $\alpha_{\overline{MS}}$. It is only for $\mu = m_e$ that the two couplings agree:

$$\alpha_{\overline{MS}}(\mu = m_e) = \alpha_{\text{phys}}.$$

76.

As usual, there is something to be gained by re-expressing the μ dependence of $\alpha_{\overline{MS}}$ of Equation 75 as a differential relation:

$$\mu^2 \frac{d\alpha_{\overline{MS}}}{d\mu^2} = +\frac{\alpha_{\overline{MS}}^2}{3\pi}.$$

77.

This is useful because the differential expression applies so long as $\alpha_{\overline{MS}} \ll 1$, whereas Equation 75 also requires $\alpha_{\overline{MS}}\log(m_e^2/\mu^2) \ll 1$. Integrating Equation 77 allows a broader inference of the μ dependence of $\alpha_{\overline{MS}}$:

$$\frac{1}{\alpha_{\overline{MS}}(\mu)} = \frac{1}{\alpha_{\overline{MS}}(\mu_0)} - \frac{1}{3\pi}\log\left(\frac{\mu^2}{\mu_0^2}\right),$$

78.

which is accurate to all orders in $(\alpha/3\pi)\log(m_e^2/\mu^2)$, so long as $(\alpha/3\pi) \ll 1$.

Equation 78 is useful because it provides a simple way to keep track of how some large logarithms appear in physical observables. For instance, consider the cross section for the scattering of electrons (plus an indeterminate number of soft photons, having energies up to $E_{\text{max}} = fE$, with $1 > f \gg m_e/E$). Such a quantity has a smooth limit as $m_e/E \to 0$ when it is expressed in terms of $\alpha_{\overline{MS}}(\mu)$ (13, 14), so on dimensional grounds we may write

$$\sigma(E, m_e, \alpha_{\text{phys}}) = \frac{1}{E^2}\left[\mathcal{F}\left(\frac{E}{\mu}, \alpha_{\overline{MS}}(\mu), f, \theta_k\right) + O\left(\frac{m_e}{E}\right)\right],$$

79.

where θ_k denote any number of dimensionless quantities (such as scattering angles) on which the observable depends, and the explicit μ dependence of $\mathcal{F}$ must cancel the μ dependence that appears implicitly through $\alpha_{\overline{MS}}(\mu)$. However, $\mathcal{F}$ is singular when $m_e/E \to 0$ when it is expressed in terms of α_{phys} because of the appearance of large logarithms. These may be included to all orders in $\alpha \log(E^2/m_e^2)$ simply by choosing $\mu = E$ in Equation 79 and using Equation 78 with Equation 76.

For the next section, it is important that Equation 77 integrates simply because the $\overline{MS}$ renormalization is a mass-independent scheme. That is, $d\alpha/d\mu$ depends only on α and does not depend explicitly on mass scales such as m_e. (On-shell renormalizations, such as where e is defined in terms of the value of a scattering amplitude at a specific momentum transfer, furnish examples of schemes that are not mass independent.)

4.1.4. Muons and the decoupling subtraction scheme.

Consider now introducing a second scale into the problem by raising the energies of interest to those above the muon mass. In this case, the underlying theory is

$$\mathcal{L} = -\frac{1}{4} F_{\mu\nu} F^{\mu\nu} - \overline{\psi}(\slashed{D} + m_e)\psi - \overline{\chi}(\slashed{D} + M_\mu)\chi, \qquad 80.$$

where χ is the Dirac spinor representing the muon, and M_μ is the muon mass. Our interest is in following how large logarithms such as $\log(M_\mu/m_e)$ appear in observables.

In this effective theory, the $\overline{MS}$ and physical electromagnetic couplings are related to one another by

$$\alpha_{\overline{MS}} = \left\{ 1 - \frac{\alpha}{3\pi} \left[\log\left(\frac{m_e^2}{\mu^2}\right) + \log\left(\frac{M_\mu^2}{\mu^2}\right) \right] \right\} \alpha_{\text{phys}}. \qquad 81.$$

This relation replaces Equation 75 of the purely electron-photon theory. Note in particular that in this theory, the physical coupling α_{phys} is no longer simply equal to $\alpha_{\overline{MS}}(\mu = m_e)$ but is now equal to $\alpha_{\overline{MS}}(\mu = \sqrt{m_e M_\mu})$. The corresponding RG equation for the running of $\alpha_{\overline{MS}}$ is

$$\mu^2 \frac{d\alpha_{\overline{MS}}}{d\mu^2} = +\frac{2\alpha^2}{3\pi}, \qquad 82.$$

with solution

$$\frac{1}{\alpha_{\overline{MS}}(\mu)} = \frac{1}{\alpha_{\overline{MS}}(\mu_0)} - \frac{2}{3\pi} \log\left(\frac{\mu^2}{\mu_0^2}\right). \qquad 83.$$

Here we see an inconvenience of the $\overline{MS}$ renormalization scheme: The right side of Equation 82 is twice as large as its counterpart, Equation 77, in the pure electron-photon theory, simply because the mass independence of the $\overline{MS}$ scheme ensures that both the electron and the muon contribute equally to the running of $\alpha_{\overline{MS}}$. The problem is that this is equally true for all μ, and it even applies at scales $\mu \ll M_\mu$, where we expect the physical influence of the muon to decouple. Of course, the physical effects of the muon indeed do decouple at scales well below the muon mass. However, this decoupling is not manifest at intermediate steps in any calculation performed with the $\overline{MS}$ scheme.

The dimensionally regularized effective Lagrangian furnishes a way to circumvent this disadvantage, by keeping the decoupling of heavy particles manifest without giving up the benefits of a mass-independent renormalization scheme. The remedy is to work with minimal subtraction, but to do so only when running couplings in an energy range between charged-particle thresholds. As the energy falls below each charged-particle threshold, a new effective theory is defined by integrating out this particle, with the couplings in the new low-energy theory found by matching to the coupling defined in the underlying theory above the relevant mass scale. The scheme defined by doing so through all particle thresholds is termed the decoupling subtraction $(\overline{DS})$ renormalization scheme (15).

For instance, for the electrodynamics of electrons and muons, the coupling constant as defined in the $\overline{MS}$ and $\overline{DS}$ schemes is identical for the full theory that describes energies greater than the muon mass, $\mu > M_\mu$. For $m_e < \mu < M_\mu$, we integrate out the muon and construct an effective theory involving only photons and electrons. This effective theory consists of the usual QED Lagrangian, plus an infinite number of higher-dimension effective interactions encoding the low-energy implications of virtual muons. Within this effective Lagrangian, the coupling constant is again defined by the coefficient Z in front of the term $F_{\mu\nu}F^{\mu\nu}$, using minimal subtraction. However, because there is no muon within this effective theory, only the electron contributes to its running.

The initial conditions for the RG equation at the muon mass are obtained by matching, as in the above sections. They are chosen to ensure that the effective theory reproduces the same predictions for all physical quantities, as does the full theory, order by order in the low-energy expansion. If there were other charged particles in the problem, each of these could be integrated out in a similar fashion as μ falls below the corresponding particle threshold.

Quantitatively, to one loop the RG equation for the $\overline{DS}$ scheme for the theory of electrons, muons, and photons becomes

$$
\begin{aligned}
\mu^2 \frac{d\alpha_{\overline{DS}}}{d\mu^2} &= \frac{2\alpha^2}{3\pi}, \quad \text{if } \mu > M_\mu, \\
&= \frac{\alpha^2}{3\pi}, \quad \text{if } m_e < \mu < M_\mu, \\
&= 0, \quad \text{if } \mu < m_e,
\end{aligned}
\tag{84.}
$$

with the boundary conditions that $\alpha_{\overline{DS}}$ should be continuous at $\mu = M_\mu$ and $\alpha_{\overline{DS}}(\mu = m_e) = \alpha_{\text{phys}}$ at $\mu = m_e$. Integrating then gives

$$
\begin{aligned}
\frac{1}{\alpha_{\overline{DS}}(E)} &= \frac{1}{\alpha_{\text{phys}}} - \frac{1}{3\pi} \log\left(\frac{E^2}{m_e^2}\right) \quad \text{for } m_e < E < M_\mu, \\
&= \frac{1}{\alpha_{\text{phys}}} - \frac{1}{3\pi} \log\left(\frac{M_\mu^2}{m_e^2}\right) - \frac{2}{3\pi} \log\left(\frac{E^2}{M_\mu^2}\right) \quad \text{for } M_\mu < E.
\end{aligned}
\tag{85.}
$$

This last expression shows how to efficiently display the various large logarithms by running a coupling with the ease of a mass-independent scheme, but with each particle explicitly decoupling as μ drops through the corresponding particle threshold.

4.2. Power-Counting Examples: QCD and Gravity

We conclude with a sketch of the utility of Equation 59 for two important examples: the interactions among pions and kaons at energies well below a GeV, and gravitational self-interactions for macroscopic systems.

4.2.1. Below the QCD scale: mesons.
We start with the interactions of pions and kaons at energies below a GeV. This represents a useful low-energy limit of the Standard Model because these mesons are Goldstone bosons for the spontaneous breaking of an approximate symmetry of the strong interactions (16). As such, their interactions are suppressed in this low-energy limit, as a general consequence of Goldstone's theorem (17). Because they interact so weakly, we represent them with fundamental scalar fields in the effective theory that applies at energies $E \ll \Lambda \sim 1$ GeV.

The resulting scalar Lagrangian has the form of Equation 51, with the constants that appear in the effective Lagrangian being $f = \sqrt{F_\pi \Lambda}$, $M = \Lambda$, and $v = F_\pi$, where $F_\pi \sim 100$ MeV defines the scale of the order parameter that describes the spontaneous breaking of the relevant approximate symmetry. In this case, the power-counting estimate of Equation 59 becomes

$$A_E(q) \sim F_\pi^2 q^2 \left(\frac{1}{F_\pi} \right)^E \left(\frac{q}{4\pi F_\pi} \right)^{2L} \left(\frac{q}{\Lambda} \right)^{\sum_{ik}(k-2)V_{ik}}, \qquad 86.$$

which is a famous result, due first to Weinberg (18). The explicit suppression of all interactions by powers of $q/\Lambda \sim q/(4\pi F_\pi)$ explicitly encodes the suppression of interactions that Goldstone's theorem requires. Because the pion mass is $m_\pi \sim 140$ MeV, this suppression is clearly relevant only for scattering at energies near threshold, $E_{cm} \sim m_\pi$.

The dominant terms in $\mathcal{L}_{eff}$ that govern the scattering at these energies correspond to choosing the smallest possible value for which $L = 0$, and $V_{ik} \neq 0$ only if $k = 2$. Because it happens that symmetries determine the effective couplings of all such terms purely in terms of F_π (in the limit of massless quarks), a great deal can be said about such low-energy meson interactions without knowing any of the dynamical details about their explicit wave functions. These predictions are consequences of the symmetry-breaking pattern, and are known as soft pion theorems. Comparisons of these predictions, including next-to-leading-order corrections and nonzero quark masses, are in good agreement with observations (19).

4.2.2. General relativity as an effective field theory.
It is instructive to repeat this power-counting analysis for the gravitational effective theory because this case furnishes a less familiar example. The result obtained also justifies the neglect of quantum effects in performing practical calculations with gravity on macroscopic scales. Even better, it permits the systematic calculation of the leading corrections in the semiclassical limit, should these ever be desired.

The field relevant for gravity (20) is the metric $g_{\mu\nu}$ (whose matrix inverse is denoted by $g^{\mu\nu}$). For applications on macroscopic scales, we use the most general effective

Lagrangian consistent with general covariance:

$$-\mathcal{L}_{\text{eff}} = \sqrt{-g}\left[\frac{1}{2}M_p^2 R + c_1 R^2 + c_2 R_{\mu\nu}R^{\mu\nu} + c_3 R_{\mu\nu\lambda\rho}R^{\mu\nu\lambda\rho} + \frac{e_1}{m_e^2}R^3 + \cdots\right], \quad 87.$$

where $g = \det(g_{\mu\nu})$, while $R_{\rho\mu\lambda\nu}$, $R_{\mu\nu} = g^{\lambda\rho}R_{\lambda\mu\rho\nu}$, and $R = g^{\mu\nu}R_{\mu\nu}$ denote the Riemann tensor, the Ricci tensor, and the Ricci scalar, respectively, each of which involves two derivatives of $g_{\mu\nu}$. For simplicity, no cosmological term is written here because this precludes a perturbative expansion about flat space, although a similar discussion could be made for perturbations about de Sitter or anti–de Sitter space. The ellipses denote terms involving at least six derivatives, one of which is displayed explicitly in Equation 87. The term linear in R is the usual Einstein-Hilbert action, with M_p denoting the usual Planck mass (whose numerical value is given in terms of Newton's constant by $8\pi G = M_p^{-2}$). The remaining effective couplings, c_k and e_k, are dimensionless, and not all of the terms written need be independent of one another. The scale m_e denotes the lightest particle (say, the electron) integrated out to obtain this effective Lagrangian.

Equation 87 has the form considered earlier, with $f = \sqrt{m_e M_p}$, $\Lambda = m_e$, and $v = M_p$. Furthermore, with these choices, the dimensionless couplings of all interactions except for the Einstein term are suppressed explicitly by the factor m_e^2/M_p^2. With these choices, the central power-counting result, Equation 59, becomes (21, 22)

$$A_E(q) \sim q^2 M_p^2 \left(\frac{1}{M_p}\right)^E \left(\frac{q}{4\pi M_p}\right)^{2L} \left(\frac{m_e^2}{M_p^2}\right)^{\sum_{i;k>2}V_{ik}} \left(\frac{q}{m_e}\right)^{\sum_{ik}(k-2)V_{ik}}$$

$$\sim q^2 M_p^2 \left(\frac{1}{M_p}\right)^E \left(\frac{q}{4\pi M_p}\right)^{2L} \left(\frac{q^2}{M_p^2}\right)^{\sum_{i;k>2}V_{ik}} \left(\frac{q}{m_e}\right)^{\sum_{i;k>4}(k-4)V_{ik}}, \quad 88.$$

where covariance requires $V_{ik} = 0$ unless $k = 2, 4, 6\ldots$, with $k = 2$ corresponding to the Einstein-Hilbert term, $k = 4$ the curvature-squared terms, and so on.

As above, the dominant term comes from choosing $L = 0$ and using only the interactions of the usual Einstein-Hilbert action: That is, $V_{ik} = 0$ for $k > 2$. The dominant contribution to gravitational physics is therefore obtained by working to tree level with the Einstein action, which is to say that one must compute the classical response of the gravitational field, using the full Einstein equations to compute this response.

The graphs responsible for the next-to-leading-order terms are also easy to determine. The minimum additional suppression by q/M_p is obtained either by working to one-loop order ($L = 1$) using the Einstein action ($V_{ik} = 0$ for $k > 2$) or by working to tree level ($L = 0$) using precisely one insertion of one of the curvature-squared interactions (i.e., with V_{i2} arbitrary but $V_{i4} = 1$ for one interaction with $k = 4$). Both cases give an additional suppression of q^2/M_p^2 relative to the leading contribution. The one-loop contribution also carries the usual additional loop factor $(1/4\pi)^2$.

Note that the derivative expansion is an expansion in q/m_e as well as in q/M_p, owing to the inverse powers of m_e that appear in the higher-curvature terms. Note also that all of the m_e dependence drops out for graphs constructed using only the Einstein and the curvature-squared terms (i.e., $V_{ik} = 0$ for $k > 4$), as it should because $1/m_e^2$ first enters Equation 87 at order curvature cubed. Although the condition $q \ll m_e$ may come as something of a surprise, it is nevertheless an excellent expansion for macroscopic applications, such as in the solar system.

DISCLOSURE STATEMENT

The author is not aware of any biases that might be perceived as affecting the objectivity of this review.

ACKNOWLEDGMENTS

This review is based on a series of lectures given in June 1995 for the Swiss Troisième Cycle in Lausanne, Switzerland, and for the University of Oslo, whose organizers I thank for their kind invitations and whose students I thank for their questions and comments. My research during the preparation of these lectures was funded in part by the Natural Sciences and Engineering Research Council (Canada), the Fonds pour la Formation de Chercheurs et l'Aide à la Recherche (Quebec), and the Killam Foundation.

LITERATURE CITED

1. Appelquist T, Carazzone J. *Phys. Rev. D* 11:2856 (1975); Buchmuller W, Wyler D. *Nucl. Phys. B* 268:621 (1986); Grinstein B, Wise MB. *Phys. Lett. B* 265:326 (1991); Meissner UG. *Rep. Prog. Phys.* 56:903 (1993); Leutwyler H. hep-ph/9406283 (1994); Manohar A. hep-ph/9606222 (1996); Pich A. hep-ph/9806303 (1998); Kaplan D. nucl-th/9506035 (1995); Georgi H. *Annu. Rev. Nucl. Part. Sci.* 43:205 (1995); Burgess CP, et al. *Phys. Rev. D* 49:6115 (1994); Rothstein IZ. hep-ph/0308266; Han Z, Skiba W. *Phys. Rev. D* 71:075009 (2005)

   These references are part of the vast literature on effective field theory.

2. Georgi H. *Weak Interactions and Modern Particle Theory*. Benjamin/Cummings (1984); Donoghue JF, Golowich E, Holstein BR. *Dynamics of the Standard Model*. Cambridge Univ. Press (1992); Weinberg S. *The Quantum Theory of Fields*, Vol. 2. Cambridge Univ. Press (1996); Burgess CP, Moore GD. *The Standard Model: A Primer*. Cambridge Univ. Press (2007)

   See these references for textbook discussions of effective field theory.

3. Polchinski J. hep-th/9210046 (1992); Shankar R. *Rev. Mod. Phys.* 66:129 (1994); Chen T, Fröhlich J, Seifert M. cond-mat/9508063 (1995); Lepage P. nucl-th/9706029 (1997); Burgess CP. *Phys. Rep. C* 330:193 (2000); Braaten E, Hammer HW. *Ann. Phys.* 322:120 (2007)

   See these references for reviews of effective field theories applied to degenerate systems.

4. Caswell WE, Lepage GP. *Phys. Lett. B* 167:437 (1986); Luke ME, Manohar AV, Rothstein IZ. *Phys. Rev. D* 61:074025 (2000); Lepage P. nucl-th/9706029; Neubert M. hep-ph/0512222 (2005)

   Applications to nonrelativistic-bound states may be found in these references.

5. Goldstone J, Salam A, Weinberg S. *Phys. Rev.* 127:965 (1962); Jona-Lasinio G. *Nuova Cim.* 34:1790 (1964)

6. Symanzik K. *Comm. Math. Phys.* 16:48 (1970); Coleman S. In *Aspects of Symmetry: Selected Erice Lectures*, p. 139. Cambridge Univ. Press. 368 pp. (1985)

7. Burgess CP, Font A, Quevedo F. *Nucl. Phys. B* 272:661 (1986)

8. Burgess CP, London D. *Phys. Rev. Lett.* 69:3428 (1992); Burgess CP, London D. *Phys. Rev. D* 48:4337 (1993)

9. Weinberg S. *The Quantum Theory of Fields*, Vol. 1. Cambridge Univ. Press (1995)

10. Furry WH. *Phys. Rev.* 51:125 (1937)

11. Aghababaie Y, Burgess CP. *Phys. Rev. D* 70:085003 (2004)

12. Itzykson C, Zuber JB. *Quantum Field Theory*. McGraw-Hill. 705 pp. (1980)

13. Weinberg S. In *Proc. Asymptot. Realms Phys.*, p. 1. Cambridge, MA: MIT Press (1983)

14. Collins J. *Renormalization*. Cambridge Univ. Press (1984)

15. Weinberg S. *Phys. Lett. B* 91:51 (1980); Ovrut BA, Schnitzer HJ. *Phys. Lett. B* 100:403 (1981); Ovrut BA, Schnitzer HJ. *Nucl. Phys. B* 179:381 (1981); Hall LJ. *Nucl. Phys. B* 178:75 (1981)

16. Weinberg S. *Phys. Rev. Lett.* 18:188 (1967); Weinberg S. *Phys. Rev.* 166:1568 (1968); Callan CG, Coleman S, Wess J, Zumino B. *Phys. Rev.* 177:2247 (1969); Gasser J, Leutwyler H. *Ann. Phys.* 158:142 (1984)

17. Goldstone J. *Nuova Cim.* 9:154 (1961); Nambu Y. *Phys. Rev. Lett.* 4:380 (1960); Goldstone J, Salam A, Weinberg S. *Phys. Rev.* 127:965 (1962)

18. Weinberg S. *Physica* 96A:327 (1979)

19. Donoghue JF, Golowich E, Holstein BR. *Dynamics of the Standard Model*. Cambridge Univ. Press (1992); Meissner UG, Bernard V. *Annu. Rev. Nucl. Part. Sci.* 57:33 (2007)

20. Weinberg S. *Gravitation and Cosmology*. Wiley & Sons (1972); Misner CW, Thorne KS, Wheeler JA. *Gravitation*. Freeman (1973); Wald R. *General Relativity*. Univ. Chicago Press (1984)

21. Burgess CP. *Living Rev. Relativ.* 7:5 (2004); Burgess CP. *Phys. Rep. C* 330:193 (2000); Donoghue JF. gr-qc/9512024 (1995)

22. Donoghue JF, Torma T. *Phys. Rev. D* 54:4963 (1996); Goldberger WD, Rothstein IZ. *Phys. Rev. D* 73:104029 (2006); Goldberger WD. hep-ph/0701129 (2007)

See this reference for a textbook discussion of photon-photon scattering.

Power counting is much more complicated for nonrelativistic sources.

Recent Developments in the Fabrication and Operation of Germanium Detectors

Kai Vetter*

Glenn T. Seaborg Institute, Lawrence Livermore National Laboratory, Livermore, California 94550; email: kvetter@llnl.gov

Annu. Rev. Nucl. Part. Sci. 2007. 57:363–404

First published online as a Review in Advance on June 28, 2007

The *Annual Review of Nuclear and Particle Science* is online at http://nucl.annualreviews.org

This article's doi:
10.1146/annurev.nucl.56.080805.140525

Copyright © 2007 by Annual Reviews.
All rights reserved

0163-8998/07/1123-0363$20.00

*The U.S. Government has the right to retain a nonexclusive, royalty-free license in and to any copyright covering this paper.

Key Words

Ge detectors, Ge detector technologies, gamma-ray tracking, Compton imaging, neutrinoless double-beta decay

Abstract

Although developed and first demonstrated more than 40 years ago, germanium detectors still represent the gold standard in detecting γ radiation for energies ranging from approximately 100 keV to 10 MeV. The combination of high-efficiency and excellent energy resolution and many recent technological developments have significantly increased the range of applications for Ge detectors. We review the state of the art in the fabrication and operation of Ge detectors by mapping recent developments in Ge detector technologies to a range of applications. These include research in nuclear physics, fundamental physics, and astrophysics, as well as applications in medical imaging and homeland security. Ge detector technologies will remain vitally important for applied and basic research instruments for high-sensitivity γ-ray detection in the future.

Contents

1. INTRODUCTION 364
    1.1. Historical Review 366
    1.2. Technical Review 371
2. RECENT TECHNICAL DEVELOPMENTS 376
    2.1. Detector Technologies and Configurations 376
    2.2. Electronics 384
    2.3. Signal Processing and Advanced Data Analysis 384
    2.4. Cryogenic Detectors 389
3. APPLICATIONS 390
    3.1. Gamma-Ray Tracking Arrays for Nuclear Physics Applications ... 390
    3.2. Gamma-Ray Imaging for National Security and Astrophysics 392
    3.3. Fundamental Physics 394
4. FUTURE DEVELOPMENTS 397
5. CONCLUSIONS 400

1. INTRODUCTION

Despite the persistent perception of germanium detectors as being outlived by other detectors for many years now, and despite significant investments in alternative detector materials and concepts, the range of applications and variety of implementations for Ge detectors have been and are still expanding. This is on the one hand due to technological advances in Ge detector and related technologies, which enable the realization of new and improved detection concepts, and on the other hand due to experimental and operational requirements, which are driving the technological advances.

Since A.J. Tavendale's review in 1967 on *Semiconductor Nuclear Radiation Detectors*, no paper has addressed the enormous advances that have been achieved in Ge detector and related technologies in this series (1). Review papers associated with the advances in Ge detector technologies were published in the 1980s and 1990s and focused on the study of specific features in nuclei (2–4). At the time of the Tavendale paper, the use of Ge detectors or silicon detectors for γ-ray detection was still in its infancy. Tavendale reports on the first lithium-drifted coaxial Ge detectors with volumes of 23 ml with, at that time, excellent energy resolution of 5.6 keV at an energy of 1368 keV. Considerable, almost inconceivable, progress in the manufacturing of large-volume detectors and processing electronics since then is the reason that, more than 40 years after the introduction of Ge as a feasible γ-ray detector material, it still provides unchallenged performance in terms of efficiency and energy resolution. Ge detectors have been built with volumes larger than 800 ml (5) and an energy resolution of 0.2 keV at low energies, which is close to the statistical limit (6). These outstanding properties are due to the unique combination of a small band gap of the elemental semiconductor Ge, very large electron and hole mobilities and lifetimes, and the fact that large crystal sizes with the highest purity and a minimum of crystal defects can be grown. The development of low-noise electronics was critical as well to achieving the high-energy resolution, particularly at low energies.

In addition to advances in the growth and purification techniques, which enable the fabrication of large-volume detectors, improved contact segmentation technologies provide the basis to measure positions and energies of individual γ-ray interactions with high accuracy and efficiency. This capability is now being used to determine the path of the γ-ray in the detector, which has a profound impact for many applications. It allows construction of arrays based on a closed shell of more than 100 of these detectors for nuclear physics experiments, providing unprecedented sensitivity in detecting many γ-rays simultaneously. In addition, the combination of high efficiency and granularity provided by segmented Ge detectors allows us to build efficient Compton-scattering-based γ-ray imaging systems, so-called Compton cameras. Compton cameras can image γ-ray sources without using heavy collimators over a large field of view, finding applications in astrophysics, biomedical imaging, nuclear nonproliferation and safeguards, and homeland security.

Complementary to the developments in Ge detector technologies themselves, advances in electronics that provide low-noise signal amplification and filtering and fast digitization were essential to realizing their demonstrated capabilities. Advances in Ge detector technologies were accompanied by advances in electronics, originally in analog signal processing primarily and more recently in digital signal processing. Digital signal processing provides higher count rate capabilities and compensation for ballistic deficit effects, two features difficult to address with analog processing. They provide more flexibility in adapting different detectors and detector configurations and in the implementation of more complex filters (e.g., adaptive and time dependent). Only the combination of two-dimensional segmentation and the digitization with high sampling rates and high resolution enabled the precise analysis of pulse shapes in Ge detectors, providing the three-dimensional position and energies in large-volume Ge detectors. Pulse-shape analysis is now employed to determine positions and energies of interactions, to distinguish between γ-rays that deposited only a fraction of the full energy, and to distinguish between γ-rays and, for example, electrons interacting in the detector by distinguishing between single and multiple interactions. On the one hand, this distinction allows the suppression of electron or beta background, for example, in astrophysics applications aimed at measuring γ-rays. On the other hand, it allows suppression of γ-rays for fundamental physics applications such as the search for the neutrinoless double-beta ($0\nu\beta\beta$) decay in isotopically enriched ^{76}Ge detectors (7).

The majority of applications for Ge detectors still relies on the γ-ray detection by measuring the ionization of Ge atoms. However, over the past 20 years, significant progress was made in measuring the heat generated by particles or photons, which allows a much improved energy resolution and sensitivities to smaller energy depositions. Ge detectors can be operated as bolometers at sub-Kelvin temperatures, measuring the number of phonons by observing the temperature rise in very low heat capacity thermistors. Although not necessarily implemented with Ge, this method provides the possibility of measuring the ionization and the heat simultaneously, which allows the discrimination, for example, of neutral particles and γ-rays.

In addition to using Ge as a particle or photon absorber, heavily doped Ge, such as neutron-transmutation-doped (NTD) Ge, is also being used as a very sensitive thermometer or thermistor owing to its very small and uniform heat capacity (8).

In the following we provide a brief historical review starting from the first demonstration of a Li-drifted Ge [Ge(Li)] detector in the 1960s to the most recent developments in large Ge detector arrays. A technical review briefly discusses some basic principles in the operation of Ge detectors. In Section 3 we discuss recent technical advances. Section 4 presents applications that drove or benefited from these advances. In Section 5 we attempt a brief outlook to the future of Ge detector and related technologies and conclude with a brief summary in Section 6. The aim of this review is not to be complete and encyclopedic but rather exemplary—to provide an overview on the breadth of developments of the recent past.

1.1. Historical Review

Since the first demonstration in the early 1960s, Ge detectors have undergone a remarkable development and now represent one of the detectors of choice for γ-ray detection. Although single crystals of Ge were grown already in 1950 (9), the first practicable Ge detector material was produced in 1960 using Li-donor compensation of p-type crystals (10, 11). These Ge(Li) detectors opened the field of high-resolution γ-ray spectroscopy. The Li drifting process was required to compensate for the remaining acceptor impurities in the Ge material, which allowed the fabrication of Ge detectors with sensitive volumes of up to 100 ml.

Owing to the much improved energy resolution as compared with NaI(Tl) detectors, Ge(Li) detectors became the state of the art for γ-ray spectroscopy in the 1960s and 1970s to study the structure of atomic nuclei. By 1970, arrangements of two or more of these detectors were used to measure coincidence relationships between γ-rays emitted from the nucleus in order to construct detailed level schemes of the excited states.

The drawback of Ge(Li) detectors is the high mobility of interstitial Li donors in Ge, which requires the detector temperature to be kept close to 77 K to prevent the Li from diffusing out of the detector, thereby removing the compensation of the acceptor impurity and making the detector inoperable, and requiring a redrifting process. The cooling requirement is a drawback not only when operating, storing, or transporting detectors, but also during the fabrication process and when transferring the detector between cryostats. In addition, the Li drifting process takes several weeks and limits the maximum size of the Ge detector. Finally, Li drifting can be performed only in p-type material, which is significantly more susceptible to radiation damage in coaxial detectors, which represent the majority of used Ge detectors. The repair of radiation damage by thermal annealing is almost impossible for Li-drifted materials.

In 1971, Hall (12) proposed the possibility of purifying and growing Ge of the high purity required without the need for subsequent Li drifting. It took approximately 10 more years before high-purity germanium (HPGe) detectors could be produced, in part owing to the seminal work by Hall at General Electric and Hansen & Haller at Lawrence Berkeley National Laboratory (13, 14). Around 1980, Ge(Li) detectors were replaced by HPGe detectors. Since then, the crystal growth process was further refined, which increased the maximum achievable diameter from approximately 60 mm in the 1980s to almost 100 mm today. P-type detectors with a diameter of 98 mm, a length of 110 mm, a corresponding volume of 830 ml, and a weight of 4.4 kg have been built (5). N-type detectors with diameters of 7 cm and lengths of 14 cm have been built for the Versatile and Efficient Gamma detector (VEGA). As a matter of fact, four of these large-volume detectors are mounted in one cryostat in a clover-like arrangement, which, by summing γ-rays from all detectors, can be treated as one large $14 \times 14 \times 14$ cm^3 or 2.5 liter Ge detector with a weight of 12 kg (15).

To improve the spectral response of Ge detectors, so-called anti-Compton shields were introduced in the 1980s. These shields are made of high-density scintillating detectors that surround the Ge detector to suppress events in which only part of the energy is deposited in the Ge. In this way, the continuous Compton background can be reduced and the overall peak-to-background ratio increased significantly. By 1985, small arrays of up to 12 Compton-suppressed Ge detectors were being widely used for nuclear physics experiments (16). The culmination of these types of arrays is Gammasphere in the United States (17) and Euroball in Europe (18). **Figure 1** shows the evolution of γ-ray detector systems in terms of the so-called resolving power. The resolving power is defined as the inverse of the sensitivity or observational limit, for example, indicating the weakest γ-ray decay channel, which can be resolved in an experiment. In addition, the figure illustrates the gain in sensitivity to measure higher spin states populated with lower intensity in the ground state rotational band and the extracted moment of inertia observed with different devices in ^{156}Dy. While the relative intensity reflects the increase of the resolving power, the moment-of-inertia plot reveals new physics associated with increased angular momentum of the nucleus (e.g., it reflects the alignments of neutron and proton angular momentum along the axis of rotation owing to Coriolis effects, the loss of suprafluid correlations and band termination, and the associated loss of quantum collectivity). Remarkably, in the ^{156}Dy case, almost every significant increase in detection sensitivity was associated with the discovery of a new phenomenon.

The current state-of-the-art detector arrays, such as Gammasphere (shown in **Figure 2**) or Euroball, comprise approximately 100 or more Compton-suppressed Ge detectors. These systems, which were built in the 1990s, have an absolute efficiency of approximately 10% for detecting the full energy of a 1-MeV γ-ray and a peak-to-total ratio of approximately 55%. Both efforts introduced technologies critical for the next-generation γ-ray spectrometer currently being built, the γ-ray-tracking-based instruments. Gammasphere introduced segmented coaxial detectors into nuclear physics applications to reduce the degradation of the intrinsic energy resolution due to Doppler broadening in in-beam experiments. Euroball introduced cryostats with seven large-volume Ge detectors—so-called cluster detectors—with each detector mounted in a hermetically sealed vacuum capsule. Mounting several detectors in one cryostat increases the efficiency and improves the peak-to-background ratio by adding the energies of all detectors within one cryostat; the encapsulation ensures the reliable operation and potentially necessary annealing procedures. In parallel to the development of large arrays of Ge detectors and

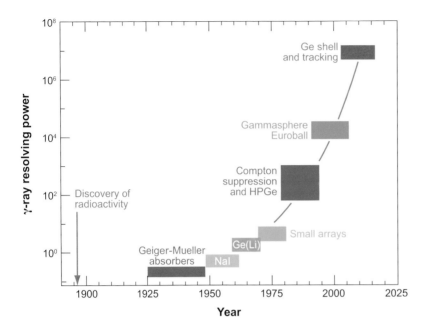

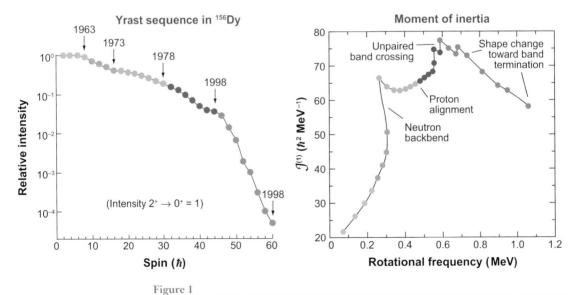

Figure 1

The evolution of γ-ray detectors in terms of resolving power. The lower left panel shows the corresponding observational limit as relative intensity measured in ^{156}Dy as a function of spin. The lower right panel shows the associated moment of inertia as a function of rotational frequency. HPGe, high-purity germanium; $\mathcal{J}^{(1)}$, moment of inertia.

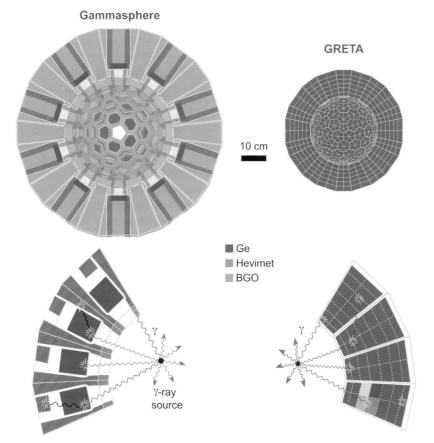

Gammasphere

GRETA

10 cm

- ■ Ge
- ■ Hevimet
- ■ BGO

γ

γ-ray
source

γ

Figure 2

Comparison between the
current state-of-the-art
detector Gammasphere
(*left*) and the proposed
Gamma-Ray Energy
Tracking Array (GRETA)
(*right*). Proposed γ-ray
tracking arrays such as
GRETA consist only of Ge
detectors; Compton shields,
for example, made of
bismuth germanate (BGO),
and absorbers are removed.
The lower panels indicate
the two different
approaches. While
anti-Compton shields
suppress cross-scattering
between Ge crystals and the
hevimet absorber prevents
direct hits of the shields to
prevent false suppression, a
γ-ray tracking array accepts
all γ-rays, significantly
increasing the efficiency.

reflecting a significant worldwide effort to study atomic nuclei far away from the valley
of stability at new radioactive facilities, smaller and more compact arrays have been
built or are currently being built. Examples of such arrays are Exogam at the Grand
Accélérateur National d'Ions Lourdes (GANIL) facility in France (19, 20), Miniball at CERN in Switzerland (21–23), the Segmented Germanium Array (SeGA) at
Michigan State University (24), and most recently the TRIUMF-ISAC Gamma-Ray
Escape Suppressed Spectrometer (TIGRESS) at TRIUMF in Canada (25).

Although both types of arrays, Gammasphere-like primarily for stable beam facilities and Exogam-like with the focus on radioactive beam facilities, represent a large
increase in efficiency and resolving power in contrast to previous arrays, they are
ultimately limited by the anti-Compton suppressor shields and the limited efficiency
of individual Ge detector systems. The removal of the Compton suppressor would
allow the solid angle coverage to increase by roughly a factor of two and, in addition,
would enable adding energies of several detectors, increasing the efficiency, for example, for the detection of a 1.3-MeV γ-ray by another factor of two. Ultimately, the
efficiency for detecting the full energy of 1.3 MeV for a closed shell built purely with
Ge detectors is approximately 60%, limited only by the finite size of the Ge crystals,

gaps, and absorption in mounting structures. However, the inability to distinguish between two γ-rays emitted and detected simultaneously and one γ-ray interacting with two detectors prevented the introduction of pure Ge shells for nuclear physics experiments. To separate multiple and simultaneous γ-rays sufficiently, more than 1000 detectors would be required, which is prohibitive owing to cost. The solution is to track the interactions of all γ-rays to identify and separate the emitted γ-rays even if several interact in one detector, as shown in **Figure 2**. Owing to the advances over the past 10 years in two-dimensional segmentation of large-volume Ge detectors and in digital signal processing, it is now possible to build detectors that can track γ-rays by measuring energies and positions of individual γ-ray interactions with high efficiency and accuracy.

The concept of γ-ray tracking in large two-dimensional segmented Ge detectors allows us to significantly increase the sensitivity in most nuclear physics experiments, ranging from radioactive beam experiments with the most exotic and rare nuclei to stable beam facilities to probe the nuclear structure at the highest attainable spins associated with the largest number of γ-rays and most complex decay schemes. In addition, it also enables us to determine the incident direction of γ-rays without using a collimator. Because it uses the Compton scattering formula to correlate the measured energies and positions with the incident angle, these systems are called Compton cameras. Because γ-rays, particularly at energies above 100 keV, cannot be focused easily, γ-ray imaging systems normally use collimator or aperture systems to project a source-location-specific pattern on a γ-ray detector. Examples are Anger cameras, consisting typically of a parallel-hole collimator in front of a position-sensitive NaI(Tl) detector, or coded-aperture or rotation-modulation systems (26–28). The latter two project the object by modulating the intensity spatially or temporally, respectively. The drawback of the collimator-based approach is the heavy collimator, which is required to generate a source-location-specific pattern on the detector. It not only makes the system heavy, particularly for higher γ-ray energies, it also introduces artifacts owing to scattering in the collimator and penetration of the collimator. Furthermore, it requires the γ-rays to enter through the collimator, limiting the solid angle that can be imaged—the so-called field of view—and, for many applications, requires heavy shielding all around the detector to reduce the background. Tracking has the additional benefits of increasing the overall sensitivity by correlating the image and the track information. In γ-ray imaging, the improvement due to γ-ray tracking is currently being explored in many areas, ranging from biomedical research (29, 30) to homeland security (31, 32) to astrophysics (33, 34). It is even being discussed for nuclear physics experiments with radioactive beams that distinguish between γ-rays of interest from the target and γ-rays from the radioactive-beam-related background. For these applications, Ge detectors provide high sensitivity owing to the high-energy and position resolution in combination with the high efficiency that can be achieved with large-volume detectors. The full γ-ray or event reconstruction associated with γ-ray tracking is also currently evaluated for fundamental physics experiments such as Majorana, which aims at observing the $0\nu\beta\beta$ decay in enriched ^{76}Ge detectors. Here, a partial or a full γ-ray reconstruction provides additional information to distinguish between the beta-decay signal and predominantly γ-ray background (7). Although these efforts are driving to a large part the developments in Ge-detector-related technologies, Ge detectors are now widely used in many other applications such as nuclear emergency response; nuclear counting facilities; safeguards applications; secondary inspection, for example, at borders or port of entries; or airborne surveillance, for example, for mining or other geological purposes. Many of these systems are now very compact, either with small liquid nitrogen dewars or operating cryogen-free by mechanically cooling the detectors and read out with fast-sampling analog-to-digital converters, where signal processing and filtering are performed digitally.

Complementary to these efforts, Ge detectors working as thermometers have been developed and are operating at much lower temperatures, for example, milli-Kelvin, providing the ability to simultaneously measure both the heat dissipation through the generations of vibrations or phonons and the ionization. The type of recoil can be identified because the ionization signal of a nuclear recoil is quenched as compared with an electron recoil. Also, because nuclear recoils are generated by heavier particles such as neutrons or potentially weakly interacting massive particles (WIMPs) and electron recoils are generated predominantly by photons, these types of interactions can be distinguished. This ability is a crucial component to reducing the background of γ-rays, for example, in the search for postulated but not yet discovered particles (35). In addition, NTD Ge thermistors have been used to fabricate the highest-resolution X-ray detectors (36, 37).

Following these discussions about the historical developments of Ge detectors and associated applications, we discuss some technical details in the operation of Ge detectors to explain why Ge detectors are still the gold standard and then illustrate some of the progress achieved.

1.2. Technical Review

Several fundamental physical properties make Ge appealing for the detection of γ radiation. The unique combination of the underlying crystal properties and the unprecedented purity enables γ-ray detection with high efficiency, good signal-to-background ratio, and good timing and excellent energy resolution. **Table 1** summarizes some of the important properties of Ge.

Gamma-ray detection is based on the effect of a γ-ray interacting with matter (38, 39). Three types of interactions are important for γ-ray detection in the range of interest for Ge detectors here, for example, from 10 keV to 20 MeV: the photoelectric absorption, the Compton effect, and the pair production. In Ge, below 150 keV the photoelectrical absorption dominates, between 150 keV and 8 MeV the Compton

Table 1 General properties of Ge

Crystal structure	Lattice constant (Å)	Z	Density (g cm^{-3})	Band gap (eV) (77° K)	Pair creation energy (eV) (77° K)	Dielectric constant	ρ at 25°C (Ω cm)	μ_e (cm^2 V^{-1} s^{-1}) (77° K)	μ_h (cm^2 Vs^{-1}) (77° K)	τ_e (s)	τ_h (s)	$\mu\tau_e$ (cm^2 V^{-1})	$\mu\tau_h$ (cm^2 V^{-1})
Cubic	5.65	32	5.32	0.7	2.96	16	47	3900 (36,000)	1900 (42,000)	>10^{-3}	10^{-3}	>1	>1

scattering process, and above 8 MeV the pair production has the largest cross section. Details on these interaction mechanisms can be found in textbooks, for example, the one by Knoll (40). A γ-ray of approximately 1 MeV typically interacts two to three times in a Ge detector via Compton scattering before it is fully absorbed by photoelectric absorption.

In the following we discuss some basic operational aspects in using Ge detectors in ionizing mode. We then discuss some specific characteristics and finally describe pulse shapes obtained and used in segmented Ge detectors.

1.2.1. Basic operation of germanium detectors. The interaction processes briefly described above generate primary charge carriers such as the photoelectrons or Compton electrons through ionization of the Ge atoms. These primary charge carriers generate electron-hole pairs while slowing down. Electrons and holes drift under an external electrical field to the respective electrodes, where they induce a displacement current that can be measured in an external circuit. To measure this current above the leakage current, Ge detectors are operated in reverse-biased diode configuration, which is based on a p-n junction extending from the so-called rectifying contact or electrode. As in a diode, a region can be formed that is depleted of free charge carriers and characterized by very low leakage current. To reduce the otherwise overwhelming leakage current due to the small band gap and associated thermal excitation of intrinsic charge carriers, Ge detectors must be cooled down to below 120 K. Depending on the electrically active net-impurity concentration N in the Ge crystal, the full crystal volume can be depleted by applying a voltage V_d. In a planar detector with thickness d, the depletion voltage can be calculated by

$$V_d = \frac{Nd^2}{2\varepsilon},$$

with ε the dielectric constant of Ge. P-type detectors are characterized by acceptors dominating the impurity concentration; in n-type detectors, the donors dominate. The type of crystal produced can be manipulated by doping the material. Two different configurations have been established for Ge detectors: a planar disk and a cylindrical or coaxial geometry.

In a planar configuration, the electrical contacts are provided on the two flat surfaces of the Ge disk. The n+ contact is normally formed by Li evaporation and diffusion into the surface of the wafer. The direct implantation of donor atoms such as phosphorus is difficult owing to a required annealing process to remove crystal damage caused by the implantation process and to make the donor atoms electrically active. In a p-type detector, the n+ contact is typically rectifying, initiating the depletion by applying positive voltage. The contact at the opposite face of the crystal must be a noninjecting or blocking contact for the minority carriers. It may consist of a p+ contact made by boron implantation or a metal-semiconductor surface barrier such as gold that acts as an electrical equivalent. The p+ contact is typically less than 1 μm in thickness, the thickness of the B contact is approximately 0.3 μm, and the n+ Li contact can be as thick as 1 mm owing to the diffusion. The latter thickness must be considered when X rays or low-energetic γ-rays are to be measured. However, this thick dead layer can serve as a passive shield for α particles or low-energy electrons or photons.

To achieve good charge collection throughout the crystal, voltages significantly higher than the depletion voltages are applied to ensure sufficiently high electrical fields to saturate the drift velocities of electrons and holes and to minimize the collection time and the detrimental effects of charge recombination and trapping. However, the maximum voltage is limited to approximately 5 kV by the leakage currents at the crystal surface and by the additional complexity to prevent high-voltage-induced breakdowns and sparks. Thus, the thickness of the detector is limited to approximately 2 cm in the planar configuration. One challenge in the fabrication of Ge detectors is the passivation of the noncontact surface areas. The goal is to be electrically neutral to establish the wanted field pattern and to withstand high voltages with low reverse-leakage current. One employed technique is to sputter hydrogenated amorphous germanium (a-Ge) onto the etched surface. Despite significant amounts of effort, the passivation of the surfaces is still not electrically neutral and electrical field lines do end on this surface, leading to incomplete charge collection (41). To circumvent the problems associated with these surfaces, so-called guard rings at the edges of the round contacts in planar detectors are introduced that separate the surface effects from the proper bulk properties.

Assuming a low impurity concentration, the electrical fields of planar detectors are almost uniform and holes and electrons drift similarly under the influence of a nearly constant field throughout the active detector volume in p-type and n-type detectors. This is quite different in coaxial detector geometries. Owing to the radial symmetry, electrical fields are nonuniform and the type of the detector has a significant influence on the details of the charge carrier motion. The advantage of the coaxial over the planar geometry is the larger depleted and active volume of a single detector that can be achieved. In addition, the capacitance of coaxial detectors is smaller, resulting in a better signal-to-noise ratio and therefore, in principal, better energy resolution if not for other detrimental effects, for example, charge trapping. While one electrode is fabricated at the outer cylindrical surface of a long Ge crystal, a second cylindrical contact is provided on the surface of an inner borehole. The so-called closed-ended configuration with the borehole reaching through approximately 80% of the detector and the front face as part of the outside contact is preferred over true-coaxial geometries. In this way, only the back side of the crystal must be passivated, with the same complications as the outside edge of planar detectors. In addition, low-energetic photons can enter the active volume, assuming a thin outside contact is being used. The disadvantage of this configuration is a nonuniform electrical field, particularly in the front corners of the detector, which are associated with the weakest electrical fields. To increase the electrical fields in these corners to improve the charge transport, the corners of the crystal as well as the hole are rounded or bulletized. In addition, the higher impurity concentration is chosen for the front of the detector to increase the electrical field as well. In coaxial detectors, the rectifying contact is typically outside to obtain the highest electrical fields outside where most of the detection volume is located. In addition to the thickness of the outside contact, other important criteria in the selection of the type of detector are the susceptibility to radiation damage and the ability to segment the outer contact. Radiation damage, for example, from neutrons emitted during in-beam nuclear physics experiments or from charged particles in space applications, predominantly generates hole traps. To minimize the hole trapping with its detrimental impact on energy resolution and, ultimately, sensitivity, the hole contribution to the detector signal must be minimized. This can be accomplished by employing n-type detectors to collect the holes outside, with the hole-collecting B-implanted p+ contact outside (42). Another advantage of n-type detectors is the ability to electrically segment the outside B-implanted contact much more easily and reliably than is possible with the thick Li contact on p-type detectors.

1.2.2. Characterization of germanium detectors. Ge detectors are characterized by their excellent energy resolution, good timing characteristic, high efficiency, and peak-to-total ratio. The excellent energy resolution is due to the small energy needed to generate information carriers—in the case of ionization electron-hole pairs—in combination with the high mobility and small charge trapping. As shown in **Table 1**, the mobility-lifetime product $\mu\tau$ is larger than 1, ensuring almost complete charge collection, which is in sharp contrast to all other semiconductor detectors to date except Si. The full width at half maximum (FWHM) of the energy resolution ΔE_{stat} in the statistical limit is defined by the number of information carriers N, which depends on the deposited energy E and the energy required to generate one information carrier ε and the Fano factor F:

$$\Delta E_{stat} = 2.35 \times \sqrt{\varepsilon \times E \times F}.$$

The Fano factor is introduced to reflect the fact that the electron-hole pairs in Ge are not created independently (43). For Ge, it is of the order 0.1. As an example, we assume a γ-ray deposits an energy of 1 MeV in the Ge detector. The average energy to produce one electron-hole pair is 2.96 eV at 80 K (44), and therefore the statistically limited energy resolution is approximately 1.3 keV. The energy to produce an electron-hole pair is approximately four times more than the band gap of 0.7 eV in Ge; that is, ~75% of the energy is converted into thermal energy in the form of phonons. The fact that these two mechanisms are coupled causes the deviation from Poisson statistics and requires the introduction of the Fano factor. In addition to the statistical uncertainty, other factors are limiting the energy resolution, such as carrier trapping and electronic noise. The electronic noise is a function of the leakage current of the detector and the amplification and readout circuitry, which typically consists of a charge-sensitive preamplifier with a field-effect transistor (FET) and a resistive or optoelectronic feedback loop. The noise characteristic of these components, the total capacitance at the FET gate lead to ground—including that of the detector, the leakage current, and any series resistance—and dielectric noise components contribute to the overall electronic noise that can be minimized by adjusting the shaping time in the following processing step. The electronic noise is independent from the energy and typically dominates the resolution at low energies. Details of noise and spectral resolution are described in the paper by Goulding & Landis (45), as well the one by Radeka (46).

Typical energy resolution values for standard Ge detectors are 0.8 keV at 60 keV and 1.8 keV at 1.33 MeV. The low-energy value is dominated by the electronics noise; the higher energy value is dominated by the statistical noise. With small devices, having small-capacity energy resolution values of 110 eV has been achieved at 5.9 keV, with energy thresholds as low as 100 eV (6).

The high purity of Ge detectors is required to deplete all parts of large crystals and provide high fields so that charge carriers can be swept out of the crystal quickly; the low density of shallow traps in combination with the high mobility provide the basis to collect all charges produced owing to the interactions. The mobility-lifetime product in Ge is much larger than the one for holes and electrons in contrast, for example, to the ternary CdZnTe semiconductor that has a mobility lifetime for holes of 10^{-4} (47). Both factors, the high purity and the large $\mu\tau$ product, in addition to the ability to apply high voltages, are necessary for depleting and achieving large volumes of HPGe detectors and therefore high efficiencies.

1.2.3. Signal generation and processing.

In this section we briefly review the generation of signals in Ge detectors. The focus here is the dependence and the parameterization of signal shapes with regard to the positions of individual γ-ray interactions. A more detailed description can be found in the review article by Radeka (46).

The motion of charge carriers in the vicinity of an electrode induces a current on this electrode that can be measured. It can be calculated by first determining the path of electrons and holes in the electrical field E_D of the detector, which is obtained by solving the Poisson equation with the proper boundary conditions and the given impurity concentration profile within the detector volume. The path of the charge carrier is calculated by moving the charges through the electrical field with the velocity $v(t)$ given the mobility $\mu_{e,b}$ with $v = \mu_{e,b} E_D$. Note that the mobilities in Ge detectors are different for holes (b) and electrons (e) and depend on the temperature and the direction of the motion relative to the crystal orientation (48). In most applications, the original charge cloud extension due to the finite range of the primary electrons such as the Compton electron is neglected, as well as the repulsion and the diffusion of the charges during the charge collection process. However, for detector configurations with segment sizes of the order of millimeters or smaller, the finite size of the charge cloud must be taken into account. Here, charge sharing occurs between segments that can be understood only by taking the underlying processes leading to an extended charge cloud into account. With segment sizes smaller than the charge cloud extension, center-of-charge methods can be applied to improve the lateral position resolution, as discussed by Radeka. These methods are implemented primarily for charged-particle tracking with thin detectors of the order 300 μm and do not require pulse-shape analysis. In applications with thicker detectors and electrode extensions of the order of or slightly larger than the charge cloud extension, this extension must be taken into account if one wishes to achieve a position resolution significantly better than the segment dimensions by pulse-shape analysis. Here, the equations of the electrical field and the motion by diffusion and repulsion must be time dependent and coupled with a finite charge distribution of electrons and holes.

As a second step, the induced current or charge on a specific electrode is deduced by calculating its weighting field. The weighting field is a measure of the electrostatic coupling between the moving charge and the sensing electrode and is calculated by applying a unity voltage to the charge-sensing electrode and zero potential to all other electrodes, with no space charge present. The result is normalized with the voltage to give a dimensionless field. The induced current $i(t)$ in the sensing electrode is then

$$i(t) = -q_m \vec{E}_w \cdot \vec{v}(t),$$

with q_m the charge in motion, E_w the weighting field, and $v(t)$ the velocity of the charge carrier. We can deduce the charge increment $\Delta q(t) = i(t)^* \Delta t$ for every time step Δt and the velocity $v(t) = \Delta r(t)/\Delta t$:

$$\Delta q(t) = -q_m \vec{E}_w \cdot \Delta \vec{r}(t).$$

The true electrical field can be quite different from the weighting field, particularly in segmented detectors. The weighting field is ultimately responsible for the dynamic range of signal shapes that enables a much more accurate position determination than given by the segment size. For example, in the 36-fold segmented prototype detector, which was developed for the Gamma-Ray Energy Tracking Array (GRETA), a position resolution of better than 1 mm at an energy of 374 keV was obtained, which was significantly better than the average segment size of approximately 20 mm (49, 50).

2. RECENT TECHNICAL DEVELOPMENTS

In this section we provide an overview about recent developments in technologies related to the basic performance and operational aspects of Ge detectors. This includes different implementations and configurations such as planar and Ge drift detectors, the encapsulation, amorphous contacts, segmentation schemes, and Ge detectors operated as bolometers. The second part briefly discusses developments in the electronics, including preamplifiers and more recently digital signal processing and pulse-shape analysis. Associated technologies that rely on some of these technologies are finally presented in the third part, such as γ-ray tracking and γ-ray imaging.

2.1. Detector Technologies and Configurations

2.1.1. Encapsulation. To increase the reliability of Ge detectors, the encapsulation technology was developed in the 1990s for nuclear physics experiments (51) and in parallel for space-flight applications (52). A vacuum-tight encapsulation provides several advantages over conventional assemblies: The crystal remains sealed at all times, avoiding contamination of the Ge surfaces, which can happen over time even with small leaks or by annealing the detectors; overall handling is greatly simplified because surfaces are protected; and annealing of radiation-damaged detectors is simplified as well because the capsule can just be put in an ordinary oven. The cold FET is outside the capsule and can be easily repaired or removed before annealing. A defective detector in a multidetector cryostat can be replaced with minimal effort and without compromising the other detectors. For nuclear physics experiments, the crystal and the capsule have a tapered, hexagonal shape. The distance between crystal surface and can is only 0.7 mm; the thickness of the can is only 0.5 mm. Electron welding of the lid and the electrical feedthroughs at the back ensure a hermetic seal. These features were essential in operating the seven Ge detector cluster detectors for Euroball and the three and four detector modules for Miniball (21). The Miniball detectors demonstrated encapsulation even for two-dimensionally segmented coaxial Ge detectors. Future γ-ray tracking arrays such as GRETA in the United States and the Advanced Gamma-Ray Tracking Array (AGATA) in Europe, consisting of three and four detector modules of (6 × 6)-fold segmented detectors, will also rely on the encapsulation technology. Without the encapsulation technology, the reliable fabrication, assembly, and operation of these complex detectors would not be possible. Encapsulation for space-based deployment of Ge detectors was developed to address the extreme challenges related to the launch, with extremely high G forces (∼100 G) and the long-term stability with lifetimes of at least five years without any possible human intervention. Although the capsules fabricated for the nuclear physics experiments are kept under vacuum, the capsules for space flights are pressurized to improve the long-term stability. A pressurized can, for example, using ultraclean nitrogen, significantly reduces the inlet of contaminants even in the case of a very small leak and therefore can extend the lifetime of a capsule significantly. It also allows the verification of the proper seal and leak integrity by measuring the nitrogen getting out of the capsule.

2.1.2. Amorphous contact. As discussed above already, Ge detectors—as well as Si detectors as the other elemental semiconductor used for radiation detection—are reverse-biased diodes operated in full depletion, with blocking contacts to maintain low leakage currents and high electrical fields for fast complete charge collection. The contacts on Ge detectors typically consist of n-type contacts formed by Li diffusion and p-type contacts formed by B ion implantation. These contacts are relatively simple to produce and are in use from the beginning of HPGe detector development. However, there are drawbacks in using the Li-diffused contact. This contact produces dead layers typically of the order of hundreds of microns, which prevents low-energy photons from reaching the sensitive detector volume. Thin n-type contacts have been developed using phosphorus implantation; however, they require elaborate processing steps to heal P-implantation-induced defects and cannot withstand high electrical fields (53). Phosphorus-implanted contacts are therefore used only in special applications such as charged-particle transmission detectors. Second, the fabrication of monolithic multi-element detectors using conventional contacts can be difficult because of the need for many processing steps. Finally, the large diffusion depths as well as the high mobility of Li in Ge at room temperature, which increases the size of the contacts, make the formation of small contact structures unreliable (54, 55).

An alternative to these conventional contacts are amorphous semiconductor contacts. The first experimental study of electrical junctions between a-Ge and crystalline Ge was published in 1964 (56). The use of a-Ge to fabricate blocking contacts on Si radiation detectors was first reported in 1971 (57). These contacts showed good bipolar blocking behavior; that is, they can block both electron and hole injection. A-Ge blocking contacts on HPGe detectors were investigated in 1977, but the devices showed large variations in leakage currents (58). Progress was made by using chemical vapor deposition of amorphous Si on Si detectors (59) and by RF sputtering of a-Ge on Ge detectors (60), instead of using vacuum evaporation deposition. Whereas the first studies used sputtered a-Ge on Ge only on one side to replace either the Li or the B contact, later the bipolar blocking characteristic was used to replace both conventional contacts. Note that the sputtering parameters used originally were identical to those used in the deposition of a-Ge for Ge detector surface passivation (61). Therefore, the a-Ge coating was and still is deposited on the total surface of the Ge. A metal layer, for example, by evaporating aluminum or gold, deposited on the top and the bottom of the detector defines the contact area; the a-Ge on the side and between contact elements serves as a surface-passivation layer. Typical thicknesses for the a-Ge and the metal layers are 300 nm and 50 nm, respectively, and therefore significantly thinner than Li-diffused contacts enabling photon detection down to a few keV that are limited by the entrance window of the cryostat and not by the detector contact. Because the a-Ge represents a true thin dead layer without charge collection, the spectral response for low γ-ray energies is improved over surface-barrier or B-implanted contacts—that is, no significant tails, steps, or generally increased background can be observed. In contrast, surface-barrier or B-implanted contacts generate a partially active layer with low or no electrical field owing to the presence of a high concentration of dopants or defects in which charge loss can take place (62). Even finely segmented contacts are relatively easy to fabricate, avoiding the problem of thick and not temperature-stable Li contacts. Owing to the bipolar blocking character, finely segmented contacts can be made on either side of p-type and n-type detectors, whether in coaxial or in planar configuration. One drawback in operating a-Ge detectors is increased leakage current for increased temperatures, which is due to the reduced barrier heights in these devices as compared with conventional p-n detectors. In addition, temperature cycling between liquid nitrogen and room temperature causes a degradation of the rectifying ability of amorphous contacts. Hull & Pehl (63) recently studied the properties of a-Ge in more detail by producing many detectors and changing sputter parameters such as temperature, time, and sputter gas composition—that is, changing the ratio of H_2/Ar gas concentrations—and found recipes that provide significantly improved performance in terms of temperature cycling and operational temperatures. A-Ge detectors can be operated up to 120 K without significant degradation owing to leakage current, similar to the conventional contacts (64; E.L. Hull, private communication).

2.1.3. Segmentation. The ability to measure the position of γ-ray interactions has many advantages in the detection of γ-rays, for example, for γ-ray tracking and associated γ-ray imaging. Position-sensitive γ-ray detection with Ge detectors was originally realized for nuclear medicine applications by employing arrays of small Ge detectors (65–67). As a more efficient and more practical approach, the segmentation of monolithic Ge detectors was introduced (54, 55, 68, 69). Segmentation of Ge detectors is now normally achieved by segmenting the contacts on the surface of the detector either by photolithography, simple masking, etching, or even by groove-cutting techniques. Accordingly, they can be more or less invasive, with varying impact on the electrical and charge collection properties in the detector. Segmentation can be implemented as either one-dimensional strips or pixels on one side, or orthogonal strips on both sides. All these implementations have been realized in planar as well as in coaxial Ge detector configurations.

2.1.3.1. Planar high-purity germanium detectors. Segmentation in Ge detectors was first demonstrated in planar HPGe detectors in one-dimensional strip configurations by several groups (54, 55, 69, 70). Photolithography techniques were employed on conventional B-implanted or Li-diffused contacts. To improve the electrical separation of adjacent strips, Protic & Riepe (69) introduced a subsequent plasma-etching process to create grooves of the order of tens of micrometers into the crystal, thereby separating the blocking contacts. With the development of a-Ge contact technologies, the limitations associated with Li-contact segmentation were circumvented and a much less complex segmentation process became possible. Hull & Pehl (63) and Luke et al. (71, 72) use a-Ge sputtering techniques; Protic & Krings (73, 74) use a-Ge evaporation. The contact and segmentation pattern is defined by the evaporation of a metal layer, that is, Au or Al, through a mask. Although Protic et al. use a-Ge contacts only on one side as n+ blocking to replace the Li contact, Luke et al. and Hull et al. coat the total surface with a-Ge, enabling the bipolar blocking characteristics of a-Ge for both contacts and as passivation of the noncontact surfaces. In this way, orthogonal strips on both sides of planar HPGe detectors, so-called double-sided strip detectors (DSSDs), have been fabricated with either n- or p-type Ge material. The ability to use either type of Ge for two-dimensional segmentation in combination with the relative simple processing of a-Ge contacts makes this approach very appealing, in addition to avoiding the limitations of Li contacts such as limited feature sizes >1 mm or long-term stability caused by the high mobility of Li in Ge (55). The a-Ge contact technologies have matured significantly over the past few years and are now available commercially; however, some detrimental effects remain. For example, incomplete charge collection is reported by Amman & Luke (72), which causes a degradation in energy resolution and photopeak efficiency. This specific problem can be avoided by employing field-shaping electrodes in between the charge-sensing electrodes as suggested by Amman, or by electrically separating the electrodes by controlled plasma etching as demonstrated by Protic (73).

Over the past 10 years, many two-dimensionally segmented planar HPGe detectors were fabricated. They range from one-sided strip and one-sided pixilated to DSSDs. **Figure 3** shows one example of a DSSD made of HPGe, with 39 strips on each side with a pitch size of 2 mm. Shown is the detector in its mount, with the readout components and the discrete preamplifiers outside the cryostat. Applications range from atomic and nuclear physics (75–78) to coded-aperture and Compton-scattering-based γ-ray imaging for astrophysics, biomedical research, nuclear medicine, nuclear safeguards, environmental remediation, and homeland security (79–87). In addition, by combining two-dimensional segmentation with the analysis of signal shapes obtained from the segments, the position resolution in all

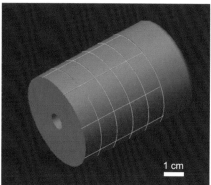

Figure 3

Two types of two-dimensionally segmented HPGe detectors. (*Left and center*) A 40-fold segmented close-ended Ge detector is shown with 5 longitudinal and 8 transversal segments. The 42 preamplifiers to read out the 40 segments, the nonsegmented front cap, and the central channels are arranged on a circular motherboard close to the vacuum feedthroughs. (*Right*) A (2×39)-fold segmented planar detector in double-sided strip configuration is shown. Each strip has a pitch width of 2 mm, resulting in 78 readout channels. As with the coaxial detector on the left, the discrete preamplifiers are operated with warm JFETs mounted close to the vacuum feedthroughs.

three dimensions within the detector can be increased significantly. This aspect is discussed in Signal Processing and Advanced Data Analysis below.

2.1.3.2. Coaxial high-purity geramium detectors. Driven by the demands to increase the sensitivity in nuclear structure experiments, the segmentation of coaxial Ge detectors was pursued from the early 1990s. The first segmented coaxial Ge detectors were fabricated and deployed for the Gammasphere spectrometer (88, 89). The purpose of the twofold segmentation of the outside contact was the increase in position resolution to reduce the impact of Doppler-shifted γ-rays. The Doppler shift of γ-ray energies emitted from excited nuclei in flight can cause a significant degradation in energy resolution owing to the finite size of Ge detectors observing the γ-ray.

Since the end of the 1990s, researchers have pursued three main approaches in segmenting and arranging coaxial HPGe detectors to build arrays of these detectors, providing high photopeak efficiency and excellent (ideally close to the intrinsic) energy resolution for nuclear physics experiments, both owing to the increased granularity provided by segmentation and advanced signal processing. The first and most advanced approach consists of two-dimensionally segmented and individually encapsulated detectors, with several of these detectors arranged in one closely packed cryostat. The aforementioned Miniball array as well as both GRETA in the United States and AGATA in Europe, γ-ray tracking arrays currently being developed, are based on this approach (23, 77, 90). Whereas the Miniball detectors mostly have only six transverse segments, the tracking arrays consist of (6×6)-fold segmented Ge detectors. The second approach consists of so-called Clover detectors, each made of four square-shaped Ge detectors arranged as a clover in one cryostat (19, 20, 91). In contrast to the Miniball detectors, Clover detectors are not encapsulated. Clover detectors are used either with no segmentation or with one- or two-dimensional segmentation. The most advanced concept for segmented Clover detectors is currently being realized at TRIUMF in Canada for the TIGRESS array. In addition to the fourfold azimuthal segmentation of each detector, a longitudinal segmentation is introduced to increase the position sensitivity in the depth of the detector as well (25). In addition, in TIGRESS, the full potential of pulse-shape analysis in this type of detector is realized (92). The third configuration of segmented coaxial HPGe detectors is the two-dimensional segmentation of HPGe detectors in the original cylindrical shape. One example is an (8 × 4)-fold segmented detector for the SEGA array currently in operation at the University of Michigan in the United States (24). In the SEGA configuration, the eight longitudinal segments, each 1 cm long, provide the granularity for Doppler-shift correction owing to the alignment of the detector symmetry axis preferentially in parallel to the beam axis. A 40-fold segmented coaxial HPGe detector shown in **Figure 3** was developed by Lawrence Livermore National Laboratory (LLNL) to demonstrate Compton imaging in this type of detector. This detector is segmented in eight azimuthal and five longitudinal segments, the latter covering only the true-coaxial part of the detector to provide maximum position sensitivity, particularly by combining the high segmentation with three-dimensional pulse-shape analysis (93). In Italy, a (6 × 4)-fold segmented coaxial HPGe detector was fabricated as a part of the European γ-ray tracking effort (94).

Note that all segmented coaxial HPGe detectors are of n-type, with external segmentation of the B-implanted contact. Except for the Gammasphere detectors, all the above mentioned detectors provide the high voltage to the central Li-diffused contact, which requires AC coupling to read out the signal from this side. In Gammasphere, the high voltage was split for the two external segments and the central core was DC coupled. The use of n-type crystals provides (a) a reliable and flexible segmentation, (b) reduced sensitivity to radiation damage, important for nuclear physics experiments, and (c) less attenuation, particularly of low-energy photons in the external contact.

Whereas most of the above mentioned applications aim at nuclear physics and ground-based experiments, one-dimensionally segmented detectors were fabricated for the Reuven Ramaty High-Energy Solar Spectroscopic Imager (RHESSI) space mission to measure X rays and γ-rays from the sun (28). The spectrometer consists of nine HPGe detectors, each with two longitudinal segments. Here the Li-diffused inner contact of the n-type detectors was segmented by cutting a groove approximately 1 cm away from the closed end of the central hole. The purpose of this segmentation was to minimize external space requirements and to separate the smaller front and pseudoplanar part and the back and nearly true-coaxial part of the detector. The front part is used primarily for low photon energies ranging from 3 keV to150 keV, providing excellent timing and energy resolution as well as fast timing characteristics necessary for the occasionally appearing high flux of low-energy photons from the Sun; the back part is used for higher photon energies with significantly reduced photon flux.

Two efforts, one in Liverpool, United Kingdom, and one for Majorana in the United States, evaluated the two-dimensional segmentation of both the external

B contact and the internal Li contact, similarly to the double-sided strip configuration of planar detectors discussed above. These two detectors were fabricated with six azimuthal and external (B-implanted) contacts and two longitudinal and internal (Li-diffused) contacts. In this way, a granularity similar to a 12-fold segmentation can be realized with only eight segments. Initial studies showed degraded performance in energy resolution. Research to characterize and potentially improve the performance of these detectors is still ongoing (A. Boston & A. Young, private communication). Finally, research is also pursued to externally segment p-type detectors by separating the Li-diffused contacts mechanically by precision cutting. By cutting grooves a few millimeters deep and a few millimeters wide, the Li can be separated sufficiently to prevent merging by migration over time. However, the impact of these large grooves remains to be evaluated, particularly with regard to pulse-shape analysis, energy resolution, and efficiency (F. Avignone, private communication).

2.1.3.3. Germanium drift detectors. Luke et al. (95) demonstrated the successful fabrication and operation of a low-capacitance and large-volume Ge detector in 1989. The configuration is similar to the more widely used implementation of Si drift detectors characterized predominantly by electrical fields in parallel to the surface of the detector. In contrast to the Si drift configuration, the Ge-based instrument is aimed at large volumes with a radius of 5 cm or larger and a depth of 5 cm or larger while maintaining the excellent noise performance owing to the small capacitance. In this configuration, the inner hole of a conventional closed-ended coaxial detector is reduced to a small disk 0.5 mm in length and 2 mm in diameter, resulting in a capacitance of approximately 1 pF at full depletion, which is much less than the typical 20 pF observed in conventional coaxial Ge detectors. The Ge drift detector was implemented using n-type material, with the p+ contact outside and the n+ contact inside. In this way, the electrons are the dominant charge carriers producing the detector signal, as almost the full signal is induced as the electrons approach the small n+ contact. The drawback of this configuration is the weak axial electrical field, which causes long electron drift times and associated susceptibility to electron trapping, degrading the energy resolution. To increase the electric field, particularly the axial component, one can either taper the detector, which implies losing a substantial amount of Ge material, or one can use materials with an axial impurity concentration gradient. This gradient is generally intrinsic to Czochralski-grown Ge crystals and more pronounced in n-type crystals owing to the segregation of impurities during the crystal growth process. In the fabricated detector, the impurity concentration was 0.7×10^{10} cm^3 at the closed end and 1.5×10^{10} cm^3 at the small electrode side. Significant electron trapping was observed, leading to an unacceptable energy resolution. By increasing the temperature of the crystal from 77 K to 125 K, the energy resolution was improved to 3.5 keV, with a long shaping time of 32 μs. The improvement of the energy resolution for higher temperatures is attributed to the more rapid release of trapped carriers; the long shaping time helps in collecting the trapped carriers over a longer time period. Although the energy resolution is degraded, the demonstrated noise level was still as low as 270 eV and could be lowered to below 100 eV by employing a better preamplifier such as a pulsed-reset preamplifier and a properly mounted FET. This noise level is significantly better than the 500 eV that can typically be achieved in coaxial Ge detectors of this size. Very recently, a Ge drift or so-called modified-electrode or point-contact HPGe detector was built in p-type configuration with significantly improved energy resolution (J. Collar, private communication). The low noise and the long drift times can be used to differentiate between single-site and multiple-site interactions in the crystal. This is due to the small pixel effect discussed above, which causes the main charge induction on the small charge-collecting electrode just before the charge carrier reaches the electrode. The low noise and therefore low trigger threshold as well as the ability to distinguish single from multiple interactions with high sensitivity and with only one electrode make it very appealing for experiments aimed at detecting low-energy particles such as WIMPs, low-energy solar neutrinos, or the $0\nu\beta\beta$ decay. The Majorana collaboration is currently considering large-volume Ge drift detectors as detector geometry to search for the $0\nu\beta\beta$ decay.

2.1.3.4. Mechanical cooling. The operating temperature of Ge detectors of approximately 100 K or less is one of its drawbacks, particularly for applications for unattended or remote detection on Earth or in space, where liquid nitrogen as coolant is not readily available. Mechanical cooling is the means to providing sufficient cryogen-free cooling for applications such as space-based missions and nuclear safeguards and customs. Mechanical cooling was originally developed in the 1990s with Joule-Thompson-based coolers with heavy, high-power compressors that could cool, in principle, any size of available Ge detectors. The next step in the development was based on the Kleemenko cycle with mixed gas refrigerants, realized in the commercially available X-COOLER by Ortec (96). Now, typically, Stirling-based cryocoolers are used, which are more efficient, more compact, and require significantly less power. However, one challenge with the Stirling coolers coupled directly to the Ge detector cryostat is the vibration-induced microphonics in the detector readout that degrades the achievable energy resolution in these systems. Complex passive and active compensation schemes or tandem coolers running in antiphase, as well as digital filtering algorithms, have been developed to minimize the impact of microphonics. Whereas the first Stirling cycle system developed by LLNL required approximately 75 W, currently available systems such as the Cryo3 requires only 15 W, weighs approximately 4.5 kg, and can be operated up to 6 hours with one set of batteries (97).

Mechanical cooling for space missions provides not only a reliable way for cooling over long periods of time, it also allows temperature cycling, including necessary annealing of radiation-induced defects, particularly important for long-duration flights such as the Mercury Surface, Space Environment, Geochemistry, and Ranging mission (MESSENGER) (98, 99). Other examples of mechanically cooled Ge detectors are RHESSI in the United States, the International Gamma-Ray Astrophysics Laboratory (INTEGRAL) in Europe, and the Selenological and Engineering Explorer (SELENE) in Japan (100–102). The MESSENGER mission represents one of the most challenging ones and aims at mapping γ-ray emissions from Mercury's surface, for example, to search for ice or water at the poles. Whereas the actual measurements in an orbit of Mercury are expected to take approximately one year, the travel to Mercury alone takes approximately seven years. The MESSENGER spacecraft was launched in 2004 and is expected to reach Mercury's orbit in 2011. A sophisticated radiation shield and Kevlar suspension were developed to maintain an operational temperature of approximately 85 K despite the extreme temperature changes expected during this mission.

2.2. Electronics

In parallel and closely related to the advances in Ge detector and other radiation detector technologies was the development of the appropriate electronics to optimize the extraction of application-specific features such as energy, time, or position. Only owing to the progress in low-noise and compact electronics was it possible to realize most of the concepts in radiation detection discussed here. The electronics developments range from analog electronics such as low-noise preamplifiers and pulse-shaping amplifiers to digital electronics with fast and high-resolution waveform digitizers and a variety of increasingly powerful digital processing units.

The function of the electronic signal processing is the determination of the energy, time, and position of the γ-ray from the detector pulse. In an analog system, separated pulse-shaping (filter) circuits optimized for energy and time measurements are used to produce pulses for an analog-to-digital converter and a time-to-digital converter. In a digital signal processing unit, the entire pulse shape from the preamplifier is digitized with a waveform digitizer, and the filtering, energy, and time determination are carried out using digital processing. The basic design principle of the filters is the same whether they are implemented in an analog circuit or in digital signal processing. Digital signal processing provides more flexibility, and the possibility of more complex processing, applications-specific analog circuits can still be more compact and less power demanding. One example is the application-specific integrated circuit (ASIC), which is essential for particle physics experiments with hundreds of thousands or even millions of readout channels. However, for high-resolution systems with high intrinsic capacities such as nonsegmented or segmented, large-volume HPGe detectors with capacities of the order 30 pF, discrete and low-noise JFETs in combination with discrete preamplifiers are still state of the art.

2.3. Signal Processing and Advanced Data Analysis

In this section we discuss recent advances in digital signal processing, which, in addition to two-dimensional segmentation, provides the ability to measure energies and three-dimensional positions of individual γ-ray interactions in large-volume Ge detectors. This enables one to track γ-rays with high efficiency, a prerequisite for the new generation of γ-ray spectrometers but also for γ-ray imaging applications, particularly for Compton imaging.

2.3.1. Signal processing in two-dimensional segmented high-purity germanium detectors. Pulse-shape analysis can be implemented in a nonsegmented coaxial detector to determine the radius of γ-ray interactions. Owing to the $1/r$ component in the electrical field, with r indicating the radius and assuming a small impurity

concentration, high electrical fields are obtained close to the inner contact in coaxial detectors associated with a large induced current or a sharp rise in the charge signal. Using these features, either the time of the maximum current or the rise time of the charge signal, one can determine the radius of a γ-ray interaction within the coaxial part and therefore the bulk of the detector. In Gammasphere, the time of the maximum current is implemented (89); a more detailed discussion on the charge signals to deduce the radius can be found in the review by Kroell et al. (103).

Segmenting the detector contacts enables one to obtain complementary position information of the γ-ray interactions and, ultimately, the full reconstruction of energies and three-dimensional positions of even multiple interactions within one or adjacent segments. In arrays with one- or two-dimensional segmentation such as Miniball, Exogam, or in the future TIGRESS, simple feature-extraction concepts similar to the radius determination discussed above are or could be implemented to obtain complementary information about the location of the first interaction. For example, the amplitude of transient-charge signals induced in adjacent azimuthal segments can be used to determine the azimuthal position with a much better accuracy than the segment size itself. The same is true for longitudinal segmentation, which can improve the accuracy of locating the depth of the interaction. However, ultimately the two-dimensional segmentation in addition to processing of the full signals allows us to deduce the energies and three-dimensional positions of even multiple interactions within one detector, enabling the concept of γ-ray tracking with high efficiency and accuracy and allowing us to significantly improve the sensitivity in the detection of γ radiation, for example, in nuclear physics and γ-ray imaging applications.

One approach to deduce the positions and energies of multiple interactions is based on the so-called signal decomposition concept, which aims at decomposing the measured signal shapes into their single-interaction components by using net-charge as well as transient-charge signals. The single-interaction components are calculated following the description from above. The measured signals are then fitted with the calculated signals, where the number of interactions, their energies, and three-dimensional positions are fit parameters. To limit the large dimension of this problem and achieve decomposition times of the order of milliseconds on a single CPU, particularly noting that at least 6–9 segment signals with typically 20–30 samples must be taken into account, the maximum number of interactions allowed is typically two per segment. **Figure 4** illustrates the signal decomposition concept in a DSSD. A γ-ray of 662 keV from a ^{137}Cs source is measured with two interactions in adjacent segments. Using the four strip signals from the top and the four strip signals from the bottom, we can reconstruct the positions and energies of both interactions shown on the left by fitting the measured signals shown on the right. To demonstrate the improvement in resolution, **Figure 5** shows position spectra of measured intensities obtained at an energy of 122 keV of a ^{57}Co source illuminating the DSSD HPGe detector with a 2-mm pitch size through a 12-mm-diameter shadow mask. To perform the full three-dimensional signal decomposition as illustrated in **Figure 4** in this detector with an 11-mm thickness, the finite extension of the charge cloud was taken into account to treat the significant amount of the observed charge sharing correctly. Position-resolution values of better than 0.5 mm in each of the three dimensions was

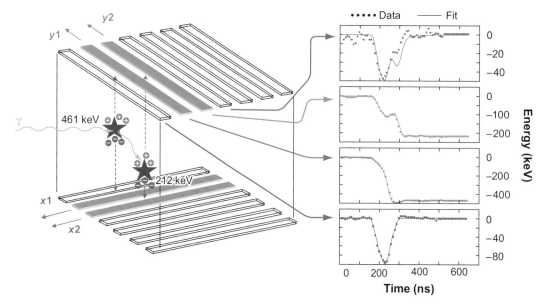

Figure 4

Demonstration of signal decomposition of two interactions in adjacent strips in a double-sided strip detector. By fitting the measured signals of eight strips (only results from the top four strips are shown on the right) with calculated signals, one can determine the energies and three-dimensional positions of individual γ-ray interactions.

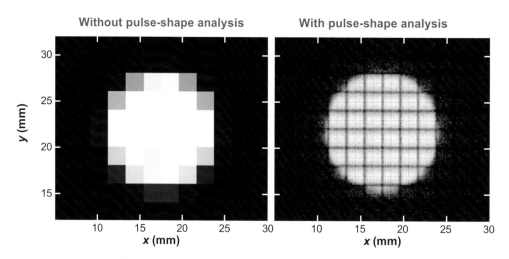

Figure 5

Demonstration of the improvement in the position resolution by applying pulse-shape analysis in a double-sided strip detector. Shown here is a γ-ray shadow of a 12-mm-diameter mask in front of a double-sided strip detector system with a 2-mm pitch size. The left image was obtained without applying pulse-shape analysis, the right image obtained with.

achieved in this detector for energies ranging from 100 keV to 400 keV. The minimum separability, for example, the minimum distance that can be resolved between two interactions, was 1–2 mm. In the GRETA prototype detectors with pixel sizes of the order 2 cm², a position resolution of approximately 2 mm was recently demonstrated (I.Y. Lee, private communication).

2.3.2. Gamma-ray tracking. To fully reconstruct or to track the γ-ray in the detector, the γ-ray interactions determined above by signal decomposition or by other means must be time ordered. Owing to the small distance between interactions—a few millimeters to a few centimeters—and a time resolution of typically 5 ns in the Ge detectors, it is not possible to determine the time sequence by measuring the time of the interactions. So-called γ-ray tracking algorithms have been developed, which rely on the characteristics of the underlying physical processes such as Compton scattering, pair production, and photoelectric absorption to order the interactions into a γ-ray path. The focus so far was on Compton scattering and the photoelectric absorption, owing to their dominance in the most interesting region of energies. Several groups have developed γ-ray tracking algorithms for γ-ray spectroscopy and Compton imaging (104–109). They typically determine a tracking figure of merit for all possible permutations of the interactions and use the scattering sequence with the lowest value. However, if the lowest value is above a certain threshold, then it is assumed that the γ-ray escaped the crystal, leaving only a fraction of its energy in the detector. In this way, not only can the scattering sequence of full-energy deposition events be determined, but one can distinguish between full and partial energy deposition and improve the peak-to-total ratio and potentially the overall sensitivity. Owing to the finite accuracy in determining positions and energies of γ-ray interactions, the peak-to-total ratio can be increased, but at the cost of full energy efficiency because some of the γ-rays that deposit their full energy are suppressed too.

These Compton-tracking algorithms typically rely on the Compton scattering kinematics, for example, energy and momentum conservation, the attenuation of γ-rays based on the γ-ray energy and distance to travel between two interactions, and the energy-dependent cross section for photoelectric absorption and Compton scattering. For applications associated with multiple γ-rays emitted and detected simultaneously, as in nuclear physics experiments employing arrays of detectors around a target or a γ-ray source, another so-called clustering algorithm must be applied to distinguish and identify individual γ-rays. As discussed by Schmid et al. (106), the full tracking process for so-called high-multiplicity or high-fold events consists of three parts: the clustering, the tracking, and the recovery. In the clustering step, interaction points within a given angular separation as viewed from the γ-ray source are grouped into a cluster. In the second step, each cluster is evaluated by tracking to determine whether it contains all interaction points belonging to a single γ-ray. Clusters with an acceptable tracking figure of merit are kept; the other clusters are evaluated further in the third step. Here, one tries to recover some of the remaining clusters by adding and subtracting individual interactions or clusters to and from other clusters. Monte Carlo simulations indicate that, given realistic assumptions on position and energy-resolution γ-ray tracking arrays consisting solely of two-dimensionally segmented coaxial Ge detectors in a closed shell around the target such as GRETA or AGATA, the sensitivity in detection of γ-rays, particularly for high-multiplicity events, can be improved substantially over anti-Compton-based arrays such as Gammasphere.

Gamma-ray tracking not only enables the reconstruction of the scattering sequence and the distinction between partial and full-energy events, it also allows us to reconstruct the full γ-ray energy for some γ-rays that left only their partial energy. Applying the Compton scattering formula, one can calculate the incident γ-ray energy by measuring the energies and positions of at least three interactions (110). Although Wulf (111) successfully demonstrated this method in thin Si detectors in DSSDs, the achievable energy resolution and efficiency are limited by the uncertainties in the energy and positions of the interaction, as well as by the finite tracking efficiency. Although very appealing in Si detectors with its lower Z and density, this concept can also be applied in Ge, particularly for higher γ-ray energies. Knowing the source location, however, allows us to reconstruct the full energy of γ-rays with at least two interactions, owing to the fact that the incident direction is known. This feature points already to the gain from correlating track with image information. How to deduce the image or the location of γ-ray sources is described in the following section.

2.3.3. Compton imaging.

The ability to determine the track of a γ-ray in the detector not only allows us to increase the sensitivity for γ-ray spectroscopy, for example, for nuclear physics experiments, but allows us to enable efficient Compton imaging. The concept of Compton imaging, or sometimes known as electronic collimation, was proposed already more than 30 years ago (112, 113). It was realized by the space-based Imaging Compton Telescope (COMPTEL) consisting of two layers of position-sensitive NaI(Tl) detectors and flown in the 1990s. The recent advances in detector segmentation and the development of fast and high-resolution digital electronics allow us now to fabricate Compton-imaging systems built from two-dimensionally segmented Ge detectors, which have significantly higher sensitivity than COMPTEL. **Figure 6** illustrates the concept of Compton imaging. An incident γ-ray interacts at least once via Compton scattering before it is stopped through photoelectric absorption. The energies and positions of these interactions are measured, and through γ-ray tracking the proper scattering sequence is deduced. Employing the Compton scatter formula

$$\cos(\theta) = 1 + \frac{511}{E_\gamma} - \frac{511}{E'_\gamma}; \quad E'_\gamma = E_\gamma - E_1,$$

with θ the scattering angle, E_γ the energy (in keV) of the incident γ-ray, and E_1 the deposited energy in the first γ-ray interaction, the incident direction of the γ-ray can be deduced to a cone with an opening angle θ and a symmetry axis defined by a line connecting the first two interactions. The projection of these cones on a sphere or a plane will overlap at the source location when many events are imaged or back projected. Without measuring the direction of the Compton electron, the incident angle can only be determined to be on a cone surface. Because in Ge the range of electrons is typically below 1 mm (for instance, a 1-MeV γ-ray generates an electron of approximately 500 keV, which has a range of approximately 0.5 mm

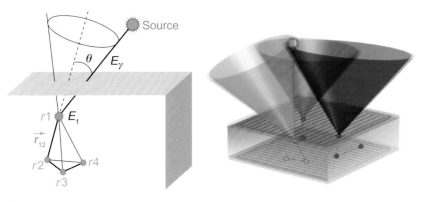

Figure 6

Concept of Compton imaging. With the measurement of individual γ-ray interactions and proper time ordering employing tracking techniques, the positions of the first two interactions as well as the energy of the first interaction can be deduced. Using the Compton scattering formula, it is possible to deduce a cone that defines a surface of possible incident directions for each γ-ray. Because all the cones intersect in the source location, we can form an image by projecting these cones onto a plane.

in Ge), it is very difficult to measure the scattering angle of the electron, particularly considering the complex slowing-down process of electrons. Only in low-Z or low-density detectors, such as gases at a pressure of approximately 1 atm, could electron vertices be measured (114). However, the efficiency to induce a γ-ray interaction at all in these instruments is extremely low. It is interesting to note that Compton imaging has opposing requirements to conventional γ-ray detectors. Although conventional γ-ray detectors aim at high efficiency with high-Z and high-density materials, a Compton imager requires at least one Compton scattering that is well separated from the second interaction to minimize the uncertainty in the direction defined by the first and second interaction for each γ-ray. To induce Compton scattering down to low energies and to maximize the distances between the first interactions, a low-Z and low-density material is required. The position and energy resolution can in principle be improved for better angular resolution, but the electron on which the photon Compton scatters carries momentum, which is inherently unknown and will limit the best angular resolution attainable (115). The most promising approach currently appears to be a hybrid design consisting of Si detectors to meet the requirements for Compton-imaging resolution, and a higher-Z material such as Ge to stop the γ-ray with high efficiency. Ge by itself can also be used as a sensitive Compton imager for γ-ray energies above 500 keV. To improve image quality, a large number of image reconstruction algorithms have been developed (116–119).

2.4. Cryogenic Detectors

The energies of photons and particles are not only lost through ionization, but predominantly by phonon production in detectors. The latter can be measured through an increase in temperatures by making use of the exceptionally small thermal capacity of materials at low temperatures. Instruments known as bolometers have been used for many years for thermal or infrared radiation in which the incident flux on a target is measured through the temperature rise as sensed by a thermistor. By reducing the temperatures to below 100 mK, it is possible to sense the momentary temperature rise caused by a single photon or incident particle. Although originally Si and Ge were used as absorbers, they are not required as such and other materials can be used as absorbers for γ-rays well. More challenging is the operation of a sensitive thermistor, which can convert the temperature rise into an electric signal as linearly as possible over the full dynamic range of interest. It turns out that NTD detectors made of Ge can meet the necessary requirements. In this way, Ge plays a role either as direct detectors, for example, in searching for cold dark matter (CDM) particles such as WIMPs, or indirectly by providing the thermistor for other materials that are sensitive to either photons or neutrons.

The need for very good energy resolution and the need to detect very small energy depositions in X-ray, γ-ray, or particle detection are the major motivations for the development of low-temperature detectors. Because the energy quanta associated with superconductors and lattice vibrations (phonons) are more than one hundred times smaller, substantial improvements have been obtained in energy resolution and in sensitivity over conventional detectors. Furthermore, these detection schemes permit tailoring of target and absorber materials to match the physics requirements. Examples can be found in particle physics experiments, such as the detection of low-energy neutrinos, search for dark matter particles, search for $0\nu\beta\beta$ decay, and β and γ spectroscopy, as well as X-ray astronomy. Silver et al. (37) report an energy resolution of 3 eV at 6 keV X rays that employ a tin absorber coupled to an NTD Ge thermistor, with a quantum efficiency of approximately 95% at 6 keV. A tin absorber with dimensions of 0.4 mm × 0.4 mm × 7 μm was used. Thermalization times, for example, signal rise times of 20 μs, and thermal recovery times, for example, decay times of 250 μs, have been achieved.

3. APPLICATIONS

In this section we discuss four applications of Ge detectors as examples that reflect the benefits of Ge detectors and some of the recent advances discussed before. These examples cover the development of γ-ray tracking arrays for nuclear physics experiments currently under construction in the United States and in Europe; Compton imaging for homeland security and astrophysics applications; the Majorana experiment, which aims at measuring the $0\nu\beta\beta$ decay in ^{76}Ge; and finally the use of Ge detectors to search for WIMPs, as currently being done in the Experience pour Detecter les WIMPs en Site Souterrain (EDELWEISS) experiment and the Cryogenic Dark Matter Search (CDMS) experiment.

3.1. Gamma-Ray Tracking Arrays for Nuclear Physics Applications

Advances in several areas led to the feasibility of large and highly efficient γ-ray tracking arrays: the two-dimensional segmentation of large-volume HPGe detectors, the pulse-shape analysis to derive energies and positions of individual γ-ray interactions, the tracking algorithms to identify and separate multiple γ-rays and to time order the interactions, the availability of fast and high-resolution waveform digitizers, and finally the availability of sufficient processing power to perform all necessary computations ranging from signal shaping to signal decomposition and γ-ray tracking in real time with event rates up to 50 kHz.

The new generation of γ-ray tracking arrays aims at answering still outstanding questions in the structure of atomic nuclei, particularly nuclei under extreme conditions, thus enabling one to distinguish between theoretical predictions. These instruments allow us to study more complex decay schemes, nuclei at higher spins, larger deformations associated with a large number of simultaneously emitted γ-rays, and collective excitation modes related to high γ-ray energies and their coupling to the normal excitation modes. Finally, these instruments will allow us to discover new excitation modes, as did Gammasphere. In addition, these instruments complement the new generation of radioactive beam facilities that are now either operational, currently being built, or will be built over the next decade. The availability of radioactive nuclei far away from the valley of stability through these facilities coupled with the high sensitivity of γ-ray tracking arrays will enable the study of the structure of nuclei at the boundary of their existence and beyond. Nuclei at extreme spin, isospin, and mass will be explored. The high position resolution and granularity in combination with the high detection efficiency will make γ-ray tracking systems the instrument of choice for both types of radioactive beam facilities, whether Isotope Separator On-Line (ISOL) type or fragmentation type. In addition to nuclear structure studies, other branches of physics will also benefit from the power of γ-ray tracking arrays such as GRETA. For example, the study of the positronium decay into four or five photons would provide higher-order tests of quantum electrodynamics and improve limits on charge conjugation symmetry-violating currents.

Simulations performed for the GRETA array consisting of 120 closed-ended Ge detectors as shown in **Figure 2** predict a full energy efficiency of 50% for a single γ-ray at 1.33 MeV. Assuming a position resolution of 2 mm (rms), which was recently demonstrated with one of the GRETA prototype detectors, the efficiency and peak-to-total ratios are still approximately 25% and 70%, respectively, even if 25 γ-rays are emitted simultaneously. These numbers must be compared with the 8% and 50% obtained for Gammasphere. Because the improvements are per γ-ray, one can understand the increase in overall sensitivity by orders of magnitude, as **Figure 1** illustrates. In addition, the efficiency for high-γ-ray energies is substantially higher, too. At 10 MeV efficiencies, up to 10% are expected, whereas Gammasphere provides only approximately 0.5% at this energy. This gain in efficiency can be explained by the full coverage of Ge detectors, which can contain the larger spatial distribution of higher-energy γ-rays. Finally, owing to the increased number of channels and shorter processing time, one expects to be able to increase the count rate per detector by a factor of five to approximately 50 kHz, reducing the necessary experiment time through larger source strengths or beam intensities, at least for a subset of experiments. The European AGATA effort aims at similar performance numbers. The main differences are the 180 detectors proposed for AGATA, the integration of three instead of four detectors in one cryostat, and the use of cold FETs inside the cryostat. However, one word of caution is justified here. Previous detector arrays such as Gammasphere relied primarily on simple features defined by the detector and the readout or generally by hardware components only, and therefore the achievable performance was fairly straightforward to predict. In contrast, γ-ray tracking arrays are not only much more complex in terms of the detector and readout hardware, but the overall performance depends strongly on the efficiency and accuracy of the data-processing algorithms, which are much more difficult to predict. Even if all data-processing steps are developed, implemented, and well benchmarked, the variation from detector to detector in terms of crystal dimensions and properties, and in particular variations in the electronic response, for example, impulse response as well as potential cross talk between segments, can not only render the characterization procedures very difficult, but impact the predictability as well. However, even if the simulations overestimate the performance by 10%–20%, the overall gain in introducing these arrays into nuclear physics experiments is substantial and will impact many facets of the atomic nucleus.

Although AGATA and GRETA represent the ultimate goal for nuclear physics experiments, both efforts are currently building smaller arrays. The precursor to AGATA is the AGATA Demonstrator, consisting of five three-detector modules; the precursor to GRETA is termed Gamma-Ray Energy Tracking for In-Beam Applications (GRETINA), consisting of seven four-detector modules. Both are compatible with the larger arrays but allow the demonstration of the overall concept and the expected gain and the initiation of a physics program that makes use of the new capabilities, particularly for the new generation of radioactive beam facilities already in operation or currently being built. For both efforts, the first multidetector module prototype detectors have been fabricated and preliminary test measurements are underway.

3.2. Gamma-Ray Imaging for National Security and Astrophysics

Ge detectors have been developed and are being used for γ-ray imaging applications, either in coded-aperture or in rotation-modulation systems. Examples are coded-aperture imagers with a planar detector in a double-sided strip configuration (120) for nuclear nonproliferation applications, a coded-aperture system combined with an array of 19 nonsegmented coaxial detectors for the space-based γ-ray observatory INTEGRAL, or a rotation-modulation imager consisting of rotating masks in front of an array of seven twofold segmented Ge detectors for the space-based RHESSI program. These imagers use heavy apertures or masks in front of the detector, which increase the weight of the instrument, limit the field of view, and are limited in complex radiation fields owing to the correlation in the data. In particular, for higher γ-ray energies, these systems are limited owing to the need for thicker and hence heavier masks. Furthermore, multiple interactions degrade the imaging performance because the first interaction cannot be determined. The DSSD-based coded-aperture system addresses this problem by placing a large-volume coaxial Ge detector closely behind the planar detector where a γ-ray first interacts in the position-sensitive DSSD and then is fully absorbed in the coaxial detector. Although this approach allows one to enhance the probability of identifying the first interaction, the main drawback is the loss in efficiency from requiring both detectors to be hit.

The ability to efficiently track γ-rays in three-dimensional position-sensitive detectors allows us to build a new generation of Compton-imaging systems. Because no collimator or space or time modulation is required, the problems associated with these collimators and associated correlations in the data can be circumvented. Many approaches are currently being pursued for Compton-imaging instruments ranging from gas- or liquid-based, time-projection chamber-like implementations (104) to semiconductor detectors. The latter includes pixelated CdZnTe and coaxial HPGe detectors (93, 121), high-purity Si- or Li-drifted Si [Si(Li)] detectors, HPGe or CdTe detectors in DSSD configuration, or a combination of HPSi or Si(Li) with HPGe or CdTe detectors (122–124).

An approach based on HPGe detectors in DSSD configuration is currently being pursued at the Space Science Laboratory in Berkeley as a potential implementation for the next-generation γ-ray observatory—the Advanced Compton Telescope (ACT)—and at LLNL for national security applications. The goal of the ACT mission is to provide unprecedented sensitivity for γ-ray energies ranging from approximately 200 keV to 10 MeV to study several astrophysical phenomena such as supernovae, galactic nucleosynthesis, γ-ray bursts, and fundamental physics (125). The design is driven by its primary science goal, the spectroscopy of the 847-keV transition in ^{56}Co from type Ia supernovae. It will allow the study of hundreds of these events over its five-year survey lifetime, with a sensitivity that is two orders of magnitude higher than that achieved by COMPTEL. Thermonuclear supernovae Ia events are used as standard candles across the Universe to measure cosmological distances and, to a large degree, responsible for the creation of our heavy elements. Yet even those near us are not well understood. The Ge-detector-based Compton-imaging concept as a key component of the instrument was adopted for ACT to maximize background rejection and signal efficiency and to utilize high spectral resolution to dramatically increase the overall sensitivity. In addition, the stringent requirements for size, weight, and power for space-based missions can be met. However, to illustrate the dimension and complexity of the challenge for ACT, an effective area of approximately 1000 cm^2 is required to meet the scientific goals. This goal requires a design for the Ge detector component of only 4 layers of 144 detectors, corresponding to 576 detectors total. Assuming 90×90 strips with a 1-mm strip pitch, approximately 100,000 electronics channels must be read out and processed. Although the lateral positions are deduced by the strips, the depths of interactions are determined by the timing of the opposing electrodes. To meet the performance and the power requirements, new low-noise ASIC technologies that can meet the specific requirements of DSSD HPGe detectors with typical strip and readout capacities of >30 pF are required.

A similar approach based on a combination of three-dimensionally position-sensitive Si and Ge detectors is being pursued at LLNL with the goal of improving the detection of illicit nuclear materials for national security applications such as nuclear nonproliferation and nuclear counterterrorism. Although γ-ray imaging is an established tool in nuclear medicine or astrophysics, only recently has the impact of γ-ray imaging for nuclear security applications been recognized. Here the goal is to provide improved capabilities to detect, localize, and characterize nuclear materials. One of the outstanding challenges in homeland security is the detection and identification of nuclear threats in a sea of nonthreat objects, which consists of legitimate radioactive objects commonly found in commerce and the environment. To mitigate the primarily nuisance alarms, it is crucial to identify the radioisotope via its characteristic γ-ray decay and to measure the location and the shape of sources through imaging. Gamma-ray imaging can potentially increase the sensitivity in finding such sources, particularly in complex and changing backgrounds owing to the ability to improve signal-to-background ratio. In particular, collimatorless Compton-imaging systems enable one to measure signals and background simultaneously and therefore provide the biggest gain in signal-to-background ratios in γ-ray imaging. Gamma-ray imagers based on position-sensitive semiconductor detectors such as HPGe detectors provide excellent imaging and spectroscopic characteristics and therefore fulfill both important requirements in national security.

At LLNL, a second-generation Compton-imaging system was recently demonstrated, consisting of two large-volume Si(Li) detectors in front of two large-volume HPGe detectors. All detectors are implemented in DSSD configuration with a 2-mm strip pitch. The 10-mm-thick Si(Li) detectors have 32 strips on each side, the 15-mm-thick HPGe detectors 37 strips. The strip signals are read out through external preamplifiers with warm FETs and fed into a digital acquisition system consisting of waveform digitizers operating at 100 MHz with an amplitude resolution of 14 bits. As pointed out above, with this implementation, a position resolution of approximately 0.5 mm in all three dimensions can be achieved. Dividing the total detector volume of approximately 244 ml into $(0.5 \text{ mm})^3$ voxels, we end up with approximately 2×10^6 voxels achieved with only 276 readout channels. This system demonstrated high efficiency, high-energy resolution, and high angular resolution and therefore an overall high γ-ray detection sensitivity for the energies of interest, ranging from 150 keV to approximately 3 MeV. Angular resolution values of approximately $3°$ at 200 keV and approximately $1°$ at 1408 keV have been achieved with this approach, and imaging efficiencies between 0.01 and 0.1 are expected for the energy range of interest with the four-detector system.

3.3. Fundamental Physics

Ge detectors have become important tools in fundamental physics studies as well. Two examples are the direct $0\nu\beta\beta$ measurements to infer the Majorana character of neutrinos and to deduce the mass scale of neutrinos and direct WIMP detection. Excellent energy resolution and large sizes of detectors to increase sensitivity in detecting the signal and the deployment deep underground are necessary but not sufficient to suppress the background. Therefore, additional degrees of freedoms to identify and suppress background events have been developed such as pulse-shape analysis and segmentation for Majorana or the correlations between ionization and phonon signals in the WIMP search. The latter rely on the fact that detectors respond differently to electron recoils and nuclear recoils, which can be measured by correlating the ionization with the thermal signal.

3.3.1. Neutrinoless double-beta decay in germanium 76.

Recent results from atmospheric, solar, and reactor-based neutrino oscillation experiments have provided compelling evidence that the neutrinos have mass and indicate already that the Standard Model is incomplete. Although the oscillatory character of neutrinos implies that neutrinos have mass, it provides only differences of neutrino masses (e.g., Δm^2). Neutrinoless double-beta decay experiments are the only proposed method of measuring neutrino mass with sufficient sensitivity for masses of interest (below 100 meV) and the only practical method to uncover the Majorana character of the neutrino. The two-neutrino double-beta ($2\nu\beta\beta$) decay is a second-order weak interaction process and has been observed in approximately 10 nuclei (126). The $2\nu\beta\beta$ decay is the transition from the nucleus (A, Z) to the nucleus (A, Z+2) and occurs when the single-beta decay to (A, Z+1) is energetically forbidden, as in most even-even nuclei, or strongly inhibited by a large change of angular momentum. In contrast, in the $0\nu\beta\beta$ decay process, no neutrino is emitted and therefore the process violates lepton number conservation, in contradiction to the Standard Model of particle physics. Whereas the $2\nu\beta\beta$ decay is already a rare process with $T_{1/2} \sim 10^{19}$ years, the $0\nu\beta\beta$ decay is even rarer, with $T_{1/2} > 10^{25}$ years. Because in this process no neutrino is emitted, both electrons always carry the full energy of the Q value. Therefore, the $0\nu\beta\beta$ decay process is characterized by a discrete energy release higher than the energy release of the $2\nu\beta\beta$ decay, which always is the dominating process.

The most sensitive experiments to date are experiments employing Ge detectors in conventional, closed-ended coaxial geometry. In one of the naturally occurring Ge isotopes, ^{76}Ge, the single-beta decay to ^{76}As is energetically forbidden and the $2\nu\beta\beta$ decay to ^{76}Se has been observed. In ^{76}Ge, the $0\nu\beta\beta$ decay is characterized by a discrete energy at 2039 keV. The most sensitive experiments to date have been the Heidelberg-Moscow experiment (127) and the International Germanium Experiment (128), both of which used Ge material enriched to approximately 86% ^{76}Ge, employing centrifuge separation, and both have been completed. Both experiments aimed at extremely low levels of background by using radio-pure materials and active and passive shielding as well as by operating in deep-underground laboratories to reduce cosmic-ray-induced backgrounds. In addition, at the later stages both experiments employed pulse-shape analysis to differentiate between the $0\nu\beta\beta$ process and background events within the Ge detectors. Although the $0\nu\beta\beta$ process is characterized by the emission of two $\sim$1 MeV electrons and therefore a localized energy deposition within approximately 2 mm—reflecting the range of these electrons in Ge—background events at 2039 keV are likely due to γ-rays interacting multiple times throughout the Ge detectors. Employing pulse-shape analysis of the central detector channels, both experiments increased the sensitivity for the $0\nu\beta\beta$ decay significantly by reducing the background rate around 2039 keV to approximately 0.1 events kg^{-1} year^{-1} keV^{-1}.

The Heidelberg-Moscow experiment has two successors, GENIUS (GErmanium in liquid NItrogen Underground Setup) and GERDA (GERmanium Detector Array) (129, 130). GENIUS consists of so-called naked enriched Ge detectors directly in liquid nitrogen. In this way, liquid nitrogen is used not only as a coolant but also as a passive shield and in addition minimizes the amount of background due to potentially contaminated support materials close to the detectors. The GENIUS test facility, with six naked and natural Ge detectors with a total Ge weight of 15 kg, has been in operation since 2004. Only recently were significant problems in the long-term behavior of this approach identified, which were associated with background from ^{222}Ra diffusing into the setup and, more importantly, an unacceptable increase of leakage current, which renders the detectors almost nonoperational.

GERDA aims at deploying two-dimensionally segmented, enriched ^{76}Ge detectors also immersed in liquid nitrogen or liquid argon. The segmentation of the Ge detectors increases the granularity of the setup over the granularity given by detectors themselves. It also enables pulse-shape analysis of the segment signals and therefore provides complementary granularity to the radius component already available in nonsegmented detectors. The advantage of liquid argon is its higher attenuation of potential background as compared to liquid argon and use as an active shield by detecting the scintillation of background events emitted from either external components or detector components. Currently, (3×6)-fold segmented prototype detectors are being built for GERDA. The Majorana collaboration is currently pursuing research and development, exploring the best approach for a ^{76}Ge-based instrument to observe the $0\nu\beta\beta$ decay based on a staged approach, aiming ultimately at a 1-ton experiment (131). The goal is to demonstrate a background rate of approximately 1 event ton^{-1} yr^{-1} keV^{-1}, which is a factor of 100 below the previous experiments. This will be achieved by reducing external as well Ge detector internal contaminations, employing ultraclean materials that are fabricated underground, and employing advanced technologies such as segmentation and pulse-shape analysis. Preliminary studies with segmented Clover detectors indicate the gain in background suppression by combining segmentation and pulse-shape analysis (132). The fundamental drawback to this approach are the additional components required to read out the additional segments, which contain some degree of radioactive contamination, thereby generating new background. An alternative to segmented detectors represents Ge drift or point-contact configurations. These detectors promise high pulse-shape sensitivity to distinguish multiple interactions with only one readout channel.

3.3.2. Detection of dark matter candidates. In contrast to the $0\nu\beta\beta$ decay discussed above with its single-site discrete energy deposition characteristics at a relatively high energy of 2039 keV, the search for CDM candidates such as WIMPs and the detection of low-energy solar neutrinos are characterized by nondiscrete deposition of energies of approximately 30 keV or less, which is experimentally even more challenging.

In our current understanding of the composition of the Universe, ~70% is contained in dark energy, ~25% can be found as nonbaryonic CDM, and only ~5% can be found as known baryonic matter. WIMPs are the leading candidate for CDM interacting only weakly or not at all with electromagnetic radiation—therefore dark matter. Owing to the mass and velocity of WIMPs relative to the observing detector, the WIMP-nucleus scattering interactions leave recoil energies in the 0–30 keV range. The first generation of experiments in the 1980s with conventional Si and Ge ionization detectors achieved large mass-lifetime exposures and low-background levels owing to underground deployment and the use of low-background materials. Although from the statistical point of view, these experiments could have been scaled to larger dimensions to achieve significantly higher sensitivity, they are limited by systematic uncertainties owing to the lack of specificity to the nuclear recoil component. Even large-mass systems built of naked Ge ionization detectors will be limited owing to the lack of specificity to background due to both distributed and local contaminations, for example, by uranium or thorium.

With the advent of cryogenic detectors operated at ~20 mK, Ge can be operated simultaneously as a bolometer that measures thermal phonons or a heat and ionization detector when operated as a conventional Ge detector in reverse-biased diode configuration. The amplitude of the phonon signal (measured as an increase in temperature) does not depend on the specific energy loss of either electrons or nuclear recoils. In contrast, in ionization measurements, measured nuclear recoil energies are suppressed or quenched to 25% relative to electron recoils. This difference in response for ionization versus phonon readout can be used to reject electron recoils to better than 99.9% above a threshold of approximately 10 keV. Because the background is due predominantly to X-ray or γ-ray interactions generating electron recoils converted into electro-hole pairs via ionization and the signal—where the scattering of WIMPs on the Ge nucleus is generated by the recoiling nucleus—these two components can be distinguished. Owing to improvements in the readout electronics, the nonthermal phonons can be detected in less than 100 μs. Cryogenic and low-noise detectors provide the necessary signal-to-noise ratio that enables the distinction between nuclear and electron recoils down to low energies, for example, 10 keV, which is critical for detecting WIMPs because the dark matter recoil energy spectrum is exponential and can therefore be confused with electronic noise or other anomalous events.

Examples of experiments using the combined readout in Ge detectors are the EDELWEISS II experiment in the Frejus tunnel, France (133), and the CDMS II experiment in the Soudan mine, United States (134), consisting of 320 g and 250 g of Ge detectors, respectively, both with the goal of deploying 20–30 kg total effective mass and an energy threshold of the order 10 keV. CDMS II also operates Si detectors in the same way to distinguish interactions via spin-dependent and spin-independent coupling.

4. FUTURE DEVELOPMENTS

Developments in Ge detector and related technologies will continue to be driven by the need for new and improved capabilities in the detection of γ-rays and particles. These developments will then enable new ideas and applications in similar or new areas in many different fields. This cycle has been demonstrated with the development of Ge detectors over the past 40 years, and there is no reason why this should not continue.

New technologies are currently being developed, such as Ge drift detectors with the promise of providing excellent energy resolution and sub-keV trigger thresholds in large-volume HPGe detectors. One could envision segmenting the small readout electrode to obtain true three-dimensional information of individual γ-ray interactions. This would be similar to two-dimensionally segmented planar or coaxial Ge detectors, however, with extremely small capacitance, thus enabling shorter shaping times and, particularly, reduced power requirements, which can be critical for large-scale applications.

The amorphous contact technology has made significant progress, already replacing the Li contact in segmented planar Ge detectors. Only the replacement of the Li n+ contact in DSSDs enabled the recent progress in Compton imaging by providing good energy and position resolutions in addition to high efficiency. In the future it may be possible to grow crystals specifically for planar detectors with just a few centimeters in length but with radii larger than 10 cm, which are currently available and dictated by the production of long coaxial detectors. For comparison, high-purity Si wafers can be grown to 15 cm now and potentially up to 20 cm. However, only up to 2-mm-thick detectors can be made because their purity is lower than that of HPGe detectors. Increasing the detector diameter will allow us to cover larger solid angles and to increase the detection probability without increasing the number of detectors and readout channels as much, effectively reducing the complexity of the instrument and reducing the power requirements. In addition, planar detectors still require guard rings of the order 3–10 mm to separate performance-degrading surface effects from the sensitive detection volume. Developing technologies to reduce or even to eliminate these guard rings will increase the effective detection volume substantially owing to the fact that most of the volume is outside. Reducing the guard ring by 5 mm will typically increase the effective volume by approximately 20%.

Current approaches for Compton imaging and for nuclear physics applications employing planar DSSD systems rely on thick Ge detectors to increase the efficiency. Fast electronics is then required to deduce the depth of interactions, and if the lateral resolution is to be improved beyond the strip size, full digital signal processing is required. Although in this way the efficiency can be maximized with the minimum number of detectors and electronics channels, it is not necessarily the most power-efficient way. Employing thin DSSDs with the resolution defined only by the thickness of the detector and the strip pitch requires slow and less power-consuming electronics to deduce the energy per strip (135). Although a factor of 5–10 more detectors and readout channels are required, the total power requirement can be reduced. In addition, simple multiplexing schemes can be applied to reduce the number of readout channels. In this way, not only can the overall power potentially be reduced, the requirements for impurities and high-voltage can be reduced too, owing to the significantly reduced depletion voltage for thinner detectors.

Over the next five years we will see the first-generation γ-ray tracking arrays, GRETINA and the AGATA Demonstrator, producing the first physics results, very likely at radioactive beam facilities, at least initially. The GERDA effort will have most of the detectors operational underground and looking for the $0\nu\beta\beta$ decay in ^{76}Ge. The Majorana experiment, with the same goal, will have explored a variety of Ge detector configurations suitable for this type of experiment and will likely have demonstrated a specific implementation that can meet the challenging background requirements with a rate of the order 1 count ton^{-1} year^{-1} in the region of interest, which is believed to be necessary for building a 1-ton scale experiment. With regard to space-based missions, MESSENGER will have measured and sent back the first spectral fingerprints from Mercury's surface. The Ge-based Next Compton Telescope will have explored the feasibility of a large-scale DSSD array for the ACT. For terrestrial applications, more compact and less power-consuming, mechanically cooled Ge detectors will be available for human-operated and remote-detection missions for national security applications worldwide. Arrays of these detectors might even be deployed in specific locations or as surveillance tools. In addition, small arrays of Ge-based Compton-imaging systems will be explored and might be on their way to deployment for homeland security and nuclear safeguard applications.

The recent advances in high-sensitivity, tracking-based γ-ray detection and the availability of fully operational prototypes such as the Si(Li)+HPGe-based Compton-imaging system will allow us to demonstrate the use of these detectors for medical applications as well. Compton cameras can be envisioned as a single-photon emission tomography system for biomedical applications with small animals or for the early detection of breast cancer and improved cancer treatment planning (136). The demonstrated angular resolution in the Si(Li)+HPGe Compton-imaging system, for example, will provide 1–2-mm spatial resolution for γ-ray energies above 300 keV for objects at 10 cm or closer, with sensitivities larger than 10^{-2}. In addition, Compton cameras will provide tomographic imaging without moving the detectors around the object, but only by moving the detection system along one side of the object. Employing large-volume, three-dimensional position-sensitive Ge detectors can also enable one to finally achieve spatial resolution of better than 1 mm for positron emission tomography system, with sensitivities of the order 10%–20% at the same time. Here the challenges are the extremely high count rates due to the amount of radioactivity applied to the patient or the small animal. However, one can estimate that with digital signal processing and adaptive filtering, assuming strip pitches of 2 mm, the count rate capabilities of Ge detectors are sufficient. The energy resolution might be degraded slightly. However, this is not critical to positron emission tomography. Finally, experiments employing the ionization and the phonon mode in Ge aimed at the detection of WIMPs will be fully operational and have collected an unprecedented amount of data, either with a detected signal or with significantly lower limits.

As in the past, the development of Ge detector technologies will be accompanied by advances in electronics and signal processing. They will enable more compact and less power-consuming, low-noise electronics essential for the development of large-scale γ-ray imaging instruments. This will be achieved by advances in analog electronics, with better matched FETs and increased integration, and in digital electronics with lower-power analog-to-digital converters, gate arrays, and processing units with ever more processing capabilities.

Predicting the developments beyond the next five years is inherently more uncertain and speculative, not only owing to the uncertain outcome of the many demonstration experiments and results by the first-generation instruments but also owing to the uncertain overall constraints for basic and applied research and the impact of potentially even more disastrous events than those that have occurred in the recent past. Nevertheless, let us try to speculate assuming basic and applied research continues to be funded as in the past. Assuming the first-generation γ-ray tracking arrays are successful and demonstrate the promised improved capabilities, the full arrays of GRETA and AGATA will be built and become operational over the next 10 years. At this time, a new generation of radioactive beam facilities above ground but potentially even underground will be available for nuclear physics studies. Gamma-ray tracking arrays will play an essential role in elucidating the structure of the most exotic nuclei at these facilities, but will also provide new insights in phenomena to be explored at stable beam facilities and will be used for complementary fundamental research experiments as well. In parallel, arrays of DSSDs with higher position resolution and imaging capabilities might be operational for specific nuclear physics experiments. The search for the $0\nu\beta\beta$ decay in ^{76}Ge might be underway with a large-scale experiment with hundreds of Ge detectors, might be in Ge drift, might be in two-dimensional segmented configuration, or might even be both combined. Depending on the future vision of NASA, an advanced Compton telescope will be built or under construction, however, not necessarily to the scale envisioned now. Larger-size Ge detector arrays might be deployed either as simple spectroscopy instruments for radioisotope identification or as Compton and coded-aperture γ-ray imaging instruments for search and surveillance applications in homeland security.

5. CONCLUSIONS

Since the introduction of Ge detectors for the detection of γ radiation approximately 40 years ago, the technology has undergone tremendous strides forward in terms of basic and operational capabilities as well as applications. In particular, over the past 10–15 years, the diversity of applications has substantially increased. Ge detectors can now be found on their way to Mercury and looking out even further. Owing to continued improvement and the development of new capabilities in Ge detector and associated technologies, Ge detectors are still playing an essential role in ongoing as well as in future experiments. On the one hand, compact and hand-held Ge detectors are now widely deployed with digital signal processing. On the other hand, large arrays with hundreds of large-volume Ge detectors are operating to look deeper into the structure of the atomic nucleus. The development of two-dimensionally segmented Ge detectors in addition to advances in signal processing enable the concept of γ-ray tracking, with unprecedented sensitivity for nuclear physics applications but also for γ-ray imaging applications, ranging from astrophysics to nuclear medicine to national security. Gamma-ray imaging with Ge detectors can be based on either spatial or temporal modulation systems or aperture- and mask-free Compton imaging. For national security, the deployment of highly efficient and high-resolution detectors is important for nuclear nonproliferation, safeguards, stockpile stewardship, and homeland security, for example, for search and surveillance applications.

Recently the large investment in room-temperature γ-ray detectors led finally to successes, for example, in the fabrication of LaBr scintillation detectors with 2%–3% relative energy resolution at 662 keV and CdZnTe detectors with energy resolution of 1% and better. Although these and other developments to come will certainly at least partially replace Ge detectors, particularly for field applications requiring compact and low-power instruments that can accept degraded energy resolution, Ge detectors will remain the workhorse for many applications for many years to come.

DISCLOSURE STATEMENT

The author is not aware of any biases that might be perceived as affecting the objectivity of this review.

ACKNOWLEDGMENTS

I would like to thank my many colleagues, who I enjoyed working with over the past few years and whose contributions to this review are too numerous to mention. I am particularly grateful to discussions with I-Yang Lee from LBNL, and Dan Chivers, Morgan Burks, and Lucian Mihailescu from LLNL. I am indebted to Paul Luke and Eugene Haller from LBNL for carefully reading this manuscript and for their sensible suggestions. This work was performed under the auspices of the U.S. Department of Energy by LLNL under contract number W-7405-Eng-48.

LITERATURE CITED

1. Tavendale AJ. *Annu. Rev. Nucl. Part. Sci.* 17:73 (1967)
2. Cline D. *Annu. Rev. Nucl. Part. Sci.* 36:637 (1996)
3. Nolan PJ, Twin PJ. *Annu. Rev. Nucl. Part. Sci.* 38:533 (1988)
4. Wu CY, Oertzen WV, Cline D, Guidry MW. *Annu. Rev. Nucl. Part. Sci.* 40:285 (1990)
5. Sangsingkeow P, et al. *Nucl. Instrum. Methods A* 505:203 (2003)
6. Darken LS, Cox CE. In *Semiconductors for Room-Temperature Detector Applications*, ed. TE Schlesinger, RB James, 43:31. Mater. Res. Soc. Symp. Proc. (1995)
7. Elliott SR, Vogel P. *Annu. Rev. Nucl. Part. Sci.* 52:115 (2002)
8. Haller EE, et al. *Proc. 4th Int. Conf. Neutron Transmutat. Doping Semiconduct. Mater.*, ed. RD Larrabee, p. 21. New York: Plenum (1984)
9. Teal GK, Little JB. *Phys. Rev. C* 78:647 (1950)
10. Pell EM. *J. Appl. Phys.* 31:291 (1960)
11. Freck DV, Wakefield J. *Nature* 193:669 (1962)
12. Hall RN, Soltys TJ. *IEEE Trans. Nucl. Sci.* 18:160 (1971)
13. Hall RN. *IEEE Trans. Nucl. Sci.* 21:62 (1974)
14. Hansen WL, Haller EE. In *Nuclear Radiation Detector Materials*, ed. EE Haller, HW Kraner, WA Higinbotham, p. 16. Amsterdam: Elsevier (1983)
15. Gerl J. *Nucl. Instrum. Methods A* 442:12238 (2000)
16. Twin PJ, et al. *Nucl. Phys. A* 409:343 (1983)
17. Deleplanque MA, Diamond RM, eds. *Gammasphere proposal preprint LBNL-5202* (1987)
18. Simpson J, et al. *Z. Phys. A* 358:139 (1997)
19. Azaiez F. *Nucl. Phys. A* 654:1003c (1999)
20. Simpson J, et al. *Acta Phys. Hung.* 11:159 (2001)
21. Eberth J, et al. *Prog. Part. Nucl. Phys.* 38:29 (1997)
22. Habs D, et al. *Prog. Part. Nucl. Phys.* 38:111 (1997)
23. Eberth J, et al. *Prog. Part. Nucl. Phys.* 46:389 (2001)
24. Mueller WF, et al. *Nucl. Instrum. Methods A* 466:492 (2001)

25. Scraggs HC, et al. *Nucl. Instrum. Methods A* 540:348 (2005)

26. Winkler C. *New Astron. Rev.* 48:183 (2004)

27. Ziock KP, et al. *IEEE Trans. Nucl. Sci.* 51:2238 (2004)

28. Lin PR, et al. *Proc. SPIE* 11:3442 (1998)

29. Conka-Nurdan T, et al. *IEEE Trans. Nucl. Sci.* 49:817 (2002)

30. Li H, Clinthorne NH. *IEEE Nucl. Sci. Symp. Rec.* 5:256 (2005)

31. Vetter K. In *Radioanalytical Methods in Interdisciplinary Research*, ed. CA Laue, KL Nash, p. 52. Washington, DC: ACS Symp. Ser. 868 (2003)

32. Vetter K, Burks M, Mihailescu L. *Nucl. Instrum. Methods A* 525:322 (2004)

33. Boggs SE, et al. *Proc. SPIE* 4851:1221 (2002)

34. Kurfess JD, et al. *New Astr. Rev.* 48:293 (2004)

35. Gaitskell RJ. *Annu. Rev. Nucl. Part. Sci.* 54:315 (2004)

36. Alessandrello A, et al. *Phys. Rev. Lett.* 82:513 (1999)

37. Silver E, et al. *Nucl. Instrum. Methods A* 545:683 (2005)

38. Evans R. *The Atomic Nucleus.* New York: McGraw-Hill (1955)

39. Siegbahn K. *Alpha-, Beta-, and Gamma-Ray Spectroscopy.* Amsterdam: North Holland (1965)

40. Knoll GF. *Radiation Detection and Measurement.* New York: Wiley & Sons. 3rd ed. (2000)

41. Hull EL, et al. *Nucl. Instrum. Methods* 385:489 (1997)

42. Pehl RH, et al. *IEEE Trans. Nucl. Sci.* 26:321 (1979)

43. Fano U. *Phys. Rev.* 72:26 (1947)

44. Pehl RG, et al. *Nucl. Instrum. Methods* 59:45 (1968)

45. Goulding FS, Landis DA. *IEEE Trans. Nucl. Sci.* 29:1125 (1982)

46. Radeka V. *IEEE Trans. Nucl. Sci.* 11:358 (1964); Radeka V. *Annu. Rev. Nucl. Part. Sci.* 38:217 (1988)

47. Squillante MR, Shah KS. In *Semiconductors for Room-Temperature Detector Applications*, ed. TE Schlesinger, RB James, 43:465. Mater. Res. Soc. Symp. Proc. (1995)

48. Mihailescu L, et al. *Nucl. Instrum. Methods A* 447:350 (2000)

49. Vetter K, et al. *Nucl. Instrum. Methods* 452:105 (2001)

50. Vetter K, et al. *Nucl. Instrum. Methods* 452:223 (2001)

51. Eberth J, et al. *Nucl. Instrum. Methods A* 369:135 (1996)

52. Cork CP, et al. *IEEE Trans. Nucl. Sci.* 43:1463 (1996)

53. Pehl RH, Luke PN, Friesel DL. *Nucl. Instrum. Methods A* 242:103 (1985)

54. Luke PN. *IEEE Trans. Nucl. Sci.* 31:312 (1984)

55. Gutknecht D. *Nucl. Instrum. Methods A* 288:13 (1990)

56. Grigorovici R, et al. Band structure and electrical conductivity in amorphous germanium. *Proc. 7th Int. Conf. Phys Semiconduct., Paris*, 1964, p. 423 (1964)

57. England JBA, Hammer VW. *Nucl. Instrum. Methods* 96:81 (1971)

58. Hansen WL, Haller EE. *IEEE Trans. Nucl. Sci.* 24:63 (1977)

59. Chiba Y, et al. *Nucl. Instrum. Methods* 299:152 (1990)

60. Luke PN, et al. *IEEE Trans. Nucl. Sci.* 39:590 (1992)

61. Hansen WL, Haller EE. *IEEE Trans. Nucl. Sci.* 27:247 (1980)

62. Luke PN, Rossington CS, Wesela MF. *IEEE Trans. Nucl. Sci.* 41:1074 (1994)

63. Hull EL, Pehl RH. *Nucl. Instrum. Methods A* 538:651 (2005)
64. Pehl RH, Haller EE, Cordi RC. *IEEE Trans. Nucl. Sci.* 20:494 (1973)
65. Detko J. *Phys. Med. Biol.* 14:245 (1969)
66. Schlosser PA, et al. *IEEE Trans. Nucl. Sci.* 21:658 (1974)
67. Kaufman L. *IEEE Trans. Nucl. Sci.* 22:395 (1975)
68. Riepe G, Protic D. *Nucl. Instrum. Methods A* 226:103 (1984)
69. Protic D, Riepe G. *IEEE Trans. Nucl. Sci.* 32:535 (1985)
70. Gerber MS, et al. *IEEE Trans. Nucl. Sci.* 24:182 (1977)
71. Luke PN, et al. *IEEE Trans. Nucl. Sci.* 47:1360 (2000)
72. Amman M, Luke PN. *Nucl. Instrum. Methods A* 452:155 (2000)
73. Protic D, Krings T. *IEEE Trans. Nucl. Sci.* 50:998 (2003)
74. Protic D, Krings T. *IEEE Trans. Nucl. Sci.* 51:1129 (2004)
75. Stoehlker T, et al. *Nucl. Instrum. Methods B* 205:210 (2003)
76. Moore EF, et al. *Nucl. Instrum. Methods A* 505:163 (2003)
77. Deleplanque MA, et al. *Nucl. Instrum. Methods A* 430:292 (1999)
78. Shimoura S. *Nucl. Instrum. Methods A* 525:188 (2004)
79. Ziock KP, et al. *IEEE Trans. Nucl. Sci.* 43:1472 (2004)
80. Inderhees SE, et al. *IEEE Trans. Nucl. Sci.* 43:1467 (1996)
81. Phlips BF, et al. *IEEE Trans. Nucl. Sci.* 43:1472 (1996)
82. Phlips BF, et al. *IEEE Trans. Nucl. Sci.* 49:597 (2002)
83. Yang YF, et al. *IEEE Trans. Nucl. Sci.* 48:656 (2001)
84. Piqueras I, et al. *Nucl. Instrum. Methods A* 525:275 (2004)
85. Vetter K, Burks M, Mihailescu L. *Nucl. Instrum. Methods A* 525:322 (2004)
86. Amrose S, et al. *Nucl. Instrum. Methods A* 505:170 (2003)
87. Boggs SE, et al. *IEEE Trans. Nucl. Sci.* 43:1472 (2002)
88. Lee IY. *Nucl. Phys A* 520:641c (1990)
89. Macchiavelli AO, et al. Performance of Gammasphere split detectors. *Proc. Conf. Phys. Large Gamma-Ray Detect. Arrays Rep.* 35687:149, Lawrence Berkeley Lab., Livermore, Calif. (1994)
90. Bazzacco D. *Nucl. Phys A* 746:248c (2004)
91. De France G. *Eur. Phys. J. A* 20:59 (2003)
92. Svensson CE, et al. *Nucl. Instrum. Methods A* 540:348 (2005)
93. Niedermayr T, et al. *Nucl. Instrum. Methods A* 553:501 (2005)
94. Kroell T, Bazzacco D. *Nucl. Instrum. Methods A* 463:227 (2001)
95. Luke PN, et al. *IEEE Trans. Nucl. Sci.* 36:926 (1989)
96. Lavietes AD, et al. *Nucl. Instrum. Methods A* 422:252 (1999)
97. Becker JA, et al. *Nucl. Instrum. Methods A* 505:167 (2003)
98. Gold RE, et al. *Planet Space Sci.* 49:1482 (2001)
99. Burks M, et al. *IEEE Trans. Nucl. Sci.* 48:656 (2004)
100. Boyle R, et al. *Cryogenics* 39:969 (1999)
101. Wilkinson, et al. *Cryogenics* 39:179 (1999)
102. Kobayashi M, et al. *Nucl. Instrum. Methods A* 548:401 (2005)
103. Kroell T, et al. *Nucl. Instrum. Methods A* 371:489 (1996)
104. Aprile E, Suzuki M. *IEEE Trans. Nucl. Sci.* 36:311 (1989)
105. Dogan N, Wehe DK, Knoll GF. *Nucl. Instrum. Methods A* 299:501 (1990)

106. Schmid GJ, et al. *Nucl. Instrum. Methods A* 430:69 (1999)

107. van der Marel J, Cederwall B. *Nucl. Instrum. Methods A* 437:538 (1999)

108. Lehner CE, He Z, Zhang F. *IEEE Trans. Nucl. Sci.* 51:1618 (2004)

109. Mihailescu, et al. *Nucl. Instrum. Methods A* 570:89 (2007)

110. Kroeger RA, et al. *IEEE Trans. Nucl. Sci.* 49:1887 (2002)

111. Wulf EA. *High resolution Compton imager for detection of shielded SNM.* Presented at Symp. Radiat. Meas. Appl., Ann Arbor (2006)

112. Todd RW, Nightingale JM, Everett DB. *Nature* 251:132 (1974)

113. Schoenfelder V, et al. *Nucl. Instrum. Methods* 107:385 (1973)

114. Bellazzini, et al. *Nucl. Instrum. Methods A* 535:477 (2004)

115. Brusa et al. *Nucl. Instrum. Methods A* 379:167 (1996)

116. Parra LC. *IEEE Trans. Nucl. Sci.* 47:1543 (2000)

117. Shepp LA, Vardi Y. *IEEE Trans. Nucl. Sci.* 1:428 (1982)

118. Wilderman SJ, et al. *IEEE Trans. Nucl. Sci.* 45:957 (1999)

119. Basko R, et al. *Phys. Med. Biol.* 43:887 (1998)

120. Ziock KP, et al. *IEEE Trans. Nucl. Sci.* 49:1737 (2002)

121. Du YF, et al. *Nucl. Instrum. Methods* 457:203 (2001)

122. Bhattacharya D, et al. *New Astr. Rev.* 48:287 (2004)

123. Takahashi T, et al. *New Astr. Rev.* 48:269 (2004)

124. Boggs SE, et al. *New Astr. Rev.* 48:251 (2004)

125. Boggs SE, et al. astro-ph/0608532 (2006)

126. Moe M, Vogel P. *Annu. Rev. Nucl. Part. Sci.* 44:247 (1994)

127. Klapdor-Kleingrothaus HV, et al. *Nucl. Instrum. Methods A* 522:371 (2004)

128. Aalseth CE, et al. *Phys. Rev. D* 65:092007 (2002)

129. Klapdor-Kleingrothaus HV, et al. *Nucl. Instrum. Methods A* 566:472 (2006)

130. Liu X, et al. *Phys. Scr.* 127:46 (2006)

131. Aalseth CE, et al. *Nucl. Phys. B* 138:217 (2005)

132. Elliott SR, et al. *Nucl. Instrum. Methods A* 588:504 (2006)

133. Navick XF, et al. *Nucl. Instrum. Methods A* 559:483 (2006)

134. Mirabolfathi N, et al. *Nucl. Instrum. Methods A* 559:417 (2006)

135. Vetter K, et al. In *Unattended Radiation Sensor Systems for Remote Applications*, ed. JI Trombka, DP Spears, PH Solomon, 632:129. AIP Conf. Proc. (2002)

136. Hartmann-Siantar CL, et al. *Cancer Biother. Radiopharm.* 17:122 (2002)

Searching for New Physics in $b \to s$ Hadronic Penguin Decays

Luca Silvestrini

Department of Physics, University of Rome, and INFN, Rome, I-00185 Rome, Italy;
email: luca.silvestrini@roma1.infn.it

Annu. Rev. Nucl. Part. Sci. 2007. 57:405–40

First published online as a Review in Advance on July 5, 2007

The *Annual Review of Nuclear and Particle Science* is online at http://nucl.annualreviews.org

This article's doi:
10.1146/annurev.nucl.57.090506.123007

Copyright © 2007 by Annual Reviews.
All rights reserved

0163-8998/07/1123-0405$20.00

Key Words

flavor physics, CP violation, B decays, supersymmetry

Abstract

We review the theoretical status of $b \to s$ hadronic penguin decays in the Standard Model and beyond. We summarize the main theoretical tools to compute branching ratios and CP asymmetries for $b \to s$ penguin-dominated nonleptonic decays and discuss the theoretical uncertainties in the prediction of time-dependent CP asymmetries in these processes. We consider general aspects of $b \to s$ transitions beyond the Standard Model. Then we present detailed predictions in supersymmetric models with new sources of flavor and CP violation.

Contents

1. INTRODUCTION 406
2. BASIC FORMALISM. 407
   2.1. Generalities 407
   2.2. Evaluation of Hadronic Matrix Elements 410
3. BRANCHING RATIOS AND CP ASYMMETRIES WITHIN THE
   STANDARD MODEL. 413
   3.1. Branching Ratios and Rate CP Asymmetries 413
   3.2. Predictions for S and ΔS in $b \rightarrow s$ Penguins 416
4. CP VIOLATION IN $b \rightarrow s$ PENGUINS BEYOND THE
   STANDARD MODEL. 420
   4.1. Model-Independent Constraints on $b \rightarrow s$ Transitions 420
   4.2. Theoretical Motivations for New Physics in $b \rightarrow s$ Transitions . . . 422
5. SUSY MODELS 423
6. NON-SUSY MODELS. 435
7. CONCLUSIONS AND OUTLOOK 435

1. INTRODUCTION

New physics (NP) can be searched for in two ways: either by raising the available energy at colliders to produce new particles and reveal them directly, or by increasing the experimental precision on certain processes involving Standard Model (SM) particles as external states. The latter option, an indirect search for NP, should be pursued using processes that are forbidden, very rare, or precisely calculable in the SM. In this respect, flavor-changing neutral current (FCNC) and CP-violating processes are among the most powerful probes of NP because in the SM they cannot arise at the tree level, and even at the loop level they are strongly suppressed by the Glashow-Iliopoulos-Maiani (GIM) mechanism. The absence of tree-level FCNCs (and the GIM mechanism) is an accidental symmetry of the SM, in the sense that it is just a consequence of renormalizability and of the field content of the SM. NP in general violates this accidental symmetry and yields too large contributions to FCNC and CP-violating processes. Furthermore, all FCNC and CP-violating processes in the quark sector are calculable in terms of the Cabibbo-Kobayashi-Maskawa (CKM) matrix, and in particular of the parameters $\bar{\rho}$ and $\bar{\eta}$ in the generalized Wolfenstein parameterization (1). Unfortunately, in many cases a deep understanding of hadronic dynamics is required to extract the relevant short-distance information from measured processes. Lattice QCD allows us to compute the necessary hadronic parameters in many processes, for example, in transitions in which the flavor quantum number F of the initial and final state differs by two units ($\Delta F = 2$ amplitudes). Indeed, the Unitarity Triangle analysis with lattice QCD input is quite successful in determining $\bar{\rho}$ and $\bar{\eta}$ and in constraining NP contributions to $\Delta F = 2$ amplitudes (2–6).

Once the CKM matrix is precisely determined by means of the Unitarity Triangle analysis (either within the SM or allowing for generic NP in $\Delta F = 2$ processes), it is possible to search for NP contributions to FCNC and CP-violating weak decays ($\Delta F = 1$ transitions). These are indeed the most sensitive probes of NP contributions to penguin operators. In particular, penguin-dominated nonleptonic B decays can reveal the presence of NP in decay amplitudes (7–9). The dominance of penguin operators is realized in $b \to s q \bar{q}$ transitions.

Thanks to the efforts of the BaBar and Belle collaborations, B factories have been able to measure CP violation with impressive accuracy in several $b \to s$ penguin-dominated channels (10). To fully exploit this rich experimental information to test the SM and look for NP, we must determine the SM predictions for each channel. As we shall see in the following, computing the uncertainty associated with the SM predictions is an extremely delicate task. Only in very few cases it is possible to control this uncertainty using only experimental data; in general, one is forced to use some dynamical information, either from flavor symmetries or from factorization. Computing CP violation in $b \to s$ penguins beyond the SM is even harder: Additional operators arise, and in many cases the dominant contribution is expected to come from new operators or from operators that are subdominant in the SM. In the near future, say before the start of the Large Hadron Collider (LHC), we can aim at establishing possible hints of NP in $b \to s$ penguins. With the advent of the LHC, two scenarios are possible. If new particles are revealed, $b \to s$ penguin decays will help us identify the flavor structure of the underlying NP model. If no new particles are seen, $b \to s$ penguins can either indirectly reveal the presence of NP, if the present hints are confirmed, or allow us to push further the lower bound on the scale of NP. In all cases, experimental and theoretical progress in $b \to s$ hadronic penguins is crucial for our understanding of flavor physics beyond the SM.

This review is organized as follows: In Section 2, we quickly review the basic formalism for $b \to s$ nonleptonic decays and the different approaches to the calculation of decay amplitudes present in the literature. In Section 3, we present the predictions for branching ratios (BRs) and CP violation within the SM, following the various approaches, and compare them with the experimental data. In Section 4, we discuss the possible sources of NP contributions to $b \to s$ penguins and how these NP contributions are constrained by experimental data on other $b \to s$ transitions. In Section 5, we concentrate on supersymmetry (SUSY) extensions of the SM, discuss the present constraints, and present detailed predictions for CP violation in $b \to s$ penguins. In Section 6, we briefly discuss $b \to s$ penguins in the context of non-SUSY extensions of the SM. Finally, in Section 7 we summarize the present status and discuss future prospects.

2. BASIC FORMALISM

2.1. Generalities

The basic theoretical framework for nonleptonic B decays is based on the operator product expansion, and renormalization group methods, which allow one to write the amplitude for a decay of a given meson $B = B_d$, B_s, B^+ into a final state F as follows:

$$\mathcal{A}(B \to F) = \langle F|\mathcal{H}_{\text{eff}}|B\rangle = \left(\frac{G_F}{\sqrt{2}}\sum_{i=1}^{12}V_i^{\text{CKM}}\,C_i(\mu) + C_i^{\text{NP}}(\mu)\right)\langle F|Q_i(\mu)|B\rangle$$

$$+ \sum_{i=1}^{N_{\text{NP}}}\tilde{C}_i^{\text{NP}}(\mu)\langle F|\tilde{Q}_i(\mu)|B\rangle. \qquad 1.$$

Here $\mathcal{H}_{\text{eff}}$ is the effective weak Hamiltonian, with Q_i denoting the relevant local operators that govern the decays in question within the SM and with $\tilde{Q}_i$ denoting the ones possibly arising beyond the SM. The CKM factors V_i^{CKM} and the Wilson coefficients $C_i(\mu)$ describe the strength with which a given operator enters the Hamiltonian; for NP contributions, we denote with $C_i^{\text{NP}}(\mu)$ and $\tilde{C}_i^{\text{NP}}(\mu)$ the Wilson coefficients arising within a given NP model, which can in general be complex. In a more intuitive language, the operators $Q_i(\mu)$ can be regarded as effective vertices and the coefficients $C_i(\mu)$ as the corresponding effective couplings. The latter can be calculated in renormalization-group-improved perturbation theory and are known to next-to-leading order in QCD within the SM and in a few SUSY models (11–13). The scale μ separates the contributions to $\mathcal{A}(B \to F)$ into short-distance contributions, with energy scales higher than μ contained in $C_i(\mu)$ and long-distance contributions with energy scales lower than μ contained in the hadronic matrix elements $\langle Q_i(\mu)\rangle$. The scale μ is usually chosen to be $O(m_b)$, but is otherwise arbitrary.

The effective weak Hamiltonian for nonleptonic $b \to s$ decays within the SM is given by

$$\mathcal{H}_{\text{eff}} = \frac{4G_F}{\sqrt{2}}\left\{V_{ub}V_{us}^*\left[C_1(\mu)\left(Q_1^u(\mu) - Q_1^c(\mu)\right) + C_2(\mu)\left(Q_2^u(\mu) - Q_2^c(\mu)\right)\right]\right.$$

$$\left. - V_{tb}V_{ts}^*\left[C_1(\mu)Q_1^c(\mu) + C_2(\mu)Q_2^c(\mu) + \sum_{i=3}^{12}C_i(\mu)Q_i(\mu)\right]\right\}, \qquad 2.$$

with

$$Q_1^{u^i} = \left(\bar{b}_L\gamma^\mu u_L^i\right)\left(\bar{u}_L^i\gamma_\mu s_L\right), \qquad Q_2^{u^i} = \left(\bar{b}_L\gamma^\mu s_L\right)\left(\bar{u}_L^i\gamma_\mu u_L^i\right),$$

$$Q_{3,5} = \sum_q(\bar{b}_L\gamma^\mu s_L)(\bar{q}_{L,R}\gamma_\mu q_{L,R}), \quad Q_4 = \sum_q(\bar{b}_L\gamma^\mu q_L)(\bar{q}_L\gamma_\mu s_L),$$

$$Q_6 = -2\sum_q(\bar{b}_L q_R)(\bar{q}_R s_L), \qquad Q_{7,9} = \frac{3}{2}\sum_q(\bar{b}_L\gamma^\mu s_L)e_q(\bar{q}_L\gamma_\mu q_L), \qquad 3.$$

$$Q_8 = -3\sum_q e_q(\bar{b}_L q_R)(\bar{q}_R s_L), \qquad Q_{10} = \frac{3}{2}\sum_q e_q(\bar{b}_L\gamma^\mu q_L)(\bar{q}_{R,L}\gamma_\mu q_{R,L}),$$

$$Q_{11} = \frac{e}{16\pi^2}m_b(\bar{b}_R\sigma^{\mu\nu}s_L)F_{\mu\nu}, \qquad Q_{12} = \frac{g}{16\pi^2}m_b(\bar{b}_R\sigma^{\mu\nu}T^a s_L)G_{\mu\nu}^a,$$

where $q_{L,R} \equiv (1 \mp \gamma_5)/2q$, $u^i = \{u, c\}$, and e_q denotes the quark electric charge ($e_u = 2/3$, $e_d = -1/3$, and so on). The sum over the quarks q runs over the active flavors at the scale μ.

Q_1 and Q_2 are the so-called current-current operators, Q_{3-6} the QCD penguin operators, Q_{7-10} the electroweak penguin operators, and $Q_{11,12}$ the (chromo)magnetic penguin operators. $C_i(\mu)$ are the Wilson coefficients evaluated at $\mu = O(m_b)$. In general, they depend upon the renormalization scheme for the operators. The scale and scheme dependence of the coefficients are canceled by the analogous dependence in the matrix elements. It is therefore convenient to identify the basic renormalization group invariant parameters (RGIs) and to express the decay amplitudes in terms of RGIs. This exercise was performed by the authors of Reference 14, who identified the RGIs and wrote the decay amplitudes for several two-body nonleptonic B decays. For our purpose, we just need to recall a few basic facts about the classification of RGIs. First, we have six nonpenguin parameters, containing only nonpenguin contractions of the current-current operators $Q_{1,2}$: emission parameters $E_{1,2}$, annihilation parameters $A_{1,2}$, and Zweig-suppressed emission-annihilation parameters $EA_{1,2}$. Second, we have four parameters containing only penguin contractions of the current-current operators $Q_{1,2}$ in the GIM-suppressed combination $Q_{1,2}^c - Q_{1,2}^u : P_1^{GIM}$ and Zweig-suppressed P_{2-4}^{GIM}. Finally, we have four parameters containing penguin contractions of current-current operators $Q_{1,2}^c$ [the so-called charming penguins (15)] and all possible contractions of penguin operators $Q_{3-12} : P_{1,2}$ and the Zweig-suppressed $P_{3,4}$.

Let us now discuss some important aspects of $b \rightarrow s$ penguin nonleptonic decays. First, we define as pure penguin channels the ones generated only by P_i and P_i^{GIM} parameters. Pure penguin $b \rightarrow s$ decays can be written schematically as

$$\mathcal{A}(B \rightarrow F) = -V_{ub}^* V_{us} \sum_i P_i^{GIM} - V_{tb}^* V_{ts} \sum_i P_i. \qquad 4.$$

Neglecting doubly Cabibbo-suppressed terms, the decay amplitude has a vanishing weak phase. Therefore, there is no direct CP violation and the coefficient S_F of the $\sin \Delta mt$ term in the time-dependent CP asymmetry (for F a CP eigenstate with eigenvalue η_F) measures the phase of the mixing amplitude: $S_F = \eta_F \, \text{Im} \, \lambda_F = -\eta_F \sin 2\phi_M$, where $\lambda_F \equiv \frac{q}{p} \frac{\bar{A}}{A} = e^{-2i\phi_M}$, $A = \mathcal{A}(B \rightarrow F)$, $\bar{A} = \mathcal{A}(\bar{B} \rightarrow F)$, and $\phi_M = \beta(-\beta_s)$ for $B_d(B_s)$ mixing. Comparing the measured S_F to the one obtained from $b \rightarrow c\bar{c}s$ transitions such as $B_{d(s)} \rightarrow J/\psi K_s(\phi)$ can reveal the presence of NP in the $b \rightarrow s$ penguin amplitude. However, to perform a precise test of the SM we need to take into account also the doubly Cabibbo-suppressed terms in Equation 4. The second term then acquires a small and calculable weak phase, leading to a small and calculable $\Delta S = -\eta_F S_F - \sin 2\phi_M$. Furthermore, we must consider the contribution from the first term, that is, the contribution of GIM penguins. An estimate of the latter requires some knowledge of penguin-type hadronic matrix elements, which can be obtained either from theory or from experimental data. Let us define this as the GIM-penguin problem; we shall return to it in the next section after introducing the necessary theoretical ingredients.

Besides pure penguins, we have $b \rightarrow s$ transitions in which emission, annihilation, or emission-annihilation parameters give a contribution to the decay amplitude. Let us call these channels penguin dominated. Then we can write schematically the decay amplitude as

$$\mathcal{A}(B \rightarrow F) = -V_{ub}^* V_{us} \sum_i (T_i + P_i^{GIM}) - V_{tb}^* V_{ts} \sum_i P_i, \qquad 5.$$

where $T_i = \{E_i, A_i, EA_i\}$. Also in this case, by neglecting doubly Cabibbo-suppressed terms, the decay amplitude has a vanishing weak phase so that $\Delta S = 0$ at this order. However, we expect $T_i > P_j$ so that the double Cabibbo suppression can be overcome by the enhancement in the matrix element, leading to a sizable ΔS. Once again, the evaluation of ΔS requires some knowledge of hadronic dynamics. Let us define this as the tree problem and return to it in the next section.

2.2. Evaluation of Hadronic Matrix Elements

The past decade has witnessed remarkable progress in the theory of nonleptonic B decays. Bjorken's color transparency argument has been put on firm ground, and there is now a wide consensus that many B two-body decay amplitudes factorize in the $m_b \to \infty$ limit and are therefore computable in this limit in terms of a few fundamental nonperturbative quantities. Three different approaches to factorization in B decays have been put forward: the so-called QCD factorization (QCDF) (16), perturbative QCD (PQCD) (17–19), and soft-collinear effective theory (SCET) (20–24). A detailed discussion of these approaches goes beyond the scope of this review; for our purpose, it suffices to quickly describe a few aspects relevant to the study of $b \to s$ penguin nonleptonic decays.

Unfortunately, as suggested in Reference 15 and later confirmed in References 25–44, subleading corrections to the infinite mass limit, being doubly Cabibbo-enhanced in $b \to s$ penguins, are very important (if not dominant) in these channels and reintroduce the strong model dependence that we had hoped to eliminate using factorization theorems. Although different approaches to factorization point to different sources of large corrections, no approach allows the computation from first principles of all the ingredients needed to test the SM in $b \to s$ penguins. Therefore, it is important to pursue, in addition to factorization studies, alternative data-driven approaches that can in some cases lead to model-independent predictions for CP violation in $b \to s$ penguins.

2.2.1. QCD factorization. The first step toward a factorization theorem was given by Bjorken's color transparency argument (45). Let us consider a decay of the B meson in two light pseudoscalars, where two light quarks are emitted from the weak interaction vertex as a fast-traveling small-size color-singlet object. In the heavy-quark limit, soft gluons cannot resolve this color dipole and therefore soft gluon exchange between the two light mesons decouples at lowest order in Λ/m_b (here and in the following, Λ denotes a typical hadronic scale of order Λ_{QCD}).

Assuming that in B decays to two light pseudoscalars, perturbative Sudakov suppression is not sufficient to guarantee the dominance of hard spectator interactions, QCDF states that all soft spectator interactions can be absorbed in the heavy-to-light form factor (16). Considering, for example, $B \to \pi\pi$ decays, the following factorization formula holds at lowest order in Λ/m_b:

$$\langle \pi(p')\pi(q)|Q_i|\bar{B}(p)\rangle = f^{B\to\pi}(q^2) \int_0^1 dx\, T_i^I(x)\phi_\pi(x)$$
$$+ \int_0^1 d\xi\, dx\, dy\, T_i^{II}(\xi, x, y)\phi_B(\xi)\phi_\pi(x)\phi_\pi(y). \qquad 6.$$

where $f^{B\to\pi}(q^2)$ is a $B \to \pi$ form factor and $\phi_\pi (\phi_B)$ are leading-twist light-cone distribution amplitudes of the pion (B meson). $T_i^{I,II}$ denote the hard scattering amplitudes. Note that T^I starts at zeroth order in α_s and at higher order contains a hard gluon exchange not involving the spectator, whereas T^{II} contains the hard interactions of the spectator and starts at order α_s.

The scheme and scale dependence of the scattering kernels $T_i^{I,II}$ match those of the Wilson coefficients, and the final result is consistently scale and scheme independent. Final-state interaction phases appear in this formalism as imaginary parts of the scattering kernels (at lowest order in Λ/m_b). These phases appear in the computation of penguin contractions and of hard gluon exchange between the two pions. This means that in the heavy-quark limit, final-state interactions can be determined perturbatively.

A few remarks are important for the discussion of CP violation in $b \to s$ penguins: Penguin contractions (including charming and GIM penguins) are found to be factorizable, at least at one loop. Subleading terms in the Λ/m_b expansion are in general nonfactorizable, so they cannot be computed from first principles. They are important for phenomenology whenever they are chirally or Cabibbo enhanced. In particular, they cannot be neglected in $b \to s$ penguin modes. This introduces a strong model dependence in the evaluation of $b \to s$ penguin BRs and CP asymmetries. Power-suppressed terms can invalidate the perturbative calculation of strong phases performed in the infinite mass limit. Indeed, in this case, subleading terms in the Λ/m_b expansion can dominate over the loop-suppressed perturbative phases arising at leading order in Λ/m_b.

2.2.2. Perturbative QCD. The basic idea underlying PQCD calculations is that the dominant process is a hard gluon exchange involving the spectator quark. PQCD adopts the three-scale factorization theorem (46) based on Lepage & Brodsky's (47) PQCD formalism, with the inclusion of the transverse momentum carried by partons inside the meson. The three different scales are the electroweak scale M_W, the scale of hard gluon exchange $t \sim O(\sqrt{\Lambda m_b})$, and the factorization scale $1/b$, where b is the conjugate variable of parton transverse momenta. The nonperturbative physics at scales below $1/b$ is encoded in process-independent meson wave functions. The inclusion of transverse momentum leads to a Sudakov form factor, which suppresses the long distance contributions in the large b region and vanishes as $b > 1/\Lambda$. This suppression renders the transverse momentum flowing into the hard amplitudes of order Λm_b. The off-shellness of internal particles then remains of order $\mathcal{O}(\Lambda m_b)$, even in the end-point region, and the singularities are removed.

Contrary to QCDF, in PQCD all contributions are assumed calculable in perturbation theory owing to the Sudakov suppression. This item remains controversial (see References 48 and 49). The dominant strong phases in this approach come from factorized annihilation diagrams. Also in this case, there is no control over subleading contributions in the Λ/m_b expansion.

2.2.3. SCET. SCET is a powerful tool to study factorization in multiscale problems. The idea is to perform a two-step matching procedure at the hard $[\mathcal{O}(m_b)]$ and hard-collinear [$\mathcal{O}(\sqrt{m_b \Lambda})$] scales. The final expression is given in terms of perturbative hard kernels, light-cone wave functions, and jet functions. For phenomenology, it is convenient to fit directly the nonperturbative parameters on data using the following expression for the decay amplitude, valid at leading order in α_s (42–44):

$$\mathcal{A}(B \to M_1 M_2) \propto f_{M_1} \zeta_J^{BM_2} \int_0^1 du \phi_{M_1}(u) T_{1J}(u) + f_{M_1} \zeta^{BM_2} T_{1\zeta} + 1 \leftrightarrow 2 + A_{cc}^{M_1 M_2},$$

where Ts are perturbative hard kernels, ζs are nonperturbative parameters, and A_{cc} denotes the charming penguin contribution. Note that charming penguins are not factorized in the infinite mass limit in this approach, contrary to what is obtained in QCDF. Phenomenological analyses are carried out at leading order in α_s and at leading power in Λ/m_b. In addition, no control is possible on power corrections to factorization.

2.2.4. SU(3) flavor symmetry.
An alternative approach pursued extensively in the literature is to use $SU(3)$ flavor symmetry to extract hadronic matrix elements from experimental data and then use them to predict $SU(3)$-related channels (50–65). In principle, in this way it is possible to eliminate all the uncertainties connected to factorization and the infinite mass limit. However, $SU(3)$ breaking must be evaluated to obtain reliable predictions.

A few comments are in order: In some fortunate cases, such as the contribution of electroweak penguin $Q_{9,10}$ to $B \to K\pi$ decays, $SU(3)$ predicts some matrix elements to vanish, so they can be assumed to be suppressed even in the presence of $SU(3)$ breaking (66). Unfortunately, explicit nonperturbative calculations of two-body nonleptonic B decays indicate that $SU(3)$-breaking corrections to B decay amplitudes can be up to 80%, thus invalidating $SU(3)$ analyses of these processes (67). To partially take into account the effects of $SU(3)$ breaking, several authors assume that symmetry breaking follows the pattern of factorized matrix elements. Although this is certainly an interesting idea, its validity for $b \to s$ penguins is questionable, given the importance of nonfactorizable contributions in these channels. It is also possible to use $SU(3)$ to obtain information on nonfactorizable terms (68).

2.2.5. General parameterizations.
The idea developed in References 28 and 69 is to write down the RGIs as the sum of their expression in the infinite mass limit, for example, using QCDF, plus an arbitrary contribution corresponding to subleading terms in the power expansion. These additional contributions are then determined by a fit to the experimental data. In $b \to s$ penguins, the dominant power-suppressed correction is given by charming penguins, and the corresponding parameter can be determined with high precision from data and is found to be compatible with a Λ/m_b correction to factorization (28). However, nondominant corrections, for example, GIM penguin parameters in $b \to s$ decays, can be extracted from data only in a few cases (for example in $B \to K\pi$ decays) (69). However, predictions for ΔS depend crucially on these corrections, so that one needs external input to constrain them. One interesting avenue is to extract the support of GIM penguins from $SU(3)$-related channels ($b \to d$ penguins) in which they are not Cabibbo suppressed and to use this support,

including a possible $SU(3)$ breaking of 100%, in the fit of $b \to s$ penguin decays. Alternatively, one can omit the calculation in factorization and fit directly the RGIs from the experimental data, instead of fitting the power-suppressed corrections (70, 71).

Note that compared with factorization approaches, general parameterizations have less predictive power but are more general and thus best suited to search for NP in a conservative way. This method has the advantage that for several channels, discussed below, the predicted ΔS decreases with the experimental uncertainty in BRs and CP asymmetries of $b \to s$ and $SU(3)$-related $b \to d$ penguins.

We conclude this section by remarking once again that neither the GIM-penguin problem nor the tree problem can be solved from first principles, and we must cope with model-dependent estimates. It then becomes very important to be able to study a variety of channels in several different approaches. In this way, we can hope to make solid predictions and to test them with high accuracy. In the following, we quickly review the present theoretical and experimental results, keeping in mind the goal of testing the SM and looking for NP.

3. BRANCHING RATIOS AND CP ASYMMETRIES WITHIN THE STANDARD MODEL

The aim of this section is to collect pre- and postdictions for BRs and CP asymmetries of $b \to s$ penguin decays obtained in the approaches briefly discussed in the previous section. The main focus will be on ΔS, but BRs and rate CP asymmetries will play a key role in assessing the reliability and the theoretical uncertainty of the different approaches.

3.1. Branching Ratios and Rate CP Asymmetries

In **Tables 1–3** we report some of the results obtained in the literature for B decay BRs and CP asymmetries. For QCDF results, the first error corresponds to variations of CKM parameters; the second to variations of the renormalization scale, quark masses, decay constants (except for transverse ones), form factors, and the $\eta - \eta'$ mixing angle. The third error corresponds to the uncertainty due to the moments in the expansion of the light-cone distribution amplitudes in Gegenbauer polynomials, and also includes the scale-dependent transverse decay constants for vector mesons. Finally, the last error corresponds to an estimate of the effect of the dominant power corrections. For PQCD results from References 40 and 41, the error includes only the variation of Gegenbauer moments, of $|V_{ub}|$, and of the CKM phase. For PQCD results from Reference 72, the errors correspond to input hadronic parameters, to scale dependence, and to CKM parameters, respectively. For SCET results, the analysis is carried out at leading order in α_s and Λ/m_b assuming exact $SU(3)$. The errors are estimates of $SU(3)$ breaking, of Λ/m_b corrections, and of the uncertainty due to SCET parameters, respectively. SCET I and SCET II denote two possible solutions for SCET parameters in the fit (44). For general parameterization (GP) results, the errors include the uncertainty on CKM parameters, form factors, quark masses, and meson decay constants, and a variation of Λ/m_b corrections up to 50% of the

Table 1   Results for CP-averaged BRs (in units of 10^{-6}) and CP asymmetries (in %) in several approaches for $B \to PP$ decays

	QCDF	PQCD	SCET	GP	Exp.
$BR(\pi^- \bar{K}^0)$	$19.3^{+1.9+11.3+1.9+13.2}_{-1.9-17.8-2.1-5.6}$	$24.5^{+13.6}_{-8.1}$	$20.8 \pm 7.9 \pm 0.6 \pm 0.7$	24.1 ± 0.7	23.1 ± 1.0
$A_{\mathrm{CP}}(\pi^- \bar{K}^0)$	$0.9^{+0.2+0.3+0.1+0.6}_{-0.3-0.3-0.1-0.5}$	0 ± 0	<5	1.2 ± 2.4	0.9 ± 2.5
$BR(\pi^0 K^-)$	$11.1^{+1.8+5.8+0.9+6.9}_{-1.7-4.0-1.0-3.0}$	$13.9^{+10.0}_{-5.6}$	$11.3 \pm 4.1 \pm 1.0 \pm 0.3$	12.6 ± 0.5	12.8 ± 0.6
$A_{\mathrm{CP}}(\pi^0 K^-)$	$7.1^{+1.7+2.0+0.8+9.0}_{-1.8-2.0-0.6-9.7}$	-1^{+3}_{-5}	$-11 \pm 9 \pm 11 \pm 2$	3.4 ± 2.4	4.7 ± 2.6
$BR(\pi^+ K^-)$	$16.3^{+2.6+9.6+1.4+11.4}_{-2.3-6.5-1.4-4.8}$	$20.9^{+15.6}_{-8.3}$	$20.1 \pm 7.4 \pm 1.3 \pm 0.6$	19.6 ± 0.5	19.4 ± 0.6
$A_{\mathrm{CP}}(\pi^+ K^-)$	$4.5^{+1.1+2.2+0.5+8.7}_{-1.1-2.5-0.6-9.5}$	-9^{+6}_{-8}	$-6 \pm 5 \pm 6 \pm 2$	-8.9 ± 1.6	-9.5 ± 1.3
$BR(\pi^0 \bar{K}^0)$	$7.0^{+0.7+4.7+0.7+5.4}_{-0.7-3.2-0.7-2.3}$	$9.1^{+5.6}_{-3.3}$	$9.4 \pm 3.6 \pm 0.2 \pm 0.3$	9.5 ± 0.4	10.0 ± 0.6
$A_{\mathrm{CP}}(\pi^0 \bar{K}^0)$	$-3.3^{+1.0+1.3+0.5+3.4}_{-0.8-1.6-1.0-3.3}$	-7^{+3}_{-3}	$5 \pm 4 \pm 4 \pm 1$	-9.8 ± 3.7	-12 ± 11
Reference(s)	(36)	(40, 41)	(44)	(69)	

Experimental averages from the Heavy Flavor Averaging Group are also shown.

leading-power emission amplitude. The values in boldface correspond to predictions (that is, the experimental value has not been used in the fit).

All approaches can reproduce the experimental BRs of $B \to PP$ penguins, although QCDF tends to predict lower BRs for $B \to P\eta'$, albeit with large uncertainties. Concerning BRs of $B \to PV$ penguins, QCDF is always on the low side and reproduces experimental BRs only when the upper range of the error due to power corrections is considered. PQCD shows similar features for K^* and ρ modes, whereas it predicts much larger values for BRs of $B \to K\omega$ decays.

The situation for rate CP asymmetries is somewhat different. Both QCDF and SCET predict $A_{\mathrm{CP}}(\bar{B}^0 \to \pi^0 \bar{K}^0) \sim -A_{\mathrm{CP}}(\bar{B}^0 \to \pi^+ K^-)$, whereas experimentally the two asymmetries have the same sign. PQCD reproduces the experimental values, although it predicts $A_{\mathrm{CP}}(\bar{B}^0 \to \pi^0 \bar{K}^0)$ on the low side of the experimental value. Note that the GP approach predicts the correct value and sign of $A_{\mathrm{CP}}(\bar{B}^0 \to \pi^0 \bar{K}^0)$

Table 2   Results for two-body $b \to s$ penguin decays to η or η' CP-averaged BRs (in units of 10^{-6}) and CP asymmetries (in %) in several approaches

	QCDF	SCET I	SCET II	Exp.
$BR(\bar{K}^0 \eta')$	$46.5^{+4.7+24.9+12.3+31.0}_{-4.4-15.4-6.8-13.5}$	$63.2 \pm 24.7 \pm 4.2 \pm 8.1$	$62.2 \pm 23.7 \pm 5.5 \pm 7.2$	64.9 ± 3.5
$A_{\mathrm{CP}}(\bar{K}^0 \eta')$	$1.8^{+0.4+0.3+0.1+0.8}_{-0.5-0.3-0.2-0.8}$	$1.1 \pm 0.6 \pm 1.2 \pm 0.2$	$-2.7 \pm 0.7 \pm 0.8 \pm 0.5$	9 ± 6
$BR(\bar{K}^0 \eta)$	$1.1^{+0.1+2.0+0.4+1.3}_{-0.1-1.3-0.5-0.5}$	$2.4 \pm 4.4 \pm 0.2 \pm 0.3$	$2.3 \pm 4.4 \pm 0.2 \pm 0.5$	<1.9
$A_{\mathrm{CP}}(\bar{K}^0 \eta)$	$-9.0^{+2.8+5.4+2.8+8.2}_{-2.1-12.6-6.2-7.8}$	$21 \pm 20 \pm 4 \pm 3$	$-18 \pm 22 \pm 6 \pm 4$	
$BR(K^- \eta')$	$49.1^{+5.1+26.5+13.6+33.6}_{-4.9-16.3-7.4-14.6}$	$69.5 \pm 27.0 \pm 4.3 \pm 7.7$	$69.3 \pm 26.0 \pm 7.1 \pm 6.3$	$69.7^{+2.8}_{-2.7}$
$A_{\mathrm{CP}}(K^- \eta')$	$2.4^{+0.6+0.6+0.3+3.4}_{-0.7-0.8-0.4-3.5}$	$-1 \pm 0.6 \pm 0.7 \pm 0.5$	$0.7 \pm 0.5 \pm 0.2 \pm 0.9$	3.1 ± 2.1
$BR(K^- \eta)$	$1.9^{+0.5+2.4+0.5+1.6}_{-0.5-1.6-0.6-0.7}$	$2.7 \pm 4.8 \pm 0.4 \pm 0.3$	$2.3 \pm 4.5 \pm 0.4 \pm 0.3$	2.2 ± 0.3
$A_{\mathrm{CP}}(K^- \eta)$	$-18.9^{+6.4+11.7+4.8+25.3}_{-6.9-17.5-8.5-21.8}$	$33 \pm 30 \pm 7 \pm 3$	$-33 \pm 39 \pm 10 \pm 4$	29 ± 11
Reference(s)	(36)	(44)	(44)	

Experimental averages from the Heavy Flavor Averaging Group are also shown.

Table 3 Results for CP-averaged BRs (in units of 10^{-6}) and CP asymmetries (in %) in several approaches for $B \to PV$ decays

	QCDF	PQCD	GP	Exp.
$BR(\pi^- \bar{K}^{*0})$	$3.6^{+0.4+1.5+1.2+7.7}_{-0.3-1.4-1.2-2.3}$	$6.0^{+2.8}_{-1.5}$	11.3 ± 0.9	10.7 ± 0.8
$A_{CP}(\pi^- \bar{K}^{*0})$	$1.6^{+0.4+0.6+0.5+2.5}_{-0.5-0.5-0.4-1.0}$	-1^{+1}_{-0}	-7 ± 6	-8.5 ± 5.7
$BR(\pi^0 K^{*-})$	$3.3^{+1.1+1.0+0.6+4.4}_{-1.0-0.9-0.6-1.4}$	$4.3^{+5.0}_{-2.2}$	7.3 ± 0.6	6.9 ± 2.3
$A_{CP}(\pi^0 K^{*-})$	$8.7^{+2.1+5.0+2.9+41.7}_{-2.6-4.3-3.4-44.2}$	-32^{+21}_{-28}	-2 ± 13	4 ± 29
$BR(\pi^+ K^{*-})$	$3.3^{+1.4+1.3+0.8+6.2}_{-1.2-1.2-0.8-1.6}$	$6.0^{+6.8}_{-2.6}$	8.5 ± 0.8	9.8 ± 1.1
$A_{CP}(\pi^+ K^{*-})$	$2.1^{+0.6+8.2+5.1+62.5}_{-0.7-7.9-5.8-64.2}$	-60^{+32}_{-19}	-4 ± 13	-5 ± 14
$BR(\pi^0 \bar{K}^{*0})$	$0.7^{+0.1+0.5+0.3+2.6}_{-0.1-0.4-0.3-0.5}$	$2.0^{+1.2}_{-0.6}$	3.1 ± 0.4	$0.0^{+1.3}_{-0.1}$
$A_{CP}(\pi^0 \bar{K}^{*0})$	$-12.8^{+4.0+4.7+2.7+31.7}_{-3.2-7.0-4.0-35.3}$	-11^{+7}_{-5}	-11 ± 15	-1 ± 27
$BR(\bar{K}^0 \rho^-)$	$5.8^{+0.6+7.0+1.5+10.3}_{-0.6-3.3-1.3-13.2}$	$8.7^{+6.8}_{-4.4}$	7.8 ± 1.1	$8.0^{+1.5}_{-1.4}$
$A_{CP}(\bar{K}^0 \rho^-)$	$0.3^{+0.1+0.3+0.2+1.6}_{-0.1-0.4-0.1-1.3}$	1 ± 1	0.02 ± 0.17	12 ± 17
$BR(K^- \rho^0)$	$2.6^{+0.9+3.1+0.8+4.3}_{-0.9-1.4-0.6-1.2}$	$5.1^{+4.1}_{-2.8}$	4.15 ± 0.50	$4.25^{+0.55}_{-0.56}$
$A_{CP}(K^- \rho^0)$	$-13.6^{+4.5+6.9+3.7+62.7}_{-5.7-4.4-3.1-55.4}$	71^{+25}_{-35}	29 ± 10	31^{+11}_{-10}
$BR(K^- \rho^+)$	$7.4^{+1.8+7.1+1.2+10.7}_{-1.9-3.6-1.1-13.5}$	$8.8^{+6.8}_{-4.5}$	10.2 ± 1.0	$15.3^{+3.7}_{-3.5}$
$A_{CP}(K^- \rho^+)$	$-3.8^{+1.3+4.4+1.9+34.5}_{-1.4-2.7-1.6-32.7}$	64^{+24}_{-30}	21 ± 10	22 ± 23
$BR(\bar{K}^0 \rho^0)$	$4.6^{+0.5+4.0+0.7+6.1}_{-0.5-2.1-0.7-2.1}$	$4.8^{+4.3}_{-2.3}$	5.2 ± 0.7	$5.4^{+0.9}_{-1.0}$
$A_{CP}(\bar{K}^0 \rho^0)$	$7.5^{+1.7+2.3+0.7+8.8}_{-2.1-2.0-0.4-8.7}$	7^{+8}_{-5}	1 ± 15	-64 ± 46
$BR(K^- \omega)$	$3.5^{+1.0+3.3+1.4+4.7}_{-1.0-1.6-0.9-1.6}$	$10.6^{+10.4}_{-5.8}$	6.9 ± 0.5	6.8 ± 0.5
$A_{CP}(K^- \omega)$	$-7.8^{+2.6+5.9+2.4+39.8}_{-3.0-3.6-1.9-38.0}$	32^{+15}_{-17}	5 ± 6	5 ± 6
$BR(\bar{K}^0 \omega)$	$2.3^{+0.3+2.8+1.3+4.3}_{-0.3-1.3-0.8-1.3}$	$9.8^{+8.6}_{-4.9}$	4.6 ± 0.5	5.2 ± 0.7
$A_{CP}(\bar{K}^0 \omega)$	$-8.1^{+2.5+3.0+1.7+11.8}_{-2.0-3.3-1.4-12.9}$	-3^{+2}_{-4}	-5 ± 11	21 ± 19
$BR(K^- \phi)$	$4.5^{+0.5+1.8+1.9+11.8}_{-0.4-1.7-2.1-13.3}$	$7.8^{+5.9}_{-1.8}$	8.39 ± 0.59	8.30 ± 0.65
$A_{CP}(K^- \phi)$	$1.6^{+0.4+0.6+0.5+3.0}_{-0.5-0.5-0.3-1.2}$	1^{+0}_{-1}	3.0 ± 4.5	3.4 ± 4.4
$BR(\bar{K}^0 \phi)$	$4.1^{+0.4+1.7+1.8+10.6}_{-0.4-1.6-1.9-13.0}$	$7.3^{+5.4}_{-1.6}$	7.8 ± 0.9	$8.3^{+1.2}_{-1.0}$
$A_{CP}(\bar{K}^0 \phi)$	$1.7^{+0.4+0.6+0.5+1.4}_{-0.5-0.5-0.3-0.8}$	3^{+1}_{-2}	1 ± 6	-1 ± 13
Reference(s)	**(36)**	**(40, 41)**	**(69)**	

Experimental averages from the Heavy Flavor Averaging Group are also shown.

in spite of the complete generality of the method. Note also that $B \to K\pi$ data in **Table 1** are perfectly reproduced in the GP approach, thus showing on general grounds the absence of any $K\pi$ puzzle, although specific dynamical assumptions may lead to discrepancies between theory and experiment (58, 73–76).

We conclude that factorization approaches in general show a remarkable agreement with experimental data, but their predictions suffer from large uncertainties. Furthermore, QCDF and SCET cannot reproduce rate asymmetries in $B \to K\pi$; this may be a hint that some delicate aspects of the dynamics of penguin decays, for example, rescattering and final-state interaction phases, are not fully under control. It is then reassuring that a more general approach such as GP can reproduce the experimental data with reasonable (but not too small) values of the Λ/m_b corrections

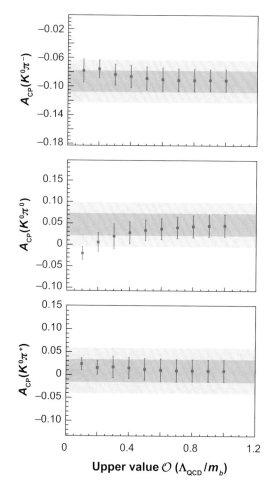

Figure 1

A_{CP} values for $B \to K\pi$ in the general parameterization (GP) approach (69) obtained varying $\mathcal{O}(\Lambda_{\mathrm{QCD}}/m_b)$ contributions in the range [0, UV], with the upper value (UV) scanned between 0 and 1 (in units of the factorized emission amplitude). For comparison, the experimental 68% (95%) probability range is given by the dark (light) band.

to factorization. To quantify this statement, we report in **Figure 1** the results of the GP fit for $A_{\mathrm{CP}}(B \to K\pi)$ as a function of the upper bound on Λ/m_b corrections (69). Clearly, imposing a too-low upper bound, of order 10%, would generate a spurious tension between theory and experiment. For the reader's convenience, we report in **Tables 4** and **5** the predictions obtained in several approaches for BRs and CP asymmetries of B_s penguin-dominated $b \to s$ decays.

3.2. Predictions for S and ΔS in $b \to s$ Penguins

Keeping in mind the results of Section 3.1, we now turn to the main topic of this review, namely, our ability to test the SM using time-dependent CP asymmetries in $b \to s$ penguin nonleptonic decays. Starting from Equation 5, we write down the expression for S_F as follows:

$$S_F = \frac{\sin(2(\beta_s + \phi_M)) + |r_F|^2 \sin(2(\phi_M + \gamma)) + 2\,\mathrm{Re}\,r_F \sin(\beta_s + 2\phi_M + \gamma)}{1 + |r_F|^2 + 2\,\mathrm{Re}\,r_F \cos(\beta_s - \gamma)}, \qquad 7.$$

Table 4 Results for CP-averaged BRs (in units of 10^{-6}) and CP asymmetries (in %) in several approaches for $B_s \to PP$ decays

	QCDF	PQCD	SCET I	SCET II
$BR(K^+K^-)$	$22.7^{+3.5+12.7+2.0+24.1}_{-3.2-18.4-2.0-19.1}$	$17.0^{+5.1+8.8+0.9}_{-4.1-5.0-0.3}$	$18.2 \pm 6.7 \pm 1.1 \pm 0.5$	
$A_{\text{CP}}(K^+K^-)$	$4.0^{+1.0+2.0+0.5+10.4}_{-1.0-2.3-0.5-11.3}$	$-25.8^{+1.1+5.2+0.9}_{-0.2-4.5-1.1}$	$-6 \pm 5 \pm 6 \pm 2$	
$BR(K^0\bar{K}^0)$	$24.7^{+2.5+13.7+2.6+25.6}_{-2.4-19.2-2.9-9.8}$	$19.6^{+6.4+10.4+0.0}_{-4.9-5.4-0.0}$	$17.7 \pm 6.6 \pm 0.5 \pm 0.6$	
$A_{\text{CP}}(K^0\bar{K}^0)$	$0.9^{+0.2+0.2+0.1+0.2}_{-0.2-0.2-0.1-0.3}$	0	<10	
$BR(\eta\eta)$	$15.6^{+1.6+9.9+2.2+13.5}_{-1.5-6.8-2.5-5.5}$	$14.6^{+4.0+8.9+0.0}_{-3.2-5.4-0.0}$	$7.1 \pm 6.4 \pm 0.2 \pm 0.8$	$6.4 \pm 6.3 \pm 0.1 \pm 0.7$
$A_{\text{CP}}(\eta\eta)$	$-1.6^{+0.5+0.6+0.4+2.2}_{-0.4-0.6-0.7-2.2}$	$-1.6^{+0.3+0.7+0.1}_{-0.3-0.6-0.1}$	$7.9 \pm 4.9 \pm 2.7 \pm 1.5$	$-1.1 \pm 5.0 \pm 3.9 \pm 1.0$
$BR(\eta\eta')$	$54.0^{+5.5+32.4+8.3+40.5}_{-5.2-22.4-6.4-16.7}$	$39.0^{+9.0+20.4+0.0}_{-7.8-13.1-0.0}$	$24.0 \pm 13.6 \pm 1.4 \pm 2.7$	$23.8 \pm 13.2 \pm 1.6 \pm 2.9$
$A_{\text{CP}}(\eta\eta')$	$0.4^{+0.1+0.3+0.1+0.4}_{-0.1-0.3-0.1-0.3}$	$-1.2^{+0.1+0.2+0.1}_{-0.0-0.1-0.1}$	$0.04 \pm 0.14 \pm 0.39 \pm 0.43$	$2.3 \pm 0.9 \pm 0.8 \pm 7.6$
$BR(\eta'\eta')$	$41.7^{+4.2+26.3+15.2+36.6}_{-4.0-17.2-8.5-15.4}$	$29.6^{+5.2+14.0+0.0}_{-5.3-8.9-0.0}$	$44.3 \pm 19.7 \pm 2.3 \pm 17.1$	$49.4 \pm 20.6 \pm 8.4 \pm 16.2$
$A_{\text{CP}}(\eta'\eta')$	$2.1^{+0.5+0.4+0.2+1.1}_{-0.6-0.4-0.3-1.2}$	$2.2^{+0.4+0.2+0.2}_{-0.4-0.4-0.1}$	$0.9 \pm 0.4 \pm 0.6 \pm 1.9$	$-3.7 \pm 1.0 \pm 1.2 \pm 5.6$
Reference(s)	(36)	(72)	(44)	(44)

The only available experimental result is $BR(B_s \to K^+K^-) = (24.4 \pm 4.8)\,10^{-6}$ (77).

Table 5 Results for CP-averaged BRs (in units of 10^{-6}) and CP asymmetries (in %) in several approaches for $B_s \to PV$ decays

Channel	QCDF	PQCD
$BR(K^+K^{*-})$	$4.1^{+1.7+1.5+1.0+9.2}_{-1.5-1.3-0.9-2.3}$	$7.4^{+2.1+1.9+0.9}_{-1.8-1.3-0.4}$
$A_{\text{CP}}(K^+K^{*-})$	$2.2^{+0.6+8.4+5.1+68.6}_{-0.7-8.0-5.9-71.0}$	$-40.6^{+2.9+2.2+1.8}_{-2.4-3.0-1.3}$
$BR(K^0\bar{K}^{*0})$	$3.9^{+0.4+1.5+1.3+10.4}_{-0.4-1.4-1.4-12.8}$	$9.1^{+3.2+2.6+0.0}_{-2.2-1.5-0.0}$
$A_{\text{CP}}(K^0\bar{K}^{*0})$	$1.7^{+0.4+0.6+0.5+1.4}_{-0.5-0.5-0.4-0.8}$	0
$BR(K^-K^{*+})$	$5.5^{+1.3+5.0+0.8+14.2}_{-1.4-2.6-0.7-13.6}$	$6.5^{+1.2+3.3+0.0}_{-1.2-1.8-0.1}$
$A_{\text{CP}}(K^-K^{*+})$	$-3.1^{+1.0+3.8+1.6+47.5}_{-1.1-2.6-1.3-45.0}$	$63.2^{+5.2+8.0+5.1}_{-5.8-10.2-2.6}$
$BR(\bar{K}^0K^{*0})$	$4.2^{+0.4+4.6+1.1+13.2}_{-0.4-2.2-0.9-3.2}$	$5.9^{+0.9+2.8+0.0}_{-1.1-1.8-0.0}$
$A_{\text{CP}}(\bar{K}^0K^{*0})$	$0.2^{+0.0+0.2+0.1+0.2}_{-0.1-0.3-0.1-0.1}$	0
$BR(\eta\omega)$	$0.012^{+0.005+0.010+0.028+0.025}_{-0.004-0.003-0.006-0.006}$	$0.10^{+0.02+0.03+0.00}_{-0.02-0.01-0.00}$
$A_{\text{CP}}(\eta\omega)$		$3.2^{+6.1+15.2+0.3}_{-3.9-11.2-0.1}$
$BR(\eta'\omega)$	$0.024^{+0.011+0.028+0.077+0.042}_{-0.009-0.006-0.010-0.015}$	$0.66^{+0.23+0.22+0.01}_{-0.18-0.21-0.03}$
$A_{\text{CP}}(\eta'\omega)$		$-0.1^{+0.7+3.9+0.0}_{-0.8-4.2-0.0}$
$BR(\eta\phi)$	$0.12^{+0.02+0.95+0.54+0.32}_{-0.02-0.14-0.12-0.13}$	$1.8^{+0.5+0.1+0.0}_{-0.5-0.2-0.0}$
$A_{\text{CP}}(\eta\phi)$	$-8.4^{+2.0+30.1+14.6+36.3}_{-2.1-71.2-44.7-59.7}$	$-0.1^{+0.2+2.3+0.0}_{-0.4-1.4-0.0}$
$BR(\eta'\phi)$	$0.05^{+0.01+1.10+0.18+0.40}_{-0.01-0.17-0.08-0.04}$	$3.6^{+1.2+0.4+0.0}_{-0.9-0.4-0.0}$
$A_{\text{CP}}(\eta'\phi)$	$-62.2^{+15.9+132.3+80.8+122.4}_{-10.2-84.2-46.8-49.9}$	$1.2^{+0.1+0.4+0.1}_{-0.0-0.6-0.1}$
Reference(s)	(36)	(72)

No experimental data are available yet.

Table 6 Predictions for S parameters in units of 10^{-2} for B decays

	PQCD	SCET I	SCET II	GP	Exp.
$S_{\pi^0 K_S}$	74^{+2}_{-3}	$80\pm2\pm2\pm1$		74.3 ± 4.4	33 ± 21
$S_{\eta' K_S}$		$70.6\pm0.5\pm0.6\pm0.3$	$71.5\pm0.5\pm0.8\pm0.2$	70.9 ± 3.9	61 ± 7
$S_{\eta K_S}$		$69\pm15\pm5\pm1$	$79\pm14\pm4\pm1$		
$S_{\phi K_S}$	71^{+1}_{-1}			71.5 ± 8.7	39 ± 18
$S_{\rho^0 K_S}$	50^{+10}_{-16}			64 ± 11	20 ± 57
$S_{\omega K_S}$	84^{+3}_{-7}			75.7 ± 10.3	48 ± 24
Reference(s)	(40, 41)	(44)	(44)	(69)	

Experimental averages from Heavy Flavor Averaging Group are also shown.

where $r_F = |V_{us}V_{ub}|/|V_{ts}V_{tb}| \times \sum(T_i + P_i^{\mathrm{GIM}})/\sum P_i$, with $T_i = 0$ for pure penguin channels. Because the angle β_s is small and very well known [$\beta_s = (2.1\pm0.1)°$], the problem is then reduced to the evaluation of $\kappa_F = \sum(T_i + P_i^{\mathrm{GIM}})/\sum P_i$ for each channel (note that $T_i = 0$ for pure penguin channels). Factorization methods have been used to provide estimates of κ_F, S_F, and ΔS_F for $b \to s$ channels. The latter two are reported in **Tables 6** and **7**. A few remarks are important: The evaluation of P_i^{GIM} relies on the factorization of penguin contractions of charm and up quarks, which is debatable even in the infinite mass limit. In addition, in factorization, P^{GIM} has a perturbative loop suppression so that it is likely to be dominated by power corrections. Furthermore, the contribution of T_i and P_i^{GIM} is particularly difficult to estimate for η and η' channels. Lastly, the determination of the sign of ΔS_F relies heavily on the determination of the sign of $\mathrm{Re}\kappa_F$. If P_i^{GIM} is dominated by power corrections, there is no guarantee that the sign given by the perturbative calculation is correct.

With the above caveat in mind, from **Tables 6** and **7** we learn the following:

- Experimentally, there is a systematic trend for negative ΔS. This might be a hint of the presence of new sources of CP violation in the $b \to s$ penguin amplitude.
- The experimental uncertainty is dominant in all channels. In addition, the GP estimate of the theoretical uncertainty, which is certainly conservative, can be reduced with experimental improvements on BRs and CP asymmetries.

Table 7 Predictions for ΔS parameters in units of 10^{-2} for B decays

	QCDF	QCDF	SCET I	SCET II	GP	Exp.
$\Delta S_{\pi^0 K_S}$	4^{+2+1}_{-3-1}	7^{+5}_{-4}	$7.7\pm2.2\pm1.8\pm1$		2.4 ± 5.9	-35 ± 21
$\Delta S_{\eta' K_S}$	0^{+0+0}_{-4-0}	1^{+1}_{-1}	$-1.9\pm0.5\pm0.6\pm0.3$	$-1.0\pm0.5\pm0.8\pm0.2$	-0.7 ± 5.4	-7 ± 7
$\Delta S_{\eta K_S}$	7^{+2+0}_{-5-0}	10^{+11}_{-7}	$-3.4\pm15.5\pm5.4\pm1.4$	$7.0\pm13.6\pm4.2\pm1.1$		
$\Delta S_{\phi K_S}$	3^{+1+1}_{-4-1}	2^{+1}_{-1}			0.4 ± 9.2	-29 ± 18
$\Delta S_{\rho^0 K_S}$	4^{+9+8}_{-10-11}	-8^{+8}_{-12}			-6.2 ± 8.4	-48 ± 57
$\Delta S_{\omega K_S}$	4^{+2+2}_{-4-1}	13^{+8}_{-8}			5.6 ± 10.7	-20 ± 24
Reference(s)	(78)	(79)	(44)	(44)	(69)	

Experimental averages from Heavy Flavor Averaging Group are also shown.

- As discussed in Section 2, the theoretical uncertainty estimated from first principles is much smaller for pure penguin decays such as $B \to \phi K_s$ than for penguin-dominated channels.
- In the model-independent GP approach, the theoretical uncertainty is smaller for $B \to \pi^0 K_s$ because the number of observables in the $B \to K\pi$ system is sufficient to efficiently constrain the hadronic parameters. This means that the theoretical error can be kept under control by improving the experimental data in these channels. However, the information on $B \to \phi K_s$ is not sufficient to bound the subleading terms, and this results in a relatively large theoretical uncertainty that cannot be decreased without additional input on hadronic parameters. Furthermore, using $SU(3)$ to constrain $\Delta S_{\phi K_s}$ is difficult because the number of amplitudes involved is very large (50, 63–65).

The ideal situation would be represented by a pure penguin decay for which the information on P_i^{GIM} is available with minimal theoretical input. Such a situation is realized by the pure penguin decays $B_s \to K^{0(*)}\bar{K}^{0(*)}$. An upper bound for the P_i^{GIM} entering this amplitude can be obtained from the $SU(3)$-related channels $B_d \to K^{0(*)}\bar{K}^{0(*)}$. Then, adding a generous 100% $SU(3)$ breaking and an arbitrary strong phase will make it possible to have full control over the theoretical error in ΔS (71). For the reader's convenience, we report in **Table 8** the predictions for the S coefficient of the time-dependent CP asymmetry for several B_s penguin-dominated decays.

Before closing this section, let us mention nonresonant three-body B decays such as $B \to K_s \pi^0 \pi^0$, $B \to K_s K_s K_s$, or $B \to K^+ K^- K_s$. In this case, a theoretical estimate of κ_F is extremely challenging, and using $SU(3)$ to constrain κ_F is difficult because of the large number of channels involved (63). Nevertheless, they are certainly helpful in completing the picture of CP violation in $b \to s$ penguins.

Table 8   Predictions for S parameters for B_s decays

	PQCD	SCET I	SCET II
$\bar{B}_s^0 \to K_S \pi^0$	$-0.46^{+0.14+0.19+0.02}_{-0.13-0.20-0.04}$	$-0.16 \pm 0.41 \pm 0.33 \pm 0.17$	
$\bar{B}_s^0 \to K_S \eta$	$-0.31^{+0.05+0.16+0.02}_{-0.05-0.17-0.03}$	$0.82 \pm 0.32 \pm 0.11 \pm 0.04$	$0.63 \pm 0.61 \pm 0.16 \pm 0.08$
$\bar{B}_s^0 \to K_S \eta'$	$-0.72^{+0.02+0.04+0.00}_{-0.02-0.03-0.00}$	$0.38 \pm 0.08 \pm 0.10 \pm 0.04$	$0.24 \pm 0.09 \pm 0.15 \pm 0.05$
$\bar{B}_s^0 \to K^- K^+$	$0.28^{+0.04+0.04+0.02}_{-0.04-0.03-0.01}$	$0.19 \pm 0.04 \pm 0.04 \pm 0.01$	
$\bar{B}_s^0 \to \pi^0 \eta$	$0.00^{+0.03+0.09+0.00}_{-0.02-0.10-0.01}$	$0.45 \pm 0.14 \pm 0.42 \pm 0.30$	$0.38 \pm 0.20 \pm 0.42 \pm 0.37$
$\bar{B}_s^0 \to \eta\eta$	$0.03^{+0.00+0.01+0.00}_{-0.00-0.01-0.00}$	$-0.026 \pm 0.040 \pm 0.030 \pm 0.014$	$-0.077 \pm 0.061 \pm 0.022 \pm 0.026$
$\bar{B}_s^0 \to \eta\eta'$	$0.04^{+0.00+0.00+0.00}_{-0.00-0.00-0.00}$	$0.041 \pm 0.004 \pm 0.002 \pm 0.051$	$0.015 \pm 0.010 \pm 0.008 \pm 0.069$
$\bar{B}_s^0 \to \eta'\eta'$	$0.04^{+0.00+0.00+0.00}_{-0.00-0.00-0.00}$	$0.049 \pm 0.005 \pm 0.005 \pm 0.031$	$0.051 \pm 0.009 \pm 0.017 \pm 0.039$
$\bar{B}_s^0 \to \omega\eta$	$0.07^{+0.00+0.04+0.00}_{-0.01-0.11-0.00}$		
$\bar{B}_s^0 \to \omega\eta'$	$-0.19^{+0.01+0.04+0.01}_{-0.01-0.04-0.03}$		
$\bar{B}_s^0 \to \phi\eta$	$0.10^{+0.01+0.04+0.01}_{-0.01-0.03-0.00}$		
$\bar{B}_s^0 \to \phi\eta'$	$0.00^{+0.00+0.02+0.00}_{-0.00-0.02-0.00}$		
$\bar{B}_s^0 \to K_S \phi$	-0.72		
Reference(s)	**(72)**	**(44)**	**(44)**

To summarize the status of $b \to s$ penguins in the SM, we can say that additional experimental data will allow us to establish whether the trend of negative ΔS shown by present data really signals the presence of NP in $b \to s$ penguins. Theoretical errors are not an issue in this respect because the estimates based on factorization can in most cases be checked using the GP approach based purely on experimental data. B_s decays will provide additional useful channels and will help considerably in assessing the presence of NP in $b \to s$ penguins.

4. CP VIOLATION IN $b \to s$ PENGUINS BEYOND THE STANDARD MODEL

There is a hint of NP in CP-violating $b \to s$ hadronic penguins. In this section, we answer two basic questions that arise when considering NP contributions to these decays:

1. What are the constraints from other processes on new sources of CP violation in $b \to s$ transitions?
2. Are NP contributions to $b \to s$ transitions well motivated from the theoretical point of view?

We consider here only model-independent aspects of these two questions and post-pone model-dependent analyses to Section 5.

4.1. Model-Independent Constraints on $b \to s$ Transitions

The past year has witnessed enormous progress in the experimental study of $b \to s$ transitions. In particular, the Tevatron experiments have provided us with the first information on the $B_s - \bar{B}_s$ mixing amplitude (80), which can be translated into constraints on the $\Delta B = \Delta S = 2$ effective Hamiltonian. In any given model, as we shall see, for example, in Section 5, these constraints can be combined with the ones from $b \to s\gamma$ and $b \to s\ell^+\ell^-$ decays to provide strong bounds on NP effects in $b \to s$ hadronic penguins.

Let us now summarize the presently available bounds on the $B_s - \bar{B}_s$ mixing amplitude, following the discussion of Reference 81. General NP contributions to the $\Delta B = \Delta S = 2$ effective Hamiltonian can be incorporated into the analysis in a model-independent way, parameterizing the shift induced in the mixing frequency and phase with two parameters, C_{B_s} and ϕ_{B_s}, equal to 1 and 0 in the SM (82–86):

$$C_{B_s} e^{2i\phi_{B_s}} = \frac{\langle B_s | \mathcal{H}_{\text{eff}}^{\text{full}} | \bar{B}_s \rangle}{\langle B_s | \mathcal{H}_{\text{eff}}^{\text{SM}} | \bar{B}_s \rangle}. \qquad 8.$$

As for the absorptive part of the $B_s - \bar{B}_s$ mixing amplitude, which is derived from the double insertion of the $\Delta B = 1$ effective Hamiltonian, it can be affected by non-negligible NP effects in $\Delta B = 1$ transitions through penguin contributions. Following References 4 and 5, we thus introduce two additional parameters, C_s^{Pen} and ϕ_s^{Pen}, which encode NP contributions to the penguin part of the $\Delta B = 1$ Hamiltonian, analogous to what C_{B_s} and ϕ_{B_s} do for the mixing amplitude.

The available experimental information includes the following: the measurement of Δm_s (80), the semileptonic asymmetry in B_s decays A_{SL}^s and the dimuon asymmetry A_{CH} from DØ (87, 88), the measurement of the B_s lifetime from flavor-specific final states (89–94), the determination of $\Delta\Gamma_s/\Gamma_s$ from the time-integrated angular analysis of $B_s \to J/\psi\phi$ decays by CDF (95), and the three-dimensional constraint on Γ_s, $\Delta\Gamma_s$, and $B_s - \bar{B}_s$ mixing phase ϕ_s from the time-dependent angular analysis of $B_s \to J/\psi\phi$ decays by DØ (96).

Making use of this experimental information, it is possible to constrain C_{B_s} and ϕ_{B_s} (5, 81, 97–100). The fourfold ambiguity for ϕ_{B_s} inherent in the untagged analysis of Reference 96 is somewhat reduced by the measurements of A_{SL}^s and A_{SL} (101), which prefer negative values of ϕ_{B_s}. The results for C_{B_s} and ϕ_{B_s}, obtained from the general analysis allowing for NP in all sectors, are (81)

$$C_{B_s} = 1.03 \pm 0.29, \quad \phi_{B_s} = (-75 \pm 14)^{\circ} \cup (-19 \pm 11)^{\circ} \cup (9 \pm 10)^{\circ}. \qquad 9.$$

Thus, the deviation from zero in ϕ_{B_s} is below the 1σ level, although clearly there is still ample room for values of ϕ_{B_s} very far from zero. The corresponding probability density function (p.d.f.) in the $C_{B_s} - \phi_{B_s}$ plane is shown in **Figure 2**.

The experimental information on $b \to s\gamma$ and $b \to s\ell^+\ell^-$ decays (102) can also be combined in a model-independent way along the lines of References 103–106. In this way, it is possible to constrain the coefficients of the $b \to s\gamma$, $b \to s\gamma^*$, and $b \to sZ$ vertices, which also contribute to $b \to s$ hadronic penguins. It turns out that order-of-magnitude enhancements of these vertices are excluded, so they are unlikely to give large effects in $b \to s$ nonleptonic decays. However, the $b \to sg$ vertex is only very weakly constrained, so it can still give large contributions to

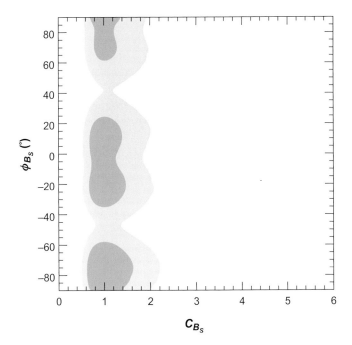

Figure 2

Constraints on the ϕ_{B_s} versus C_{B_s} plane (81). Darker (lighter) regions correspond to 68% (95%) probability.

$b \rightarrow s$ hadronic penguins (107). Finally, the information contained in Equation 9 can be used to constrain NP effects in $b \rightarrow s$ hadronic decays only within a given model, as a connection between $\Delta B = 2$ and $\Delta B = 1$ effective Hamiltonians is possible only once the model is specified. We shall return to this point in Section 5.

4.2. Theoretical Motivations for New Physics in $b \rightarrow s$ Transitions

We now turn to the second question formulated at the beginning of this section, namely, whether on general grounds it is natural to expect NP to show up in $b \rightarrow s$ transitions. The general picture emerging from the generalized Unitarity Triangle analysis performed in References 4, 5, and 81 and from the very recent data on $D-\bar{D}$ mixing (108–111) is that no new sources of CP violation are present in B_d, K, and D mixing amplitudes. Conversely, large NP contributions to $s \rightarrow dg$, $b \rightarrow dg$, and $b \rightarrow sg$ transitions are not at all excluded. Therefore, although the idea of minimal flavor violation is phenomenologically appealing (11, 112–117), an equally possible alternative is that NP is contributing more to $\Delta F = 1$ transitions than to $\Delta F = 2$ ones. Within the class of $\Delta F = 1$ transitions, (chromo)magnetic vertices are peculiar because they require a chirality flip to take place, which leads to a down-type quark mass suppression within the SM. However, NP models can weaken this suppression if they contain additional heavy fermions and/or additional sources of chiral mixing. In this case, they can lead to spectacular enhancements for the coefficients of (chromo)magnetic operators. Furthermore, if the relevant new particles are colored, they can naturally give a strong enhancement of chromomagnetic operators, whereas magnetic operators may be only marginally modified. The electric dipole moment of the neutron puts strong constraints on new sources of CP violation in chirality-flipping flavor-conserving operators involving light quarks, but this does not necessarily imply the suppression of flavor-violating operators, especially those involving b quarks. Therefore, assuming that NP is sizable in hadronic $b \rightarrow s$ penguins is perfectly legitimate given the present information available on flavor physics.

From a theoretical point of view, a crucial observation is the strong breaking of the SM $SU(3)^5$ flavor symmetry by the top quark Yukawa coupling. This breaking necessarily propagates in the NP sector so that in general it is very difficult to suppress NP contributions to CP violation in B decays, and these NP contributions could be naturally larger in $b \rightarrow s$ transitions than in $b \rightarrow d$ ones. This is indeed the case in several flavor models (see, e.g., Reference 118).

Another interesting argument is the connection between quark and lepton flavor violation in grand unified models (119–122). The idea is very simple: The large flavor mixing present in the neutrino sector, if generated mainly by Yukawa couplings, should be shared by right-handed down-type quarks that sit in the same $SU(5)$ multiplet with left-handed leptons. Once again, one expects in this case large NP contributions to $b \rightarrow s$ transitions.

We conclude that the possibility of large NP effects in $b \rightarrow s$ penguin hadronic decays is theoretically well motivated on general grounds. The arguments sketched above can of course be put on firmer ground in the context of specific models, and we refer the reader to the rich literature on this subject.

5. SUSY MODELS

Let us now focus on SUSY and discuss the phenomenological effects of the new sources of flavor and CP violation in $b \to s$ processes that arise in the squark sector (123–147). In general, in the Minimal Supersymmetric Standard Model, squark masses are neither flavor universal nor aligned to quark masses, so they are not flavor diagonal in the super-CKM basis in which quark masses are diagonal, and all neutral current vertices are flavor diagonal. The ratios of off-diagonal squark mass terms to the average squark mass define four new sources of flavor violation in the $b \to s$ sector: the mass insertions $(\delta_{23}^d)_{AB}$, with A, B = L, R referring to the helicity of the corresponding quarks. In general, these δs are complex, so they also violate CP. One can think of them as additional CKM-type mixings arising from the SUSY sector. Assuming that the dominant SUSY contribution comes from the strong interaction sector, that is, from gluino exchange, all FCNC processes can be computed in terms of the SM parameters plus the four δs plus the relevant SUSY parameters: the gluino mass $m_{\tilde{g}}$, the average squark mass $m_{\tilde{q}}$, $\tan \beta$, and the μ parameter. The impact of additional SUSY contributions such as chargino exchange has been discussed in detail in Reference 138. We consider only the case of small or moderate $\tan \beta$ because for large $\tan \beta$, the constraints from $B_s \to \mu^+ \mu^-$ and Δm_s preclude the possibility of having large effects in $b \to s$ hadronic penguin decays (148).

Barring accidental cancellations, one can consider one single δ parameter, fix the SUSY masses, and study the phenomenology. The constraints on δs come at present from $B \to X_s \gamma$, $B \to X_s l^+ l^-$, and from the $B_s - \bar{B}_s$ mixing amplitude as given in Equation 9. We refer the reader to References 149 and 150 for all the details of this analysis.

Fixing as an example $m_{\tilde{g}} = m_{\tilde{q}} = |\mu| = 350$ GeV and $\tan \beta = 3$ or 10, one obtains the constraints on δs reported in **Figures 3–5**. We plot in light green the allowed region, considering only the constraint from the C_{B_s} versus ϕ_{B_s} p.d.f. of **Figure 2**, in light blue the allowed region considering only the constraint from $b \to s \ell^+ \ell^-$, and in violet the allowed region considering only the constraint from $b \to s \gamma$. The dark blue region is the one selected after imposing all constraints simultaneously.

Several comments are in order at this point. Only $(\delta_{23}^d)_{LL,LR}$ generate amplitudes that interfere with the SM in rare decays. Therefore, the constraints from rare decays for $(\delta_{23}^d)_{RL,RR}$ are symmetric around zero, whereas the interference with the SM produces the circular shape of the $B \to X_s \gamma$ constraint on $(\delta_{23}^d)_{LL,LR}$. We recall that LR and RL mass insertions generate much larger contributions to the (chromo)magnetic operators, as the necessary chirality flip can be performed on the gluino line ($\propto m_{\tilde{g}}$) rather than on the quark line ($\propto m_b$). Therefore, the constraints from rare decays are much more effective on these insertions, so the bound from $B_s - \bar{B}_s$ has no impact in this case. The $\mu \tan \beta$ flavor-conserving LR squark mass term generates, together with a flavor-changing LL mass insertion, an effective $(\delta_{23}^d)_{LR}^{\text{eff}}$ that contributes to $B \to X_s \gamma$. For positive (negative) μ, we have $(\delta_{23}^d)_{LR}^{\text{eff}} \propto +(-)(\delta_{23}^d)_{LL}$, and therefore the circle determined by $B \to X_s \gamma$ in the LL and LR cases lies on the same (opposite) side of the origin (see **Figures 3** and **4**).

For $\tan \beta = 3$, we see from the upper panels of **Figure 3** that the bound on $(\delta_{23}^d)_{LL}$ from $B_s - \bar{B}_s$ mixing is competitive with the one from rare decays, whereas

LL

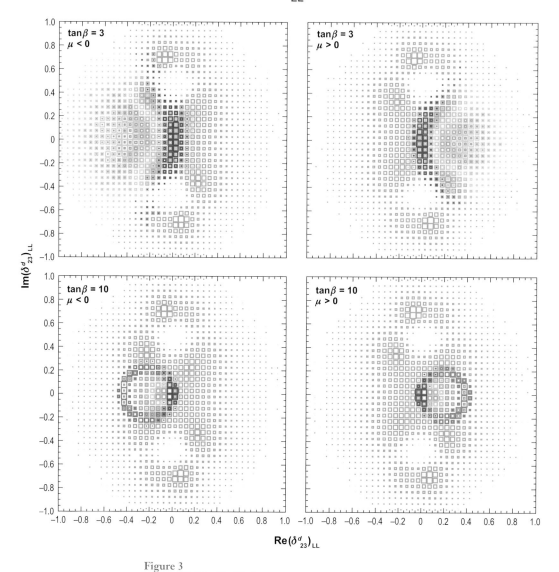

Figure 3

Allowed region in the $\mathrm{Re}(\delta_{23}^{d})_{\mathrm{LL}}$-$\mathrm{Im}(\delta_{23}^{d})_{\mathrm{LL}}$ plane. In the plots on the left (right), negative (positive) μ is considered. Plots in the upper (lower) panels correspond to $\tan\beta = 3$ ($\tan\beta = 10$). See text for details.

for $\tan\beta = 10$, rare decays give the strongest constraints (lower panels of **Figure 3**). The bounds on all other δs do not depend upon the sign of μ or the value of $\tan\beta$ for this choice of SUSY parameters. For LL and LR cases, $B \rightarrow X_s\gamma$ and $B \rightarrow X_s l^+ l^-$ produce bounds with different shapes on the $\mathrm{Re}\delta$-$\mathrm{Im}\delta$ plane (violet and light blue regions in **Figures 3** and **4**). Only a much smaller region around the origin is allowed

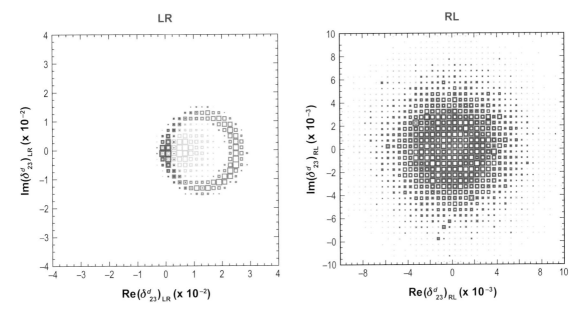

Figure 4

Allowed region in the $\mathrm{Re}\big(\delta_{23}^{d}\big)_{\mathrm{LR}}$-$\mathrm{Im}\big(\delta_{23}^{d}\big)_{\mathrm{LR}}$ (*left panel*) and $\mathrm{Re}\big(\delta_{23}^{d}\big)_{\mathrm{RL}}$-$\mathrm{Im}\big(\delta_{23}^{d}\big)_{\mathrm{RL}}$ (*right panel*) plane. Results do not depend upon the sign of μ or upon the value of $\tan\beta$.

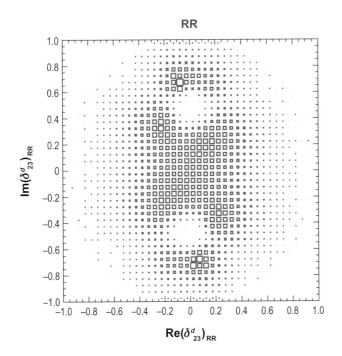

Figure 5

Allowed region in the $\mathrm{Re}\big(\delta_{23}^{d}\big)_{\mathrm{RR}}$-$\mathrm{Im}\big(\delta_{23}^{d}\big)_{\mathrm{RR}}$ plane. Results do not depend upon the sign of μ or upon the value of $\tan\beta$. See text for details.

by both constraints (dark blue regions in **Figures 3** and **4**). This shows the key role played by rare decays in constraining new sources of flavor and CP violation in the squark sector. For the RR case, the constraints from rare decays are very weak, so the only significant bound comes from $B_s - \bar{B}_s$ mixing. If $(\delta_{23}^d)_{LL}$ and $(\delta_{23}^d)_{RR}$ insertions are simultaneously nonzero, they generate chirality-breaking contributions that are strongly enhanced over chirality-conserving ones so that the product $(\delta_{23}^d)_{LL}(\delta_{23}^d)_{RR}$ is severely bounded. In **Figure 6** we report the allowed region obtained in the case $(\delta_{23}^d)_{LL} = (\delta_{23}^d)_{RR}$. For $(\delta_{23}^d)_{LL} \neq (\delta_{23}^d)_{RR}$, this constraint can be interpreted as a bound on $\sqrt{(\delta_{23}^d)_{LL}(\delta_{23}^d)_{RR}}$. We observe a very interesting interplay between the constraints from rare decays and the one from $B_s - \bar{B}_s$ mixing. Increasing $\tan \beta$ from 3 to 10, the bound from rare decays becomes tighter, but $B_s - \bar{B}_s$ mixing still plays a relevant role. All constraints scale approximately linearly with squark and gluino masses.

Having determined the p.d.f.s for the four δs, we now turn to the evaluation of the time-dependent CP asymmetries. As discussed in Section 2, the uncertainty in the calculation of SUSY effects is even larger than that for SM. Furthermore, we cannot use the GP approach because to estimate the SUSY contribution we must evaluate the hadronic matrix elements explicitly. Following Reference 149, we use QCDF, enlarging the range for power-suppressed contributions to annihilation chosen in Reference 36, as suggested in Reference 28. We warn the reader about the large theoretical uncertainties that affect this evaluation.

In **Figures 7–10**, we present the results for $S_{\phi K_S}$, $S_{\pi^0 K_S}$, $S_{\eta' K_S}$ and $S_{\omega K_S}$. They do not show a sizable dependence upon the sign of μ or $\tan \beta$ for the chosen range of SUSY parameters. We see that deviations from the SM expectations (*a*) are possible in all channels and the present experimental central values can be reproduced, and (*b*) are more easily generated by LR and RL insertions owing to the enhancement mechanism discussed above. As noted in References 151 and 152, the correlation between ΔS_{PP} and ΔS_{PV} depends upon the chirality of the NP contributions. For example, we show in **Figure 11** the correlation between $\Delta S_{K_S \phi}$ and $\Delta S_{K_S \pi^0}$ for the four possible choices for mass insertions. We see that the $\Delta S_{K_S \phi}$ and $\Delta S_{K_S \pi^0}$ are correlated for LL and LR mass insertions, and anticorrelated for RL and RR mass insertions.

An interesting issue is the scaling of SUSY effects in ΔS with squark and gluino masses. We have noted above that the constraints from other processes scale linearly with the SUSY masses. The dominant SUSY contribution to ΔS, the chromomagnetic one, also scales linearly with SUSY masses as long as $m_{\tilde{g}} \sim m_{\tilde{q}} \sim \mu$. This means that there is no decoupling of SUSY contributions to ΔS as long as the constraint from other processes can be saturated for $\delta < 1$. From **Figures 3–5**, we see that the bounds on LL and RR mass insertions quickly reach the physical boundary at $\delta = 1$, whereas LR and RL are safely below that bound. Additional bounds on chirality-flipping LR and RL insertions can be derived by studying the scalar potential of the Minimal Supersymmetric Standard Model, imposing the absence of charge and color-breaking minima and unbounded-from-below directions (153). However, it is easy to check that the flavor bounds given above are stronger for SUSY masses up to (and above) the TeV scale. We conclude that LR and RL mass insertions can give

LL = RR

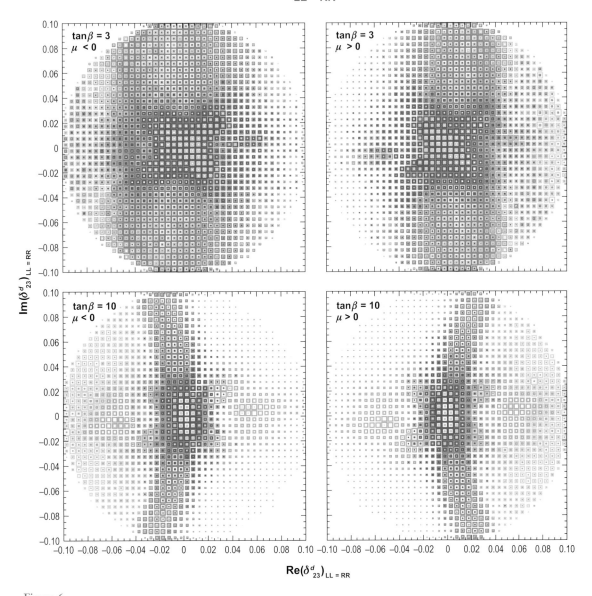

Figure 6

Allowed region in the $\mathrm{Re}(\delta_{23}^d)_{\mathrm{LL=RR}}$-$\mathrm{Im}(\delta_{23}^d)_{\mathrm{LL=RR}}$ plane. In the plots on the left (right), negative (positive) μ is considered. Plots in the upper (lower) panels correspond to $\tan\beta = 3$ ($\tan\beta = 10$). See text for details.

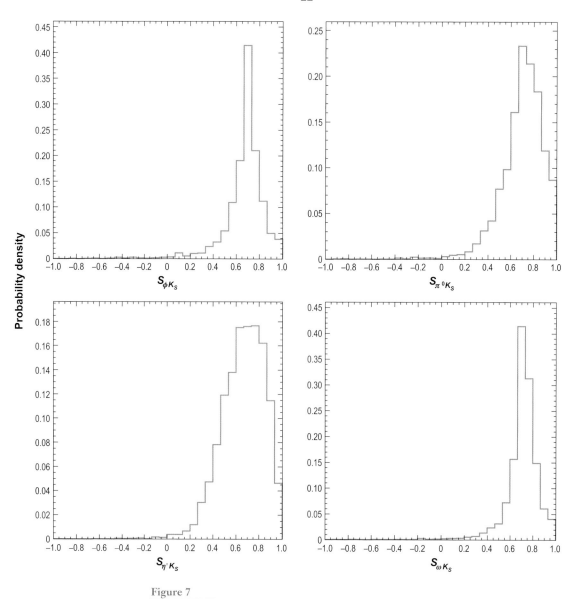

Figure 7

Probability density functions for $S_{\phi K_s}$, $S_{\pi^0 K_s}$, $S_{\eta' K_s}$, and $S_{\omega K_s}$ induced by $(\delta^d_{23})_{LL}$.

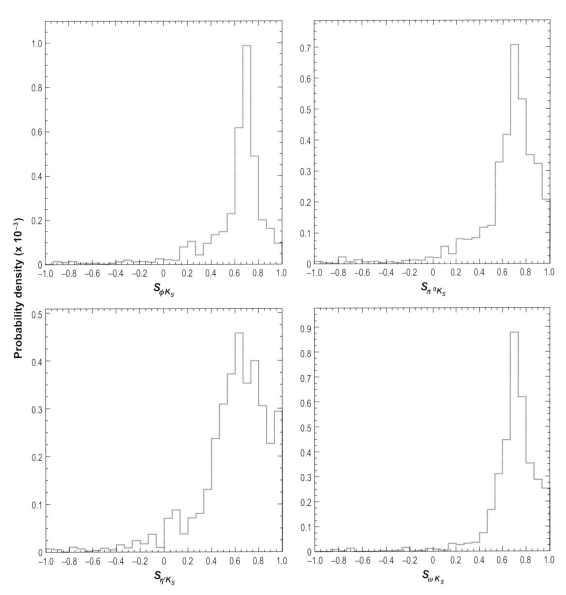

Figure 8

Probability density functions for $S_{\phi K_s}$, $S_{\pi^0 K_s}$, $S_{\eta' K_s}$, and $S_{\omega K_s}$ induced by $(\delta_{23}^d)_{\text{LR}}$.

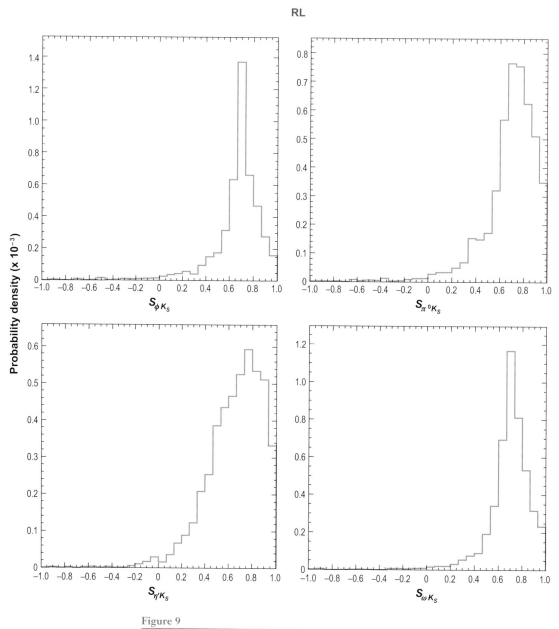

Figure 9

Probability density functions for $S_{\phi K_s}$, $S_{\pi^0 K_s}$, $S_{\eta' K_s}$, and $S_{\omega K_s}$ induced by $(\delta_{23}^d)_{RL}$.

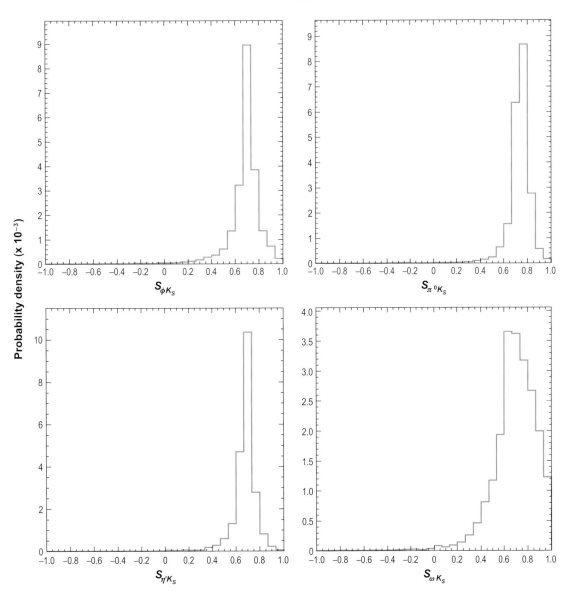

Figure 10

Probability density functions for $S_{\phi K_s}$, $S_{\pi^0 K_s}$, $S_{\eta' K_s}$, and $S_{\omega K_s}$ induced by $(\delta_{23}^d)_{RR}$.

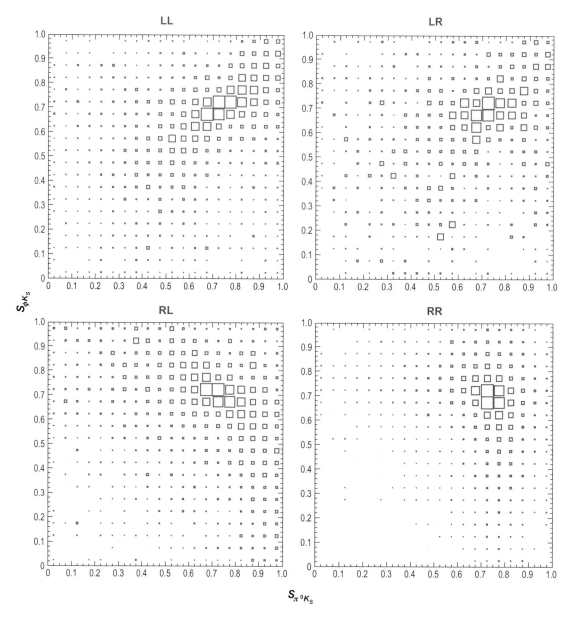

Figure 11

Correlation between $S_{\phi K_s}$ and $S_{\pi^0 K_s}$ for LL, LR, RL, and RR mass insertions.

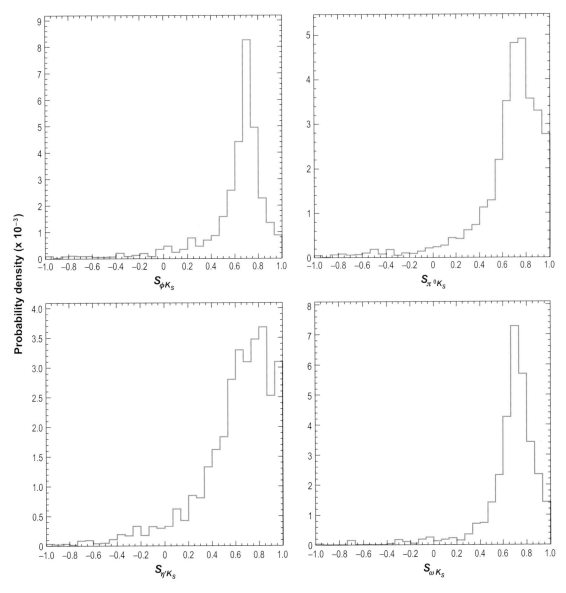

Figure 12

Probability density functions for $S_{\phi K_s}$, $S_{\pi^0 K_s}$, $S_{\eta' K_s}$, and $S_{\omega K_s}$ induced by $(\delta_{23}^d)_{LR}$ for $m_{\tilde{g}} = m_{\tilde{q}} = \mu = 1$ TeV.

RL

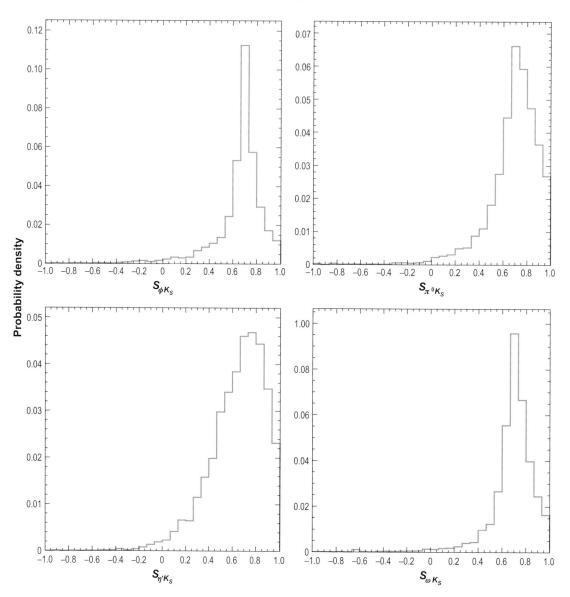

Figure 13

Probability density functions for $S_{\phi K_s}$, $S_{\pi^0 K_s}$, $S_{\eta' K_s}$, and $S_{\omega K_s}$ induced by $(\delta_{23}^d)_{\mathrm{RL}}$ for $m_{\tilde{g}} = m_{\tilde{q}} = \mu = 1$ TeV.

434   *Silvestrini*

observable effects for SUSY masses within the reach of the LHC and above. This is shown explicitly in **Figures 12** and **13**, where we present the p.d.f.s for $S_{\phi K_s}$, $S_{\pi^0 K_s}$, $S_{\eta' K_s}$, and $S_{\omega K_s}$ for SUSY masses of 1 TeV.

6. NON-SUSY MODELS

In general, one expects sizable values of ΔS in all models in which new sources of CP violation are present in $b \to s$ penguins. In particular, models with a fourth generation, both vector-like and sequential, models with warped extra dimensions in which the flavor structure of the SM is obtained using localization of fermion wave functions, and models with Z' gauge bosons can all give potentially large contributions to $b \to s$ penguins (154–158).

In any given NP model, it is possible to perform a detailed analysis along the lines of Section 5, considering the constraints from $B_s - \bar{B}_s$ mixing and from rare B decays, plus the constraints from all other sectors if they are correlated with $b \to s$ transitions in the model. On general grounds, the dominant contributions to $b \to s$ hadronic decays are expected to come from electroweak or chromomagnetic penguins. The correlation between the induced ΔS_{PP} and ΔS_{PV} can give a handle on the chirality of the NP-generated operators. NP effects in electroweak penguin contributions are in general correlated with effects in $b \to s l^+ l^-$, in $b \to s\gamma$, and possibly in $Z \to b\bar{b}$. Depending upon the flavor structure of NP, other effects may be seen in $K \to \pi \nu \bar{\nu}$ or in ε'/ε. NP effects in the chromomagnetic penguin may also show up in $b \to s\gamma$, in $B_s - \bar{B}_s$ mixing, and, if there is a correlation between the B and K sectors, in ε'/ε.

7. CONCLUSIONS AND OUTLOOK

We have reviewed the theoretical status of hadronic $b \to s$ penguin decays and have shown that, in spite of the theoretical difficulties in the evaluation of hadronic matrix elements, in the SM it is possible to obtain sound theoretical predictions for the coefficient S_F of time-dependent CP asymmetries, using either models of hadronic dynamics or data-driven approaches. Experimental data show an interesting trend of deviations from the SM predictions that definitely deserves further theoretical and experimental investigation.

From the point of view of NP, the recent improvements in the experimental study of other $b \to s$ processes such as $B_s - \bar{B}_s$ transitions or $b \to s\gamma$ and $b \to s l^+ l^-$ have considerably restricted the NP parameter space. However, there are still several NP models, in particular SUSY with new sources of $b \to s$ mixing in squark mass matrices, that can produce deviations from the SM in the ballpark of experimental values. In any given model, the study of hadronic $b \to s$ penguins and of the correlation with other FCNC processes in B and K physics is a very powerful tool for unraveling the flavor structure of NP.

Any NP model with new sources of CP violation and new particles within the mass reach of the LHC can potentially produce sizable deviations from the SM in $b \to s$ penguins. It will be exciting to combine the direct information from the LHC and the indirect one from flavor physics to identify the physics beyond the SM that has been hiding behind the corner for the past decades. In this respect, future facilities for *B* physics will provide us with an invaluable tool for studying the origin of fermion masses and of flavor symmetry breaking, two aspects of elementary particle physics that remain obscure despite the theoretical and experimental efforts in flavor physics.

DISCLOSURE STATEMENT

The author is not aware of any biases that might be perceived as affecting the objectivity of this review.

ACKNOWLEDGMENTS

I am grateful to M. Ciuchini, E. Franco, and M. Pierini for their careful reading of this manuscript and useful discussions. I acknowledge partial support from the Research Training Networks European contracts MRTN-CT-2004-503369, MRTN-CT-2006-035482, and MRTN-CT-2006-035505.

LITERATURE CITED

1. Wolfenstein L. *Phys. Rev. Lett.* 51:1945 (1983); Buras AJ, Lautenbacher ME, Ostermaier G. *Phys. Rev. D* 50:3433 (1994)
2. Bona M, et al. (UTfit Collab.) *JHEP* 07:028 (2005)
3. Bona M, et al. (UTfit Collab.) *JHEP* 10:081 (2006)
4. Bona M, et al. (UTfit Collab.) *JHEP* 03:080 (2006)
5. Bona M, et al. (UTfit Collab.) *Phys. Rev. Lett.* 97:151803 (2006)
6. Charles J, et al. *Eur. Phys. J. C* 41:1 (2005)
7. Grossman Y, Worah MP. *Phys. Lett. B* 395:241 (1997)
8. Ciuchini M, et al. *Phys. Rev. Lett.* 79:978 (1997)
9. London D, Soni A. *Phys. Lett. B* 407:61 (1997)
10. Chen KF, et al. (Belle Collab.) *Phys. Rev. Lett.* 98:031802 (2007); Aubert B, et al. (BaBar Collab.) hep-ex/0607112 (2006); Aubert B, et al. (BaBar Collab.) *Phys. Rev. Lett.* 98:031801 (2007); Aubert B, et al. (BaBar Collab.) hep-ex/0702046 (2007); Abe K, et al. (Belle Collab.) hep-ex/0609006 (2006); Aubert B, et al. (BaBar Collab.) hep-ex/0607096 (2006); Aubert B, et al. (BaBar Collab.) *Phys. Rev. Lett.* 98:051803 (2007); Aubert B, et al. (BaBar Collab.) hep-ex/0607101 (2006); Aubert B, et al. (BaBar Collab.) hep-ex/0408095 (2004); Aubert B, et al. (BaBar Collab.) hep-ex/0702010 (2007); Aubert B, et al. (BaBar Collab.) hep-ex/0507016 (2005)
11. Ciuchini M, Degrassi G, Gambino P, Giudice GF. *Nucl. Phys. B* 534:3 (1998)
12. Bobeth C, Misiak M, Urban J. *Nucl. Phys. B* 567:153 (2000)
13. Degrassi G, Gambino P, Giudice GF. *JHEP* 12:009 (2000)
14. Buras AJ, Silvestrini L. *Nucl. Phys. B* 569:3 (2000)
15. Ciuchini M, Franco E, Martinelli G, Silvestrini L. *Nucl. Phys. B* 501:271 (1997)

16. Beneke M, Buchalla G, Neubert M, Sachrajda CT. *Phys. Rev. Lett.* 83:1914 (1999); Beneke M, Buchalla G, Neubert M, Sachrajda CT. *Nucl. Phys. B* 591:313 (2000); Beneke M, Neubert M. *Nucl. Phys. B* 651:225 (2003); Beneke M, Jager S. *Nucl. Phys. B* 751:160 (2006); Beneke M, Jager S. *Nucl. Phys. B* 768:51 (2007)

17. Li Hn, Yu HL. *Phys. Rev. Lett.* 74:4388 (1995)

18. Li Hn, Yu HL. *Phys. Lett. B* 353:301 (1995)

19. Li Hn, Yu HL. *Phys. Rev. D* 53:2480 (1996)

20. Bauer CW, Pirjol D, Stewart IW. *Phys. Rev. D* 65:054022 (2002)

21. Bauer CW, et al. *Phys. Rev. D* 66:014017 (2002)

22. Bauer CW, Pirjol D, Stewart IW. *Phys. Rev. Lett.* 87:201806 (2001)

23. Chay J, Kim C. *Phys. Rev. D* 68:071502 (2003)

24. Chay J, Kim C. *Nucl. Phys. B* 680:302 (2004)

25. Ciuchini M, et al. *Nucl. Phys. B* 512:3 (1998)

26. Ciuchini M, et al. *Nucl. Instrum. Methods A* 408:28 (1998)

27. Isola C, et al. *Phys. Rev. D* 64:014029 (2001)

28. Ciuchini M, et al. *Phys. Lett. B* 515:33 (2001)

29. Isola C, et al. *Phys. Rev. D* 65:094005 (2002)

30. Ciuchini M, et al. hep-ph/0208048 (2002)

31. Isola C, Ladisa M, Nardulli G, Santorelli P. *Phys. Rev. D* 68:114001 (2003)

32. Zenczykowski P. *Phys. Lett. B* 590:63 (2004)

33. Ciuchini M, et al. hep-ph/0407073 (2004)

34. Cheng HY, Yang KC. *Phys. Rev. D* 64:074004 (2001)

35. Beneke M, Buchalla G, Neubert M, Sachrajda CT. *Nucl. Phys. B* 606:245 (2001)

36. Beneke M, Neubert M. *Nucl. Phys. B* 675:333 (2003)

37. Beneke M, Rohrer J, Yang D. hep-ph/0612290 (2006)

38. Keum YY, Li Hn, Sanda AI. *Phys. Rev. D* 63:054008 (2001)

39. Li Xq, Yang Yd. *Phys. Rev. D* 72:074007 (2005)

40. Li Hn, Mishima S, Sanda AI. *Phys. Rev. D* 72:114005 (2005)

41. Li Hn, Mishima S. *Phys. Rev. D* 74:094020 (2006)

42. Bauer CW, Pirjol D, Rothstein IZ, Stewart IW. *Phys. Rev. D* 70:054015 (2004)

43. Bauer CW, Rothstein IZ, Stewart IW. *Phys. Rev. D* 74:034010 (2006)

44. Williamson AR, Zupan J. *Phys. Rev. D* 74:014003 (2006)

45. Bjorken JD. *Nucl. Phys. Proc. Suppl.* 11:325 (1989)

46. Chang CHV, Li Hn. *Phys. Rev. D* 55:5577 (1997)

47. Lepage GP, Brodsky SJ. *Phys. Rev. D* 22:2157 (1980)

48. Descotes-Genon S, Sachrajda CT. *Nucl. Phys. B* 625:239 (2002)

49. Keum YY, Li Hn, Sanda AI. *AIP Conf. Proc.* 618:229 (2002)

50. Zeppenfeld D. *Zeit. Phys. C* 8:77 (1981); Savage MJ, Wise MB. *Phys. Rev. D* 39:3346 (1989); Chau LL, et al. *Phys. Rev. D* 43:2176 (1991); Gronau M, Hernandez OF, London D, Rosner JL. *Phys. Rev. D* 50:4529 (1994); Gronau M, Hernandez OF, London D, Rosner JL. *Phys. Rev. D* 52:6374 (1995); Gronau M, Rosner JL. *Phys. Rev. D* 53:2516 (1996); Grinstein B, Lebed RF. *Phys. Rev. D* 53:6344 (1996); Chiang CW, et al. *Phys. Rev. D* 69:034001 (2004); Chiang CW, Gronau M, Rosner JL, Suprun DA. *Phys. Rev. D* 70:034020 (2004); Chiang CW, Zhou YF. *JHEP* 12:027 (2006)

51. Fleischer R. *Phys. Rev. D* 58:093001 (1998)
52. Fleischer R. *Phys. Lett. B* 435:221 (1998)
53. Buras AJ, Fleischer R. *Eur. Phys. J. C* 11:93 (1999)
54. Fleischer R. *Phys. Lett. B* 459:306 (1999)
55. Buras AJ, Fleischer R. *Eur. Phys. J. C* 16:97 (2000)
56. Fleischer R, Matias J. *Phys. Rev. D* 61:074004 (2000)
57. Fleischer R, Matias J. *Phys. Rev. D* 66:054009 (2002)
58. Buras AJ, Fleischer R, Recksiegel S, Schwab F. *Eur. Phys. J. C* 32:45 (2003)
59. Buras AJ, Fleischer R, Recksiegel S, Schwab F. *Phys. Rev. Lett.* 92:101804 (2004)
60. Buras AJ, Fleischer R, Recksiegel S, Schwab F. *Acta Phys. Pol.* B36:2015 (2005)
61. Buras AJ, Fleischer R, Recksiegel S, Schwab F. *Eur. Phys. J. C* 45:701 (2006)
62. Fleischer R, Recksiegel S, Schwab F. *Eur. Phys. J. C* 51:55 (2007)
63. Engelhard G, Nir Y, Raz G. *Phys. Rev. D* 72:075013 (2005)
64. Engelhard G, Raz G. *Phys. Rev. D* 72:114017 (2005)
65. Raz G. hep-ph/0509125 (2005)
66. Neubert M, Rosner JL. *Phys. Rev. Lett.* 81:5076 (1998); Gronau M, Pirjol D, Yan TM. *Phys. Rev. D* 60:034021 (1999); Neubert M. *JHEP* 9902:014 (1999); Gronau M, Pirjol D, Yan TM. *Phys. Rev. D* 69:119901 (2004)
67. Khodjamirian A, Mannel T, Melcher M. *Phys. Rev. D* 68:114007 (2003)
68. Descotes-Genon S, Matias J, Virto J. *Phys. Rev. Lett.* 97:061801 (2006); Matias J. hep-ph/0701116 (2007); Descotes-Genon S, Matias J, Virto J. arxiv:0705.0477 (2007)
69. Ciuchini M, et al. *Proc. 4th Int. Workshop CKM Unitarity Triangle KEK Rep.,* Nagoya, Japan. In press (2007)
70. Ciuchini M, Pierini M, Silvestrini L. *Phys. Rev. Lett.* 95:221804 (2005)
71. Ciuchini M, Pierini M, Silvestrini L. hep-ph/0703137 (2007)
72. Ali A, et al. hep-ph/0703162 (2007)
73. Baek S, et al. *Phys. Rev. D* 71:057502 (2005)
74. Kim CS, Oh S, Yu C. *Phys. Rev. D* 72:074005 (2005)
75. Baek S, London D. hep-ph/0701181 (2007)
76. Fleischer R. hep-ph/0701217 (2007)
77. Morello M. (CDF Collab.) hep-ex/0612018 (2006)
78. Cheng HY, Chua CK, Soni A. *Phys. Rev. D* 72:014006 (2005)
79. Beneke M. *Phys. Lett. B* 620:143 (2005)
80. Abulencia A, et al. (CDF Collab.) *Phys. Rev. Lett.* 97:242003 (2006)
81. Bona M, et al. (UTfit Collab.) arxiv:0707.0636
82. Soares JM, Wolfenstein L. *Phys. Rev. D* 47:1021 (1993)
83. Deshpande NG, Dutta B, Oh S. *Phys. Rev. Lett.* 77:4499 (1996)
84. Silva JP, Wolfenstein L. *Phys. Rev. D* 55:5331 (1997)
85. Cohen AG, Kaplan DB, Lepeintre F, Nelson AE. *Phys. Rev. Lett.* 78:2300 (1997)
86. Grossman Y, Nir Y, Worah MP. *Phys. Lett. B* 407:307 (1997)
87. Abazov VM, et al. (D0 Collab.) *Phys. Rev. Lett.* 98:151801 (2007)
88. Abazov VM, et al. (D0 Collab.) *Phys. Rev. D* 74:092001 (2006)
89. Buskulic D, et al. (ALEPH Collab.) *Phys. Lett. B* 377:205 (1996)
90. Abe F, et al. (CDF Collab.) *Phys. Rev. D* 59:032004 (1999)

91. Abreu P, et al. (DELPHI Collab.) *Eur. Phys. J. C* 16:555 (2000)
92. Ackerstaff K, et al. (OPAL Collab.) *Phys. Lett. B* 426:161 (1998)
93. Abazov VM, et al. (D0 Collab.) *Phys. Rev. Lett.* 97:241801 (2006)
94. Barberio E, et al. (HFAG) hep-ex/0603003 (2006)
95. Acosta D, et al. (CDF Collab.) *Phys. Rev. Lett.* 94:101803 (2005)
96. Abazov VM, et al. (D0 Collab.) *Phys. Rev. Lett.* 98:121801 (2007)
97. Ligeti Z, Papucci M, Perez G. *Phys. Rev. Lett* 97:101801 (2006)
98. Ball P, Fleischer R. *Eur. Phys. J. C* 48:413 (2006)
99. Grossman Y, Nir Y, Raz G. *Phys. Rev. Lett.* 97:151801 (2006)
100. Lenz A, Nierste U. hep-ph/0612167 (2006)
101. Aubert B, et al. (BaBar Collab.) *Phys. Rev. Lett.* 96:251802 (2006)
102. Ishikawa A, et al. (Belle Collab.) *Phys. Rev. Lett.* 91:261601 (2003); Koppenburg P, et al. (Belle Collab.) *Phys. Rev. Lett.* 93:061803 (2004); Iwasaki M, et al. (Belle Collab.) *Phys. Rev. D* 72:092005 (2005); Ishikawa A, et al. *Phys. Rev. Lett.* 96:251801 (2006); Aubert B, et al. (BaBar Collab.) *Phys. Rev. Lett.* 93:081802 (2004); Aubert B, et al. (BaBar Collab.) *Phys. Rev. D* 72:052004 (2005); Aubert B, et al. (BaBar Collab.) *Phys. Rev. D* 73:092001 (2006)
103. Ali A, Giudice GF, Mannel T. *Z. Phys.* C67:417 (1995)
104. Ali A, Lunghi E, Greub C, Hiller G. *Phys. Rev. D* 66:034002 (2002)
105. Hiller G, Kruger F. *Phys. Rev. D* 69:074020 (2004)
106. Gambino P, Haisch U, Misiak M. *Phys. Rev. Lett.* 94:061803 (2005)
107. Ciuchini M, Gabrielli E, Giudice GF. *Phys. Lett. B* 388:353 (1996)
108. Aubert B, et al. (BaBar Collab.) *Phys. Rev. Lett.* 98:211802 (2007)
109. Abe K, et al. (Belle Collab.) *Phys. Rev. Lett.* 98:211803 (2007)
110. Abe K, et al. (Belle Collab.) arXiv:0704.1000 (2007)
111. Ciuchini M, et al. hep-ph/0703204 (2007)
112. Gabrielli E, Giudice GF. *Nucl. Phys. B* 433:3 (1995)
113. Misiak M, Pokorski S, Rosiek J. *Adv. Ser. Direct High Energy Phys.* 15:795 (1998)
114. Buras AJ, et al. *Phys. Lett. B* 500:161 (2001)
115. D'Ambrosio G, Giudice GF, Isidori G, Strumia A. *Nucl. Phys. B* 645:155 (2002)
116. Bobeth C, et al. *Nucl. Phys. B* 726:252 (2005)
117. Blanke M, Buras AJ, Guadagnoli D, Tarantino C. *JHEP* 10:003 (2006)
118. Masiero A, Piai M, Romanino A, Silvestrini L. *Phys. Rev. D* 64:075005 (2001)
119. Baek S, Goto T, Okada Y, Okumura Ki. *Phys. Rev. D* 63:051701 (2001)
120. Harnik R, Larson DT, Murayama H, Pierce A. *Phys. Rev. D* 69:094024 (2004)
121. Hisano J, Shimizu Y. *Phys. Lett. B* 565:183 (2003)
122. Huang CS, Li Tj, Liao W. *Nucl. Phys. B* 673:331 (2003)
123. Gabbiani F, Gabrielli E, Masiero A, Silvestrini L. *Nucl. Phys. B* 477:321 (1996)
124. Barbieri R, Strumia A. *Nucl. Phys. B* 508:3 (1997)
125. Kagan AL, Neubert M. *Phys. Rev. D* 58:094012 (1998)
126. Abel SA, Cottingham WN, Whittingham IB. *Phys. Rev. D* 58:073006 (1998)
127. Kagan A. hep-ph/9806266 (1997)
128. Fleischer R, Mannel T. *Phys. Lett. B* 511:240 (2001)
129. Besmer T, Greub C, Hurth T. *Nucl. Phys. B* 609:359 (2001)
130. Lunghi E, Wyler D. *Phys. Lett. B* 521:320 (2001)

131. Causse MB. hep-ph/0207070 (2002)

132. Hiller G. *Phys. Rev. D* 66:071502 (2002)

133. Khalil S, Kou E. *Phys. Rev. D* 67:055009 (2003)

134. Kane GL, et al. *Phys. Rev. D* 70:035015 (2004)

135. Baek S. *Phys. Rev. D* 67:096004 (2003)

136. Agashe K, Carone CD. *Phys. Rev. D* 68:035017 (2003)

137. Cheng JF, Huang CS, Wu Xh. *Phys. Lett. B* 585:287 (2004)

138. Chakraverty D, Gabrielli E, Huitu K, Khalil S. *Phys. Rev. D* 68:095004 (2003)

139. Khalil S, Kou E. *Phys. Rev. Lett.* 91:241602 (2003)

140. Khalil S, Kou E. hep-ph/0307024 (2003)

141. Cheng JF, Huang CS, Wu Xh. *Nucl. Phys. B* 701:54 (2004)

142. Khalil S, Kou E. *Phys. Rev. D* 71:114016 (2005)

143. Gabrielli E, Huitu K, Khalil S. *Nucl. Phys. B* 710:139 (2005)

144. Khalil S. *Mod. Phys. Lett.* A19:2745 (2004)

145. Khalil S. *Phys. Rev. D* 72:035007 (2005)

146. Baek S, London D, Matias J, Virto J. *JHEP* 0602:027 (2006)

147. Baek S, London D, Matias J, Virto J. *JHEP* 0612:019 (2006)

148. Isidori G, Retico A. *JHEP* 11:001 (2001); Buras AJ, Chankowski PH, Rosiek J, Slawianowska L. *Nucl. Phys. B* 659:3 (2003); Foster J, Okumura Ki, Roszkowski L. *JHEP* 03:044 (2006); Foster J, Okumura Ki, Roszkowski L. *Phys. Lett. B* 641:452 (2006); Isidori G, Paradisi P. *Phys. Lett. B* 639:499 (2006); Isidori G, Mescia F, Paradisi P, Temes D. hep-ph/0703035 (2007)

149. Ciuchini M, Franco E, Masiero A, Silvestrini L. *Phys. Rev. D* 67:075016 (2003)

150. Ciuchini M, Silvestrini L. *Phys. Rev. Lett.* 97:021803 (2006)

151. Kagan AL. hep-ph/0407076 (2004)

152. Endo M, Mishima S, Yamaguchi M. *Phys. Lett. B* 609:95 (2005)

153. Casas JA, Dimopoulos S. *Phys. Lett. B* 387:107 (1996)

154. Atwood D, Hiller G. hep-ph/0307251 (2003)

155. Burdman G. *Phys. Lett. B* 590:86 (2004)

156. Buchalla G, Hiller G, Nir Y, Raz G. *JHEP* 09:074 (2005)

157. Agashe K, Papucci M, Perez G, Pirjol D. hep-ph/0509117 (2005)

158. Hou WS, Li Hn, Mishima S, Nagashima M. *Phys. Rev. Lett.* 98:131801 (2007)

Quantum Communication

Judy Jackson[1] and Neil Calder[2]

[1] Fermi National Accelerator Laboratory, Batavia, Illinois 60540;
email: jjackson@fnal.gov

[2] Stanford Linear Accelerator Center, Stanford University, Stanford,
California 94025; email: neil.calder@slac.stanford.edu

Annu. Rev. Nucl. Part. Sci. 2007. 57:441–62

First published online as a Review in Advance on
July 3, 2007

The *Annual Review of Nuclear and Particle Science* is
online at http://nucl.annualreviews.org

This article's doi:
10.1146/annurev.nucl.57.090506.123030

Copyright © 2007 by Annual Reviews.
All rights reserved

0163-8998/07/1123-0441$20.00

Key Words

particle physics, collaboration, international, SLAC, Fermilab,
symmetry, interactions

Abstract

Few would dispute that the science of particle physics in the United
States has reached a crossroads. Policies, decisions, and events of the
coming decade will be critical in determining whether the United
States continues to carry out a competitive program of leadership in
this field of fundamental science. The field of particle physics has
responded to this reality by fundamentally changing its model of
communication from "business as usual" to a strategic and collaborative method that is clearly focused on achieving a healthy future
for the science. Over the past half-dozen years, the particle physics
community has gone from being an oft-cited example of how not
to communicate effectively, to a frequently cited—and emulated—
model for science communication. This review outlines the new
approach toward communication in particle physics and then goes
into detail about three case studies.

Contents

THE DILEMMA OF PARTICLE PHYSICS 442
    Strategic Communication 443
WHY? 444
    Who Are the Audiences? 444
THERE'S NOTHING LIKE DATA 446
COLLABORATION 448
CASE STUDIES 450
    Seeking Symmetry 450
QUANTUM UNIVERSE 454
DISCOVERING THE QUANTUM UNIVERSE 458
CONCLUSION 462

THE DILEMMA OF PARTICLE PHYSICS

Twenty-first-century particle physics in the United States faces a future that is as exciting and promising as at any time in its history—and at least as challenging. On the one hand, the prospects for particle physics have never been brighter or more fascinating. As advances in recent years have revealed, an unexplored Universe— far stranger and more wonderful than scientists ever imagined—awaits discovery. Illuminating the nature of dark matter and dark energy, finding the origin of mass, searching for possible evidence of extra dimensions, and the unification of forces will produce a Copernican-scale revolution in the human understanding of the Universe and its physical laws. Particle physics discoveries will be fundamental to this new science of the Universe. An undiscovered Universe beckons—a true paradise for particle physics.

On the other hand, there looms the sheer scale of the technology required to embark on this great twenty-first-century scientific adventure. Experimental particle physics requires multibillion-dollar particle accelerators; ever-larger, more complex, and costlier detectors; and giant experimental collaborations to make advances in a field of science that is difficult to comprehend, has few immediate applications, and must compete with other scientific disciplines and compelling societal needs at a time of constrained national budgets.

Furthermore, because it operates at such an enormous scale, the particle physics research of the future can only be achieved with a level of international cooperation and collaboration that is unprecedented, even in the field of particle physics, where scientists long ago became accustomed to carrying out their research across national borders. Few if any models exist to show how global scientific collaboration on this scale might succeed. Twenty-first-century particle physics will demand not only national but international long-term commitment to very large and expensive projects, perhaps on foreign shores, whose payoffs are for the most part unknown.

Confronted simultaneously with such extraordinary scientific opportunity and such formidable obstacles, what will the future hold for particle physics? This is the central question for those engaged in particle physics communication in today's environment. To create the kind of robust and competitive program of research that would allow the United States to remain a leader in particle physics, the particle physics community must respond successfully to the greatest communication challenge the field has ever faced. In this unprecedented context, the familiar models of communication have not proved effective. In the face of this extraordinary—and imminent—challenge to its future, the field has transformed the way it communicates. Particle physics has adopted a new model of strategic, collaborative communication designed specifically to meet clearly defined goals for the field. Moreover, this transformation has occurred in an environment in which resources for communication are few and the most effective use must be made of every communication dollar.

Strategic Communication

Particle physicists have long understood the need to communicate better. For decades, the subject of improved communication constituted an obligatory session at every physics conference, motivating countless well-meaning attempts to get the word out via a variety of media about the nature and value of this field of science. Yet there was little evidence that such efforts to communicate for the sake of communicating had borne fruit in the form of increased support for particle physics—or that they had even reached the audiences who influence science policy. One physics communicator refers to this approach as the "let's do a brochure" school of communication—the faith that producing a full-color booklet, a clever give-away item, or a half-hour CD will somehow lead to some ill-defined but desirable result. Too often, the brochures sat on the shelf, the gizmos came and went, and the CDs went unviewed. However well intentioned, such efforts were seldom effective because they started at the wrong end of the process: choosing particular tactics before the would-be communicators had defined their communication goals and answered crucial questions.

Designing effective communication can be compared to designing a particle detector. No experimental physics collaboration would design a detector by saying "let's have a drift chamber," or "a gas calorimeter might be good," or "how about some pixels?" Before experimenters begin the process of designing a detector, they ask fundamental strategic questions: What is the science that we want to do? What are the key scientific questions that we seek to answer? What particle interactions do we want to observe? And finally, what are the most effective—and most cost-effective—technological means to obtaining our scientific objectives? Only after systematically answering such questions in depth and detail do experimenters begin the task of designing individual detector elements and combining them into an effective scientific instrument. As detector design proceeds, a continuous review process examines every aspect of the project to make sure that it stays on track to achieve its scientific objectives. Course corrections are a normal occurrence.

Like successful detector design, effective communication results from clearly defined goals and from posing the right set of fundamental questions. Clear goals and basic questions allow the development of a communication strategy that is systematically constructed to reach the desired set of goals. Only by methodically asking and responding to such questions is it possible to determine what the elements—the communication equivalents of the drift chamber or the pixels—of the communication plan should be and how to combine them to achieve the desired result.

Just as detector design depends on asking the right set of questions, it is also critical to get the questions right in designing effective communication. What is the communication goal? Who are the key audiences? What are the messages that should be communicated to the audiences? Only after answering these questions is it possible to find rational answers to the question of how. What means of communication are most likely to succeed in communicating the message to the key audiences to achieve the goal? And, as with detectors, frequent review and feedback—with changes in tactics if necessary—are essential to the process.

WHY?

The goal of particle physics communication in the United States is to strengthen support for particle physics to ensure a strong and healthy future for the field, so that the nation can continue in its historic role as a leader in this fundamental field of science. There may be secondary, strategic goals, but they all follow from this central goal, the measuring stick for evaluating every communication strategy or tactic. Does this brochure, this publication, this Web site, or this video help strengthen particle physics? The way to answer those questions is to ask the next set of questions.

Who Are the Audiences?

Consideration of the communication goal—strengthening support for particle physics—determines the key audiences for particle physics communication. Who are the people whose views, decisions, and actions influence the level and the nature of support for the field and thus determine its future? In "The Conceptualization and Measurement of Policy Leadership" (1), science communication researcher Jon D. Miller asks the question, "Who are the people who exercise—or who have the opportunity to exercise—influence on the definition of science and technology policy?" These policy makers and opinion leaders are a relatively small, well-defined, and well-characterized group.

Notwithstanding the widespread perception that in order to influence science policy it is necessary to gain the support of the general public, Miller argues in "Communication Strategies for Meeting the Information Needs of Science and Energy Policy Leaders" (2) that "[d]espite the substantial impact of science and technology on life in the United States, there is no evidence that any congressional or senatorial election has ever been determined by differences over a science or technology issue." Instead, science policy is defined as a low-salience issue. That is, rather than being made through the electoral process, science policy is made by a fairly small number of government officials in the executive and legislative branches of government who turn to a somewhat larger group of nongovernmental opinion leaders for information, advice, and guidance. These opinion leaders typically include leading scientists in research universities, national laboratories, and science-based corporations; members of the National Academy of Sciences and the National Academy of Engineering; the leadership of major universities, corporations, national laboratories, and other organizations active in scientific work; scientific and engineering professional societies; and certain science journalists for national or international publications.

Besides policy makers and opinion leaders, there are other audiences important to the achievement of the communication goal. Particle physics cannot thrive without attracting and inspiring successive generations of young physicists. Students and teachers are therefore a significant audience for education and outreach. The physics-interested public (Miller estimates the science-interested public at approximately 12% of the American population; the particle-physics-interested public is presumably some fraction of that number) may in some circumstances constitute a key audience. Particularly for national laboratories, the support and participation of residents in neighboring communities are critical to long-term success; they are thus an important audience, as is the internal community of employees and university researchers at national laboratories.

To succeed in influencing science policy, then, effective science communication must provide these key audiences with information that they need and want from sources that they trust via media that they use. By asking yet another set of questions about each audience, the opportunity exists to develop communication strategies that have a high probability of success.

What do audiences want? What do audiences need and want to know? Where do they get their information? What sources do they trust? Which messages are effective and which messages are not? What media, topics, subjects, articles, and formats interest and engage them? Answers to these questions are likely to vary (although our experience indicates that the variation may be less than expected) for different audiences. There is really only one way to get accurate answers to these critical questions: to ask the audiences themselves. Too often in the past, the particle physics community approached the process of communication abstractly, using what could be described as a theoretical approach, devising the answers from a set of assumptions and first principles without verifying their accuracy. As a result, past communications often missed their targets and failed to achieve the desired results. Reports gathered dust, or in a few extreme cases inspired actual animosity in their target audiences. Exhibit posters dazed their viewers with acres of fine print. Rather than taking a theoretical approach, particle physics communicators recognized the need to follow an experimental mode, basing their conclusions on real data.

Benedict and Miller's 2002 survey of science-policy makers and opinion leaders provides useful information about where science-policy leaders obtain information and about the level of confidence they express in various information sources. More than half of survey respondents said that they consult the Internet for information on matters related to scientific policy. Much smaller percentages said that they relied on government agencies, newspapers, magazines, television, or radio. Nearly half reported contacting a national laboratory, suggesting that science-policy leaders find national laboratories a useful source of information. When asked about their level of confidence in various sources of information, policy leaders listed reports from the National Academy of Sciences, articles in *Science* or *Nature*, and reports from national laboratories as the top three sources of scientific policy in which they had the most confidence. Their confidence in newspaper stories, reports from federal agencies, and television news reports was significantly lower.

THERE'S NOTHING LIKE DATA

Although such survey results are instructive, to get accurate and timely data, there is no substitute for face-to-face conversations with members of key audiences. Meeting with science-policy makers and opinion leaders to ask them what information they need and want and how they would like to receive it is essential to developing strategies, tactics, and materials that meet communication goals. Just as in the case of experimental science, the results of such a data-driven approach to science communication are often surprising. For example, it is perhaps obvious—or should be obvious—that "Science first!" is a key message to communicate to particle physics policy makers and opinion leaders. Nevertheless, it took forceful and repeated delivery of the concept, "Before you tell me what you want to build, tell me the questions that you want to answer," by key policy makers before "Science first!" reached the top of the list of key messages from the particle physics community.

More unexpected was a finding that, not only for policy makers and opinion leaders but also for many science journalists, the classified advertisements ranked among the favorite departments of *FermiNews*, a long-time publication of Fermi National Accelerator Laboratory. Why? Because, these audiences said, the classifieds gave them a glimpse into the day-to-day life of a national laboratory, something very few felt they understood. "I wanted to see what the physicists were buying and selling," one reader said.

This surprising feedback that key *FermiNews* audiences liked reading the classified advertisements helped shape editorial policy for *FermiNews'* successor publication, *symmetry* magazine, the U.S. particle physics magazine of Fermilab and Stanford Linear Accelerator Center. Feature articles on subjects such as the meals devised for survival by poverty-stricken graduate students, what world-traveling physicists pack in their luggage, and how families of physicists cope with extreme travel schedules are among the most-cited of readers' favorite articles. Such stories provide *symmetry* readers with a window into the daily lives of the human beings who do particle physics. Direct feedback from readers shows that readers in all audiences value the opportunity to make such human connections.

Based in part on published survey results of policy makers and opinion leaders and in part on direct feedback from these audiences and others, particle physics communication makes heavy use of the Internet. The dedicated Web sites for *Quantum Universe* publications and *symmetry* magazine (whose archive of one-minute explanations of particle physics topics receives some 60,000 visits per month) have significantly extended the reach of the print versions of these publications. Electronic tools make it particularly easy to track the number of visits to these sites and to determine which features are of most value and interest to audiences.

Similarly, evaluating the effectiveness of communication initiatives requires data, usually obtained by seeking direct and regular feedback from key audiences. Before beginning the process of writing and producing the High Energy Physics Advisory Panel's (HEPAP) publication *Quantum Universe*, the committee charged with producing the report met with the policy makers who had requested the report to solicit their views on its content, messages, and overall approach. Midway through the process of preparing the report, the committee returned to these same stakeholders to present their progress and direction. The response showed that many members of these key audiences found significant flaws in the committee's approach. The result was a significant change in direction for the report and ultimately an extremely positive reception for *Quantum Universe*.

In another example, the Fermilab and SLAC communication team that jointly publishes *symmetry* engaged in a lengthy debate about the cover for the magazine's third issue, devoted to the World Year of Physics 2005, celebrating the centennial of Einstein's *annus mirabilis*. The proposed cover showed a pajama-clad little girl, mounting the stairs to bed, dragging behind her a stuffed Einstein teddy bear (**Figure 1**).

Was this cover too cute? the publishers worried. Would it diminish the stature of a publication attempting to establish itself as the particle physics magazine of the

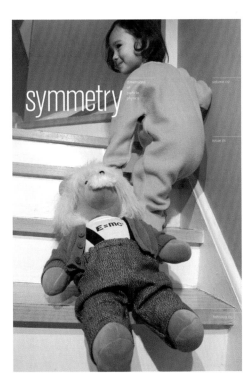

Figure 1

Cover of *symmetry* magazine, February 2005. Figure adapted with permission from *symmetry* magazine.

United States? Should the publication instead run a more conventional and traditional cover? The editors obtained what feedback they could and decided, with some trepidation, to go ahead with the teddy bear. The response from readers was immediate and unequivocal. Key audiences loved it—and not only ordered hundreds of extra copies but wondered where they could get Einstein teddy bears of their own. The teddy bear cover won over sectors of the audience who had hitherto been skeptical, and remains among the most popular covers that *symmetry* has ever published. It was an important strategic lesson for the publishers.

COLLABORATION

Besides requiring a strategic model, successful particle physics communication also depends on a high degree of collaboration. The particle physics community operates with a level of global collaboration among physicists, universities, and laboratories worldwide that is unmatched by that of any other field. For decades, such collaboration has been a fact of life for particle physics experimenters and theorists alike. More recently, even the construction of particle accelerators, traditionally the responsibility of the host nation alone, has become a shared endeavor across national and regional boundaries. Moreover, the nature of particle physics research in the twenty-first century means that every advance, every discovery made by one experiment or laboratory, rests upon the achievements of other experiments and other laboratories—and affects future discoveries at other institutions. Truly, no particle physics experiment or laboratory is an island.

Despite this long-standing tradition of global collaboration and the interwoven nature of advances in the field, particle physics communication followed no such collaborative model. Until 2001, each laboratory communicated as an independent entity, apparently oblivious to, and, in the worst cases, at cross-purpose with, other laboratories engaged in this worldwide scientific endeavor. Competitiveness, suspicion, and one-upmanship characterized the policy and practice of particle physics communication.

The first tentative movements toward a more collaborative model of communication among communication directors of the world's particle physics laboratories were propelled rapidly forward by the attacks on the World Trade Center on September 11, 2001. On September 12, Petra Folkerts, then communication director of the Deutsches Elektronen Synchrotron in Hamburg, Germany, emailed one of the authors:

> I want to say that we all are with you in these days. I myself can't find the right words to express my feeling and my mind after this terrible September 11. From my point of view NOW it's absolutely important that we HEP Outreach people [a]round the world will meet as soon as possible in the United States. Not only to figure out how to help international particle physics stay alive but also how we, in our field of activity, can set visible footprints for the significance of peaceful collaboration across all borders. What do you think about that? (If you would prefer to meet in Europe, I could organize a meeting at DESY.)

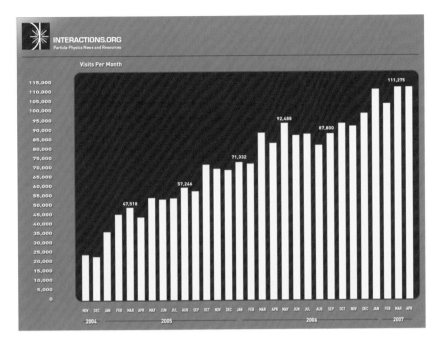

Figure 2

Interactions.org, established
as central resource for
particle physics information.

Because foreign travel to the United States was difficult in the immediate aftermath of September 11, a meeting of particle physics communicators from U.S. and European laboratories was organized at DESY in Hamburg on December 6 and 7, 2001, and the InterAction Collaboration for particle physics communication was born. The collaboration's mission statement came directly from the original message: "Not only to strengthen the international science of particle physics but to set visible footprints for peaceful collaboration across all borders."

Since 2001, the InterAction Collaboration has grown to include communication directors from 23 particle physics organizations in Europe, North America, and Asia. The collaboration jointly developed and maintains the Interactions.org Web site, with a full-time dedicated Web master, as a communication resource provided by the world's particle physics laboratories. In the three years since the site's appearance, the number of unique monthly visits to the site has reached approximately 110,000.

Some 2200 subscribers worldwide receive particle physics press releases via the Interactions News Wire (**Figure 2**). Most importantly, the InterAction Collaboration has fundamentally changed the model for particle physics communication (**Figure 3**).

From the writing and dissemination of press releases to the design of Web sites and publications, the exchange of personnel, and the near-daily consultation with colleagues, the communication of particle physics has become as thoroughly and globally collaborative as the science itself. *symmetry* magazine, the presence of a full-time U.S. communicator in the CERN Press Office, and the global communication team for the proposed International Linear Collider (ILC) offer a few high-profile examples.

A second significant aspect of collaboration is that between professional communicators and scientists. Historically, communicators may be asked to edit or illustrate

Figure 3

Meeting of the InterAction Collaboration at Stanford Linear Accelerator Center, April 2007.

reports prepared by scientists, as a final step before publication. Scientists may be consulted to check the accuracy of a communication project. However, the two groups seldom worked side by side as part of the same communication team. The *Quantum Universe* project changed that model, for the first time creating a committee comprising both physicists and communicators from the outset. That approach has become the norm in U.S. particle physics communication projects. The subsequent *Discovering the Quantum Universe* report on the science of the Large Hadron Collider (LHC) and the ILC, the *Deep Science* report on the science of a deep underground science laboratory, and the accompanying document for the *Reference Design Report* of the Global Design Effort for the ILC have all followed this model, combining the skills and perspectives of scientists and communicators to shape the content, voice, appearance, and dissemination of particle physics communication.

CASE STUDIES

Seeking Symmetry

As has been mentioned several times, the landscape of particle physics, and of particle physics communication, has changed, and although Fermilab and SLAC each had a print publication that served multiple audiences and goals, neither publication served the most important strategic aims of the laboratories. The Fermilab publication *FermiNews* was seen as too focused on local interests for a broader audience, and the SLAC publication *Beamline* was too technical for nonspecialists.

As particle physics entered a more global and collaborative phase, in a tightening budget situation in the United States, Fermilab and SLAC decided to combine resources on a publication that would better serve their needs. The key requirements of the new publication, and now its explicit goals, were to support the progress of particle physics in the United States, including the hoped-for ILC project specifically. The publication would show how Fermilab and SLAC fit into this global scenario.

Owing to the nature of policy making in the United States and what would be required to sustain particle physics research, the laboratories decided that a publication should target policy makers and policy leaders. The publication should reflect the exciting new developments in and changing nature of particle physics and present an image of the field more appealing to policy leaders than that from either a local news publication or a technical digest. Secondary audiences include the particle physics research community, the local communities around Fermilab and SLAC, and the science-interested general public.

Reaching the primary audience imposed many requirements on the publication. We had assumptions on what the target audience wanted, but one of our first actions was to visit Washington, DC, and ask the main funding agencies, the Department of Energy (DOE) and the National Science Foundation (NSF), what sort of publication would be most useful for them. We also visited the Office of Management and Budget, the Office of Science and Technology Policy, and various staffers on Capitol Hill. Although studies show that policy leaders are comfortable reading on the Web, the discussions with members of the prospective target audience convinced us that a print publication was also necessary, so that readers could carry it with them to look at in a few spare minutes and to have a physical presence among the deluge of information that policy leaders typically receive.

To reach the audience, the publication needed to work on various levels. It needed to have an immediate visual appeal, so that it would not be discarded as just another annual report or overly glossy advertising brochure. It needed to stand out from other publications competing for the reader's time, but still be sufficiently clear in purpose so that even a reader who only flicked through the publication would retain some kind of positive impression of the field of particle physics. The editorial team developed a list of adjectives that it would like people to use when describing the publication. Although each of these needs clarification, the simple list of adjectives included approachable, clean, artful/stylish, novel, credible, alive, and engaging (**Figure 4**).

The content of the magazine needed to be appropriate for the readers. The editorial committee determined that the primary audience would be well educated and intelligent, but would not necessarily have any background knowledge of particle physics. Many readers would need to read the publication for work reasons, but the magazine needed to engage readers without assuming they were required to read the publication. These conditions meant that the publication had to present stories about people and places, as well as about the science, and that the language had to

Figure 4

Cover of *symmetry*
magazine, January 2007.

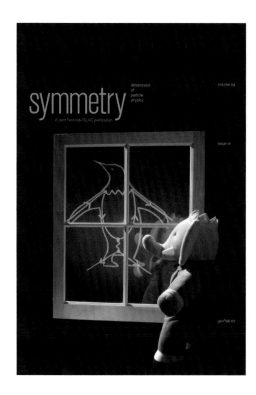

be at a level similar to popular science magazines or science sections of newspapers. The editorial team was determined that the magazine would not include unexplained jargon or look like a technical publication with many references and equations. On the contrary, the content should take a variety of forms so that there is something for everybody, in the style of any commercial magazine. With these requirements in mind, and after conducting many discussions with representative members of our audience, the editorial team determined that a magazine, named *symmetry*, appearing 10 times per year, in full color with high production values, would best suit the target audience.

Two years into publication, these choices seem to be paying off, based on many discussions with members of our target audiences. Policy makers are aware of *symmetry* magazine, speak highly of it, participate by contributing opinion pieces and story ideas, and supply no-strings-attached funding. Individual members of the U.S. Congress who we had targeted as playing major roles in the science-policy process requested that *symmetry* be sent to all its members, as they deem it worthwhile and interesting.

Circulation consists of approximately 15,000 print copies, going to our primary audience and the particle physics community. The online edition of the magazine reaches approximately another 45,000 unique readers each month. Readers on the Web are often driven by considerable buzz in the blogosphere about specific

Figure 5

"Explain it in 60 Seconds" has proved one of the most popular pages in *symmetry* magazine.

articles, particularly the "Explain it in 60 Seconds" archive, a constant favorite (**Figure 5**).

Reader feedback is consistently positive, with researchers happy with the accuracy of the content, policy makers happy with the accessibility, and nonscientific general readers particularly liking the people-oriented stories. Even so, the symmetry team continually asks for more feedback from our readers and is always tweaking its approach to best address the strategic needs of the U.S. particle physics research community.

An additional benefit of the magazine is that the effort we put into design and production provides other advantages. The physics community can reuse many images for reports, internal communications, public talks, and posters. The text is often useful for quickly generating summaries of research projects and decision-making processes for press releases and other documents. Images, stories, and reporting from the magazine have been reprinted or translated and used in other publications around the world, ranging from the *New York Times* and *Die Zeit*, to *Newton Magazine* in Italy and the pop-lifestyle magazine *Revue* in Germany (**Figure 6**).

symmetry magazine has demonstrated to us that combining resources and embedding a communications project in the global field of particle physics, rather than taking an overly parochial, laboratory-based or even region-based view, are useful strategies. These have allowed *symmetry* to reach many audiences in a way to which they seem to respond and in a way that appears to help the strategic aims of the

Figure 6

Bling Einstein. Graphic in *symmetry* subsequently published in Germany's leading weekly *Die Zeit*.

publication. The authors would like to recognize the contribution of each member of the *symmetry* team, but in particular to highlight the vision and tenacity of the magazine's editor-in-chief, David Harris.

QUANTUM UNIVERSE

The hunting call "Higgs, Higgs, Higgs, Higgs!" gives a fair representation of the worldwide justification for funding particle physics at the energy frontier for the past three decades. The Large Electron-Positron Collider (LEP), the Superconducting Super Collider, the LEPII, the Tevatron Run II, and the LHC—all huge and costly initiatives—have the same goal: find the Higgs boson. Meanwhile, enormous changes had been taking place in physics. Results from a diverse range of experiments, in space, on the ground, and underground, had produced a coherent but radically different vision of the Universe. There are many new parameters to this vision, but the blockbuster is that all visible matter makes up only 4% of the total mass of the Universe. What is the rest? Answering this question is the greatest scientific challenge of our time:

> Or like stout Cortez, when with eagle eyes
> He stared at the Pacific—and all his men
> Looked at each other with a wild surmise—
> Silent on a peak in Darien.

> John Keats: *On First Looking into Chapman's Homer*

As Cortez, so the particle physics community. There was tremendous excitement, but what to do next was to some extent governed by "wild surmise." On October 22, 2003, Raymond L. Orbach, Director of the DOE Office of Science, and Michael Turner, Assistant Director of the Mathematical and Physical Sciences division of the National Science Foundation, addressed a letter to HEPAP stating the following:

> Recent scientific discoveries at the energy frontier and in the far reaches of the Universe have redefined the scientific landscape for cosmology, astrophysics and high energy physics, and revealed new and compelling mysteries.... We are writing to ask the High Energy Physics Advisory Panel (HEPAP) to take the lead in producing a report which will illuminate the issues, and provide the funding and science-policy agencies with a clear picture of the connected, complementary experimental approaches to the truly exciting scientific questions of this century.

This letter was the genesis of the *Quantum Universe* report.

A committee was rapidly formed under Persis Drell's, Deputy Director of SLAC, leadership (**Table 1**). A breakthrough in the composition of the committee was the inclusion of two professional communicators—the authors. Great credit for the successful outcome of *Quantum Universe* must be given to Persis Drell's inspirational leadership. She immediately established an expectation of hard work and succeeded in catalyzing and focusing the excitement felt by all the members of the group. All involved will certainly agree with Persis' words in the *Stanford Report*, "It was a blast, the most fun committee I have been on."

As we described above, understanding the needs of the audience is a key, but frequently overlooked, factor in effective communication. Therefore, a trip to

Table 1   The *Quantum Universe* committee

Sam Aronson	Brookhaven National Laboratory
Jon Bagger	Johns Hopkins University
Keith Baker	Hampton University
Neil Calder	Stanford Linear Accelerator Center
Persis Drell	Stanford Linear Accelerator Center, Chair
Andreas Albrecht	University of California, Davis
Evalyn Gates	University of Chicago
Fred Gilman	Carnegie Mellon University
Judy Jackson	Fermi National Accelerator Laboratory
Steve Kahn	Stanford University/Stanford Linear Accelerator Center
Rocky Kolb	Fermi National Accelerator Laboratory
Joe Lykken	Fermi National Accelerator Laboratory
Hitoshi Murayama	University of California, Berkeley
Hamish Robertson	University of Washington
Jim Siegrist	University of California, Berkeley
Simon Swordy	University of Chicago
John Womersley	Fermi National Accelerator Laboratory

Washington, DC, was quickly organized. Most of the committee members attended meetings on November 18, 2003, with Jack Marburger, Mike Holland, and Pat Looney from the Office of Science and Technology Policy; Joel Parriott from the Office of Management and Budget; Jo Dehmer and Mike Turner from the NSF; and Ray Orbach and Robin Staffin from the DOE. This is what we heard: Lead with the science! Clarify the large array of tools that are seemingly unconnected. Articulate the questions that are driving the field. Outline the tool kit needed to answer these questions and show how the different tools are connected. Show how scientific questions map onto experimental space.

The committee took this input very seriously and constantly returned to it during the initial writing stage. They got to work with weekly teleconferences from mid-October 2003 through June 2004. There was also a face-to-face meeting at SLAC in January 2004. Three main themes were quickly established: What is the nature of the Universe and what is it made of? What are matter, energy, space, and time? How did we get here and where are we going? These themes were then refined into three chapters with nine interrelated questions:

1. Einstein's dream of unified forces: Are there undiscovered principles of nature: new symmetries, new physical laws? How can we solve the mystery of dark energy? Are there extra dimensions of space? Do all the forces become one?

2. The particle world: Why are there so many kinds of particles? What is dark matter? How can we make it in the laboratory? What are neutrinos telling us?

3. The birth of the Universe: How did the Universe come to be? What happened to the antimatter?

These questions guide a coherent campaign on a variety of technical fronts, and one of the aims of *Quantum Universe* is to make it clear that traditional particle physics will advance in symbiosis with particle astrophysics. Determining the fundamental nature of the new Universe requires a diverse tool kit consisting of accelerator-based experiments, underground laboratories, space probes, and ground-based telescopes. The unknowns of the Universe would undergo a two-pronged attack using astrophysical observations to determine the parameters of the Universe and accelerator experiments to search for a quantum explanation.

Writing *Quantum Universe* presented challenges on many levels: to agree on the scientific content but then, perhaps the greater challenge, to communicate the science with precision, accuracy, and clarity. Throughout the writing process, the rallying cry was, "Remember the audience." Having communicators embedded in the process certainly helped, as we were able to remind our colleagues of this when the discussion became too esoteric. We were also determined that *Quantum Universe* should look and feel different from any previous report. The science was so extraordinary and we wanted the printed report to generate the same kind of astonishment. Therefore, a great deal of thought and care went into the design. It had to be different and engaging but avoid accusations of being glitzy.

The biggest break from tradition was the cover (**Figure 7**). We chose to mimic the almost Dickensian design of a laboratory log book, as this generated an immediate tension between twenty-first-century physics in a nineteenth-century cover. The

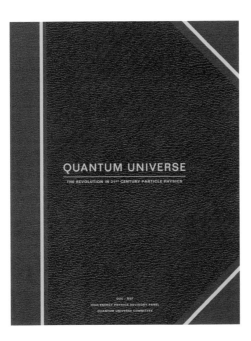

Figure 7

Cover of *Quantum Universe*.

choice also allowed great creativity inside the report. It became a physicist's personal log book, complete with coffee cup stains, glued-in Polaroids, and playful details such as a fake sticky note on the back cover. The approach to the layout was an integral part of the concentration on "Remember the audience." We were determined that the final layout and style of the document would enhance the science. We wanted people to read the whole document and really appreciate the science, so the overall rhythm of the document was broken up with graphics, sidebars, and cartoons.

The reaction to *Quantum Universe* when it was published in June 2004 was overwhelmingly positive. The greatest compliments came precisely from the audiences we were most eager to impress. Senator Ted Stevens, then head of the Senate Committee on Appropriations, read the report from cover to cover and asked that it be distributed to every member of the committee. The accessibility of *Quantum Universe* allowed our well wishers in Washington, DC, to promote the field, as they had at their disposal a scientific report that, perhaps for the first time, intrigued rather than intimidated.

Quantum Universe describes a revolution in particle physics, but it was also a revolution in particle physics communication. The strategy that was adopted for *Quantum Universe* has become the template for communication initiatives. Here are the lessons learned:

- Embed communicators with the scientists.
- Find out what your audience needs.
- Never forget your audience during the preparation of the report.
- Be daring with graphics and design.

- Be international; a particle physics report can no longer talk only about U.S. efforts.
- Communicate enthusiasm.

Quantum Universe and its dedicated Web site **www.quantumuniversereport.org** have brought a major change in particle physics communication. The science is wonderful, but that in itself is not enough. The people who support and fund the science have to be convinced. *Quantum Universe* is the product of serious strategic thinking on the optimum manner of reaching this key audience.

DISCOVERING THE QUANTUM UNIVERSE

The success of the *Quantum Universe* report rapidly generated another request from the DOE and the NSF to HEPAP.

> Dear Professor Gilman:
>
> We wish to congratulate you on the widely successful *Quantum Universe* report that with clarity and elegance expresses the great discovery opportunities in particle physics today. It has made a positive impact in Washington, DC, in the Nation, and abroad in conveying the drivers of the coming scientific revolution. As funding agencies and advisors of the Nation's research portfolio in this field, our ability to bring clarity and focus to outstanding scientific issues is an important responsibility. You have succeeded well with *Quantum Universe*.
>
> To educate us and the public, and to clarify the matter more generally, we would like HEPAP to form a committee to write a document that addresses the following:
>
> - In the context of already known physics, i.e., our current understanding of the electroweak symmetry breaking sector, what are the synergies and complementarities of these two machines? How would an LC be utilized in understanding a Standard Model Higgs, or whatever fulfills its role in the electroweak interaction?
> - In the context of physics discoveries beyond the Standard Model (supersymmetry, extra dimensions or other new physics) that are assumed to be made at the Tevatron or early at the LHC, what would be the role of a TeV Linear Collider in making additional and unique contributions to these discoveries, in distinguishing between models, and in establishing connections to cosmological observations?

Quantum Universe had established the revolution in particle physics and the logical next step was to write a follow-up that outlined with similar clarity the role of the future colliders, the LHC and the ILC, in discovering the quantum Universe. *Discovering the Quantum Universe* was quickly decided upon as a suitable title. Throughout the conception and writing of *Discovering the Quantum Universe*, the committee adhered strictly to the lessons learned during the process of writing the *Quantum Universe*.

HEPAP quickly appointed a committee to prepare the report. This committee was chaired jointly by Jim Siegrist from Lawrence Berkeley National Laboratory and Joe Lykken from Fermilab, and included nine members of the original *Quantum Universe*

Table 2   The *Discovering the Quantum Universe* committee

Jim Siegrist	Lawrence Berkeley National Laboratory, Co-Chair
Joe Lykken	Fermi National Accelerator Laboratory, Co-Chair
Jonathan Bagger	Johns Hopkins University
Barry Barish	California Institute of Technology
Neil Calder	Stanford Linear Accelerator Center
Albert de Roeck	CERN
Jonathan L. Feng	University of California, Irvine
Fred Gilman	Carnegie Mellon University
JoAnne Hewett	Stanford Linear Accelerator Center
John Huth	Harvard University
Judy Jackson	Fermi National Accelerator Laboratory
Young-Kee Kim	University of Chicago
Rocky Kolb	Fermi National Accelerator Laboratory
Konstantin Matchev	University of Florida
Hitoshi Murayama	Lawrence Berkeley National Laboratory
Rainer Weiss	Massachusetts Institute of Technology

committee. This overlap of experience proved invaluable in focusing on the areas of the report that were to prove most difficult and in convincing the new members that adopting an adventurous graphic design should not be feared. The committee members are listed in **Table 2**.

The job of writing *Discovering the Quantum Universe* was soon complicated by the realization that in fact two different reports were needed. Concurrently with the preparation of *Discovering the Quantum Universe*, a National Research Council committee was preparing an extremely important report for the future of particle physics in the United States. The report, *Revealing the Hidden Nature of Space and Time: Charting the Course for Elementary Particle Physics*, known as EPP2010, was published in April 2006 and was charged with laying out a 15-year plan for the field that was realistic in the current funding climate and in line with the big physics questions of the day. Providing clear and convincing input to the EPP2010 panel became one of the principle goals of the *Discovering the Quantum Universe* committee. Our guidance from EPP2010 is that they were looking for a white paper on the physics related to the LHC and ILC. This fulfilled a significant part but not all of our charge from HEPAP. Thus, we decided to produce two reports: the first for EPP2010 and a later document for a wider audience. The feedback from the EPP2010 committee was that we had delivered what they were hoping for and had answered all of their questions. A very important achievement of *Discovering the Quantum Universe* was the invaluable input that was given to the EPP2010 panel. Their final report was extremely positive about the future of particle physics and was taken very seriously by government policy makers. This was precisely the audience we were trying to influence with *Discovering the Quantum Universe*, and our ability to feed EPP2010 with well-written and well-documented information may in the long run be the most influential action of the committee.

Discovering the Quantum Universe was a more politically sensitive communication challenge than its older sister *Quantum Universe*, as it inevitably involved comparison and contrast of the two colliders, the LHC and the ILC. Any clumsiness could have catalyzed rifts and dissent between the two communities. We were therefore extremely careful to get as much input as possible from both groups during the preparation of the report. For example, Albert de Roeck, a CERN employee, was a very active member of the committee, and drafts of the report were circulated to leaders of both the LHC and ILC communities.

A primary communication goal of the report was to dispel the misconception that if the LHC discovers more and measures more, then there is less motivation for the ILC. Our report made it clear that the opposite is true. There was also a misconception that once the LHC discovers a Higgs particle, the rest is details. Our report made it clear that the discovery of a Higgs particle would raise urgent questions, leading to even greater discoveries. It is also a misconception that the only thing colliders do is discover particles. *Discovering the Quantum Universe* explained how particles are the tools that physicists use to resolve mysteries and to discover new laws of nature.

A lot of time was spent on condensing the overall message of the report in one opening paragraph. The idea was to convey the excitement and message of the whole report on the first spread (**Figure 8**):

Right now is a time of radical change in particle physics. Recent experimental evidence demands a revolutionary new vision of the universe. Discoveries are at hand that will stretch the imagination with new forms of matter, new forces of nature, new dimensions of space and time. Breakthroughs will come from the next generation of particle accelerators—the Large Hadron Collider, now under construction in Europe, and the proposed International Linear Collider. Experiments at these accelerators will revolutionize your concept of the universe.

Figure 8

Opening spread of *Discovering the Quantum Universe*.

As with *Quantum Universe*, we put the physics first, basing the report on the nine great questions from *Quantum Universe*. These were mapped into the three basic physics themes most relevant for the LHC and the ILC:

1. Mysteries of the Terascale: The LHC should discover the Higgs. It should also discover supersymmetry or some other new principle that explains the Higgs' existence. The linear collider would resolve the hidden messages of the Higgs, the superpartners, and their Terascale relatives.
2. Light on dark matter: Most theories of Terascale physics contain new massive particles with the right properties to contribute to dark matter. Such particles would first be produced at the LHC. Experiments at the LHC would establish whether they are actually dark matter.
3. Einstein's telescope: Most theories of Terascale physics contain new massive particles with the right properties to contribute to dark matter. Such particles would first be produced at the LHC. Experiments at the linear collider would establish whether they are actually dark matter.

An interesting spin off from *Discovering the Quantum Universe* is the word Terascale, which we coined to describe the TeV energy regime. The Terascale has since become the standard terminology for the discovery range of the two colliders. The three physics themes mentioned above were then broken down into nine discovery scenarios: the Higgs is different, a shortage of antimatter, mapping the dark Universe, exploring extra dimensions, dark matter in the laboratory, supersymmetry, matter unification, unknown forces, and concerto for strings.

The very successful laboratory log book design of *Quantum Universe* was echoed in the new report, as we felt very strongly that the two volumes should be seen as two

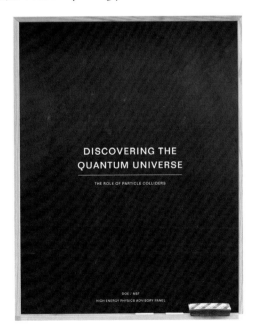

Figure 9

Cover of *Discovering the Quantum Universe*.

parts of the same communications exercise. The graphic theme used in *Discovering the Quantum Universe* was the blackboard and corkboard, both everyday tools for physicists (**Figure 9**). The accompanying Web site, **http://www.interactions.org/ quantumuniverse/qu2006/index.html**, also followed the same design concept.

Concentration on the needs of the audience, high-quality writing, innovative graphic design, and pervasive enthusiasm are the hallmarks of the *Quantum Universe* reports. They have set a completely new standard for high-energy physics communication that all subsequent communications initiatives will have to match.

CONCLUSION

The traditional model of communication used by the major physics laboratories in the United States and in fact around the world was inadequate. The time had come to think differently. A more professional attitude has been adopted that concentrates on a clear definition of the goals of the communication, the needs of the audience, and the most efficient tools to be used. Concurrently, communicators in the laboratories have been encouraged to work together to support the field as a whole, in contrast to the previous model whereby each institute or laboratory communicated independently, frequently to the detriment of other physics centers. Great progress has been made, but real proof of the effectiveness of this new approach can only be given by raising funding for particle physics in the coming years.

DISCLOSURE STATEMENT

The authors are not aware of any biases that might be perceived as affecting the objectivity of this review.

LITERATURE CITED

1. Miller D. *The conceptualization and measurement of policy leadership*. Presented at Annu. Meet. Am. Assoc. Adv. Sci., Denver (2003)
2. Miller D. *Communication strategies for meeting the information needs of science and energy policy leaders*. Presented at Annu. Meet. Am. Assoc. Adv. Sci., Denver (2003)

Primordial Nucleosynthesis in the Precision Cosmology Era

Gary Steigman

Departments of Physics and Astronomy, The Ohio State University, Columbus, Ohio 43210; email: steigman.1@osu.edu

Annu. Rev. Nucl. Part. Sci. 2007. 57:463–91

The *Annual Review of Nuclear and Particle Science* is online at http://nucl.annualreviews.org

This article's doi: 10.1146/annurev.nucl.56.080805.140437

Copyright © 2007 by Annual Reviews. All rights reserved

0163-8998/07/1123-0463$20.00

Key Words

Big Bang Nucleosynthesis, baryon density, early-Universe expansion rate, neutrino asymmetry

Abstract

Primordial nucleosynthesis probes the Universe during its early evolution. Given the progress in exploring the constituents, structure, and recent evolution of the Universe, it is timely to review the status of Big Bang Nucleosynthesis (BBN) and confront its predictions, and the constraints that emerge from them, with those derived from independent observations of the Universe at much later epochs. Following an overview of the key physics that controls the synthesis of the elements in the early Universe, the predictions of BBN in the standard (and some nonstandard) models of cosmology and particle physics are presented. The observational data used to infer the primordial abundances are described, with an emphasis on the distinction between *precision* and *accuracy*. These are compared with the predictions, testing the internal consistency of BBN and enabling a comparison of the BBN-inferred constraints with those derived from the cosmic microwave background radiation and large scale structure data.

Contents

1. INTRODUCTION 464
2. A SYNOPSIS OF BIG BANG NUCLEOSYNTHESIS 465
    2.1. Big Bang Nucleosynthesis Chronology 468
    2.2. Standard Big Bang Nucleosynthesis–Predicted Abundances 470
    2.3. Nonstandard Big Bang Nucleosynthesis: $S \neq 1$ 472
    2.4. Nonstandard Big Bang Nucleosynthesis: $\xi_e \neq 0$ 474
    2.5. Nonstandard Big Bang Nucleosynthesis: $S \neq 1$ and $\xi_e \neq 0$ 475
3. THE RELIC NUCLIDES OBSERVED 475
    3.1. Deuterium: The Baryometer of Choice 477
    3.2. Helium-3 478
    3.3. Helium-4 480
    3.4. Lithium-7 484
    3.5. Adopted Primordial Abundances 486
4. CONFRONTATION OF THEORY WITH DATA 486
    4.1. Standard Big Bang Nucleosynthesis 487
    4.2. Nonstandard Big Bang Nucleosynthesis 488
5. SUMMARY 489

1. INTRODUCTION

As the Universe evolved from its early, hot, dense beginnings (the Big Bang) to its present, cold, dilute state, it passed through a brief epoch when the temperature (average thermal energy) and density of its nucleon component were such that nuclear reactions were effective in building complex nuclei. Because the nucleon content of the Universe is small (in a sense to be described below) and because the Universe evolved through this epoch very rapidly, only the lightest nuclides (D, ^{3}He, ^{4}He, and ^{7}Li) could be synthesized in astrophysically interesting abundances. The abundances of these relic nuclides provide probes of the conditions and contents of the Universe at a very early epoch in its evolution (the first few minutes) that would otherwise be hidden from our view. The standard model of cosmology subsumes the standard model of particle physics (specifically, three families of very light left-handed neutrinos, and their right-handed antineutrinos) and uses general relativity (e.g., the Friedman equation) to track the time evolution of the universal expansion rate and its matter and radiation contents. Big Bang Nucleosynthesis (BBN) begins in earnest when the Universe is a few minutes old and ends less than half an hour later when the nuclear reactions are quenched by the low temperatures and densities. The BBN-predicted abundances depend on the conditions at those times (e.g., temperature, nucleon density, expansion rate, neutrino content, neutrino-antineutrino asymmetry, etc.) and are largely independent of the detailed processes that established them. Consequently, BBN can test the standard models of cosmology and particle physics and constrain their parameters, as well as probe nonstandard physics or cosmology that may alter the conditions at BBN.

This review begins with a synopsis of the physics relevant to a description of the early evolution of the Universe at the epoch of primordial nucleosynthesis, and with an outline of the specific nuclear and weak-interaction physics of importance for BBN—in the standard model as well as in the context of some very general extensions of the standard models of cosmology and/or particle physics. Having established the framework, we then present our predictions of the relic abundances of the nuclides synthesized during BBN. Next, the current status of the observationally inferred relic abundances is reviewed, with an emphasis on the uncertainties associated with these determinations. The predicted and observed abundances are then compared to test the internal consistency of the standard model and to constrain extensions beyond the standard model. Observations of the cosmic microwave background radiation (CMB) temperature fluctuations and of the large scale structure (LSS) provide probes of the physics associated with the later evolution of the Universe, complementary to that provided by BBN. The parameter estimates/constraints from BBN and the CMB are compared, again testing the consistency of the standard model and probing or constraining some classes of nonstandard models.

2. A SYNOPSIS OF BIG BANG NUCLEOSYNTHESIS

The Universe is expanding and filled with radiation. All wavelengths, those of the CMB photons as well as the de Broglie wavelengths of all freely expanding massive particles, are stretched along with this expansion. Therefore, although the present Universe is cool and dilute, during its early evolution the Universe was hot and dense. The combination of high temperature (average energy) and density ensures that collision rates are very high during early epochs, guaranteeing that all particles, with the possible exception of those that have only gravitational strength interactions, were in kinetic and thermal equilibrium at sufficiently early times. As the Universe expands and cools, interaction rates decline and, depending on the strength of their interactions, different particles depart from equilibrium at different epochs. For the standard, active neutrinos (ν_e, ν_μ, ν_τ), this departure from equilibrium occurs when the Universe is only a few tenths of a second old and the temperature of the photons, $e^\pm$ pairs, and neutrinos (the only relativistic, standard model particles populated at that time) is a few MeV. Note that departure from equilibrium is not sharp and collisions among neutrinos and other particles continue to occur. When the temperature drops below $T \lesssim 2$–3 MeV, the interaction rates of neutrinos with CMB photons and $e^\pm$ pairs become slower than the universal expansion rate (as measured by the Hubble parameter H), and the neutrinos effectively decouple from the CMB photons and $e^\pm$ pairs. However, the electron neutrinos (and antineutrinos) continue to interact with the baryons (nucleons) via the charged-current weak interactions until the Universe is a few seconds old and the temperature has dropped below ~0.8 MeV. This decoupling, too, is not abrupt; the neutrinos do not "freeze out" (see **Figure 1**).

Indeed, two-body reactions among neutrons, protons, $e^\pm$, and ν_e ($\bar{\nu}_e$) continue to influence the ratio of neutrons to protons, albeit not rapidly enough to allow the neutron-to-proton (n/p) ratio to track its equilibrium value of n/p $= \exp(-\Delta m/T)$, where the neutron-proton mass difference is $\Delta m = m_n - m_p = 1.29\,\text{MeV}$. As a result,

Figure 1

The time-temperature evolution of the neutron-to-proton (n/p) ratio. The solid red curve indicates the true variation. The steep decline at a few hundred seconds is the result of the onset of BBN. The dashed blue curve indicates the equilibrium n/p ratio $\exp(-\Delta m/T)$, and the dotted gray curve indicates free-neutron decay $\exp(-t/\tau_n)$.

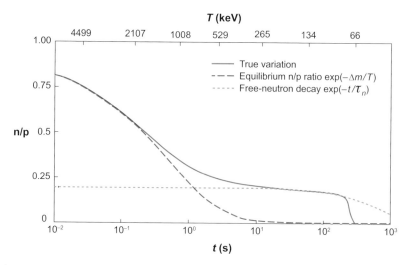

the n/p ratio continues to decrease from $\sim 1/6$ at freeze out to $\sim 1/7$ when BBN begins at ~ 200 s ($T \approx 80$ keV). Because the neutrinos are extremely relativistic during these epochs, they may influence BBN in several ways. The universal expansion rate (the Hubble parameter H) is determined through the Friedman equation by the total energy density, which, during these early epochs, is dominated by massless particles in addition to those massive particles that are extremely relativistic at these epochs: CMB photons, $e^\pm$ pairs, and neutrinos. The early Universe is radiation dominated and the neutrinos constitute a significant component of the radiation. In addition, through their charged-current weak interactions, the electron-type neutrinos help control the fraction of free neutrons available, which, as we see below, effectively limits the primordial abundance of ^{4}He.

Although the $e^\pm$ pairs annihilate during the first few seconds when $T \lesssim m_e$, the surviving electrons (the excess of electrons over positrons), equal in number to the protons to ensure charge neutrality, remain coupled to the CMB photons via Compton scattering. Only after the electrons and the nuclides (mainly protons and alphas) combine to form neutral atoms at "recombination" are the CMB photons released from the grasp of the electrons to propagate freely. This occurs when the Universe is some 400,000 years old, and the relic photons—redshifted to the currently observed black body radiation at $T = 2.725$ K—provide a snapshot of the Universe at this early epoch. At this relatively late stage (in contrast to BBN) in the early evolution of the Universe, the freely propagating, relativistic neutrinos contribute to the radiation density, influencing the magnitude of the universal expansion rate (e.g., the time-temperature relation). Note that if the neutrino masses are sufficiently large, the neutrinos will have become nonrelativistic prior to recombination, and their free streaming would have the potential to damp density fluctuations in the baryon fluid. The important topic of neutrino masses is not addressed here (for recent reviews see Reference 1); for our analysis, it is sufficient to assume that $m_\nu \lesssim$ few eV.

The primordial abundances of the relic nuclei produced during BBN depend on the baryon (nucleon) density and on the early-Universe expansion rate. The amplitude and angular distribution of the CMB temperature fluctuations depend on these same parameters (as well as on several others). The universal abundance of baryons may be quantified by comparing the number density of baryons (nucleons) to the number density of CMB photons,

$$\eta_{10} \equiv 10^{10}(n_B/n_\gamma).\qquad\qquad 1.$$

As the Universe expands, the densities of baryons and photons both decrease, whereas, according to the standard model, after $e^\pm$ annihilation, the numbers of baryons and CMB photons in a comoving volume are unchanged. As a result, in the standard model, η_{10} measured at present, at recombination, and at BBN are the same. Any departure would be a sign of new physics/cosmology beyond the standard models. This is one of the key cosmological tests. Because the baryon mass density ($\rho_B \equiv \Omega_B \rho_c$, where $\rho_c = 3 H_0^2/8\pi G_N$ is the present critical mass density, H_0 is the present value of the Hubble parameter, and G_N is the Newton constant) plays a direct role in the growth of perturbations, it is equally convenient to quantify the baryon abundance using a combination of Ω_B and h, the present value of the Hubble parameter measured in units of 100 km s^{-1} Mpc^{-1},

$$\eta_{10} = 274\, \omega_B \equiv 274\, \Omega_B h^2.\qquad\qquad 2.$$

Until very recently, the comparison between the observationally inferred and the BBN-predicted primordial abundances, especially the relic abundance of D, provided the most accurate determination of η_{10}. However, this has now changed as a result of the very high-quality data from a variety of CMB experiments (2), complemented by observations of LSS (3). At present, the best non-BBN value (see References 2 and 3) is $\Omega_B h^2 = 0.0223 \pm 0.0007$, corresponding to $\eta_{10} = 6.11 \pm 0.20$. This value is used below to predict the relic abundances for comparison with the observational data (and vice versa).

The Hubble parameter, $H = H(t)$, measures the expansion rate of the Universe. Deviations from the standard model ($H \to H'$) may be parameterized by an expansion rate parameter, $S \equiv H'/H$. In the standard model, during the early radiation-dominated evolution, H is determined by the energy density in relativistic particles so that deviations from the standard cosmology ($S \neq 1$) may be quantified equally well by the "equivalent number of neutrinos," $\Delta N_\nu \equiv N_\nu - 3$:

$$\rho_R \to \rho_R' \equiv \rho_R + \Delta N_\nu \rho_\nu.\qquad\qquad 3.$$

Prior to $e^\pm$ annihilation, these two parameters are related by

$$S = (1 + 7\Delta N_\nu/43)^{1/2}.\qquad\qquad 4.$$

ΔN_ν and S are equivalent ways to quantify any deviation from the standard model expansion rate; ΔN_ν is not necessarily related to extra (or fewer) neutrinos. For example, if the value of the gravitational constant, G_N, was different in the early Universe from its value at present, $S = (G_N'/G_N)^{1/2}$, and $\Delta G_N/G_N = 7\Delta N_\nu/43$ (4, 5). Changes (from the standard model) in the expansion rate at BBN will modify the neutron abundance and the time available for nuclear production/destruction, changing the BBN-predicted relic abundances.

Although most models of particle physics beyond the standard model adopt (or impose) a lepton asymmetry of the same order of magnitude as the (very small) baryon asymmetry ($\frac{n_{\rm B} - n_{\bar{\rm B}}}{n_\gamma} = 10^{-10}\eta_{10}$), lepton (neutrino) asymmetries that are orders of magnitude larger are currently not excluded by any experimental data. In analogy with η_{10}, which provides a measure of the baryon asymmetry, the lepton (neutrino) asymmetry, $L = L_\nu \equiv \Sigma_\alpha L_{\nu_\alpha}$ ($\alpha \equiv e, \mu, \tau$), may be quantified by the ratios of the neutral lepton chemical potentials to the temperature (in energy units) $\xi_{\nu_\alpha} \equiv \mu_{\nu_\alpha}/kT$, where

$$L_{\nu_\alpha} \equiv \left(\frac{n_{\nu_\alpha} - n_{\bar{\nu}_\alpha}}{n_\gamma} \right) = \frac{\pi^2}{12\zeta(3)} \left(\frac{T_{\nu_\alpha}}{T_\gamma} \right)^3 \xi_{\nu_\alpha} \left(1 + \left(\frac{\xi_{\nu_\alpha}}{\pi} \right)^2 \right). \qquad 5.$$

Although any neutrino degeneracy always increases the energy density in the neutrinos, resulting in an effective $\Delta N_\nu > 0$, the range of $|\xi_{\nu_\alpha}|$ of interest to BBN is limited to sufficiently small values that the increase in S arising from a nonzero ξ_{ν_α} is negligible and $T_{\nu\alpha} = T_\gamma$ prior to $e^\pm$ annihilation (6). However, a small asymmetry between electron-type neutrinos and antineutrinos ($|\xi_e| \sim 10^{-2}$), although very large compared to the baryon asymmetry, can have a significant impact on BBN by modifying the pre-BBN n/p ratio. Because any charged lepton asymmetry is limited by the baryon asymmetry to be very small, any non-negligible lepton asymmetry must be among the neutral lepton, the neutrinos. For this reason, in the following, the subscript ν will be dropped ($\xi_{\nu_\alpha} \equiv \xi_\alpha$).

2.1. Big Bang Nucleosynthesis Chronology

When the Universe is only a fraction of a second old, at temperatures above several MeV, thermodynamic equilibrium has already been established among the key BBN players: neutrinos, $e^\pm$ pairs, photons, and nucleons. Consequently, without loss of generality, discussion of BBN can begin at this epoch. Early in this epoch, the charged-current weak interactions proceed sufficiently rapidly to keep the n/p ratio close to its equilibrium value $(n/p)_{eq} = e^{-\Delta m/T}$. Note that if there is an asymmetry between the numbers of ν_e and $\bar{\nu}_e$, the equilibrium n/p ratio is modified to $(n/p)_{eq} = \exp(-\Delta m/T - \mu_e/T) = e^{-\xi_e}(n/p)^0_{eq}$. As the Universe continues to expand and cool, the lighter protons are favored over the heavier neutrons and the n/p ratio decreases, initially tracking the equilibrium expression. However, as the temperature decreases below $T \sim 0.8$ MeV, when the Universe is $\sim$1 s old, the weak interactions have become too slow to maintain equilibrium, and the n/p ratio, while continuing to fall, deviates from (exceeds) the equilibrium value (see **Figure 1**). Because the n/p ratio depends on the competition between the weak-interaction rates and the early-Universe expansion rate (as well as on a possible neutrino asymmetry), deviations from the standard model (e.g., $S \neq 1$ and/or $\xi_e \neq 0$) will change the relative numbers of neutrons and protons available for building the complex nuclides.

Simultaneously, nuclear reactions among the neutrons and protons (e.g., $n + p \leftrightarrow D + \gamma$) are proceeding very rapidly ($\Gamma_{\rm nuc} \gg H$). However, any D synthesized by this reaction finds itself bathed in a background of energetic γ rays (the CMB blueshifted to average photon energies $E_\gamma \sim 3T_\gamma \gtrsim$ few MeV), and before another neutron or proton can be captured by the deuteron to begin building more complex nuclei, the

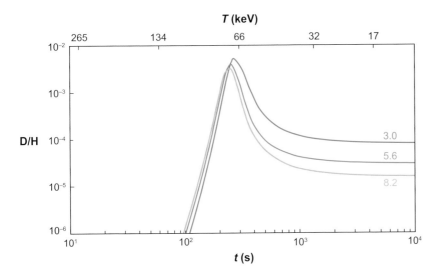

Figure 2

The time-temperature
evolution of the deuteron
abundance (the D/H ratio).
The curves are labeled by
the corresponding baryon
abundances η_{10}.

deuteron is photodissociated and BBN remains stillborn. This bottleneck to BBN
persists until the temperature drops even further below the deuteron binding energy,
when there are too few sufficiently energetic photons to photodissociate the deuteron
before it captures additional nucleons, launching BBN in earnest. This transition
(smooth, but rapid) occurs after $e^{\pm}$ annihilation, when the Universe is now a few
minutes old and the temperature has dropped below $\sim$80 keV. In **Figure 2**, the time
evolution of the ratio of deuterons to protons (D/H) is shown as a function of time;
note the very rapid rise of the deuteron abundance. Once BBN begins, neutrons and
protons quickly combine to build D, ^{3}H, ^{3}He, and ^{4}He. The time evolution of the
^{4}He (α-particle) abundance is shown in **Figure 3**; note that the rapid rise in ^{4}He
follows that in D. Because ^{4}He is the most tightly bound of the light nuclides and

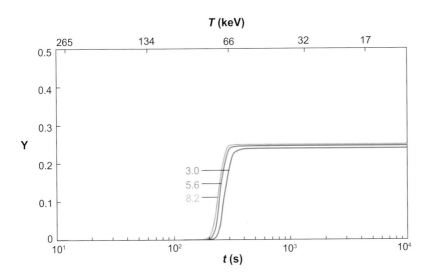

Figure 3

The time-temperature
evolution of the ^{4}He
abundance $Y \equiv 4y/(1 + 4y)$,
where $y \equiv n_{He}/n_H$. The
curves are labeled by the
corresponding baryon
abundances η_{10}.

Figure 4

The time-temperature
evolution of the mass-7
abundances. The solid
curves are for direct
production of ^{7}Be, and the
dashed curves are for direct
production of ^{7}Li. The
curves are labeled by the
corresponding baryon
abundances, η_{10}.

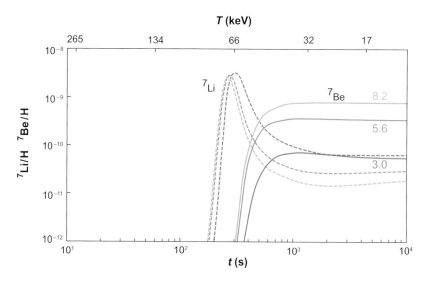

because there are no stable mass-5 nuclides, a new bottleneck appears at ^{4}He. As a
result, the nuclear reactions quickly incorporate all available neutrons into ^{4}He, and
the ^{4}He relic abundance is limited by the availability of neutrons at BBN. As may be
seen in **Figure 3**, the ^{4}He abundance is very insensitive to the baryon abundance.

To jump the gap at mass-5 requires Coulomb-suppressed reactions of ^{4}He with
D, ^{3}H, or ^{3}He, guaranteeing that the abundances of the heavier nuclides are severely
depressed below that of ^{4}He (and even that of D and ^{3}He). The few reactions that
do manage to bridge the mass-5 gap lead mainly to mass-7, producing ^{7}Li and ^{7}Be.
Later, when the Universe has cooled further, ^{7}Be captures an electron and decays to
^{7}Li. As seen in **Figure 4**, direct production of ^{7}Li by ^{3}H(α, γ) ^{7}Li reactions dominates
at low baryon abundance ($\eta_{10} \lesssim 3$), whereas direct production of ^{7}Be via ^{3}He(α, γ)
^{7}Be reactions dominates at higher baryon abundance ($\eta_{10} \gtrsim 3$). For the range of η_{10} of
interest, the BBN-predicted abundance of ^{6}Li is more than three orders of magnitude
below that of the more tightly bound ^{7}Li. Finally, there is another gap at mass-
8, ensuring that during BBN no heavier nuclides are produced in astrophysically
interesting abundances.

The primordial nuclear reactor is short lived. As seen from **Figures 2–4**, as the
temperature drops below $T \lesssim 30$ keV, when the Universe is $\sim$20 min old, Coulomb
barriers and the absence of free neutrons (almost all those present when BBN began
have been incorporated into ^{4}He) abruptly suppress all nuclear reactions. Afterward,
until the first stars form, no relic primordial nuclides are destroyed (except that ^{3}H
and ^{7}Be are unstable and decay to ^{3}He and ^{7}Li, respectively) and no new nuclides are
created. In $\sim$1000 s, BBN has run its course.

2.2. Standard Big Bang Nucleosynthesis–Predicted Abundances

For standard Big Bang Nucleosynthesis (SBBN), where $S = 1$ ($N_\nu = 3$) and $\xi_e = 0$,
the BBN-predicted primordial abundances depend on only one free parameter, the

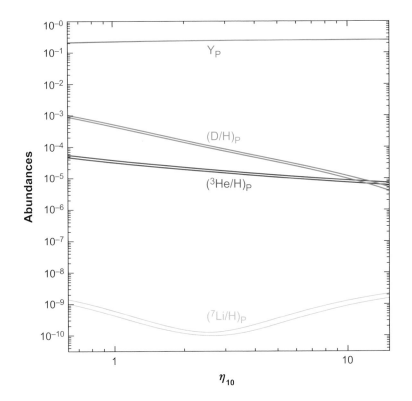

The SBBN-predicted primordial abundances of D, ^{3}He, and ^{7}Li (relative to hydrogen by number), and the ^{4}He mass fraction (Y_P), as functions of the baryon abundance parameter η_{10}. The widths of the bands (including the band for Y_P!) reflect the uncertainties in the nuclear and weak-interaction rates.

Figure 5

baryon abundance η_{10}. In **Figure 5**, the SBBN-predicted relic abundances of D, ^{3}He, ^{4}He, and ^{7}Li as a function of η_{10} are shown. The trends revealed in **Figure 5** are easily understood on the basis of the preceding discussion. For example, D, ^{3}H, and ^{3}He are burned to ^{4}He, and the higher the baryon abundance, the faster are D and ^{3}He burned away and the smaller are their surviving abundances. Because the ^{4}He abundance is limited by the abundance of neutrons, the primordial ^{4}He mass fraction is very insensitive to η_{10}, $Y_P \equiv 4y/(1 + 4y) \approx \frac{2(n/p)_{BBN}}{1+(n/p)_{BBN}} \approx 1/4$ (see **Figure 3**), where $y \equiv n_{He}/n_H$. Of course, defined this way, Y_P is not really the mass fraction because this expression adopts precisely 4 for the ^{4}He-to-H mass ratio. However, the reader should be warned that Y_P defined this way is conventionally referred to by cosmologists as the ^{4}He mass fraction. The residual dependence of Y_P on η_{10} results from the fact that the higher the baryon abundance, the earlier the D bottleneck is breached—at a higher temperature, where the n/p ratio is slightly higher. As a result, Y_P increases, but only logarithmically, with η_{10}. The valley shape of the ^{7}Li abundance curve is a reflection of the two paths to mass-7 (see **Figure 4**). At low baryon abundance, the directly produced ^{7}Li is burned away as the baryon abundance increases, whereas at higher baryon abundance, ^{7}Be is synthesized more rapidly as the baryon abundance increases in the range of interest. Eventually, at much higher η_{10}, the ^{7}Be will also be burned away as the baryon abundance increases.

Over the years, many have written computer codes to integrate the coupled set of differential equations that track element production/destruction to solve for the BBN-predicted abundances of the light nuclides. Once the cosmology is defined, the time-temperature-density relations are known and the only uncertain inputs are the nuclear and weak-interaction cross sections and rates. It is not surprising, indeed it is required, that with the same input the codes should predict the same abundances as a function of the one free parameter, the baryon abundance η_{10}. Now that observations of the CMB temperature fluctuations and of the distribution of LSS have become sufficiently precise, the range of η_{10} of interest is considerably restricted (3): with $\sim 95\%$ confidence, $5.7 \lesssim \eta_{10} \lesssim 6.5$. Within this limited range of η_{10}, there is no need to have access to a sophisticated, state-of-the-art BBN code, as the following simple fits are quite accurate (to within the nuclear and weak rate uncertainties) and are consistent with the published results of many independent BBN codes (7):

$$y_D \equiv 10^5 (D/H) = 2.68(1 \pm 0.03)(6/\eta_{10})^{1.6}, \qquad 6.$$

$$y_3 \equiv 10^5 ({}^3He/H) = 1.06(1 \pm 0.03)(6/\eta_{10})^{0.6}, \qquad 7.$$

$$Y_P = 0.2483 \pm 0.0005 + 0.0016(\eta_{10} - 6), \qquad 8.$$

$$y_{Li} \equiv 10^{10}({}^7Li/H) = 4.30(1 \pm 0.1)(\eta_{10}/6)^2. \qquad 9.$$

Note that the lithium abundance is often expressed logarithmically as $[Li] \equiv 12 + \log(Li/H) = 2 + \log y_{Li}$. The expression for Y_P corresponds to the currently accepted value of the neutron lifetime, $\tau_n = 885.7 \pm 0.8$ s (8), which normalizes the strength of the charged-current weak interactions responsible for neutron-proton interconversions. Changes in the currently accepted value may be accounted for by adding $2 \times 10^{-4}(\tau_n - 885.7)$ to the right side of the expression for Y_P. The above abundances have been calculated under the assumption that the three flavors of active neutrinos were populated in the pre-BBN Universe and had decoupled before $e^{\pm}$ annihilation, prior to BBN. Although this latter is a good approximation, it is not entirely accurate. Mangano et al. (9) relaxed the assumption of complete decoupling at the time of $e^{\pm}$ annihilation, finding that in the post-$e^{\pm}$ annihilation Universe (relevant for comparisons with the CMB and LSS) the relativistic energy density is modified (increased) in contrast to the complete decoupling limit. This can be accounted for by N_ν (post-$e^{\pm}$ ann.) $= 3 \rightarrow 3.046$ (9). Mangano et al. find that the effect on the BBN yields is small, well within the uncertainties quoted above. For ${}^4He$, the predicted relic abundance (Equation 8) increases by $\sim 2 \times 10^{-4}$.

As an example to be explored more carefully below, if the CMB/LSS result for the baryon abundance is adopted ($\eta_{10} = 6.11 \pm 0.20$ at $\sim 1\sigma$), these analytic fits predict $y_D = 2.60 \pm 0.16$, $y_3 = 1.05 \pm 0.04$, $Y_P = 0.2487 \pm 0.0006$, and $[Li] = 2.65^{+0.05}_{-0.06}$. The corresponding 95% confidence ranges are $2.3 \lesssim y_D \lesssim 2.9$, $0.97 \lesssim y_3 \lesssim 1.12$, $0.247 \lesssim Y_P \lesssim 0.250$, and $2.53 \lesssim [Li] \lesssim 2.74$.

2.3. Nonstandard Big Bang Nucleosynthesis: $S \neq 1$

From the preceding description of BBN, it is straightforward to anticipate the changes in the BBN-predicted abundances when $S \neq 1$. Because D and ${}^3He$ are being burned to ^{4}He, a faster-than-standard expansion ($S > 1$) leaves less time for D and ^{3}He destruction and their relic abundances increase, and vice versa for a slower-than-standard expansion. The ^{4}He relic abundance is tied tightly to the neutron abundance at BBN. A faster expansion provides less time for neutrons to transform into protons, and the higher neutron fraction results in an increase in Y_P. The effect on the mass-7 abundance when $S \neq 1$ is a bit more complex. At low baryon abundance, ^{7}Li is being destroyed and $S > 1$ inhibits its destruction, increasing the relic abundance of mass-7. In contrast, for high baryon abundance ($\eta_{10} \gtrsim 3$), direct production of ^{7}Be dominates and a faster-than-standard expansion provides less time to produce mass-7, reducing its relic abundance. Isoabundance curves for D and ^{4}He in the $S - \eta_{10}$ plane are shown in **Figure 6**. Although, in general, access to a BBN code is necessary to track the primordial abundances as functions of η_{10} and S, Kneller & Steigman (10) have identified extremely simple but quite accurate analytic fits over a limited range in these variables ($4 \lesssim \eta_{10} \lesssim 8$, $0.85 \lesssim S \lesssim 1.15$, corresponding to $1.3 \lesssim N_\nu \lesssim 5.0$). For D and ^{4}He [including the Mangano et al. (9) correction], these are

$$y_D \equiv 46.5(1 \pm 0.03)\eta_D^{-1.6}; \quad Y_P \equiv (0.2386 \pm 0.0006) + \eta_{He}/625, \qquad 10.$$

where

$$\eta_D \equiv \eta_{10} - 6(S - 1); \quad \eta_{He} \equiv \eta_{10} + 100(S - 1). \qquad 11.$$

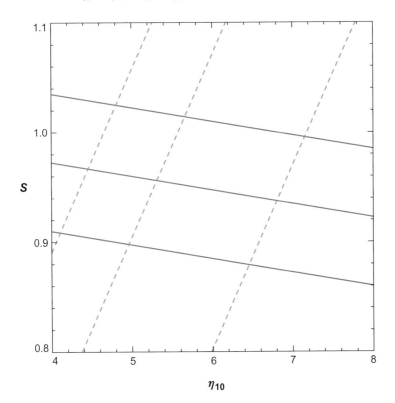

Figure 6

Isoabundance curves for D (*dashed blue lines*) and ^{4}He (*solid red lines*) in the expansion rate factor (S)–baryon abundance (η_{10}) plane. The ^{4}He curves, from bottom to top, are for $Y_P = 0.23$, 0.24, and 0.25. The D curves, from left to right, are for $y_D = 4.0$, 3.0, and 2.0.

Over the same high-η_{10} range, the fit for ^{7}Li is

$$y_{\mathrm{Li}} \equiv \frac{(1 \pm 0.1)}{8.5} \eta_{\mathrm{Li}}^2, \qquad\qquad 12.$$

where

$$\eta_{\mathrm{Li}} \equiv \eta_{10} - 3(S - 1). \qquad\qquad 13.$$

Note that Kneller & Steigman (10) chose the coefficients in these fits to maximize the goodness of fit over the above ranges in η_{10} and S, and not to minimize the difference between the fits and the more accurate results from the BBN code with $S = 1$. As a result, although these fits differ very slightly from those in Section 2.2, they do agree with them within the quoted 1σ errors.

As seen from **Figure 6** and the above equations, D (and ^{7}Li) is (are) most sensitive to η_{10}, whereas Y_P is most sensitive to S. Observations that constrain the abundances of ^{4}He and D (or ^{7}Li) have the potential to constrain the cosmology/particle physics parameters η_{10} and S. Note also that because η_D and η_{Li} depend very similarly on η_{10} and S, constraining either the D or ^{7}Li abundance has the potential to predict or constrain the other:

$$\eta_{\mathrm{Li}} = \eta_D + 3(S - 1) \approx \eta_D. \qquad\qquad 14.$$

2.4. Nonstandard Big Bang Nucleosynthesis: $\xi_e \neq 0$

Although the most popular models for generating the baryon asymmetry in the Universe tie it to the lepton asymmetry, suggesting that $|L_{\nu_e}| \sim \xi_{\nu_e} \sim \eta_{10}$, this assumption is untested. An asymmetry in the electron-type neutrinos will modify the n/p ratio at BBN that affects the primordial abundance of ^{4}He, with smaller effects on the other light-nuclide relic abundances. As already noted, for $\xi_{\nu_e} > 0$, the n/p ratio increases over its SBBN value, leading to an increase in the ^{4}He yield. Isoabundance curves for D and ^{4}He in the $\xi_e - \eta_{10}$ plane are shown in **Figure 7**. In this case too, Kneller & Steigman (10) identified simple analytic fits that are quite accurate over limited ranges in η_{10} and ξ_e ($4 \lesssim \eta_{10} \lesssim 8$, $-0.1 \lesssim \xi_e \lesssim 0.1$):

$$\eta_D \equiv \left(\frac{46.5(1 \pm 0.03)}{y_D} \right)^{1/1.6} = \eta_{10} + 5\xi_e/4, \qquad\qquad 15.$$

$$\eta_{\mathrm{He}} \equiv 625(Y_P - 0.2386 \pm 0.0006) = \eta_{10} - 574\xi_e/4, \qquad\qquad 16.$$

$$\eta_{\mathrm{Li}} \equiv (8.5(1 \pm 0.1)y_{\mathrm{Li}})^{1/2} = \eta_{10} - 7\xi_e/4. \qquad\qquad 17.$$

As for the case of a nonstandard expansion rate, here the ^{4}He abundance provides the most sensitive probe of lepton asymmetry, whereas D and ^{7}Li mainly constrain the baryon abundance and their abundances are strongly correlated:

$$\eta_{\mathrm{Li}} = \eta_D - 3\xi_e \approx \eta_D. \qquad\qquad 18.$$

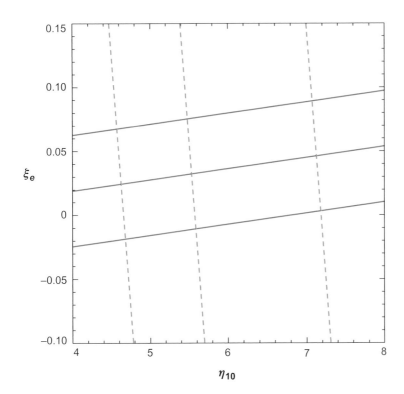

Figure 7

Isoabundance curves for D (*dashed blue lines*) and ^{4}He (*solid red lines*) in the lepton asymmetry (ξ_e)–baryon abundance (η_{10}) plane. The ^{4}He curves, from top to bottom, are for $Y_P = 0.23$, 0.24, and 0.25. The D curves, from left to right, are for $y_D = 4.0$, 3.0, and 2.0.

2.5. Nonstandard Big Bang Nucleosynthesis: $S \neq 1$ and $\xi_e \neq 0$

The two cases of nonstandard BBN discussed in the preceding sections may be combined using the linear relations among the three parameters derived from observations of the relic abundances (η_D, η_{He}, η_{Li}) and the three cosmological/particle physics parameters (η_{10}, S, ξ_e). However, although in principle observations of the relic abundances of D, ^{4}He, and ^{7}Li can constrain η_{10}, S, and ξ_e, in practice the observational uncertainties are too large at present to lead to useful constraints. Therefore, the strategy in the following discussion is to use the inferred relic abundances of D and ^{4}He to constrain the pairs $\{\eta_{10}, S\}$ or $\{\eta_{10}, \xi_e\}$ and to use the results to predict the BBN abundance of ^{7}Li.

3. THE RELIC NUCLIDES OBSERVED

In the precision era of cosmology, systematic errors have come to dominate over statistical uncertainties. This is true, with a vengeance, of the relic abundances of the BBN-produced light nuclides inferred from current observational data. Although the bad news is that these largely uncontrolled systematic errors affect D, ^{3}He, ^{4}He, and ^{7}Li, the good news is that each of these nuclides is observed in completely different astronomical objects by entirely different observational techniques so that these errors are uncorrelated. Furthermore, the post-BBN evolution of these nuclides, with its potential to modify the relic abundances, is different for each of them.

For example, owing to its very weak binding, any D incorporated into stars is completely burned away during the collapse of the prestellar nebula. For the same reason, any D synthesized in stellar interiors is burned away, to ^{3}He and beyond, before it can be returned to the interstellar medium (ISM). As a result, the post-BBN evolution of D is simple and monotonic: It is only destroyed, with the consequence being that D observed anywhere at any time in the post-BBN evolution of the Universe provides a lower bound to the primordial D abundance (11). In contrast, the post-BBN evolution of the more tightly bound ^{3}He nucleus is more complex. Prestellar D is burned to ^{3}He, which, along with any prestellar ^{3}He, may be preserved in the cooler outer layers of stars or burned away in the hotter interiors. In addition, hydrogen burning in the relatively cooler interiors of lower-mass stars may result in the net production of new ^{3}He. If this material survives the later stages of stellar evolution, it may result in stellar-produced ^{3}He being returned to the ISM. The bottom line is that stellar and galactic evolution models are necessary to track the post-BBN evolution of ^{3}He in regions containing stellar-processed material, with the result being that the relic abundance inferred from such observations is model dependent (12).

As gas is cycled through successive generations of stars, hydrogen is burned to helium (^{4}He) and beyond, with the net effect being that the post-BBN abundance of ^{4}He increases along with the abundances of the heavier elements, "metals" such as C, N, and O. This contamination of the relic ^{4}He abundance by stellar-produced helium is non-negligible, although by tracking its dependence on the metallicity (e.g., oxygen abundance) observed in the same astrophysical sites provides a means to estimate and correct for such pollution. In general, however, the extrapolation to zero metallicity introduces uncertainties in the inferred ^{4}He primordial abundance.

Finally, although ^{7}Li is, like D, a very weakly bound nucleus, its post-BBN evolution is more complicated than that of D. Whereas any ^{7}Li in the interiors of stars is burned away, some ^{7}Li in the cooler outer layers of the lowest-mass, coolest stars, where the majority of the spectra used to infer the lithium abundances are formed, may survive. Furthermore, observations of enhanced lithium in some evolved stars (super-lithium-rich red giants) suggest that ^{7}Li formed in the hotter interiors of some stars may be transported to the cooler exteriors before being destroyed. If this lithium-rich material is returned to the ISM before being burned away, these (and other) stars may be net lithium producers. In addition, ^{7}Li is synthesized (along with ^{6}Li, ^{9}Be, and ^{10,11}B) in nonstellar processes involving collisions of cosmic ray nuclei (protons, alphas, and CNO nuclei) with their counterparts in the ISM. Because stellar-synthesized CNO nuclei are necessary for spallation-produced ^{7}Li, this provides a correlation between ^{7}Li and the metallicity, which may help track this component of post-BBN lithium synthesis. However, production of ^{7}Li via α-α fusion may occur prior to the production of the heavier elements, mimicking relic production (13).

Below, the observations relevant to inferring the BBN abundances of each of these nuclides are critically reviewed. We emphasize the distinction between statistical precision and the limited accuracy—which results from difficult-to-quantify systematic uncertainties in the observational data, within the context of our understanding of the post-BBN evolution of these relic nuclei and of the astrophysical sites where they are observed.

3.1. Deuterium: The Baryometer of Choice

Deuterium's simple post-BBN evolution, combined with its sensitivity to the baryon abundance ($y_D \propto \eta_{10}^{-1.6}$), singles it out among the relic nuclides as the baryometer of choice. Although there are observations of D in the solar system (14) and the ISM of the Galaxy (15), which provide interesting lower bounds to its primordial abundance, any attempt to employ those data to infer the primordial abundance introduces model-dependent, galactic evolution uncertainties. From the discussion above, the D abundance is expected to approach its BBN value in systems of very low metallicity and/or those observed at earlier epochs (high redshift) in the evolution of the Universe. As the metallicity (Z) decreases and/or the redshift (z) increases, the corresponding D abundances should approach a plateau at the BBN-predicted abundance. Access to D at high z and low Z is provided by observations of the absorption by neutral gas of light emitted by distant QSOs. By comparing observations of absorption due to D with the much larger absorption by hydrogen in these QSO absorption-line systems (QSOALS), the nearly primordial D/H ratio may be inferred.

The identical absorption spectra of D I and H I (modulo the $\sim$81 km s^{-1} velocity offset that results from the isotope shift) are a severe liability, creating the potential for confusion of D I absorption with that of an H I interloper masquerading as D I (16). This is exacerbated by the fact that there are many more H I absorbers at low column density than at high column density. As a result, it is necessary to select very carefully those QSOALS with simple, well-understood velocity structures. This selection process is telescope intensive, leading to the rejection of D/H determinations of many potential QSOALS targets identified from low-resolution spectra, after having invested the time and effort to obtain the necessary high-resolution spectra. This has drastically limited the number of useful targets in the otherwise vast Lyman-α forest of QSO absorption spectra (see Reference 17 for further discussion).

The higher H I column-density absorbers (e.g., damped Lyman-α absorbers) have advantages over the lower H I column-density absorbers (Lyman limit systems) by enabling observations of many lines in the Lyman series. However, a precise determination of the damped Lyman-α H I column density using the damping wings of the H I absorption requires an accurate placement of the continuum, which could be compromised by H I interlopers, leading to the potential for systematic errors in the inferred H I column density. These complications are real, and the path to primordial D using QSOALS has been fraught with obstacles, with some abundance claims having had to be withdrawn or revised. As a result, despite much work utilizing some of the largest telescopes, through 2006 there have been only six sufficiently simple QSOALS with D detections that lead to reasonably robust abundance determinations (see Reference 18 and further references therein). These are shown in **Figure 8**, in addition to, for comparison, the corresponding solar system and ISM D abundances. It is clear from **Figure 8** that there is significant dispersion, in excess of the claimed observational errors, among the derived D abundances at low metallicity, which, so far, masks the anticipated primordial D plateau. This large dispersion suggests that systematic errors, whose magnitudes are hard to estimate, may have contaminated the determinations of (at least some of) the D I and/or H I column densities.

Figure 8

Deuterium abundances
derived from observations
of high-redshift,
low-metallicity QSOALS as
a function of the
corresponding metallicities
(shown relative to solar on a
log scale). Also shown for
comparison are the D
abundances derived from
solar system observations
(SUN), as well as the range
in D and oxygen
abundances inferred from
observations of the ISM.

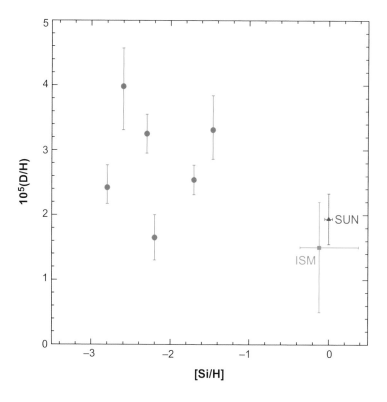

O'Meara et al. (18) choose the mean of the log of the six individual D abundances as an estimator of the primordial D abundance, finding $\log(y_{DP}) = 0.454 \pm 0.036$. This choice, corresponding to $y_{DP} = 2.84^{+0.25}_{-0.23}$, has $\chi^2 = 18.5$ for five degrees of freedom. This author is unconvinced that the mean of the log of the individual D abundances is the best estimator of the relic D abundance. If, instead, the χ^2 of the individual y_D determinations is minimized, $\langle y_D \rangle = 2.68$ and $\chi^2_{min} = 19.3$ (for five dof). Although the χ^2 for either of these estimators is excessive, we prefer to use the individual y_D determinations to find $\langle y_D \rangle$. In an attempt to compensate for the large dispersion among the individual y_D values, the individual errors are inflated by a factor of $(\chi^2_{min}/\text{dof})^{1/2} = 1.96$, which leads to the following estimate of the primordial D abundance that is adopted for the subsequent discussion:

$$y_{DP} = 2.68^{+0.27}_{-0.25}. \qquad \qquad 19.$$

3.2. Helium-3

In contrast to D, which is observed in neutral gas via absorption, ^{3}He is observed in emission from regions of ionized gas, H$_{II}$ regions. The ^{3}He nucleus has a net spin so that for singly ionized ^{3}He, the analog of the 21-cm spin-flip transition in neutral hydrogen occurs at 3.46 cm, providing the ^{3}He observational signature. The

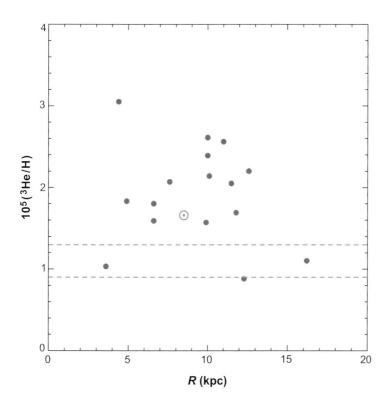

Figure 9

The ^{3}He abundances (by number relative to hydrogen) derived from observations of H$_{II}$ regions in the Galaxy (19) as a function of the corresponding distances from the galactic center (R). The blue solar symbol indicates the ^{3}He abundance for the presolar nebula (14). The dashed gray lines show the 1σ band adopted by Bania et al. (19) for an upper limit to the primordial ^{3}He abundance.

emission is quite weak, with the result that ^{3}He is observed only (outside of the solar system) in H$_{II}$ regions and in a few planetary nebulae in the Galaxy. The latter confirm the net stellar production of ^{3}He in at least some stars. In the Galaxy, there is a clear gradient of metallicity with distance from the center, with higher metallicity— more stellar processing—in the center and less in the suburbs. If in the course of the chemical evolution of the Galaxy, the abundance of ^{3}He increases (net production) or decreases (net destruction), a gradient in ^{3}He abundance (with metallicity and/or with distance) is also expected. However, as is clear from **Figure 9**, the data (14, 19) reveal no statistically significant correlation between the ^{3}He abundance and location (or metallicity) in the Galaxy. This suggests a very delicate balance between post-BBN net production and net destruction of ^{3}He. For a review of the current status of ^{3}He evolution, see Romano et al. (20). Although the absence of a gradient suggests that the mean ^{3}He abundance in the Galaxy ($y_3 \approx 1.9 \pm 0.6$) may provide a good estimate of the primordial abundance, Bania et al. (19) instead prefer to adopt as an upper limit to the primordial abundance the ^{3}He abundance inferred from observations of the most distant (from the galactic center), most metal-poor, galactic H$_{II}$ region, $y_{3P} \lesssim 1.1 \pm 0.2$ (see **Figure 9**). For purposes of this review, the Bania et al. (19) value is adopted as an estimate of the primordial abundance of ^{3}He.

$$y_{3P} = 1.1 \pm 0.2. \qquad\qquad 20.$$

Given that the post-BBN evolution of ^{3}He, involving competition among stellar production, destruction, and survival, is considerably more complex and model dependent than that of D, and that observations of ^{3}He are restricted to the chemically evolved solar system and the Galaxy, the utility of ^{3}He as a baryometer is limited. In the subsequent comparison of predictions and observations, ^{3}He will mainly be employed to provide a consistency check.

3.3. Helium-4

The post-BBN evolution of ^{4}He is quite simple. As gas cycles through successive generations of stars, hydrogen is burned to ^{4}He (and beyond), increasing the ^{4}He abundance above its primordial value. The ^{4}He mass fraction in the present Universe, Y_0, has received significant contributions from post-BBN stellar nucleosynthesis, so that $Y_0 > Y_P$. However, because some elements such as oxygen are produced by short-lived massive stars and ^{4}He is synthesized (to a greater or lesser extent) by all stars, at very low metallicity the increase in Y should lag that in O/H, resulting in a ^{4}He plateau, with $Y \rightarrow Y_P$ as O/H $\rightarrow$ 0. Therefore, although ^{4}He is observed in the Sun and in galactic H$_{II}$ regions, to minimize model-dependent evolutionary corrections, the key data for inferring its primordial abundance are provided by observations of helium and hydrogen emission lines generated from the recombination of ionized hydrogen and helium in low-metallicity extragalactic H$_{II}$ regions. The present inventory of such regions studied for their helium content exceeds 80 [see Izotov & Thuan (IT) (21)]. Because for such a large data set even modest observational errors can result in an inferred primordial abundance whose formal statistical uncertainty may be quite small, special care must be taken to include hitherto ignored or unaccounted for systematic corrections. It is the general consensus that the present uncertainty in Y_P is dominated by the latter, rather than by the former, errors.

Although astronomers have generally been long aware of important sources of potential systematic errors (22), attempts to account for them have often been unsystematic or entirely absent. When it comes to using published estimates of Y_P, *caveat emptor*. The current conventional wisdom that the accuracy of the data demands the inclusion of systematic errors has led to recent attempts to account for some of them (21–28). In **Figure 10**, a sample of Y_P determinations from 1992 to 2006 is shown (29). Most of these observationally inferred estimates (largely uncorrected for systematic errors or corrected unsystematically for some) fall below the SBBN-predicted primordial abundance (see Section 2.2), hinting either at new physics or at the need to account more carefully and consistently for all known systematic corrections.

Keeping in mind that more data do not necessarily translate to higher accuracy, the largest observed, reduced, and consistently analyzed data set of helium abundance determinations from low-metallicity extragalactic H$_{II}$ regions is that from IT 2004 (21). For their full sample of more than 80 H$_{II}$ regions, IT infer $Y_P^{IT} = 0.243 \pm 0.001$. From an analysis of an a posteriori–selected subset of seven of these H$_{II}$ regions, in which they attempt to account for some of the systematic errors, IT derive a consistent, slightly smaller value of $Y_P = 0.242 \pm 0.002$. Both of these Y_P values are shown in **Figure 10**. The IT analysis of this subset is typical of many of the recent attempts

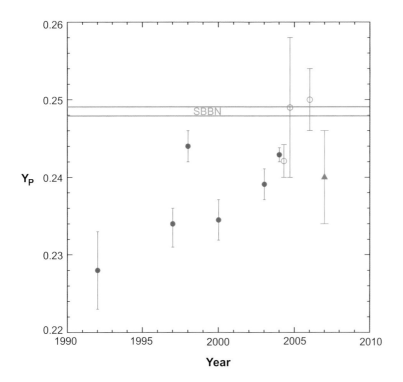

Figure 10

A sample of the observationally inferred primordial ^{4}He abundances published from 1992 to 2006 (see text). The error bars are the quoted 1σ uncertainties. The open red circles are derived from various a posteriori–selected subsets of the IT 2004 H$_{II}$ regions. The filled blue triangle is the value of Y_P adopted in this review. Also shown is the 1σ band for the SBBN-predicted relic abundance.

to include systematic corrections: Almost always only a subset of the known sources of systematic errors is analyzed, and almost always these analyses are applied to a very limited set (typically 1–7) of H$_{II}$ regions, which have usually been selected a posteriori.

For example, Luridiana et al. (24) use photoionization models to account for the effect of collisional excitation of Balmer lines for three (out of a total of five) H$_{II}$ regions, and extrapolate their results for the individual helium and oxygen abundances to zero oxygen abundance using the slope of the Y-versus-O relation derived from chemical evolution models. Their result, $Y_P = 0.239 \pm 0.002$, is shown in **Figure 10**.

The Olive & Skillman (OS) (23) and Fukugita & Kawasaki (FK) (27) analyses of the IT data attempt to account for the effect of underlying stellar absorption on the helium and hydrogen emission lines. Following criteria outlined in their 2001 paper (23), OS found they could apply their analysis to only 7 of the 82 IT H$_{II}$ regions. This small data set, combined with its restricted range in metallicity (oxygen abundance), severely limits the statistical significance of the OS conclusions. For example, there is no evidence from the seven OS H$_{II}$ regions that $\Delta Y \equiv Y^{OS} - Y^{IT}$ is correlated with metallicity, and the weighted mean correction and the error in its mean are $\Delta Y = 0.0029 \pm 0.0032$ (the average correction and its average error are $\Delta Y = 0.0009 \pm 0.0095$), consistent with zero at 1σ. If the weighted mean offset is applied to the IT-derived primordial abundance of $Y_P^{IT} = 0.243 \pm 0.001$, the corrected primordial value is $Y_P^{IT/OS} = 0.246 \pm 0.004$ (where, to be conservative, the errors have been added linearly). In contrast, OS prefer to force a fit of the seven data points to a linear Y-versus-O/H relation and from it derive the primordial abundance.

It is not surprising that for only seven data points spanning a relatively narrow range in metallicity, their linear fit, $Y_7^{OS} = 0.2495 \pm 0.0092 + (54 \pm 187)$ (O/H), is not statistically significant. In particular, the large error in Y_P is dominated by the very large uncertainty in the slope of the Y-versus-O relation (which includes unphysical, negative slopes). Indeed, the OS linear fit is not preferred statistically over the simple weighted mean of the seven OS helium abundances because the reduced χ^2 (χ^2/dof) is actually higher for the linear fit. All eight HII regions reanalyzed by OS are consistent with a weighted mean, along with the error in the mean, of $\langle Y \rangle = 0.250 \pm 0.002$ (χ^2/dof $= 0.51$). This result is of interest in that it provides an upper bound to primordial helium, $Y_P \leq \langle Y \rangle \leq 0.254$ at $\sim 2\sigma$.

Fukugita & Kawasaki (27) performed a very similar analysis to that of OS using 30 of the IT HII regions and the Small Magellanic Cloud HII region from Peimbert et al. (24). For this larger data set, Fukugita & Kawasaki (27) found an anticorrelation between their correction, ΔY, and the oxygen abundance. This flattens the FK-inferred Y-versus-O/H relation to the extent that they too found no evidence for a statistically significant correlation of He and oxygen abundances (dY/dZ $= 1.1 \pm 1.4$), leading to a zero-metallicity intercept, $Y_P^{FK} = 0.250 \pm 0.004$. As with the OS analysis, it seems that the most robust conclusion that can be drawn from the FK analysis is to use the weighted mean, along with the error in the mean of their 31 HII regions (χ^2/dof $= 0.58$) to provide an upper bound to primordial helium: $Y_P < \langle Y \rangle = 0.253 \pm 0.001 \leq 0.255$ at $\sim 2\sigma$.

Very recently, Peimbert, Luridiana, & Peimbert (PLP07) (28), using new atomic data, reanalyzed five HII regions spanning a factor of six in metallicity. Four of their five HII regions were in common with those analyzed by OS. PLP07, too, found no support from the data for a positive correlation between helium and oxygen. In **Figure 11**, the OS and PLP07 results for He/H and O/H (the ratios by number for helium to hydrogen and for oxygen to hydrogen, respectively) are shown. Although PLP07 force a model-dependent linear correlation to their data that, when extrapolated to zero metallicity, leads to $Y_P = 0.247 \pm 0.003$, their data are better fit by the weighted mean (χ^2/dof $= 0.07$) and the error in the mean: $\langle Y \rangle = 0.251 \pm 0.002$, leading to an upper bound to $Y_P < \langle Y \rangle \leq 0.255$ at $\sim 2\sigma$.

There are other sources of systematic errors that have not been included in (some of) these and other analyses. For example, because hydrogen and helium recombination lines are used, the observations are blind to any neutral helium or hydrogen. Estimates of the ionization correction factor (icf), although model dependent, are large (26). Using models of HII regions ionized by the distribution of stars of different masses and ages and comparing them to the IT 1998 data, Gruenwald et al. (GSV) (26) concluded that the IT analysis overestimated the primordial ^{4}He abundance by ΔY^{GSV}(icf) $\approx 0.006 \pm 0.002$; Sauer & Jedamzik (26) found an even larger correction. If the GSV correction is applied to the OS-revised, IT primordial abundance, the icf-corrected value becomes $Y_P^{IT/OS/GSV} = 0.240 \pm 0.006$ (as above, the errors have been added linearly).

The lesson from our discussion (and from **Figure 10**) is that although recent attempts to determine the primordial abundance of ^{4}He may have achieved high precision, their accuracy remains in question. The latter is limited by our understanding

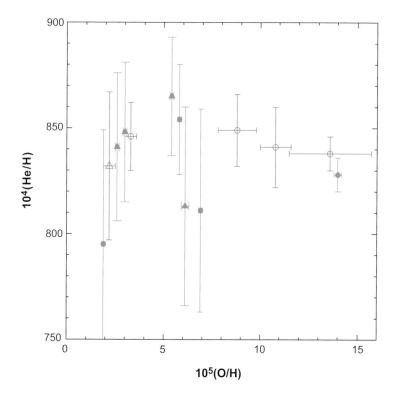

Figure 11

The HII region
helium-to-hydrogen
(He/H) ratios (by number)
as a function of the
oxygen-to-hydrogen (O/H)
ratios (by number) from the
analyses by Olive &
Skillman (*blue filled triangles
and circles*) (23) and by
Peimbert et al. (*red open
triangles and circles*) (28). The
filled and open circles are
for the HII regions common
to the two analyses.

of and ability to account for systematic corrections and their errors, not by the statistical uncertainties. The good news is that carefully planned and targeted studies of the lowest-metallicity extragalactic HII regions may go a long way toward a truly accurate determination of Y_P. In the best of all worlds, a team of astronomers would develop a well-defined observing strategy designed to acquire the data necessary to address all the known sources of systematic error, identify an a priori–selected set of target HII regions, and carry out the observations and analyses. The bad news is that too few astronomers and telescope time-allocation committee members are either aware or convinced that this is interesting and important and worth their effort and dedicated telescope time.

In the opinion of this author, the question of the observationally determined value of Y_P (and its error) is currently unresolved. The most recent analyses (OS, FK, PLP07) fail to find evidence for the anticipated correlation between the helium and oxygen abundances, calling into question the model-dependent extrapolations to zero metallicity often employed in the quest for primordial helium. Perhaps the best that can be done at present is to adopt a defensible value for Y_P and, especially, its uncertainty. To this end, the $Y_P^{\mathrm{IT/OS/GSV}}$ estimate is chosen for the subsequent discussion:

$$Y_P = 0.240 \pm 0.006. \qquad\qquad 21.$$

The adopted error is an attempt to account for the systematic as well as the statistical uncertainties. Although the central value of Y_P is low, it is only slightly more than 1σ below the SBBN-predicted central value (see **Figure 10**). Because systematic errors dominate, it is unlikely that the errors are Gaussian distributed.

Alternatively, the recent analyses (e.g., OS, FK, PLP07), although differing among themselves in detail, are in agreement on the weighted mean of the post-BBN abundance, which can be used to provide an upper bound to Y_P. To this end, the PLP07 result is adopted as an alternate constraint on (an upper bound to) Y_P: $Y_P < 0.251 \pm 0.002$.

3.4. Lithium-7

Outside of the Sun, the solar system, and the local ISM (which are all chemically evolved), lithium has been observed only in the absorption spectra of very old, very metal-poor stars in the halo of the Galaxy or in similarly metal-poor galactic globular cluster (GGC) stars. These metal-poor targets are, of course, ideal for probing the primordial abundance of lithium. Even though lithium is easily destroyed in the hot interiors of stars, theoretical expectations supported by observational data suggest that although lithium may have been depleted in many stars, the overall trend is that its galactic abundance has increased with time (see Section 3). Therefore, to probe the BBN yield of ^{7}Li, the key data are from the oldest, most metal-poor halo or GGC stars in the Galaxy (expected to form a plateau in a plot of Li/H versus Fe/H) such as those shown at low metallicity in **Figure 12**.

As for ^{4}He, the history of the relic ^{7}Li abundance determinations is an interesting and, perhaps, cautionary tale. For example, using a set of the lowest-metallicity halo stars, Ryan et al. (30) claimed evidence for a 0.3 dex increase in the lithium abundance, $[Li] \equiv 12 + \log(Li/H)$, as the metallicity (measured logarithmically by the iron abundance relative to solar) increased over the range $-3.5 \leq [Fe/H] \leq -1$. From this trend, they derived a primordial abundance of $[Li]_P \approx 2.0-2.1$. This abundance is low compared to an earlier estimate of Thorburn's (31), who found $[Li]_P \approx 2.25 \pm 0.10$. One source of systematic errors is the stellar temperature scale, which plays a key role in the connection between the observed equivalent widths and the inferred ^{7}Li abundance.

Studies of halo and GGC stars employing the infrared flux method effective temperature scale suggested a higher lithium plateau abundance (32), $[Li]_P = 2.24 \pm 0.01$, similar to Thorburn's (31) value. Melendez & Ramirez (33) reanalyzed 62 halo dwarfs using an improved infrared flux method effective temperature scale, failing to confirm the [Li]-[Fe/H] correlation claimed by Ryan et al. (30) and finding a higher relic lithium abundance, $[Li]_P = 2.37 \pm 0.05$. In a very detailed and careful reanalysis of extant observations, with great attention to systematic uncertainties and the error budget, Charbonnel & Primas (34) also found no convincing evidence for a lithium trend with metallicity, and derived $[Li]_P = 2.21 \pm 0.09$ for their full sample and $[Li]_P = 2.18 \pm 0.07$ when they restricted their sample to unevolved (dwarf) stars. They suggested the Melendez-Ramirez value should be corrected downward

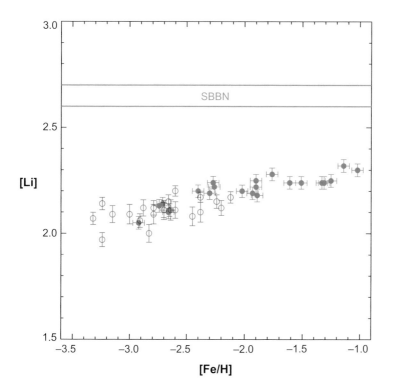

Figure 12

Lithium abundances,
[Li] ≡ 12+ log(Li/H),
versus metallicity (on a log
scale relative to solar) from
Reference 30 (*open red
circles*) and Reference 35
(*filled blue circles*). The
SBBN-predicted 1σ band
(see Section 2.2) is shown
for comparison.

by 0.08 dex to account for different stellar atmosphere models, thereby bringing it into closer agreement with their results.

More recently, Asplund et al. (35) obtained and analyzed data for 24 metal-poor halo stars (filled circles in **Figure 12**), taking special care to address many stellar and atomic physics issues related to the path from the data to the abundances. The great surprise of their results is the apparent absence of a lithium plateau; at least from their data, the lithium abundance appears to continue to decrease with decreasing metallicity. Inclusion of Ryan et al.'s data at low metallicity flattens this trend, suggesting that $[Li]_P \approx 2.1 \pm 0.1$. There is clearly tension (if not outright conflict) between this estimate and the SBBN-predicted relic abundance of $[Li]_P = 2.65^{+0.05}_{-0.06}$ (see Section 2.2). Asplund et al. identify and discuss in some detail several possible sources of systematic errors: systematic errors in the abundance analysis; dilution and destruction of ^{7}Li in the very old, very metal-poor stars; and uncertain or erroneous nuclear reaction rates (resulting in an incorrect SBBN-predicted abundance). Asplund et al. argue that it is unlikely that the abundance analysis errors can be large enough to bridge the gap, and Cyburt et al. (36) have used the observed solar neutrino flux to limit the nuclear reaction rate uncertainty, removing that too as a likely explanation. However, because the low-metallicity halo or GGC stars used to constrain primordial lithium are among the oldest in the Galaxy, they have had the most time to alter their surface lithium abundances by dilution and/or destruction, as is known to be important for many younger, higher-metallicity stars such as the Sun. Although mixing stellar surface material to the interior (or interior material to the surface) would destroy or dilute any prestellar lithium, the small dispersion observed among the low-metallicity halo star lithium abundances (in contrast to the very large spread for the higher-metallicity stars) suggests this correction may not be large enough ($\lesssim 0.1$–0.2 dex at most) to bridge the gap between theory and observation; see, for example, Pinsonneault et al. (37) and further references therein.

In this context, Korn et al.'s (38) recent observations, coupled with stellar modeling, are of special interest. Korn et al. have observed stars in the globular cluster NGC 6397. The advantage of this approach is that these stars are (or should be) the same age and metallicity. They compare their observations of lithium and iron to models of stellar diffusion, finding evidence that both lithium and iron have settled out of the atmospheres of these old stars. Applying their stellar models to the data they infer for the unevolved abundances, [Fe/H] = −2.1 and [Li] = 2.54 ± 0.10, in excellent agreement with the SBBN prediction. More such data are eagerly anticipated.

3.5. Adopted Primordial Abundances

The discussion above reveals that large uncertainties in the relic abundances, inferred from the observational data, persist in this era of precision cosmology. Much work remains to be done by observers and theorists alike. At present, the relic abundance of D (the baryometer of choice) appears to be quite well constrained. Because observations of ^{3}He are limited to the chemically evolved Galaxy, uncertain corrections to the zero-metallicity abundance are the largest source of uncertainty in its primordial abundance. Although there are a wealth of data on the abundances of ^{4}He and ^{7}Li in metal-poor astrophysical sites, systematic corrections are the sources of the largest uncertainties for these nuclides. For this reason, an alternate abundance is considered for each of them (shown in parentheses below). With this in mind, the following primordial abundances are adopted for the comparison between the observations and the theoretical predictions:

$$10^5(\text{D/H})_{\text{P}} \equiv y_{\text{DP}} = 2.68^{+0.27}_{-0.25}, \qquad\qquad 22.$$

$$10^5(^3\text{He/H})_{\text{P}} \equiv y_{3\text{P}} = 1.1 \pm 0.2, \qquad\qquad 23.$$

$$Y_{\text{P}} = 0.240 \pm 0.006 \; (<0.251 \pm 0.002), \qquad\qquad 24.$$

$$12 + \log(\text{Li/H})_{\text{P}} \equiv 2 + \log(y_{\text{Li}})_{\text{P}} \equiv [\text{Li}]_{\text{P}} = 2.1 \pm 0.1(2.5 \pm 0.1). \qquad 25.$$

4. CONFRONTATION OF THEORY WITH DATA

In the context of the standard models of cosmology and particle physics, the BBN-predicted abundances of the light nuclides depend on only one free parameter, the baryon abundance η_{10}. Consistency of SBBN requires there be a unique value (or a limited range) of η_{10} for which the predicted and observationally inferred abundances of D, ^{3}He, ^{4}He, and ^{7}Li agree. If they agree, then in this era of precision cosmology it is interesting to ask if this value/range of η_{10} is in agreement with that inferred from non-BBN-related observations of the CMB and the growth of LSS. If they do not agree, the options may be limited only by the creativity of cosmologists and physicists. It is, of course, interesting to ask if any challenges to the standard model(s) may be resolved through a nonstandard expansion rate ($S \neq 1$) or a lepton asymmetry ($\xi_e \neq 0$).

4.1. Standard Big Bang Nucleosynthesis

Let us first put SBBN to the test. In **Figure 13**, the SBBN-predicted values of η_{10} (and the 1σ ranges) corresponding to the observationally inferred abundances adopted in Section 3.5 are shown. Although the SBBN-predicted lithium abundance is a double-valued function of η_{10} (see **Figure 5**), only the higher-η_{10} branch is plotted here. From D, the baryometer of choice, we find $\eta_D = 6.0 \pm 0.4$. Although this value, accurate to $\sim6\%$, is in excellent agreement with that inferred from the less well-constrained abundance of ^{3}He ($\eta_3 = 5.6^{+2.2}_{-1.4}$), the abundances of ^{4}He and ^{7}Li correspond to very different—much smaller—values of the baryon abundance ($\eta_{He} = 2.7^{+1.2}_{-0.9}$, $\eta_{Li} = 4.0 \pm 0.6$). Because the corresponding values of η_{10} are outside the ranges of applicability of the fits described in Section 2.2, the values of η_{He} and η_{Li} in **Figure 12** are from a numerical BBN code. However, for the central value of the D-predicted baryon abundance, the SBBN-predicted helium abundance is $Y_P = 0.248$, only 1.3σ away from the observationally inferred value. The lithium abundance poses

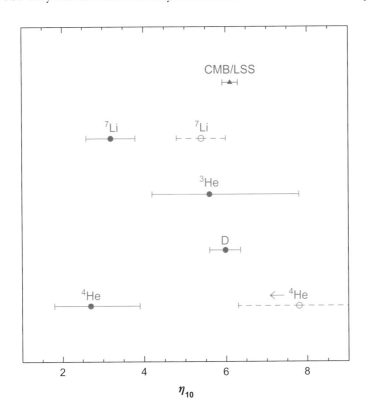

Figure 13

The SBBN-predicted values of η_{10}, and their 1σ uncertainties (*red filled circles*), corresponding to the primordial abundances adopted in Section 3.5, and the non-BBN value inferred from cosmic microwave background radiation (CMB) and large scale structure (LSS) data (*blue triangle*). The open circles and dashed lines correspond to the alternate abundances proposed for ^{4}He and ^{7}Li in Section 3.5.

a greater challenge; for $\eta_{10} \equiv \eta_D = 6.0$, the SBBN-predicted lithium abundance, $[\text{Li}]_P = 2.6$, is far from the observationally inferred value of $[\text{Li}]_P = 2.1 \pm 0.1$.

It is interesting that the two nuclides that may pose challenges to SBBN are those for which systematic corrections may change the inferred relic abundances by the largest amounts. The alternate choices considered in Section 3.5, shown by the open circles in **Figure 13**, are in much better agreement with the D and ^{3}He determined baryon abundance ($\eta_{He} < 7.8^{+1.9}_{-1.5}$, $\eta_{Li} = 5.4 \pm 0.6$).

Observations of the small temperature fluctuations in the CMB and of the LSS they seeded currently provide the tightest constraints on the universal abundance of baryons: $\eta_{CMB/LSS} = 6.11 \pm 0.20$ (2, 3), as shown in **Figure 13**. The observationally inferred relic abundances of D and ^{3}He are in excellent agreement with the SBBN predictions for this value/range of η_{10}. Depending on the outcome of various systematic corrections for the primordial abundances of ^{4}He and ^{7}Li, the observationally inferred abundances may or may not pose a challenge to SBBN.

4.2. Nonstandard Big Bang Nucleosynthesis

Whether or not the relic abundances of ^{4}He and/or ^{7}Li are consistent with SBBN, the relic abundances, especially of D and ^{4}He, can provide bounds on the parameters of some general classes of nonstandard physics. The data may be used in two ways to reveal such constraints. For example, ignoring the non-BBN (CMB/LSS) estimate for η_{10}, the primordial abundances of D and ^{4}He may be used to constrain $S(N_\nu)$ or ξ_e:

$$\eta_{He} - \eta_D = -106(S-1) = -\frac{579}{4}\xi_e. \qquad 26.$$

For the abundances choices (alternate) in Section 3.5,

$$S = 0.952 \pm 0.036 \ (<1.017 \pm 0.013), \ \xi_e = 0.034 \pm 0.026 \ (>-0.012 \pm 0.009). \quad 27.$$

The constraint on S corresponds to $N_\nu = 2.42^{+0.43}_{-0.41}$ ($N_\nu < 3.21 \pm 0.16$ for the alternate ^{4}He abundance), whereas the 2σ range is $1.61 \leq N_\nu \leq 3.30$ ($N_\nu < 3.54$). In either case, the central values are less than $\sim 1.5\sigma$ away from SBBN. Indeed, it is interesting that the 2σ upper bound to the effective number of neutrinos at BBN is $N_\nu \leq 3.3$ ($N_\nu < 3.5$), constraining the presence at BBN of even one additional active or fully mixed neutrino. $N_\nu \leq 3.3(<3.5)$ also precludes the existence at that time of a thermalized, light (relativistic) scalar for which the equivalent number of neutrinos corresponds to $\Delta N_\nu = 4/7$. For $N_\nu = 3$, the constraint on S bounds the ratio of the early Universe gravitational constant to its present value $G'_N/G_N = 0.91 \pm 0.07$ or, at 2σ, $0.77 \leq G'_N/G_N \leq 1.05$ (< 1.09). Note that although allowing $S \neq 1$ and/or $\xi_e \neq 0$ can reconcile the BBN-predicted and observed relic abundances of D and ^{4}He, even if all three (η_{10}, S, ξ_e) are allowed to vary, there is no combination that can resolve the lithium problem. As seen from Equations 14 and 18, the non-SBBN-predicted abundance of lithium remains very close to its SBBN-predicted value.

Very similar, nearly identical, constraints are found if in place of the D abundance to constrain η_{10}, the CMB/LSS value of η_{10} is employed. Here, too, the non-SBBN value of the lithium abundance remains very close to its SBBN-predicted value.

5. SUMMARY

Some 20 years ago, an article, "Big Bang Nucleosynthesis: Theories and Observations," by A.M. Boesgaard and the current author appeared in the 1985 issue of the *Annual Review of Astronomy and Astrophysics* (39). It is interesting to compare the conclusions from that review to those reached in 2006. At that time, the only data on D came from the solar system and the local ISM of the Galaxy. As a result, potentially large and uncertain evolutionary corrections were a barrier to constraining its primordial abundance, leading to the very large range in the relic abundance adopted there: $1-2 < y_{DP} < 20$. With such a large uncertainty in the baryometer of choice, all that could be inferred about the baryon abundance was that $\eta_{10} < 7-10$, consistent with our present estimate. Although there were no observations of ^{3}He outside of the solar system at that time, Boesgaard & Steigman (39) noted, following Yang et al. (40), that solar system observations of D and ^{3}He, in addition to some very general assumptions about the post-BBN evolution of D and ^{3}He, suggested a bound on the sum of the relic abundances of D and ^{3}He: $y_{DP} + y_{3P} < 6-10$, leading to a lower bound on the baryon abundance of $\eta_{10} > 3-4$, again consistent with our present estimate. As for ^{4}He, we argued that all high-quality data were consistent with $Y_P = 0.24 \pm 0.02$, and we cautioned that "systematic effects—and not statistical uncertainties—dominate." *Plus ça change*.... Then, as now, the new (at that time) data on lithium in metal-poor stars favored $y_{Li} = 0.7-1.8$, corresponding to a very low SBBN baryon abundance of $1.6 \lesssim \eta_{10} \lesssim 4.0$. Twenty years ago, the neutron lifetime was quite uncertain and the central value and range $[900 \lesssim \tau_n(1985) \lesssim 935]$ were significantly different from the currently favored value $[\tau_n = 885.7 \pm 0.8\ (8)]$. Accounting for this difference, the Boesgaard-Steigman-quoted bound (39) on the effective number of neutrinos, $N_\nu < 3.8$ (for an assumed upper bound of $Y_P < 0.254$ and a lower bound on the baryon abundance of $\eta_{10} > 3$), translates to a present-day limit of $N_\nu < 4.0$.

The past 20 years have seen dramatic increases in the amount and the precision of cosmological data. Non-BBN data from the CMB and LSS (2, 3) constrain the baryon abundance to high accuracy ($\eta_{10} = 6.11 \pm 0.20$), leading to very accurate SBBN predictions of the relic abundances of D, ^{3}He, ^{4}He, and ^{7}Li (see Section 2.2). A wealth of observational data has extended our reach on the primordial abundances of these light nuclides. These high-precision data reinforce the importance of accounting for systematic uncertainties if they are to be transformed into accurate relic abundance estimates. At present, the SBBN-predicted and observationally inferred primordial abundances of D and ^{3}He are in excellent agreement, providing support for the standard models of cosmology and particle physics. The standard cosmological model provides a concordant description of the Universe at a few minutes and 400,000 years. Although it has profited from more high-precision data, the ^{4}He abundance continues to be dominated by systematic uncertainties. Nonetheless, the predicted and observed ^{4}He abundances agree within 2σ, leading to strong constraints on new physics $[1.6 \lesssim N_\nu \lesssim 3.3\ (N_\nu \leq 3.5); -0.027 \lesssim \xi_e \lesssim 0.086\ (\xi_e \geq -0.030)$; see Section 4.2]. BBN continues to provide a unique window on the early evolution of the Universe and on its particle content.

DISCLOSURE STATEMENT

The author is not aware of any biases that might be perceived as affecting the objectivity of this review.

ACKNOWLEDGMENTS

I am appreciative of the insights and advice provided over the years by my many collaborators on the topic of this review. This review profited from helpful discussions with Manuel Peimbert, Marc Pinsonneault, Donatella Romano, and Monica Tosi. I wish to thank Mandeep Gill for a careful reading of this manuscript. I owe a debt of gratitude to the copyeditor and the referee for useful suggestions that improved the accuracy and clarity of this manuscript. This research is supported at The Ohio State University by a grant from the U.S. Department of Energy (DE-FG02-91ER40690).

LITERATURE CITED

1. Lesgourgues J, Pastor S. *Phys. Rep.* 429:307 (2006); Hannestad S. *Annu. Rev. Nucl. Part. Sci.* 56:137 (2006)
2. Spergel DN, et al. *Astrophys. J. Suppl.* 170:377 (2007)
3. Tegmark M, Eisenstein DJ, Strauss MA, Weinberg DH. *Phys. Rev. D* 74:123507 (2006)
4. Steigman G. *Nature* 261:479 (1976)
5. Yang J, Schramm DN, Steigman G, Rood RT. *Astrophys. J.* 227:696 (1979)
6. Kang H-S, Steigman G. *Nucl. Phys. B* 372:494 (1992)
7. Kernan P, Krauss L. *Phys. Rev. Lett.* 72:3309 (1994); Copi C, Schramm DN, Turner MS. *Science* 267:192 (1995); Coc A, et al. *Astrophys. J.* 600:544 (2004); Cyburt RH. *Phys. Rev. D* 70:023505 (2004); Serpico PD, et al. *JCAP* 0412:010 (2004)
8. Yao W-M, et al. *J. Phys. G Nucl. Part. Phys.* 33:1 (2006)
9. Mangano G, et al. *Nucl. Phys. B* 729:221 (2005)
10. Kneller JP, Steigman G. *New J. Phys.* 6:117 (2004)
11. Epstein RJ, Lattimer J, Schramm DN. *Nature* 263:198 (1976)
12. Rood RT, Steigman G, Tinsley BM. *Astrophys. J. Lett.* 207:L57 (1976); Olive KA, et al. *Astrophys. J.* 444:680 (1995); Dearborn DSP, Steigman G, Tosi M. *Astrophys. J.* 465:887 (1996). Erratum. *Astrophys. J.* 473:570 (1996); Galli D, Stanghellini L, Tosi M, Palla F. *Astrophys. J.* 477:218 (1997); Olive KA, Schramm DN, Scully ST, Truran J. *Astrophys. J.* 479:752 (1997); Palla F, et al. *Astron. Astrophys.* 355:69 (2000); Chiappini C, Renda A, Matteucci F. *Astron. Astrophys.* 395:789 (2002)
13. Steigman G, Walker TP. *Astrophys. J. Lett.* 385:L13 (1992)
14. Geiss J, Gloeckler JG. *Space Sci. Rev.* 84:239 (1998)
15. Linsky J, et al. *Astrophys. J.* 647:1106 (2006)
16. Steigman G. *MNRAS* 269:L53 (1994)
17. Kirkman D, et al. *Astrophys. J. Suppl.* 149:1 (2003)
18. O'Meara JM, Burles S, Prochaska JX, Prochter GE. *Astrophys. J.* 649:L61 (2006)
19. Bania TM, Rood RT, Balser DS. *Nature* 415:54 (2002)

20. Romano D, Tosi M, Matteucci F, Chiappini C. *MNRAS* 346:295 (2003)
21. Izotov YI, Thuan TX. *Astrophys. J.* 500:188 (1998); Izotov YI, Thuan TX. *Astrophys. J.* 602:200 (2004)
22. Davidson K, Kinman TD. *Astrophys. J.* 58:321 (1985)
23. Olive KA, Skillman ED. *New Astron.* 6:119 (2001); Olive KA, Skillman ED. *Astrophys. J.* 617:29 (2004)
24. Peimbert M, Peimbert A, Ruiz MT. *Astrophys. J.* 541:688 (2000); Peimbert A, Peimbert M, Luridiana V. *Astrophys. J.* 565:668 (2002); Luridiana V, Peimbert A, Peimbert M, Cerviño M. *Astrophys. J.* 592:846 (2003)
25. Steigman G, Viegas SM, Gruenwald R. *Astrophys. J.* 490:187 (1997)
26. Viegas SM, Gruenwald R, Steigman G. *Astrophys. J.* 531:813 (2000); Gruenwald R, Steigman G, Viegas SM. *Astrophys. J.* 567:931 (2002); Sauer D, Jedamzik K. *Astron. Astrophys.* 381:361 (2002)
27. Fukugita M, Kawasaki M. *Astrophys. J.* 646:691 (2006)
28. Peimbert M, Luridiana V, Peimbert A, Carigi L. astro-ph/0701313 (2007); Peimbert M, Luridiana V, Peimbert A. astro-ph/0701580 (2007)
29. Pagel BEJ, Simonson EA, Terlevich RJ, Edmunds MJ. *MNRAS* 255:325 (1992); Olive KA, Skillman ED, Steigman G. *Astrophys. J.* 489:1006 (1997); Izotov YI, Thuan TX. *Astrophys. J.* 500:188 (1998); Olive KA, Steigman G, Walker TP. *Phys. Rep.* 333:389 (2000); Luridiana V, Peimbert A, Peimbert M, Cerviño M. *Astrophys. J.* 592:846 (2003); Izotov YI, Thuan TX. *Astrophys. J.* 602:200 (2004); Olive KA, Skillman ED. *Astrophys. J.* 617:29 (2004); Fukugita M, Kawasaki M. *Astrophys. J.* 646:691 (2006)
30. Ryan SG, Norris JE, Beers TC. *Astrophys. J.* 523:654 (1999); Ryan SG, et al. *Astrophys. J.* 530:L57 (2000)
31. Thorburn JA. *Astrophys. J.* 421:318 (1994)
32. Bonifacio P, Molaro P. *MNRAS* 285:847 (1997); Bonifacio P, Molaro P, Pasquini L. *MNRAS* 292:L1 (1997)
33. Melendez J, Ramirez I. *Astrophys. J.* 615:L33 (2004)
34. Charbonnel C, Primas F. *Astron. Astrophys.* 442:961 (2005)
35. Asplund M, et al. *Astrophys. J.* 644:229 (2006)
36. Cyburt RH, Fields BD, Olive KA. *Phys. Rev. D* 69:123519 (2004)
37. Pinsonneault MH, Steigman G, Walker TP, Narayanan VK. *Astrophys. J.* 574:398 (2002)
38. Korn AJ, et al. *Nature* 442:657 (2006)
39. Boesgaard AM, Steigman G. *Annu. Rev. Astron. Astrophys.* 23:319 (1985)
40. Yang J, et al. *Astrophys. J.* 276:92

Cumulative Indexes

Contributing Authors, Volumes 48–57

A

Acosta DE, 49:389–434
Adelberger EG, 53:77–121
Alford M, 51:131–60
Armbruster P, 50:411–79
Asztalos SJ, 56:293–326
Aumann T, 48:351–99

B

Baier R, 50:37–69
Baldo Ceolin M, 52:1–21
Barker AR, 50:249–97
Beck DH, 51:189–217
Bedaque PF, 52:339–96
Bernard V, 57:33–60
Bertulani CA, 55:271–310
Blazey GC, 49:633–85
Blessing SK, 49:389–434
Bloch P, 53:123–61
Blumenhagen R, 55:71–139
Bortignon PF, 48:351–99
Boyanovsky D, 56:441–500
Bowman CD, 48:505–56
Browder TE, 53:353–86
Brown GE, 51:1–22
Buchmüller W, 55:311–55
Bunce G, 50:525–75
Burdman G, 53:431–99
Burgess CP, 57:329–62

C

Cabibbo N, 53:39–75
Calder N, 57:441–62
Camp JB, 54:525–77
Carlson CE, 57:171–204
Chakraborty D, 53:301–51
Clark RM, 50:1–36
Collon P, 54:39–67
Cornish NJ, 54:525–77
Courant ED, 53:1–37
Cvetiĉ M, 55:71–139

D

Das SR, 50:153–206
Dasgupta M, 48:401–61
Davier M, 54:115–40
Dawson S, 54:269–314
de Jager K, 54:217–67
Denef F, 57:119–44
Deshpande A, 55:165–228
de Vega HJ, 56:441–500
Diehl HT, 50:71–117
Dokshitzer YL, 54:487–523
Douglas DR, 51:413–50;
    53:387–429
Douglas MR, 57:119–44
Drechsel D, 54:69–114

E

El-Khadra AX, 52:201–51
Elliott SR, 52:115–52
Ellison J, 48:33–80
Emling H, 48:351–99

F

Faccini R, 53:353–86
Fisher P, 49:481–527
Flaugher BL, 49:633–85
Formaggio JA, 54:361–412
Frankfurt L, 55:403–65
Franzini P, 56:207–51
Frauendorf SG, 51:219–59
Froidevaux D, 56:375–440

G

Gaisser TK, 52:153–99
Gaitskell RJ, 54:315–59
Gay C, 50:577–641
Geesaman DF, 56:53–92
Gelbke CK, 56:53–92
Gerschel C, 49:255–301
Gibbons LK, 48:121–74
Glasmacher T, 48:1–31
Glenzinski DA, 50:207–48
Gomez-Cadenas JJ, 52:253–302
Guet C, 51:219–59

H

Hagiwara K, 48:463–504
Hannestad S, 56:137–61
Hansen PG, 53:219–61
Harris DA, 52:253–302
Harrison M, 52:425–71
Hashimoto S, 54:451–86
Haxton W, 51:261–93
Heckel BR, 53:77–121

Heinrich J, 57:145–69
Heintz U, 50:207–48
Heinz U, 49:529–79
Heiselberg H, 50:481–524
Herrmann N, 49:581–632
Hertzog DW, 54:141–74
Heßberger FP, 54:175–215
Hewett J, 52:397–424
Hinchliffe I, 50:643–78
Hinde DJ, 48:401–61
Höcker A, 56:501–67
Honda M, 52:153–99
Hüfner J, 49:255–301
Hughes EW, 49:303–39
Hughes V, 50:i–xxxvii
Huovinen P, 56:163–206
Hyde-Wright CE, 54:217–67

J

Jacak BV, 49:529–79
Jackson J, 57:441–62
Jackson JD, 49:1–33
Janesick JR, 53:263–300
Janssens RVF, 56:53–92
Jenkins E, 48:81–119
Ji X, 54:413–50
Jung C, 51:451–88

K

Kachru S, 57:119–44
Kado MM, 52:65–114
Kajita T, 51:451–88
Kamionkowski M, 49:77–123
Kappeler F, 48:175–251
Kayser B, 49:481–527
Kettell SH, 50:249–97
Kharzeev DE, 54:487–523
Klein JR, 55:141–63
Klein SR, 55:271–310
Konigsberg J, 53:301–51
Kosowsky A, 49:77–123
Krafft GA, 51:413–50;
    53:387–429
Kubodera K, 54:19–37
Kutschera W, 54:39–67

L

Laermann E, 53:163–98
Langacker P, 55:71–139

Lattimer J, 51:295–344
Learned J, 50:679–749
LeCompte T, 50:71–117
Lee T-SH, 52:23–64
Leemann CW, 51:413–50
Leino M, 54:175–215
Ligeti Z, 56:501–67
Lingel K, 48:253–306
Lisa MA, 55:357–402
Lu Z-T, 54:39–67
Luke M, 52:201–51
Lundberg B, 53:199–218
Lyons L, 57:145–69

M

Macchiavelli AO, 50:1–36
Mangano ML, 55:555–88
Mann T, 51:451–88
Mannheim K, 50:679–749
Manohar A, 50:643–78
Marciano WJ, 54:115–40
Margetis S, 50:299–342
Martoff CJ, 54:361–412
Masiero A, 51:161–87
Mathur SD, 50:153–206
McFarland KS, 49:481–527
McGaughey PL, 49:217–53
McGrew C, 51:451–88
McKeown RD, 51:189–217
Meißner U-G, 57:33–60
Merminga L, 53:387–429
Meyer H-O, 57:1–31
Mezzacappa A, 55:467–515
Miller GA, 56:253–92
Miller ML, 57:205–43
Milner R, 55:165–228
Mohapatra RN, 56:569–
    628
Morse WM, 54:141–74
Moskalenko IV, 57:285–327
Moss JM, 49:217–53
Moulson M, 56:207–51
Mrówczyński S, 57:61–94
Müller B, 56:93–135

N

Nagle JL, 56:93–135
Nelson AE, 53:77–121
Neuberger H, 51:23–52
Nico JS, 55:27–69

Niwa K, 53:199–218
Nystrand J, 55:271–310

O

Onogi T, 54:451–86
Opper AK, 56:253–92
Oreglia M, 54:269–314

P

Page D, 56:327–74
Page SA, 56:1–52
Pandharipande V, 50:481–524
Paolone V, 53:199–218
Park T-S, 54:19–37
Peccei RD, 55:311–55
Peggs S, 52:425–71
Peng JC, 49:217–53
Perkins DH, 55:1–26
Phelps ME, 52:303–38
Philipsen O, 53:163–98
Pieper S, 51:53–90
Poppitz E, 48:307–50
Prakash M, 51:295–344
Pratt S, 55:357–402
Ptuskin VS, 57:285–327
Putnam G, 53:263–300

R

Raffelt GG, 49:163–216
Rainwater D, 53:301–51
Ramsey-Musolf MJ, 56:1–52
Reddy S, 56:327–74
Redwine RP, 52:23–64
Rehm E, 51:91–129
Reygers K, 57:205–43
Riotto A, 49:35–75
Roodman A, 55:141–63
Rosenberg LJ, 56:293–326
Roser T, 52:425–71
Rowley N, 48:401–61
Rowson P, 51:345–412
Ruuskanen PV, 56:163–206

S

Šafařík S, 50:299–342
Saito NS, 50:525–75
Samtleben D, 57:245–83

Sanders SJ, 57:205–43
Sauli F, 49:341–87
Savard G, 50:119–52
Sawyer R, 51:295–344
Schiff D, 50:37–69
Schmaltz M, 55:229–70
Schwarz DJ, 56:441–500
Sharma A, 49:341–87
Sherrill BM, 56:53–92
Shintake T, 49:125–62
Shipsey I, 53:431–99
Shiu G, 55:71–139
Sikivie P, 56:293–326
Silvestrini L, 57:405–40
Skwarnicki T, 48:253–306
Smirnov AY, 56:569–628
Smith JG, 48:253–306
Smith MS, 51:91–129
Snow WM, 55:27–69
Soffer J, 50:525–75
Soltz R, 55:357–402
Son DT, 57:95–118
Sphicas P, 56:375–440
Spiropulu M, 52:397–424
Staggs S, 57:245–83
Stankus P, 55:517–53
Starinets AO, 57:95–118
Stefanini AM, 48:401–61
Steigman G, 57:463–91
Steinberg P, 57:205–43
Stelzer TJ, 55:555–88

Stephenson EJ, 56:253–92
Strikman M, 55:403–65
Strong AW, 57:285–327
Su D, 51:345–412
Swallow EC, 53:39–75

T

Tauscher L, 53:123–61
Tenenbaum P, 49:125–62
Thielemann F, 48:175–251
Thoma MH, 57:61–94
Tiator L, 54:69–114
Tollefson K, 49:435–79
Tostevin JA, 53:219–61
Trivedi SP, 48:307–50
Trodden M, 49:35–75
Tucker-Smith D, 55:229–70
Tully CG, 52:65–114

V

van Bibber K, 56:293–326
Vanderhaeghen M, 57:171–204
van Kolck U, 52:339–96
Varnes EW, 49:435–79
Venugopalan V, 55:165–228
Verbaarschot JJM, 50:343–410
Vetter K, 57:363–404
Villalobos Baillie O, 50:299–342
Vives O, 51:161–87

Vogel P, 52:115–52
Vogelsang W, 50:525–75;
  55:165–228
Volkas R, 51:295–344
Voss R, 49:303–39

W

Weiss C, 55:403–65
Werth G, 50:119–52
Wessels JP, 49:581–632
Wettig T, 50:343–410
Wiedemann U, 55:357–402
Wieman C, 51:261–93
Wienold T, 49:581–632
Wiescher M, 48:175–251
Willocq S, 51:345–412
Winstein B, 57:245–83
Winston R, 53:39–75
Wiringa R, 51:53–90
Wolfenstein L, 54:1–17
Wudka J, 48:33–80

Y

Yanagida T, 55:311–55

Z

Zakharov BG, 50:37–69
Zioutas K, 56:293–326

Chapter Titles, Volumes 48–57

Prefatory Chapters

Snapshots of a Physicist's Life	JD Jackson	49:1–33
Various Researches in Physics	VW Hughes	50:i–xxxvii
The Discreet Charm of the Nuclear Emulsion Era	M Baldo Ceolin	52:1–21
Accelerators, Colliders, and Snakes	ED Courant	53:1–37
The Strength of the Weak Interactions	L Wolfenstein	54:1–17
From Pions to Proton Decay: Tales of the Unexpected	DH Perkins	55:1–26

Accelerators

Coulomb Excitation at Intermediate Energies	T Glasmacher	48:1–31
Measurement of Small Electron-Beam Spots	P Tenenbaum, T Shintake	49:125–62
Physics Opportunities at Neutrino Factories	JJ Gomez-Cadenas, DA Harris	52:253–302
The RHIC Accelerator	M Harrison, S Peggs, T Roser	52:425–71
Accelerators, Colliders, and Snakes	ED Courant	53:1–37
High-Current Energy-Recovering Electron Linacs	L Merminga, DR Douglas, GA Krafft	53:387–429
Physics Opportunities with a TeV Linear Collider	S Dawson, M Oreglia	54:269–314
Study of the Fundamental Structure of Matter with an Electron-Ion Collider	A Deshpande, R Milner, R Venugopalan, W Vogelsang	55:165–228
The Indiana Cooler: A Retrospective	H-O Meyer	57:1–31

Astrophysics

Current Quests in Nuclear Astrophysics and
   Experimental Approaches   F Käppeler,   48:175–251
       F-K Thielemann,
       M Wiescher

Recent Progress in Baryogenesis   A Riotto, M Trodden   49:35–75
The Cosmic Microwave Background
   and Particle Physics   M Kamionkowski,   49:77–123
       A Kosowsky

Particle Physics From Stars   GG Raffelt   49:163–216
The Quantum Physics of Black Holes: Results
   from String Theory   SR Das, SD Mathur   50:153–206
High-Energy Neutrino Astronophysics   JG Learned,   50:679–749
       K Mannheim

Flux of Atmospheric Neutrinos   TK Gaisser,   52:153–99
       M Honda

The Solar *hep* Process   K Kubodera,   54:19–37
       T-S Park

Direct Detection of Dark Matter   RJ Gaitskell   54:315–59
Gravitational Wave Astronomy   JB Camp, NJ Cornish   54:525–77
Leptogenesis as the Origin of Matter   W Buchmüller,   55:311–55
       RD Peccei,
       T Yanagida

Ascertaining the Core Collapse Superrnova
   Mechanism: The State of the Art
   and the Road Ahead   A Mezzacappa   55:467–515
Primordial Neutrinos   S Hannestad   56:137–61
Charge Symmetry Breaking and QCD   GA Miller,   56:253–92
       AK Opper,
       EJ Stephenson

Searches for Astrophysical and Cosmological
   Axions   SJ Asztalos,   56:293–326
       LJ Rosenberg,
       K van Bibber,
       P Sikivie,
       K Zioutas

Phase Transitions in the Early and Present
   Universe   D Boyanovsky,   56:441–500
       HJ de Vega,
       DJ Schwarz

The Cosmic Microwave Background for
   Pedestrians: A Review for Particle
   and Nuclear Physicists   D Samtleben,   57:245–83
       S Staggs,
       B Winstein

Cosmic-Ray Propagation and Interactions
   in the Galaxy   AW Strong,   57:285–327
       IV Moskalenko,
       VS Ptuskin

Primordial Nucleosynthesis in the Precision
   Cosmology Era   G Steigman   57:463–91

Instrumentation and Techniques

Measurement of Small Electron-Beam Spots	P Tenenbaum, T Shintake	49:125–62
Micropattern Gaseous Detectors	F Sauli, A Sharma	49:341–87
The CDF and DØ Upgrades for Run II	T LeCompte, HT Diehl	50:71–117
Molecular Imaging with Positron Emission Tomography	ME Phelps	52:303–38
The RHIC Accelerator	M Harrison, S Peggs, T Roser	52:425–71
Tests of the Gravitational Inverse-Square Law	EG Adelberger, BR Heckel, AE Nelson	53:77–121
Developments and Applications of High-Performance CCD and CMOS Imaging Arrays	JR Janesick, G Putnam	53:263–300
Backgrounds to Sensitive Experiments Underground	JA Formaggio, CJ Martoff	54:361–412
Blind Analysis in Nuclear and Particle Physics	JR Klein, A Roodman	55:141–63
Tools for the Simulation of Hard Hadronic Collisions	ML Mangano, TJ Stelzer	55:555–88
General-Purpose Detectors for the Large Hadron Collider	D Froidevaux, P Sphicas	56:375–440
Systematic Errors	J Heinrich, L Lyons	57:145–69
Recent Developments in the Fabrication and Operation of Germanium Detectors	K Vetter	57:363–404

Nuclear Applications

Accelerator-Driven Systems for Nuclear Waste Transmutation	CD Bowman	48:505–56
Molecular Imaging with Positron Emission Tomography	ME Phelps	52:303–38
Tracing Noble Gas Radionuclides in the Environment	P Collon, W Kutschera, Z-T Lu	54:39–67
Physics of a Rare Isotope Accelerator	DF Geesaman, CK Gelbke, BM Sherrill, RVF Janssens	56:53–92

Nuclear Experiment

Direct Reactions with Exotic Nuclei	PG Hansen, JA Tostevin	53:219–61
Physics of Ultra-Peripheral Nuclear Collisions	CA Bertulani, SR Klein, J Nystrand	55:271–310

Femtoscopy in Relativistic Heavy Ion
    Collisions: Two Decades of Progress   MA Lisa, S Pratt,   55:357–402
                                                    R Soltz,
                                                    U Wiedemann
Direct Photon Production in Relativistic
    Heavy-Ion Collisions   P Stankus   55:517–53
Results from the Relativistic Heavy Ion
    Collider   B Müller, JL Nagle   56:93–135

Nuclear Reaction Mechanisms–Heavy Particles
Measuring Barriers to Fusion   M Dasgupta,   48:401–61
                                                    DJ Hinde,
                                                    N Rowley,
                                                    AM Stefanini
Charmonium Suppression in Heavy-Ion
    Collisions   C Gerschel, J Hüfner   49:255–301
Two-Particle Correlations in Relativistic
    Heavy-Ion Collisions   U Heinz, BV Jacak   49:529–79
Collective Flow in Heavy-Ion Collisions   N Herrmann,   49:581–632
                                                    JP Wessels,
                                                    T Wienold
Energy Loss in Perturbative QCD   R Baier, D Schiff,   50:37–69
                                                    BG Zakharov
Strangeness Production in Heavy-Ion
    Collisions   S Margetis, K Šafařík,   50:299–342
                                                    O Villalobos Baillie
The RHIC Accelerator   M Harrison, S Peggs,   52:425–71
                                                    T Roser

Nuclear Reaction Mechanisms–Light Particles
Multiphonon Giant Resonances in Nuclei   T Aumann,   48:351–99
                                                    PF Bortignon,
                                                    H Emling
Pion-Nucleus Interactions   T-SH Lee,   52:23–64
                                                    RP Redwine
Two-Photon Physics in Hadronic Processes   CE Carlson,   57:171–204
                                                    M Vanderhaeghen

Nuclear Structure
High-Energy Hadron-Induced Dilepton
    Production from Nucleons and Nuclei   PL McGaughey,   49:217–53
                                                    JM Moss, JC Peng
Spin Structure Functions   E Hughes, R Voss   49:303–39
The Shears Mechanism in Nuclei   RM Clark,   50:1–36
                                                    AO Macchiavelli
Precision Nuclear Measurements
    with Ion Traps   G Savard, G Werth   50:119–52
On the Production of Superheavy Elements   P Armbruster   50:411–79
Pion-Nucleus Interactions   T-SH Lee,   52:23–64
                                                    RP Redwine

Effective Field Theory for Few-Nucleon Systems	PF Bedaque, U van Kolck	52:339–96
The Nuclear Structure of Heavy-Actinide and TransActinide Nuclei	M Leino, FP Heßberger	54:175–215
Electromagnetic Form Factors of the Nucleon and Compton Scattering	CE Hyde-Wright, K de Jager	54:217–67

Nuclear Theory

Random Matrix Theory and Chiral Symmetry in QCD	JJM Verbaarschot, T Wettig	50:343–410
Recent Progress in Neutron Star Theory	H Heiselberg, V Pandharipande	50:481–524
Prospects for Spin Physics at RHIC	G Bunce, N Saito, J Soffer, W Vogelsang	50:525–75
Effective Field Theory for Few-Nucleon Systems	PF Bedaque, U van Kolck	52:339–96
Lattice QCD at Finite Temperature	E Laermann, O Philipsen	53:163–98
The Solar *hep* Process	K Kubodera, T-S Park	54:19–37
Hydrodynamic Models for Heavy Ion Collisions	P Huovinen, PV Ruuskanen	56:163–206
Charge Symmetry Breaking and QCD	GA Miller, AK Opper, EJ Stephenson	56:253–92
Chiral Perturbation Theory	V Bernard, U-G Meißner	57:33–60
What Do Electromagnetic Plasmas Tell Us about the Quark-Gluon Plasma?	S Mrówczyński, MH Thoma	57:61–94
Two-Photon Physics in Hadronic Processes	M Vanderhaeghen, CE Carlson	57:171–204
Glauber Modeling in High-Energy Nuclear Collisions	ML Miller, K Reygers, SJ Sanders, P Steinberg	57:205–43

Particle Interactions at High Energies

Study of Trilinear Gauge-Boson Couplings at the Tevatron Collider	J Ellison, J Wudka	48:33–80
High-Energy Hadron-Induced Dilepton Production from Nucleons and Nuclei	PL McGaughey, JM Moss, JC Peng	49:217–53

Spin Structure Functions	E Hughes, R Voss	49:303–39
Leptoquark Searches at HERA and the Tevatron	DE Acosta, SK Blessing	49:389–434
Direct Measurement of the Top Quark Mass	K Tollefson, EW Varnes	49:435–79
Inclusive Jet and Dijet Production at the Tevatron	GC Blazey, BL Flaugher	49:633–85
Precision Measurments of the W Boson Mass	DA Glenzinski, U Heintz	50:207–48
Developments in Rare Kaon Decay Physics	AR Barker, SH Kettell	50:249–97
Flux of Atmospheric Neutrinos	TK Gaisser, M Honda	52:153–99
Physics Opportunities at Neutrino Factories	JJ Gomez-Cadenas, DA Harris	52:253–302
Particle Physics Probes of Extra Spacetime Dimensions	J Hewett, M Spiropulu	52:397–424
Lattice QCD at Finite Temperature	E Laermann, O Philipsen	53:163–98
Top-Quark Physics	D Chakraborty, J Konigsberg, D Rainwater	53:301–51
Physics Opportunities with a TeV Linear Collider	S Dawson, M Oreglia	54:269–314
Small-x Physics: From HERA to LHC and Beyond	L Frankfurt, M Strikman, C Weiss	55:403–65
Searching for New Physics in $b{\to}s$ Hadronic Penguin Decays	L Silvestrini	57:405–40

Particle Spectroscopy

Large-N_c Baryons	E Jenkins	48:81–119
High-Energy Hadron-Induced Dilepton Production from Nucleons and Nuclei	PL McGaughey, JM Moss, JC Peng	49:217–53
Charmonium Suppression in Heavy-Ion Collisions	C Gerschel, J Hüfner	49:255–301
Direct Measurement of the Top Quark Mass	K Tollefson, EW Varnes	49:435–79
The Mass of the b Quark	AX El-Khadra, M Luke	52:201–51
Top-Quark Physics	D Chakraborty, J Konigsberg, D Rainwater	53:301–51

Particle Theory

Dynamical Supersymmetry Breaking	E Poppitz, SP Trivedi	48:307–50
The QCD Coupling Constant	I Hinchliffe, A Manohar	50:643–78
The Mass of the b Quark	AX El-Khadra, M Luke	52:201–51

Effective Field Theory for Few-Nucleon Systems	PF Bedaque, U van Kolck	52:339–96
Particle Physics Probes of Extra Spacetime Dimensions	J Hewett, M Spiropulu	52:397–424
Semileptonic Hyperon Decays	N Cabibbo, EC Swallow, R Winston	53:39–75
Tests of the Gravitational Inverse-Square Law	EG Adelberger, BR Heckel, AE Nelson	53:77–121
Lattice QCD at Finite Temperature	E Laermann, O Philipsen	53:163–98
Top-Quark Physics	D Chakraborty, J Konigsberg, D Rainwater	53:301–51
The Strength of the Weak Interactions	L Wolfenstein	54:1–17
The Solar *hep* Process	K Kubodera, T-S Park	54:19–37
The Gerasimov-Drell-Hearn Sum Rule and the Spin Structure of the Nucleon	D Drechsel, L Tiator	54:69–114
The Theoretical Prediction for the Muon Anomalous Magnetic Moment	M Davier, WJ Marciano	54:115–40
Physics Opportunities with a TeV Linear Collider	S Dawson, M Oreglia	54:269–314
Generalized Parton Distributions	X Ji	54:413–50
Heavy Quarks on the Lattice	S Hashimoto, T Onogi	54:451–86
The Gribov Conception of Quantum Chromodynamics	YL Dokshitzer, DE Kharzeev	54:487–523
Toward Realistic Intersecting D-Brane Models	R Blumenhagen, M Cvetic, P Langacker, G Shiu	55:71–139
Little Higgs Theories	M Schmaltz, D Tucker-Smith	55:229–70
Neutrino Mass and New Physics	RN Mohapatra, AY Smirnov	56:569–628
Chiral Perturbation Theory	V Bernard, U-G Meißner	57:33–60
What Do Electromagnetic Plasmas Tell Us about the Quark-Gluon Plasma?	S Mrówczyński, MH Thoma	57:61–94
Viscosity, Black Holes, and Quantum Field Theory	DT Son, AO Starinets	57:95–118
Physics of String Flux Compactifications	F Denef, MR Douglas, S Kachru	57:119–144
An Introduction to Effective Field Theory	CP Burgess	57:329–362
Searching for New Physics in $b \rightarrow s$ Hadronic Penguin Decays	L Silvestrini	57:405–40

Weak and Electromagnetic Interactions

Measurement of the CKM Matrix Element V_{ub} Exclusive $B\pi\nu$ and $B\rho\nu$ Decays	LK Gibbons	48:121–74
Penguin Decays of B Mesons	K Lingel, T Skwarnicki, JG Smith	48:253–306
Electroweak Studies at Z Factories	K Hagiwara	48:463–504
Recent Progress in Baryogenesis	A Riotto, M Trodden	49:35–75
Direct Measurement of the Top Quark Mass	K Tollefson, EW Varnes	49:435–79
Neutrino Mass and Oscillation	P Fisher, B Kayser, KS McFarland	49:481–527
B Mixing	C Gay	50:577–641
The Searches for Higgs Bosons at LEP	MM Kado, CG Tully	52:65–114
Double Beta Decay	SR Elliott, P Vogel	52:115–52
The Mass of the b Quark	AX El-Khadra, M Luke	52:201–51
Semileptonic Hyperon Decays	N Cabibbo, EC Swallow, R Winston	53:39–75
Tests of Discrete Symmetries with CPLEAR	P Bloch, L Tauscher	53:123–61
Observation of the Tau Neutrino	B Lundberg, K Niwa, V Paolone	53:199–218
Top-Quark Physics	D Chakraborty, J Konigsberg, D Rainwater	53:301–51
Establishment of CP Violation in B Decays	TE Browder, R Faccini	53:353–86
$D^0 - \bar{D}^0$ Mixing and Rare Charm Decays	G Burdman, I Shipsey	53:431–99
The Strength of the Weak Interactions	L Wolfenstein	54:1–17
The Brookhaven Muon Anomalous Magnetic Moment Experiment	DW Hertzog, WM Morse	54:141–74
Fundamental Neutron Physics	JS Nico, WM Snow	55:27–69
Hadronic Parity Violation: A New View Through the Looking Glass	MJ Ramsey-Musolf, SA Page	56:1–52
The Physics of DAΦNE and KLOE	P Franzini, M Moulson	56:207–51
CP Violation and the CKM Matrix	A Höcker, Z Ligeti	56:501–67

Special Topics

Quantum Communication	J Jackson, N Calder	57:441–62

RETURN TO: PHYSICS-ASTRONOMY LIBRARY
351 LeConte Hall   510-642-3122

LOAN PERIOD 1	2	3
1-MONTH		
4	5	6

ALL BOOKS MAY BE RECALLED AFTER 7 DAYS.
Renewable by telephone.

DUE AS STAMPED BELOW.

This book will be held
in PHYSICS LIBRARY
until   JUL 1 0 2008

FORM NO. DD 22   UNIVERSITY OF CALIFORNIA, BERKELEY
2M 7-07   Berkeley, California 94720–6000

ANNUAL REVIEWS

Intelligent Synthesis of the Scientific Literature

Annual Reviews – Your Starting Point for Research Online
http://arjournals.annualreviews.org

- Over 1150 Annual Reviews volumes—more than 26,000 critical, authoritative review articles in 35 disciplines spanning the Biomedical, Physical, and Social sciences—available online, including all Annual Reviews back volumes, dating to 1932

- Current individual subscriptions include seamless online access to full-text articles, PDFs, Reviews in Advance (as much as 6 months ahead of print publication), bibliographies, and other supplementary material in the current volume and the prior 4 years' volumes

- All articles are fully supplemented, searchable, and downloadable — see http://nucl.annualreviews.org

- Access links to the reviewed references (when available online)

- Site features include customized alerting services, citation tracking, and saved searches

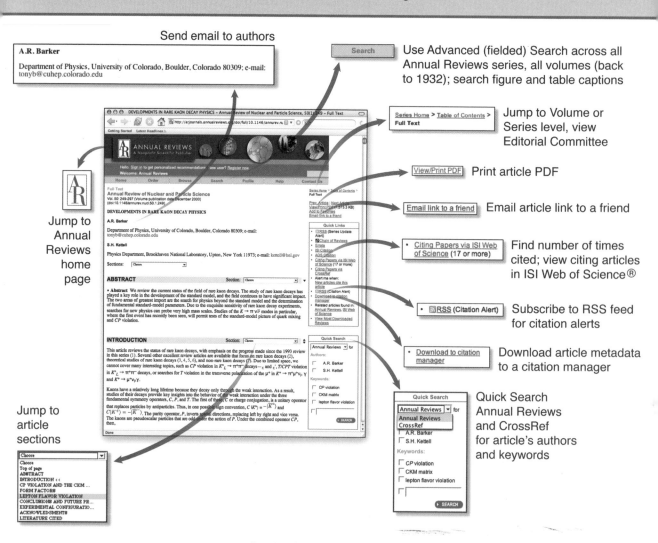

Send email to authors

Use Advanced (fielded) Search across all Annual Reviews series, all volumes (back to 1932); search figure and table captions

Jump to Volume or Series level, view Editorial Committee

Print article PDF

Email article link to a friend

Find number of times cited; view citing articles in ISI Web of Science®

Subscribe to RSS feed for citation alerts

Download article metadata to a citation manager

Quick Search Annual Reviews and CrossRef for article's authors and keywords

Jump to Annual Reviews home page

Jump to article sections

Copyright © 2007 Annual Reviews, Nonprofit Publisher of the *Annual Review of* Series